The Hutchinson

Concise
Science Dictionary

CONTRIBUTORS

Dougal Dixon
Ian Kingston
Peter Lafferty
Carol Lister
Graham Littler
David Penfold
Ian Ridpath
Julian Rowe
Jack Schofield
Steve Smyth
Catherine Thompson

EDITORS

Editor	Catherine Thompson
Managing Editor	Sara Jenkins-Jones
Editorial Director	Michael Upshall
Proofreader	Helen Maxey
Production	Tony Ballsdon
Art Editor	Terence Caven
Design	Roger Walker
Page makeup	TechType

The Hutchinson

Concise
Science Dictionary

Helicon

First published 1995

Copyright © Helicon Publishing Ltd 1995

All rights reserved

Helicon Publishing Ltd
42 Hythe Bridge Street
Oxford OX1 2EP

Printed and bound in Great Britain by Cox & Wyman Ltd, Reading, Berkshire

ISBN 1 85986 062 1

British Cataloguing in Publication Data

A catalogue record of this book is available from the British Library

A in physics, symbol for ◊*ampere*, a unit of electrical current.

abdomen in vertebrates, the part of the body below the ◊thorax, containing the digestive organs; in insects and other arthropods, it is the hind part of the body. In mammals, the abdomen is separated from the thorax by the diaphragm, a sheet of muscular tissue; in arthropods, commonly by a narrow constriction. The female reproductive organs are in the abdomen. In insects and spiders, it is characterized by the absence of limbs.

aberration of starlight apparent displacement of a star from its true position, due to the combined effects of the speed of light and the speed of the Earth in orbit around the Sun (about 30 km per sec/18.5 mi per sec).

aberration, optical any of a number of defects that impair the image in an optical instrument. Aberration occurs because of minute variations in lenses and mirrors, and because different parts of the light ◊spectrum are reflected or refracted by varying amounts.

In *chromatic aberration* the image is surrounded by coloured fringes, because light of different colours is brought to different focal points by a lens. In *spherical aberration* the image is blurred because different parts of a spherical lens or mirror have different focal lengths. In *astigmatism* the image appears elliptical or cross-shaped because of an irregularity in the curvature of the lens. In *coma* the images appear to be progressively elongated towards the edge of the field of view.

abiotic factor a nonorganic variable within the ecosystem, affecting the life of organisms. Examples include temperature, light, and soil structure. Abiotic factors can be harmful to the environment, as when sulphur dioxide emissions from power stations produce acid rain.

abrasive substance used for cutting and polishing or for removing small amounts of the surface of hard materials. There are two types: natural and artificial abrasives, and their hardness is measured using the ◊Mohs' scale. Natural abrasives include quartz, sandstone, pumice, diamond, and corundum; artificial abrasives include rouge, whiting, and carborundum.

abscissa in ◊coordinate geometry, the *x*-coordinate of a point – that is, the horizontal distance of that point from the vertical or *y*-axis. For example, a point with the coordinates (4, 3) has an abscissa of 4. The *y*-coordinate of a point is known as the ◊ordinate.

abscissin or *abscissic acid* plant hormone found in all higher plants. It is involved in the process of ◊abscission and also inhibits stem elongation, germination of seeds, and the sprouting of buds.

abscission in botany, the controlled separation of part of a plant from the main plant body – most commonly, the falling of leaves or the dropping of fruit controlled by ◊abscissin. In ◊deciduous plants the leaves are shed before the winter or dry season, whereas ◊evergreen plants drop their leaves continually throughout the year. Fruitdrop, the abscission of fruit while still immature, is a naturally occurring process.

absolute value or *modulus* in mathematics, the value, or magnitude, of a number irrespective of its sign. The absolute value of a number n is written $|n|$ (or sometimes as mod n), and is defined as the positive square root of n^2. For example, the numbers -5 and 5 have the same absolute value:

$$|5| = |-5| = 5$$

absolute zero lowest temperature theoretically possible, zero Kelvin (0K), equivalent to $-273.15°C/-459.67°F$, at which molecules are motionless. Although the third law of ◊thermodynamics indicates the impossibility of reaching absolute zero exactly, a temperature of $7 \times 10^{-6}K$ (seven millionths of a degree above absolute zero) was produced 1993 at Lancaster University, England. Near absolute zero, the physical properties of some materials change substantially (see ◊cryogenics); for example, some metals lose their electrical resistance and become superconductive.

absorption in science, the taking up of one substance by another, such as a liquid by a solid (ink by blotting paper) or a gas by a liquid (ammonia by water). In biology, absorption describes the passing of nutrients or medication into and through tissues such as intestinal walls and blood vessels. In physics, absorption is the phenomenon by which a substance retains radiation of particular wavelengths; for example, a piece of blue glass absorbs all visible light except the wavelengths in the blue part of the spectrum; it also refers to the partial loss of energy resulting from light and other electromagnetic waves passing through a medium. In nuclear physics, absorption is the capture by elements, such as boron, of neutrons produced by fission in a reactor.

absorption spectroscopy or *absorptiometry* in analytical chemistry, a technique for determining the identity or amount present of a chemical substance by measuring the amount of electromagnetic radiation the substance absorbs at specific wavelengths; see ◊spectroscopy.

abzyme in biotechnology, an artificially created antibody that can be used like an enzyme to accelerate reactions.

AC in physics, abbreviation for ◊*alternating current*.

acceleration rate of change of the velocity of a moving body. It is usually measured in metres per second per second (m s^{-2}) or feet per second per second (ft s^{-2}). Because velocity is a ◊vector quantity (possessing both magnitude and direction) a body travelling at constant speed may be said to be accelerating if its direction of motion changes. According to Newton's second law of motion, a body will accelerate only if it is acted upon by an unbalanced, or resultant, ◊force.

Acceleration due to gravity is the acceleration of a body falling freely under the influence of the Earth's gravitational field; it varies slightly at different latitudes and altitudes. The value adopted internationally for gravitational acceleration is 9.806 m s^{-2}/32.174 ft s^{-2}.

acceleration, secular in astronomy, the continuous and nonperiodic change in orbital velocity of one body around another, or the axial rotation period of a body.

acceleration, uniform in physics, acceleration in which the velocity of a body changes by equal amounts in successive time intervals. Uniform acceleration is represented by a straight line on a ◊speed–time graph.

accelerator in physics, a device to bring charged particles (such as protons and electrons) up to high speeds and energies, at which they can be of use in industry, medicine, and pure physics. At low energies, accelerated particles can be used to produce the image on a television screen and generate X-rays (by means of a ◊cathode-ray tube), destroy tumour cells, or kill bacteria. When high-energy particles collide with other particles, the fragments formed reveal the nature of the fundamental forces of nature.

The first accelerators used high voltages (produced by ◊van de Graaff generators) to generate a strong, unvarying electric field. Charged particles were accelerated as they passed through the electric field. However, because the voltage produced by a generator is limited, these accelerators were replaced by machines where the particles passed through regions of alternating electric fields, receiving a succession of small pushes to accelerate them. The first of these accelerators was the ◊linear accelerator or *linac*.

Another way of making repeated use of an electric field is to bend the path of a particle into a circle so that it passes repeatedly through the same electric field. The first accelerator to use this idea was the *cyclotron*. This consists of an electromagnet with two hollow metal semicircular structures, called dees, supported between the poles of an electromagnet. Particles such as protons are introduced at the centre of the machine and travel outwards in a spiral path, being accelerated by an oscillating electric field each time they pass through the gap between the dees. Cyclotrons can accelerate particles up to energies of 25 MeV (25 million electron volts).

In the ◊synchrotron, particles travel in a circular path of constant radius, guided by electromagnets. The strengths of the electromagnets are varied to keep the particles on an accurate path. Electric fields at points around the path accelerate the particles.

accelerometer apparatus, either mechanical or electromechanical, for measuring ◊acceleration or deceleration – that is, the rate of increase or decrease in the ◊velocity of a moving object.

access time or *reaction time* in computing, the time taken by a computer, after an instruction has been given, to read from or write to ◊memory.

acclimation or *acclimatization* the physiological changes induced in an organism by exposure to new environmental conditions. When humans move to higher altitudes, for example, the number of red blood cells rises to

accomodation *The mechanism by which the shape of the lens in the eye is changed so that clear images of objects, whether distant or near, can be focused on the retina.*

increase the oxygen-carrying capacity of the blood in order to compensate for the lower levels of oxygen in the air.

accommodation in biology, the ability of the ◊eye to focus on near or far objects by changing the shape of the lens.

For an object to be viewed clearly its image must be precisely focused on the retina, the light-sensitive layer of cells at the rear of the eye. Close objects can be seen when the lens takes up a more spherical shape, far objects when the lens is flattened. These changes in shape are caused by the movement of ligaments attached to a ring of ciliary muscles lying beneath the iris.

accumulator in electricity, a storage ◊battery – that is, a group of rechargeable secondary cells. A familiar example is the lead–acid car battery.

acesulfame-K non-carbohydrate sweetener that is up to 300 times as sweet as sugar. It is used in soft drinks and desserts.

acetabulum deep socket in the hip bone in which the head of the thigh bone (◊femur) is contained.

acetaldehyde common name for ◊ethanal.

acetate common name for ◊ethanoate.

acetic acid common name for ◊ethanoic acid.

acetone common name for ◊propanone.

acetylcholine (ACh) chemical that serves as a ◊neurotransmitter, communicating nerve impulses between the cells of the nervous system. It is largely associated with the transmission of impulses across the ◊synapse (junction) between nerve and muscle cells, causing the muscles to contract.

Postmortems on sufferers of Alzheimer's disease show a deficit of ACh. This has lead to the hypothesis 1993 that loss of ACh contributes to memory loss.

acetylene common name for ◊ethyne.

acetylsalicylic acid chemical name for the painkilling drug ◊aspirin.

achene dry, one-seeded ◊fruit that develops from a single ◊ovary and does not split open to disperse the seed. Achenes commonly occur in groups – for example, the fruiting heads of buttercup *Ranunculus* and clematis. The outer surface may be smooth, spiny, ribbed, or tuberculate, depending on the species.

An achene with part of the fruit wall extended to form a membranous wing is called a *samara*; an example is the pendulous fruit of the ash *Fraxinus*. A ◊*caryopsis*, another type of achene, is formed when the ◊carpel wall becomes fused to the seed coat and is typical of grasses and cereals. A *cypsela* is derived from an inferior ovary and is characteristic of the daisy family (Compositae). It often has a ◊pappus of hairs attached, which aids its dispersal by the wind, as in the dandelion.

Achilles tendon tendon at the back of the ankle attaching the calf muscles to the heel bone. It is one of the largest tendons in the human body.

achromatic lens combination of lenses made from materials of different refractive indexes, constructed in such a way as to minimize chromatic aberration (which in a single lens causes coloured fringes around images because the lens diffracts the different wavelengths in white light to slightly different extents).

acid compound that, in solution in an ionizing solvent (usually water), gives rise to hydrogen ions (H⁺ or protons). In modern chemistry, acids are defined as substances that are proton donors and accept electrons to form ◊ionic bonds. Acids react with ◊bases to form salts, and they act as solvents. Strong acids are corrosive; dilute acids have a sour or sharp taste, although in some organic acids this may be partially masked by other flavour characteristics.

Acids can be detected by using coloured indicators such as ◊litmus and methyl orange. The strength of an acid is measured by its hydrogen-ion concentration, indicated by the ◊pH value. Acids are classified as monobasic, dibasic, tribasic, and so forth, according to the number of hydrogen atoms, replaceable by bases, in a molecule. The first known acid was vinegar (ethanoic or acetic acid). Inorganic acids include boric, carbonic, hydrochloric, hydrofluoric, nitric, phosphoric, and sulphuric. Organic acids include acetic, benzoic, citric, formic, lactic, oxalic, and salicylic, as well as complex substances such as ◊nucleic acids and ◊amino acids.

acid–base balance essential balance between the amount of carbonic acid and bicarbonate (an alkali) in the blood. It must be kept constant so that the hydrogen ion concentration in the blood plasma is in turn kept constant. Any deviation in acid–base balance can have profound effects on physiological functioning.

acidic oxide oxide of a ◊nonmetal. Acidic oxides are covalent compounds. Those that dissolve in water, such as sulphur dioxide, give acidic solutions.

$$SO_2 + H_2O \rightleftharpoons H_2SO_{3(aq)} \rightleftharpoons H^+_{(aq)} + HSO_{3(aq)}^-$$

All acidic oxides react with alkalis to form salts.

$$CO_2 + NaOH \rightarrow NaHCO_3$$

acid rain acidic precipitation thought to be caused principally by the release into the atmosphere of sulphur dioxide (SO_2) and oxides of nitrogen. Sulphur dioxide is formed by the burning of fossil fuels, such as coal, that contain high quantities of sulphur; nitrogen oxides are contributed from various industrial activities and from car exhaust fumes.

acid salt chemical compound formed by the partial neutralization of a dibasic or tribasic ◊acid (one that contains two or three hydrogen atoms). Although a salt, it contains replaceable hydrogen, so it may undergo the typical reactions of an acid. Examples are sodium hydrogen sulphate ($NaHSO_4$) and acid phosphates.

acoustic coupler device that enables computer data to be transmitted and received through a normal telephone handset; the handset rests on the coupler to make the connection. A small speaker within the device is used to convert the computer's digital output data into sound signals, which are then picked up by the handset and transmitted through the telephone system. At the receiving telephone, a second acoustic coupler or modem converts the sound signals back into digital data for input into a computer.

acoustic ohm c.g.s. unit of acoustic impedance (the ratio of the sound pressure on a surface to the sound flux through the surface). It is analogous to the ohm as the unit of electrical ◊impedance.

acoustics in general, the experimental and theoretical science of sound and its transmission; in particular, that branch of the science that has to do with the phenomena of sound in a particular space such as a room or theatre. In architecture, the sound-reflecting character of a room.

acquired character feature of the body that develops during the lifetime of an individual, usually as a result of repeated use or disuse, such as the enlarged muscles of a weightlifter.

French naturalist Jean Baptiste Lamarck's (1744–1829) theory of evolution assumed that acquired characters were passed from parent to offspring. Modern evolutionary theory does not recognize the inheritance of acquired characters because there is no reliable scientific evidence that it occurs, and because no mechanism is known whereby bodily changes can influence the genetic material. See also ◊central dogma.

acquired immune deficiency syndrome full name for the disease ◊AIDS.

acre-foot unit sometimes used to measure large volumes of water, such as the capacity of a reservoir (equal to its area in acres multiplied by its average depth in feet). One acre-foot equals 1,233.5 cu m/43,560 cu ft or the amount of water covering one acre to a depth of one foot.

acridine $C_{13}H_9N$ organic compound that occurs in coal tar. It is extracted by dilute acids but can also be obtained synthetically. It is used to make dyes and drugs.

acrylic acid common name for ◊propenoic acid.

ACTH (abbreviation for *adrenocorticotrophic hormone*) ◊hormone secreted by the anterior lobe of the ◊pituitary gland. It controls the production of corticosteroid hormones by the ◊adrenal gland and is commonly produced as a response to stress.

actinide any of a series of 15 radioactive metallic chemical elements with atomic numbers 89 (actinium) to 103 (lawrencium). Elements 89 to 95 occur in nature; the rest of the series are synthesized elements only. Actinides are grouped together because of their chemical similarities (for example, they are all bivalent), the properties differing only slightly with atomic number. The series is set out in a band in the ◊periodic table of the elements, as are the ◊lanthanids.

actinium white, radioactive, metallic element, the first of the actinide series, symbol Ac, atomic number 89, relative atomic mass 227; it is a weak emitter of high-energy alpha particles.

Actinium occurs with uranium and radium in ◊pitchblende and other ores, and can be synthesized by bombarding radium with neutrons. The longest-lived isotope, Ac-227, has a half-life of 21.8 years (all the other isotopes have very short half-lives).

actinium K original name given to the radioactive element ◊francium 1939 by its discoverer, the French scientist Marguerite Perey (1909–1975).

action potential in biology, a change in the potential difference (voltage) across the membrane of a nerve cell when an impulse passes along it. A change in potential (from about –60 to +45 millivolts) accompanies the passage of sodium and potassium ions across the membrane.

activation analysis in analytical chemistry, a technique used to reveal the presence and amount of minute impurities in a substance or element. A sample of a material that may contain traces of a certain element is irradiated with ◊neutrons, as in a reactor. Gamma rays emitted by the material's radioisotope have unique energies and relative intensities, similar to the spectral lines from a luminous gas. Measurements and interpretation of the gamma-ray spectrum, using data from standard samples for comparison, provide information on the amount of impurities present.

activation energy in chemistry, the energy required in order to start a chemical reaction. Some elements and compounds will react together merely by bringing them into contact (spontaneous reaction). For others it is necessary to supply energy in order to start the reaction, even if there is ultimately a net output of energy. This initial energy is the activation energy.

active transport in cells, the use of energy to move substances, usually molecules or ions, across a membrane.

Energy is needed because movement occurs against a concentration gradient, with substances being passed into a region where they are already present in significant quantities. Active transport thus differs from diffusion, the process by which substances move towards a region where they are in lower concentration, as when oxygen passes into the blood vessels of the lungs. Diffusion requires no input of energy.

activity in physics, the number of particles emitted in one second by a radioactive source. The term is used to describe the radioactivity or the potential danger of that source. The unit of activity is the becquerel (Bq).

activity series in chemistry, alternative name for ◊reactivity series.

Ada high-level computer-programming language, developed and owned by the US Department of Defense, designed for use in situations in which a computer directly controls a process or machine, such as a military aircraft. The language took more than five years to specify, and became commercially available only in the late 1980s. It is named after English mathematician Ada Augusta Byron (1815–1851).

adaptation in biology, any change in the structure or function of an organism that allows it to survive and reproduce more effectively in its environment. In ◊evolution, adaptation is thought to occur as a result of random variation in the genetic make-up of organisms coupled with ◊natural selection. Species become extinct when they are no longer adapted to their environment – for instance, if the climate suddenly becomes colder.

adaptive radiation in evolution, the formation of several species, with adaptations to different ways of life, from a single ancestral type. Adaptive radiation is likely to occur whenever members of a species migrate to a new habitat with unoccupied ecological niches. It is thought that the lack of competition in such niches allows sections of the migrant population to develop new adaptations, and eventually to become new species.

The colonization of newly formed volcanic islands has led to the development of many unique species. The 13 species of Darwin's finch on the Galápagos Islands, for example, are all descended from a single species from the South American mainland. The parent stock evolved into different species that now occupy a range of diverse niches.

ADC in electronics, abbreviation for ◊*analogue-to-digital converter*.

addition polymerization ◊polymerization reaction in which a single monomer gives rise to a single polymer, with no other reaction products.

addition reaction chemical reaction in which the atoms of an element or compound react with a double bond or triple bond in an organic compound by opening up one of the bonds and becoming attached to it, for example:

$$CH_2=CH_2 + HCl \rightarrow CH_3CH_2Cl$$

An example is the addition of hydrogen atoms to ◊unsaturated compounds in vegetable oils to produce margarine.

additive in food, any natural or artificial chemical added to prolong the shelf life of processed foods (salt or nitrates), alter the colour or flavour of food, or improve its food value (vitamins or minerals). Many chemical additives are used and they are subject to regulation, since individuals may be affected by constant exposure even to traces of certain additives and may suffer side effects ranging from headaches and hyperactivity to cancer. Food companies in many countries are now required by law to list additives used in their products. Within the European Union, approved additives are given an official ◊E number.

address in a computer memory, a number indicating a specific location. At each address, a single piece of data can be stored. For microcomputers, this normally amounts to one ◊byte (enough to represent a single character, such as a letter or digit).

adenoids masses of lymphoid tissue, similar to ◊tonsils, located in the upper part of the throat, behind the nose. They are part of a child's natural defences against the entry of germs but usually shrink and disappear by the age of ten.

adenosine triphosphate compound present in cells. See ◊ATP.

adenovirus any of a group of DNA-containing viruses that cause respiratory infections and conjunctivitis.

ADH in biology, abbreviation for *antidiuretic hormone*, part of the system maintaining a correct salt/water balance in vertebrates.

Its release is stimulated by the ◊hypothalamus in the brain, which constantly receives information about salt concentration from receptors situated in the neck. In conditions of water shortage increased ADH secretion from the brain will cause more efficient conservation of water in the kidney, so that fluid is retained by the body. When an animal is able to take in plenty of water, decreased ADH secretion will cause the urine to become dilute and plentiful. The system allows the body to compensate for a varying fluid intake and maintain a correct balance.

adhesive substance that sticks two surfaces together. Natural adhesives (glues) include gelatin in its crude industrial form (made from bones, hide fragments, and fish offal) and vegetable gums. Synthetic adhesives include thermoplastic and thermosetting resins, which are often stronger than the substances they join; mixtures of ◊epoxy resin and hardener that set by chemical reaction; and elastomeric (stretching) adhesives for flexible joints. Superglues are fast-setting adhesives used in very small quantities.

adiabatic in physics, a process that occurs without loss or gain of heat, especially the expansion or contraction of a gas in which a change takes place in the pressure or volume, although no heat is allowed to enter or leave.

adipose tissue type of ◊connective tissue of vertebrates that serves as an energy reserve, and also pads some organs. It is commonly called fat tissue, and consists of large spherical cells filled with fat. In mammals, major layers are in the inner layer of skin and around the kidneys and heart.

adolescence in the human life cycle, the period between the beginning of puberty and adulthood.

ADP (abbreviation for *adenosine diphosphate*) in biology, used in the manufacture of ◊ATP, the molecule used by all cells to drive their chemical reactions.

adrenal gland or *suprarenal gland* triangular gland situated on top of the kidney. The adrenals are soft and yellow, and consist of two parts: the cortex and medulla. The *cortex* (outer part) secretes various steroid hormones, controls salt and water metabolism, and regulates the use of carbohydrates, proteins, and fats. The *medulla* (inner part) secretes the hormones adrenaline and noradrenaline which, during times of stress, cause the heart to beat faster and harder, increase blood flow to the heart and muscle cells, and dilate airways in the lungs, thereby delivering more oxygen to cells throughout the body and in general preparing the body for 'fight or flight'.

adrenaline or *epinephrine* hormone secreted by the medulla of the ◊adrenal glands.

adrenocorticotrophic hormone hormone secreted by the anterior lobe of the pituitary gland; see ◊ACTH.

adsorption taking up of a gas or liquid at the surface of another substance, usually a solid (for example, activated charcoal adsorbs gases). It involves molecular attraction at the surface, and should be distinguished from ◊absorption (in which a uniform solution results from a gas or liquid being incorporated into the bulk structure of a liquid or solid).

advanced gas-cooled reactor (AGR) type of ◊nuclear reactor widely used in W Europe. The AGR uses a fuel of enriched uranium dioxide in stainless-steel cladding and a moderator of graphite. Carbon dioxide gas is pumped through the reactor core to extract the heat produced by the ◊fission of the uranium. The heat is transferred to water in a steam generator, and the steam drives a turbogenerator to produce electricity.

adventitious in botany, arising in an abnormal position, as in roots developing on the stem of a cutting or buds developing on roots.

aerenchyma plant tissue with numerous air-filled spaces between the cells. It occurs in the stems and roots of many aquatic plants where it aids buoyancy and facilitates transport of oxygen around the plant.

aerial or *antenna* in radio and television broadcasting, a conducting device that radiates or receives electromagnetic waves. The design of an aerial depends principally on the wavelength of the signal. Long waves (hundreds of metres in wavelength) may employ long wire aerials; short waves (several centimetres in wavelength) may employ rods and dipoles; microwaves may also use dipoles – often with reflectors arranged like a toast rack – or highly directional parabolic dish aerials. Because microwaves travel in straight lines, giving line-of-sight communication, microwave aerials are usually located at the tops of tall masts or towers.

aerial oxidation in chemistry, a reaction in which air is used to oxidize another substance, as in the contact process for the manufacture of sulphuric acid, and in the ◊souring of wine.

aerobic in biology, descriptive of those living organisms that require oxygen (usually dissolved in water) for the efficient release of energy contained in food molecules, such as glucose. They include almost all living organisms (plants as well as animals) with the exception of certain bacteria, yeasts, and internal parasites.

Aerobic reactions occur inside every cell and lead to the formation of energy-rich ◊ATP, subsequently used by the cell for driving its metabolic processes. Oxygen is used to convert glucose to carbon dioxide and water, thereby releasing energy.

aerodynamics branch of fluid physics that studies the forces exerted by air or other gases in motion – for example, the airflow around bodies (such as land vehicles, bullets, rockets, and aircraft) moving at speed through the atmosphere. For maximum efficiency, the aim is usually to design the shape of an object to produce a streamlined flow, with a minimum of turbulence in the moving air.

aerogel light, transparent, highly porous material composed of more than 90% air. Such materials are formed from silica, metal oxides, and organic chemicals, and are produced by drying gels – networks of linked molecules suspended in a liquid – so that air fills the spaces previously occupied by the liquid. They are excellent heat insulators and have unusual optical, electrical, and acoustic properties.

aeronautics science of travel through the Earth's atmosphere, including aerodynamics, aircraft structures, jet and rocket propulsion, and aerial navigation.

In *subsonic aeronautics* (below the speed of sound), aerodynamic forces increase at the rate of the square of the speed.

Transsonic aeronautics covers the speed range from just below to just above the speed of sound and is crucial to aircraft design. Ordinary sound waves move at about 1,225 kph/760 mph at sea level, and air in front of an aircraft moving slower than this is 'warned' by the waves so that it can move aside. However, as the flying speed approaches that of the sound waves, the warning is too late for the air to escape, and the aircraft pushes the air aside, creating shock waves, which absorb much power and create design problems. On the ground the shock waves give rise to a ◊sonic boom. It was once thought that the speed of sound was a speed limit to aircraft, and the term ◊sound barrier came into use.

Supersonic aeronautics concerns speeds above that of sound and in one sense may be considered a much older study than aeronautics itself, since the study of the flight of bullets, known as ◊ballistics, was undertaken soon after the introduction of firearms. *Hypersonics* is the study of airflows and forces at speeds above five times that of sound (Mach 5); for example, for guided missiles, space rockets, and advanced concepts such as HOTOL (horizontal takeoff and landing). For all flight speeds streamlining is necessary to reduce the effects of air resistance.

aeroplane (US *airplane*) powered heavier-than-air craft supported in flight by fixed wings. Aeroplanes are propelled by the thrust of a jet engine or airscrew (propeller). They must be designed aerodynamically, since streamlining ensures maximum flight efficiency.

aerosol particles of liquid or solid suspended in a gas. Fog is a common natural example. Aerosol cans contain a substance such as scent or cleaner packed under pressure with a device for releasing it as a fine spray.

Most aerosols used chlorofluorocarbons (CFCs) as propellants until these were found to cause destruction of the ◊ozone layer in the stratosphere.

aestivation in zoology, a state of inactivity and reduced metabolic activity, similar to ◊hibernation, that occurs during the dry season in species such as lungfish and snails. In botany, the term is used to describe the way in which flower petals and sepals are folded in the buds. It is an important feature in ◊plant classification.

aether alternative form of ◊ether, the hypothetical medium once believed to permeate all of space.

affinity in chemistry, the force of attraction (see ◊bond) between atoms that helps to keep them in combination in a molecule. The term is also applied to attraction between molecules, such as those of biochemical significance (for example, between ◊enzymes and substrate molecules). This is the basis for affinity ◊chromatography, by which biologically important compounds are separated.

afterbirth in mammals, the placenta, umbilical cord and ruptured membranes, which become detached from the uterus and expelled soon after birth. In the natural world it is often eaten.

afterburning method of increasing the thrust of a gas turbine (jet) aeroplane engine by spraying additional fuel into the hot exhaust duct between the turbojet and the tailpipe where it ignites. Used for short-term increase of power during takeoff, or during combat in military aircraft.

afterimage persistence of an image on the retina of the eye after the object producing it has been removed. This leads to persistence of

aerosol *The aerosol can produces a fine spray of liquid particles, called an aerosol. When the top button is pressed, a valve is opened, allowing the pressurized propellant in the can to force out a spray of the liquid contents. As the liquid sprays from the can, the small amount of propellant dissolved in the liquid vaporizes, producing a fine spray of small droplets.*

vision, a necessary phenomenon for the illusion of continuous movement in films and television. The term is also used for the persistence of sensations other than vision.

after-ripening process undergone by the seeds of some plants before germination can occur. The length of the after-ripening period in different species may vary from a few weeks to many months.

It helps seeds to germinate at a time when conditions are most favourable for growth. In some cases the embryo is not fully mature at the time of dispersal and must develop further before germination can take place. Other seeds do not germinate even when the embryo is mature, probably owing to growth inhibitors within the seed that must be leached out or broken down before germination can begin.

agar jellylike carbohydrate, obtained from seaweeds. It is used mainly in microbiological experiments as a culture medium for growing bacteria and other microorganisms. The agar is resistant to breakdown by microorganisms, remaining a solid jelly throughout the course of the experiment.

agate banded or cloudy type of ▷chalcedony, a silica, SiO_2, that forms in rock cavities. Agates are used as ornamental stones and for art objects.

ageing in common usage, the period of deterioration of the physical condition of a living organism that leads to death; in biological terms, the entire life process.

Agent Orange selective weedkiller, notorious for its use in the 1960s during the Vietnam War by US forces to eliminate ground cover which could protect enemy forces. It was subsequently discovered to contain highly poisonous ▷dioxin. Thousands of US troops who had handled it later developed cancer or fathered deformed babies.

Agent Orange, named after the distinctive orange stripe on its packaging, combines equal parts of 2,4-D (2,4–dichlorophenoxyacetic acid) and 2,4,5-T (2,4,5–trichlorophenoxyacetic acid), both now banned in the USA.

age–sex graph graph of the population of an area.

agglutination in biology, the clumping together of ▷antigens, such as red blood cells or bacteria, to form larger, visible clumps, under the influence of ▷antibodies. As each antigen clumps only in response to its particular antibody, agglutination provides a way of determining ▷blood groups and the identity of unknown bacteria.

aggression in biology, behaviour used to intimidate or injure another organism (of the same or of a different species), usually for the purposes of gaining territory, a mate, or food. Aggression often involves an escalating series of threats aimed at intimidating an opponent without having to engage in potentially dangerous physical contact. Aggressive signals include roaring by red deer, snarling by dogs, the fluffing up of feathers by birds, and the raising of fins by some species of fish.

AGR abbreviation for ▷*advanced gas-cooled reactor*, a type of nuclear reactor.

AI abbreviation for ▷*artificial intelligence*.

AIDS (acronym for *acquired immune deficiency syndrome*) the gravest of the sexually transmitted diseases. It is caused by the human immunodeficiency virus (HIV), now known to be a ▷retrovirus, an organism first identified 1983. HIV is transmitted in body fluids, mainly blood and genital secretions.

air the Earth's ▷atmosphere.

air conditioning system that controls the state of the air inside a building or vehicle. A complete air-conditioning unit controls the temperature and humidity of the air, removes dust and odours from it, and circulates it by means of a fan.

air mass large body of air with particular characteristics of temperature and humidity. An air mass forms when air rests over an area long enough to pick up the conditions of that area. For example, an air mass formed over the Sahara will be hot and dry. When an air mass moves to another area it affects the weather of that area, but its own characteristics become modified in the process. For example, a Saharan air mass becomes cooler as it moves northwards.

air pollution contamination of the atmosphere caused by the discharge, accidental or deliberate, of a wide range of toxic airborne substances. Often the amount of the released substance is relatively high in a certain locality, so the harmful effects become more noticeable. The cost of preventing any discharge of pollutants into the air is prohibitive, so attempts are more usually made to reduce gradually the amount of discharge and to disperse this as quickly as possible by using a very tall chimney, or by intermittent release.

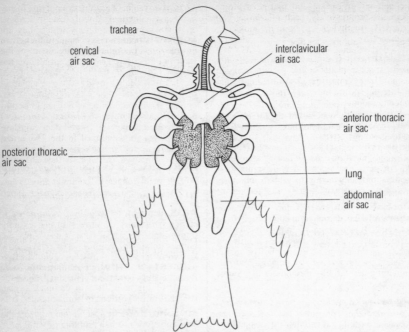

air sac *The air sacs of birds are thin-walled extensions of the lungs. They lie in the abdomen and thorax, and extend even into some of the bones. Compression and extension of the sacs assist ventilation of the lungs and respiratory system.*

air sac in birds, a thin-walled extension of the lungs. There are nine of these and they extend into the abdomen and bones, effectively increasing lung capacity. In mammals, it is another name for the alveoli in the lungs, and in some insects, for widenings of the trachea.

alabaster naturally occurring fine-grained white or light-coloured translucent form of ◊gypsum, often streaked or mottled.

albedo the fraction of the incoming light reflected by a body such as a planet. A body with a high albedo, near 1, is very bright, while a body with a low albedo, near 0, is dark. The Moon has an average albedo of 0.12, Venus 0.65, Earth 0.37.

albinism rare hereditary condition in which the body has no tyrosinase, one of the enzymes that form the pigment ◊melanin, normally found in the skin, hair, and eyes. As a result, the hair is white and the skin and eyes are pink. The skin and eyes are abnormally sensitive to light, and vision is often impaired. The condition occurs among all human and animal groups

albumin or *albumen* any of a group of sulphur-containing ◊proteins. The best known is in the form of egg white; others occur in milk, and as a major component of serum. They are soluble in water and dilute salt solutions, and are coagulated by heat.

alcohol any member of a group of organic chemical compounds characterized by the presence of one or more aliphatic OH (hydroxyl) groups in the molecule, and which form ◊esters with acids. The main uses of alcohols are as solvents for gums, resins, lacquers, and varnishes; in the making of dyes; for essential oils in perfumery; and for medical substances in pharmacy. Alcohol (ethanol) is produced naturally in the ◊fermentation process and is consumed as part of alcoholic beverages.

Alcohols may be liquids or solids, according to the size and complexity of the molecule. The five simplest alcohols form a series in which the number of carbon and hydrogen atoms increases progressively, each one having an extra CH_2 (methylene) group in the molecule: methanol or wood spirit (methyl alcohol, CH_3OH); ethanol (ethyl alcohol, C_2H_5OH); propanol (propyl alcohol, C_3H_7OH); butanol (butyl alcohol, C_4H_9OH); and pentanol (amyl alcohol, $C_5H_{11}OH$). The lower alcohols are liquids that mix with water; the higher alcohols, such as pentanol, are oily liquids immiscible with water; and the highest are waxy solids for example, hexadecanol (cetyl alcohol, $C_{16}H_{33}OH$) and melissyl alcohol ($C_{30}H_{61}OH$), which occur in sperm-whale oil and beeswax respectively. Alcohols containing the CH_2OH group are primary; those containing CHOH are secondary; while those containing COH are tertiary.

alcoholic solution solution produced when a solute is dissolved in ethanol.

Aldebaran or *Alpha Tauri* brightest star in the constellation Taurus and the 14th brightest star in the sky; it marks the eye of the 'bull'. Aldebaran is a red giant 60 light years away, shining with a true luminosity of about 100 times that of the Sun.

aldehyde any of a group of organic chemical compounds prepared by oxidation of primary alcohols, so that the OH (hydroxyl) group loses its hydrogen to give an oxygen joined by a double bond to a carbon atom (the aldehyde group, with the formula CHO).

The name is made up from *al*cohol *dehyd*rogenation – that is, alcohol from which hydrogen has been removed. Aldehydes are usually liquids and include methanal (formaldehyde), ethanal (acetaldehyde), and benzaldehyde.

alexandrite rare gemstone variety of the mineral chrysoberyl (beryllium aluminium oxide $BeAl_2O_4$), which is green in daylight but appears red in artificial light.

algae (singular *alga*) diverse group of plants (including those commonly called seaweeds) that shows great variety of form, ranging from single-celled forms to multicellular seaweeds of considerable size and complexity.

Algae were formerly included within the division Thallophyta, together with fungi and bacteria. Their classification changed with increased awareness of the important differences existing between the algae and Thallophyta, and also between the groups of algae themselves; many botanists now place each algal group in a separate class or division of its own.

algebra system of arithmetic applying to any set of non-numerical symbols (usually letters), and the axioms and rules by which they are combined or operated upon; sometimes known as *generalized arithmetic*.

alginate salt of alginic acid, $(C_6H_8O_6)_n$, obtained from brown seaweeds and used in textiles, paper, food products, and pharmaceuticals.

ALGOL (acronym for *algorithmic language*) in computing, an early high-level programming language, developed in the 1950s and 1960s for scientific applications. A general-purpose language, ALGOL is best suited to mathematical work and has an algebraic style. Although no longer in common use, it has greatly influenced more recent languages, such as Ada and PASCAL.

Algol or *Beta Persei* ◊eclipsing binary, a pair of rotating stars in the constellation Perseus, one of which eclipses the other every 69 hours, causing its brightness to drop by two-thirds.

alimentary canal in animals, the tube through which food passes; it extends from the mouth to the anus. It is a complex organ, adapted for ◊digestion. In human adults, it is about 9 m/30 ft long, consisting of the mouth cavity, pharynx, oesophagus, stomach, and the small and large intestines.

aliphatic compound any organic chemical compound in which the carbon atoms are joined in straight chains, as in hexane (C_6H_{14}), or in branched chains, as in 2-methylpentane ($CH_3CH(CH_3)CH_2CH_2CH_3$).

Aliphatic compounds have bonding electrons localized within the vicinity of the bonded atoms. ◊Cyclic compouns that do not have delocalized electrons are also aliphatic, as in the alicyclic compound cyclohexane (C_6H_{12}) or the heterocyclic piperidine ($C_5H_{11}N$). Compare ◊aromatic compound.

alkali in chemistry, a compound classed as a ◊base that is soluble in water. Alkalis neutralize acids and are soapy to the touch.

The hydroxides of metals are alkalis; those of sodium and potassium being chemically powerful; both were historically derived from the ashes of plants.

alkali metal any of a group of six metallic elements with similar chemical bonding properties: lithium, sodium, potassium, rubidium, caesium, and francium. They form a linked group in the ◊periodic table of the elements. They are univalent (have a valency of one) and of very low density (lithium, sodium, and potassium float on water); in general they are reactive, soft, low-melting-point metals. Because of their reactivity they are only found as compounds in nature.

alkaline-earth metal any of a group of six metallic elements with similar bonding properties: beryllium, magnesium, calcium, strontium, barium, and radium. They form a linked group in the ◊periodic table of the elements. They are strongly basic, bivalent (have a valency of two), and occur in nature only in compounds.

alkaloid any of a number of physiologically active and frequently poisonous substances contained in some plants. They are usually organic bases and contain nitrogen. They form salts with acids and, when soluble, give alkaline solutions.

Substances in this group are included by custom rather than by scientific rules. Examples include morphine, cocaine, quinine, caffeine, strychnine, nicotine, and atropine. In 1992, epibatidine, a chemical extracted from the skin of an Ecuadorian frog, was identified as a member of an entirely new class of alkaloid. It is an organochlorine compound, which is rarely found in animals, and a powerful painkiller, about 200 times as effective as morphine.

alkane member of a group of ◊hydrocarbons having the general formula C_nH_{2n+2}, commonly known as *paraffins*. Lighter alkanes, such as methane, ethane, propane, and butane, are colourless gases; heavier ones are liquids or solids. In nature they are found in natural gas and petroleum. As alkanes contain only single ◊covalent bonds, they are said to be saturated.

alkene member of the group of ◊hydrocarbons having the general formula C_nH_{2n}, formerly known as *olefins*. Lighter alkenes, such as ethene and propene, are gases, obtained from the ◊cracking of oil fractions. Alkenes are unsaturated compounds, characterized by one or more double bonds between adjacent carbon atoms. They react by addition, and many useful compounds, such as poly(ethene) and bromoethane, are made from them.

alkyne member of the group of ◊hydrocarbons with the general formula C_nH_{2n-2}, formerly known as the *acetylenes*. They are unsaturated compounds, characterized by one or more triple bonds between adjacent carbon atoms. Lighter alkynes, such as ethyne, are gases; heavier ones are liquids or solids.

allele one of two or more alternative forms of a ◊gene at a given position (locus) on a chromosome, caused by a difference in the ◊DNA. Blue and brown eyes in humans are determined by different alleles of the gene for eye colour.

Organisms with two sets of chromosomes (diploids) will have two copies of each gene. If the two alleles are identical the individual is said to be ◊homozygous at that locus; if different, the individual is ◊heterozygous at that locus. Some alleles show ◊dominance over others.

allergy special sensitivity of the body that makes it react with an exaggerated response of

name	molecular structure	structural formula
methane	CH_4	H–C–H with H above and H below

uses: domestic fuel (natural gas)

| ethane | C_2H_6 | H₃C–CH₃ (H–C–C–H with H's) |

uses: industrial fuel and chemical feedstock

| propane | C_3H_8 | H–C–C–C–H with H's |

uses: bottled gas (camping gas)

| butane | C_4H_{10} | H–C–C–C–C–H with H's |

uses: bottled gas (lighter fuel, camping gas)

alkane *The alkanes, or paraffins, burn in air to produce carbon dioxide and water, and react with halogens, otherwise they are unreactive. They are used as fuels and to make other organic substances, such as plastics.*

the natural immune defence mechanism to the introduction of an otherwise harmless foreign substance (*allergen*).

allometry in biology, a regular relationship between a given feature (for example, the size of an organ) and the size of the body as a whole, when this relationship is not a simple proportion of body size. Thus, an organ may increase in size proportionately faster, or slower, than body size does. For example, a human baby's head is much larger in relation to its body than is an adult's.

allotropy property whereby an element can exist in two or more forms (allotropes), each possessing different physical properties but the same state of matter (gas, liquid, or solid). The allotropes of carbon are diamond and graphite. Sulphur has several different forms (flowers of sulphur, plastic, rhombic, and monoclinic). These solids have different crystal structures, as do the white and grey forms of tin and the black, red, and white forms of phosphorus.

alloy metal blended with some other metallic or nonmetallic substance to give it special qualities, such as resistance to corrosion, greater hardness, or tensile strength. Useful alloys include bronze, brass, cupronickel, duralumin, German silver, gunmetal, pewter, solder, steel, and stainless steel.

alluvial deposit layer of broken rocky matter, or sediment, formed from material that has been carried in suspension by a river or stream and dropped as the velocity of the current changes. River plains and deltas are made entirely of alluvial deposits, but smaller pockets can be found in the beds of upland torrents.

alpha in computing, a 64-bit ◊RISC chip launched in 1993 by Digital Equipment (DEC). It is seen as a rival to Intel's ◊Pentium chip.

Alpha Centauri or *Rigil Kent* brightest star in the constellation Centaurus and the third brightest star in the sky. It is actually a triple star (see ◊binary star); the two brighter stars orbit each other every 80 years, and the third, Proxima Centauri, is the closest star to the Sun, 4.2 light years away, 0.1 light years closer than the other two.

alpha decay disintegration of the nucleus of an atom to produce an ◊alpha particle. See also ◊radioactivity.

alpha particle positively charged, high-energy particle emitted from the nucleus of a

common alloys

name	approximate composition	uses
brass	35–10% zinc, 65–90% copper	decorative metalwork, plumbing fittings, industrial tubing
bronze		
common	2% zinc, 6% tin, 92% copper	machinery, decorative work
aluminium	10% aluminium, 90% copper	machinery castings
coinage	1% zinc, 4% tin, 95% copper	coins
cast iron	2–4% carbon, 96–98% iron	decorative metalwork, engine blocks, industrial machinery
dentist's amalgam	30% copper, 70% mercury	dental fillings
duralumin	0.5 % magnesium, 0.5% manganese, 5% copper, 95% aluminium	framework of aircraft
gold		
coinage	10% copper, 90% gold	coins
dental	14–28% silver, 14–28% copper, 58% gold	dental fillings
lead battery plate	6% antimony, 94% lead	car batteries
manganin	1.5% nickel, 16% manganese, 82.5% copper	resistance wire
nichrome	20% chromium, 80% nickel	heating elements
pewter	20% lead, 80% tin	utensils
silver		
coinage	10% copper, 90% silver	coins
solder	50% tin, 50% lead	joining iron surfaces
steel		
stainless	8–20% nickel, 10–20% chromium, 60–80% iron	kitchen utensils
armour	1–4% nickel, 0.5–2% chromium, 95–98% iron	armour plating
tool	2–4% chromium, 6–7% molybdenum, 90–95% iron	tools

radioactive atom. It is one of the products of the spontaneous disintegration of radioactive elements (see ◊radioactivity) such as radium and thorium, and is identical with the nucleus of a helium atom – that is, it consists of two protons and two neutrons. The process of emission, *alpha decay*, transforms one element into another, decreasing the atomic (or proton) number by two and the atomic mass (or nucleon number) by four.

Because of their large mass alpha particles have a short range of only a few centimetres in air, and can be stopped by a sheet of paper. They have a strongly ionizing effect (see ◊ionizing radiation) on the molecules that they strike, and are therefore capable of damaging living cells. Alpha particles travelling in a vacuum are deflected slightly by magnetic and electric fields.

Altair or *Alpha Aquilae* brightest star in the constellation Aquila and the 12th brightest star in the sky. It is a white star 16 light years away and forms the so-called Summer Triangle with the stars Deneb (in the constellation Cygnus) and Vega (in Lyra).

alternating current (AC) electric current that flows for an interval of time in one direction and then in the opposite direction, that is, a current that flows in alternately reversed directions through or around a circuit. Electric energy is usually generated as alternating current in a power station, and alternating currents may be used for both power and lighting.

alternation of generations typical life cycle of terrestrial plants and some seaweeds, in which there are two distinct forms occurring alternately: *diploid* (having two sets of chromosomes) and *haploid* (one set of chromosomes). The diploid generation produces haploid spores by ◊meiosis, and is called the sporophyte, while the haploid generation produces gametes (sex cells), and is called the gametophyte. The gametes fuse to form a diploid ◊zygote which develops into a new sporophyte; thus the sporophyte and gametophyte alternate.

alternator electricity ◊generator that produces an alternating current.

altimeter instrument used in aircraft that measures altitude, or height above sea level. The common type is a form of aneroid ◊barometer, which works by sensing the differences in air pressure at different altitudes. This must continually be recalibrated because of the change in air pressure with changing weather conditions. The ◊radar altimeter measures the height of the aircraft above the ground, measuring the time it takes for radio pulses emitted by the aircraft to be reflected. Radar altimeters are essential features of automatic and blind-landing systems.

altruism in biology, helping another individual of the same species to reproduce more effectively, as a direct result of which the altruist may leave fewer offspring itself. Female honey bees (workers) behave altruistically by rearing sisters in order to help their mother, the queen bee, reproduce, and forgo any possibility of reproducing themselves.

ALU abbreviation for ◊*arithmetic and logic unit*.

alum any double sulphate of a monovalent metal or radical (such as sodium, potassium, or ammonium) and a trivalent metal (such as aluminium or iron). The commonest alum is the double sulphate of potassium and aluminium, $K_2Al_2(SO_4)_4.24H_2O$, a white crystalline powder that is readily soluble in water. It is used in curing animal skins. Other alums are used in papermaking and to fix dye in the textile industry.

alumina or *corundum* Al_2O_3 oxide of aluminium, widely distributed in clays, slates, and shales. It is formed by the decomposition of the feldspars in granite and used as an abrasive. Typically it is a white powder, soluble in most strong acids or caustic alkalis but not in water. Impure alumina is called 'emery'. Rubies and sapphires are corundum gemstones.

aluminium lightweight, silver-white, ductile and malleable, metallic element, symbol Al, atomic number 13, relative atomic mass 26.9815. It is the third most abundant element (and the most abundant metal) in the Earth's crust, of which it makes up about 8.1% by mass. It is an excellent conductor of electricity and oxidizes easily, the layer of oxide on its surface making it highly resistant to tarnish. Commercially, it is prepared by the electrolysis of ◊bauxite. In its pure state aluminium is a weak metal, but when combined with elements such as copper, silicon, or magnesium it forms alloys of great strength. Because of its light weight (specific gravity 2.70), aluminium is widely used in the shipbuilding and aircraft industries.

aluminium chloride $AlCl_3$ white solid made by direct combination of aluminium and chlorine.

$$2Al + 3Cl_2 \rightarrow 2AlCl_3$$

aluminium *The extraction of aluminium from bauxite by electrolysis.*

The anhydrous form is a typical covalent compound (see ◊covalency).

aluminium hydroxide or *alumina cream* Al(OH)₃ gelatinous ◊precipitate formed when a small amount of alkali solution is added to a solution of an aluminium salt.

$$Al_{(aq)}^{3+} + 3OH_{(aq)} \rightarrow Al(OH)_{3(s)}$$

It is an ◊amphoteric compound as it readily reacts with both acids and alkalis.

aluminium ore raw material from which aluminium is extracted. The main ore is bauxite, a mixture of minerals, found in economic quantities in Australia, Guinea, West Indies, and several other countries.

alveolus (plural *alveoli*) one of the many thousands of tiny air sacs in the ◊lungs in

alveolus *The tiny air sacs, called alveoli, in the lungs are covered with a network of blood capillaries, allowing oxygen to enter the blood. The alveoli of an adult have a total surface area of 70 sq m/750 sq ft.*

which exchange of oxygen and carbon dioxide takes place between air and the bloodstream.

AM abbreviation for ◊*amplitude modulation*.

amalgam any alloy of mercury with other metals. Most metals will form amalgams, except iron and platinum. Amalgam is used in dentistry for filling teeth, and usually contains copper, silver, and zinc as the main alloying ingredients. This amalgam is pliable when first mixed and then sets hard, but the mercury leaches out and may cause a type of heavy-metal poisoning.

Amalgamation, the process of forming an amalgam, is a technique sometimes used to extract gold and silver from their ores. The ores are treated with mercury, which combines with the precious metals.

amatol explosive consisting of ammonium nitrate and TNT (trinitrotoluene) in almost any proportions.

amber fossilized resin from coniferous trees of the Middle Tertiary period. It is often washed ashore on the Baltic coast with plant and animal specimens preserved in it; many extinct species have been found preserved in this way. It ranges in colour from red to yellow, and is used to make jewellery.

American National Standards Institute (ANSI) US national standards body. It sets official procedures in (among other areas) computing and electronics.

americium radioactive metallic element of the ◊actinide series, symbol Am, atomic number 95, relative atomic mass 243.13; it was first synthesized 1944. It occurs in nature in minute quantities in ◊pitchblende and other uranium ores, where it is produced from the decay of neutron-bombarded plutonium, and is the element with the highest atomic number that occurs in nature. It is synthesized in quantity only in nuclear reactors by the bombardment of plutonium with neutrons. Its longest-lived isotope is Am-243, with a half-life of 7,650 years.

amethyst variety of ◊quartz, SiO₂, coloured violet by the presence of small quantities of impurities such as manganese or iron; used as a semiprecious stone. Amethysts are found chiefly in the Ural Mountains, India, the USA, Uruguay, and Brazil.

amide any organic chemical derived from a fatty acid by the replacement of the hydroxyl group (–OH) by an amino group (–NH₂). One of the simplest amides is acetamide (CH₃CONH₂), which has a strong mousy odour.

amino acid Amino acids are natural organic compounds that make up proteins and can thus be considered the basic molecules of life. There are 20 different amino acids. They consist mainly of carbon, oxygen, hydrogen, and nitrogen. Each amino acid has a common core structure (consisting of two carbon atoms, two oxygen atoms, a nitrogen atom, and four hydrogen atoms) to which is attached a variable group, known as the R group. In glycine, the R group is a single hydrogen atom; in alanine, the R group consists of a carbon and three hydrogen atoms.

amine any of a class of organic chemical compounds in which one or more of the hydrogen atoms of ammonia (NH_3) have been replaced by other groups of atoms.

amino acid water-soluble organic ◊molecule, mainly composed of carbon, oxygen, hydrogen, and nitrogen, containing both a basic amino group (NH_2) and an acidic carboxyl (COOH) group. When two or more amino acids are joined together, they are known as ◊peptides; ◊proteins are made up of interacting polypeptides (peptide chains consisting of more than three amino acids) and are folded or twisted in characteristic shapes.

Many different proteins are found in the cells of living organisms, but they are all made up of the same 20 amino acids, joined together in varying combinations (although other types of amino acid do occur infrequently in nature). Eight of these, the *essential amino acids*, cannot be synthesized by humans and must be obtained from the diet.

ammeter instrument that measures electric current, usually in ◊amperes.

ammonia NH_3 colourless pungent-smelling gas, lighter than air and very soluble in water. It is made on an industrial scale by the ◊Haber process, and used mainly to produce nitrogenous fertilizers, some explosives, and nitric acid.

In aquatic organisms and some insects, nitrogenous waste (from breakdown of amino acids and so on) is excreted in the form of ammonia, rather than urea as in mammals.

ammoniacal solution in chemistry, a solution produced by dissolving a solute in aqueous ammonia.

ammonite extinct marine ◊cephalopod mollusc of the order Ammonoidea, related to the modern nautilus. The shell was curled in a plane spiral and made up of numerous gas-filled chambers, the outermost containing the body of the animal. Many species flourished between 200 million and 65 million years ago, ranging in size from that of a small coin to 2 m/6 ft across.

ammonium NH_4^+ ion formed when ammonia accepts a proton H^+ from an acid. It is the only positive ion that is not a metal and that forms a series of salts. When heated with an alkali it produces ammonia gas.

ammonium carbonate $(NH_4)_2CO_3$ white, crystalline solid that readily sublimes at room temperature into its constituent gases: ammonia, carbon dioxide, and water. It was formerly used in ◊smelling salts.

ammonium chloride or *sal ammoniac* NH₄Cl a volatile salt that forms white crystals around volcanic craters. It is prepared synthetically for use in 'dry-cell' batteries, fertilizers, and dyes.

ammonium nitrate NH₄NO₃ colourless, crystalline solid, prepared by ◊neutralization of nitric acid with ammonia; the salt is crystallized from the solution. It sublimes on heating.

amniocentesis sampling the amniotic fluid surrounding a fetus in the womb for diagnostic purposes. It is used to detect Down's syndrome and other genetic abnormalities.

amnion innermost of three membranes that enclose the embryo within the egg (reptiles and birds) or within the uterus (mammals). It contains the amniotic fluid that helps to cushion the embryo.

amoeba (plural *amoebae*) one of the simplest living animals, consisting of a single cell and belonging to the ◊protozoa group. The body consists of colourless protoplasm. Its activities are controlled by the nucleus, and it feeds by flowing round and engulfing organic debris. It reproduces by ◊binary fission. Some species of amoeba are harmful parasites.

amp in physics, abbreviation for ampere, a unit of electrical current.

ampere SI unit (abbreviation amp, symbol A) of electrical current. Electrical current is measured in a similar way to water current, in terms of an amount per unit time; one ampere represents a flow of about 6.28×10^{18} ◊electrons per second, or a rate of flow of charge of one coulomb per second. It is named after the French scientist André Ampère.

Ampère's rule rule developed by French physicist André Ampère connecting the direction of an electric current and its associated magnetic currents. It states that if a person were travelling along a current-carrying wire in the direction of conventional current flow (from the positive to the negative terminal), and carrying a magnetic compass, then the north pole of the compass needle would be deflected to the left-hand side.

amphetamine or *speed* powerful synthetic stimulant. Indications for its use today are very restricted because of severe side effects, including addiction. It is a sulphate or phosphate form of $C_9H_{13}N$.

amphibian member of the vertebrate class Amphibia, which generally spend their larval (tadpole) stage in fresh water, transferring to

amoeba *The amoebae are among the simplest living organisms, consisting of a single cell. Within the cell, there is a nucleus, which controls cell activity, and many other microscopic bodies and vacuoles (fluid-filled spaces surrounded by a membrane) with specialized functions. Amoebae eat by flowing around food particles, engulfing the particle; a process called phagocytosis.*

land at maturity (after ◊metamorphosis) and generally returning to water to breed. Like fish and reptiles, they continue to grow throughout life, and cannot maintain a temperature greatly differing from that of their environment. The class includes caecilians (wormlike in appearance), salamanders, frogs, and toads.

amphibole any one of a large group of rock-forming silicate minerals with an internal structure based on double chains of silicon and oxygen, and a general formula $X_2Y_5Si_8O_{22}(OH)_2$. It is closely related to ◊pyroxene. Amphiboles form orthorhombic, monoclinic, and triclinic ◊crystals.

amphoteric term used to describe the ability of some chemical compounds to behave either as an ◊acid or as a ◊base depending on their environment. For example, the metals aluminium and zinc, and their oxides and hydroxides, act as bases in acidic solutions and as acids in alkaline solutions.

amplifier electronic device that magnifies the strength of a signal, such as a radio signal. The ratio of output signal strength to input signal strength is called the *gain* of the amplifier. As well as achieving high gain, an amplifier should be free from distortion and able to operate over a range of frequencies. Practical amplifiers are usually complex circuits, although simple amplifiers can be built from single transistors or valves.

amplitude maximum displacement of an oscillation from the equilibrium position. For a wave motion, it is the height of a crest (or the depth of a trough). With a sound wave, for example, amplitude corresponds to the intensity (loudness) of the sound. In AM (amplitude modulation) radio broadcasting, the required audio-frequency signal is made to modulate (vary slightly) the amplitude of a continuously transmitted radio carrier wave.

amplitude or *argument* in mathematics, the angle in an ◊Argand diagram between the line that represents the complex number and the real (positive horizontal) axis. If the complex number is written in the form $r(\cos \theta + i \sin \theta)$, where r is radius and $i = \sqrt{-1}$, the amplitude is the angle θ.

amplitude modulation (AM) method by which radio waves are altered for the transmission of broadcasting signals. AM waves are constant in frequency, but the amplitude of the transmitting wave varies in accordance with the signal being broadcast.

ampulla in the inner ◊ear, a slight swelling at the end of each semicircular canal, able to sense the motion of the head.

The sense of balance depends largely on sensitive hairs within the ampullae responding to movements of fluid within the canal.

amylase one of a group of ◊enzymes that break down starches into their component molecules (sugars) for use in the body. It occurs widely in both plants and animals. In humans, it is found in saliva and in pancreatic juices.

anabolic steroid any ◊hormone of the ◊steroid group that stimulates tissue growth. Its use in medicine is limited to the treatment of some anaemias and breast cancers; it may help to break up blood clots. Side effects include aggressive behaviour, masculinization in women, and, in children, reduced height.

anabolism process of building up body tissue, promoted by the influence of certain hormones. It is the constructive side of ◊metabolism, as opposed to ◊catabolism.

anaemia condition caused by a shortage of haemoglobin, the oxygen-carrying component of red blood cells. The main symptoms are fatigue, pallor, breathlessness, palpitations, and poor resistance to infection. Treatment depends on the cause.

Anaemia arises either from abnormal loss or defective production of haemoglobin. Excessive loss occurs, for instance, with chronic slow bleeding or with accelerated destruction (◊haemolysis) of red blood cells. Defective production may be due to iron deficiency, vitamin B_{12} deficiency (pernicious anaemia), certain blood diseases (sickle-cell disease and thalassaemia), chronic infection, kidney disease, or certain kinds of poisoning.

anaerobic (of living organisms) not requiring oxygen for the release of energy from food molecules such as glucose. Anaerobic organisms include many bacteria, yeasts, and internal parasites.

Obligate anaerobes such as certain primitive bacteria cannot function in the presence of oxygen; but *facultative anaerobes*, like the fermenting yeasts and most bacteria, can function with or without oxygen. Anaerobic organisms release much less of the available energy from their food than do ◊aerobic organisms.

anaesthetic drug that produces loss of sensation or consciousness; the resulting state is ***anaesthesia***, in which the patient is insensitive to stimuli. Anaesthesia may also happen as a result of nerve disorder.

analgesic agent for relieving ◊pain. ◊Opiates alter the perception or appreciation of pain and are effective in controlling 'deep' visceral (internal) pain. Non-opiates, such as aspirin, paracetamol, and NSAIDs (non-steroidal anti-inflammatory drugs), relieve musculoskeletal pain and reduce inflammation in soft tissues.

analogous in biology, term describing a structure that has a similar function to a structure in another organism, but not a similar evolutionary path. For example, the wings of bees and of birds have the same purpose – to give powered flight – but have different origins. Compare ◊homologous.

analogue (of a quantity or device) changing continuously; by contrast a ◊digital quantity or device varies in series of distinct steps. For example, an analogue clock measures time by means of a continuous movement of hands around a dial, whereas a digital clock measures time with a numerical display that changes in a series of discrete steps.

analogue computer computing device that performs calculations through the interaction of continuously varying physical quantities, such as voltages (as distinct from the more common ◊digital computer, which works with discrete quantities). An analogue computer is said to operate in real time (corresponding to time in the real world), and can therefore be used to monitor and control other events as they happen.

analogue signal in electronics, current or voltage that conveys or stores information, and varies continuously in the same way as the information it represents (compare ◊digital signal). Analogue signals are prone to interference and distortion.

analogue-to-digital converter (ADC) electronic circuit that converts an analogue signal into a digital one. Such a circuit is needed to convert the signal from an analogue device into a digital signal for input into a computer. For example, many ◊sensors designed to measure physical quantities, such as temperature and pressure, produce an analogue signal in the form of voltage and this must be passed through an ADC before computer input and processing. A ◊digital-to-analogue converter performs the opposite process.

analysis in chemistry, the determination of the composition of substances; see ◊analytical chemistry.

analytical chemistry branch of chemistry that deals with the determination of the chemical composition of substances. *Qualitative analysis* determines the identities of the substances in a given sample; *quantitative analysis* determines how much of a particular substance is present.

Simple qualitative techniques exploit the specific, easily observable properties of elements or compounds – for example, the flame test makes use of the different flame-colours produced by metal cations when their compounds are held in a hot flame. More sophisticated methods, such as those of ◊spectroscopy, are required where substances are present in very low concentrations or where several substances have similar properties.

Most quantitative analyses involve initial stages in which the substance to be measured is extracted from the test sample, and purified. The final analytical stages (or 'finishes') may involve measurement of the substance's mass (gravimetry) or volume (volumetry, titrimetry), or a number of techniques initially developed for qualitative analysis, such as fluorescence and absorption spectroscopy, chromatography, electrophoresis, and polarography. Many modern methods enable quantification by means of a detecting device that is integrated into the extraction procedure (as in gas-liquid chromatography).

anatomy study of the structure of the body and its component parts, especially the ◊human body, as distinguished from physiology, which is the study of bodily functions.

andalusite aluminium silicate, Al_2SiO_5, a white to pinkish mineral crystallizing as square- or rhombus-based prisms. It is common in metamorphic rocks formed from clay sediments under low pressure conditions. Andalusite, kyanite, and sillimanite are all polymorphs of Al_2SiO_5.

andesite volcanic igneous rock, intermediate in silica content between rhyolite and basalt. It is characterized by a large quantity of feldspar ◊minerals, giving it a light colour. Andesite erupts from volcanoes at destructive plate margins (where one plate of the Earth's surface moves beneath another; see ◊plate tectonics), including the Andes, from which it gets its name.

AND gate in electronics, a type of ◊logic gate.

androecium male part of a flower, comprising a number of ◊stamens.

androgen general name for any male sex hormone, of which ◊testosterone is the most important. They are all ◊steroids and are principally involved in the production of male ◊secondary sexual characteristics (such as beard growth).

Andromeda major constellation of the northern hemisphere, visible in autumn. Its main feature is the Andromeda galaxy. The star Alpha Andromedae forms one corner of the Square of Pegasus. It is named after the princess of Greek mythology.

Andromeda galaxy galaxy 2.2 million light years away from Earth in the constellation Andromeda, and the most distant object visible to the naked eye. It is the largest member of the ◊Local Group of galaxies.

Like the Milky Way, it is a spiral orbited by several companion galaxies but contains about twice as many stars. It is about 200,000 light years across. In 1993 US astronomers detected two components at the centre of Andromeda, indicating that it may have a double nucleus, and a black hole with a mass tens of millions times that of the Sun.

AND rule rule used for finding the combined probability of two or more independent events both occurring. If two events E_1 and E_2 are independent (have no effect on each other) and the probabilities of their taking place are p_1 and p_2, respectively, then the combined probability p that both E_1 *and* E_2 will happen is given by:

$$p = p_1 \times p_2$$

anechoic chamber room designed to be of high sound absorbency. All surfaces inside the chamber are covered by sound-absorbent materials such as rubber. The walls are often covered with inward-facing pyramids of rubber, to minimize reflections. It is used for experiments in ◊acoustics and for testing audio equipment.

anemometer device for measuring wind speed and liquid flow. The most basic form, the *cup-type anemometer*, consists of cups at the ends of arms, which rotate when the wind blows. The speed of rotation indicates the wind speed.

anemophily type of ◊pollination in which the pollen is carried on the wind. Anemophilous flowers are usually unscented, have either very reduced petals and sepals or lack them altogether, and do not produce nectar. In some species they are borne in ◊catkins. Male and female reproductive structures are commonly found in separate flowers. The male flowers have numerous exposed stamens, often on long filaments; the female flowers have long, often branched, feathery stigmas.

aneroid barometer kind of ◊barometer.

angiosperm flowering plant in which the seeds are enclosed within an ovary, which ripens to a fruit. Angiosperms are divided into ◊monocotyledons (single seed leaf in the embryo) and ◊dicotyledons (two seed leaves in the embryo). They include the majority of flowers, herbs, grasses, and trees except conifers.

There are over 250,000 different species of angiosperm, found in a wide range of habitats. Like ◊gymnosperms, they are seed plants, but differ in that ovules and seeds are protected within the carpel. Fertilization occurs by male gametes passing into the ovary from a pollen tube. After fertilization the ovule develops into the seed while the ovary wall develops into the fruit.

angle in mathematics, the amount of turn or rotation; it may be defined by a pair of rays (half-lines) that share a common endpoint but do not lie on the same line. Angles are measured in ◊degrees (°) or ◊radians (rads) – a complete turn or circle being 360° or 2π rads.

Angles are classified generally by their degree measures: *acute angles* are less than 90°; *right angles* are exactly 90° (a quarter turn); *obtuse angles* are greater than 90° but less than 180°; *reflex angles* are greater than 180° but less than 360°.

angle of declination angle at a particular point on the Earth's surface between the direction of the true or geographic North Pole and the magnetic north pole. The angle of declination has varied over time because of the slow drift in the position of the magnetic north pole.

angle of dip or *angle of inclination* angle at a particular point on the Earth's surface between the direction of the Earth's magnetic field and the horizontal; see ◊magnetic dip.

angle of incidence angle between a ray of light striking a mirror incident ray) and the normal to that mirror. It is equal to the ◊angle of reflection.

angle of reflection angle between a ray of light reflected from a mirror and the normal to that mirror. It is equal to the ◊angle of incidence.

angle of refraction angle between a refracted ray of light and the normal to the surface at which ◊refraction occurred. When a ray passes from air into a denser medium such as glass, it is bent towards the normal so that the angle of refraction is less than the ◊angle of incidence.

angstrom unit (symbol Å) of length equal to 10^{-10} metre or one-ten-millionth of a millimetre, used for atomic measurements and the wavelengths of electromagnetic radiation. It is named after the Swedish scientist A J Ångström.

angular momentum in physics, a type of ◊momentum.

anhydride chemical compound obtained by the removal of water from another compound; usually a dehydrated acid. For example, sulphur(VI) oxide (sulphur trioxide, SO_3) is the anhydride of sulphuric acid (H_2SO_4).

anhydrite naturally occurring anhydrous calcium sulphate ($CaSO_4$). It is used commercially for the manufacture of plaster of Paris and builders' plaster.

anhydrous of a chemical compound that contains no water. If the water of crystallization is removed from blue crystals of copper(II) sulphate, a white powder (anhydrous copper sulphate) results. Liquids from which all traces of water have been removed are also described as being anhydrous.

aniline $C_6H_5NH_2$ or *phenylamine* one of the simplest aromatic chemicals (a substance related to benzene, with its carbon atoms joined in a ring). When pure, it is a colourless oily liquid; it has a characteristic odour, and turns brown on contact with air. It occurs in coal tar, and is used in the rubber industry and to make drugs and dyes. It is highly poisonous.

animal or *metazoan* member of the kingdom Animalia, one of the major categories of living things, the science of which is *zoology*. Animals are all ◊heterotrophs (they obtain their energy from organic substances produced by other organisms); they have eukaryotic cells (the genetic material is contained within a distinct nucleus) bounded by a thin cell membrane rather than the thick cell wall of plants. Most animals are capable of moving around for at least part of their life cycle.

animal behaviour scientific study of the behaviour of animals, either by comparative psychologists (with an interest mainly in the psychological processes involved in the control of behaviour) or by ethologists (with an interest in the biological context and relevance of behaviour; see ◊ethology).

anion ion carrying a negative charge. During electrolysis, anions in the electrolyte move towards the anode (positive electrode).

ankle hinge joint connecting the lower leg and the foot. It is formed by the ankle bone (talus) that fits into a socket provided by the lower ends of the ◊tibia and ◊fibula.

annealing process of heating a material (usually glass or metal) for a given time at a given temperature, followed by slow cooling, to increase ductility and strength. It is a common form of ◊heat treatment.

annelid any segmented worm of the phylum Annelida. Annelids include earthworms, leeches, and marine worms such as lugworms.

annihilation in nuclear physics, a process in which a particle and its *antiparticle* ('mirror image' particle) collide and disappear, with the creation of a burst of energy.

The energy created is equivalent to the mass of the colliding particles in accordance with the ◊mass–energy equation. For example, an electron and a positron annihilate to produce a burst of high-energy X-rays.

Not all particle–antiparticle interactions result in annihilation; the exception concerns the group called ◊mesons, which are composed of ◊quarks and their antiquarks. See ◊antimatter.

annual plant plant that completes its life cycle within one year, during which time it germinates, grows to maturity, bears flowers, produces seed, and then dies.

annual rings or *growth rings* concentric rings visible on the wood of a cut tree trunk or other woody stem. Each ring represents a period of growth when new ◊xylem is laid down to replace tissue being converted into wood (secondary xylem). The wood formed from xylem produced in the spring and early summer has larger and more numerous vessels than the wood formed from xylem produced in autumn when growth is slowing down. The result is a clear boundary between the pale spring wood and the denser, darker autumn wood. Annual rings may be used to estimate the age of the plant (see ◊dendrochronology), although occasionally more than one growth ring is produced in a given year.

anode in chemistry, the positive electrode of an electrolytic ◊cell, towards which negative particles (anions), usually in solution, are attracted. See ◊electrolysis.

anodizing process that increases the resistance to ◊corrosion of a metal, such as aluminium, by building up a protective oxide layer on the surface. The natural corrosion resistance of aluminium is provided by a thin film of aluminium oxide; anodizing increases the thickness of this film and thus the corrosion protection.

anomalous expansion of water expansion of water as it is cooled from 4°C to 0°C. This behaviour is unusual because most substances contract when they are cooled. It means that water has a greater density at 4°C than at 0°C. Hence ice floats on water, and the water at the bottom of a pool in winter is warmer than at the surface. As a result large lakes freeze slowly in winter and aquatic life is more likely to survive.

anoxia or *hypoxia* in biology, deprivation of oxygen, a condition that rapidly leads to collapse or death, unless immediately reversed.

ANSI (abbreviation for *American National Standards Institute*) US national standards body. It sets official procedures in (among other areas) computing and electronics.

antacid any substance that neutralizes stomach acid, such as sodium bicarbonate or magnesium hydroxide ('milk of magnesia').

Antacids are weak ◊bases, swallowed as solids or emulsions. They may be taken between meals to relieve symptoms of hyperacidity, such as pain, bloating, nausea, and 'heartburn'. Excessive or prolonged need for antacids should be investigated medically.

antagonist in medicine, a drug or body chemical with the reverse effect of another drug or chemical. The drug naloxone is an antagonist of ◊morphine.

antagonistic muscles in the body, a pair of muscles allowing coordinated movement of the skeletal joints. The extension of the arm, for example, requires one set of muscles to relax, while another set contracts. The individual components of antagonistic pairs can be classified into extensors (muscles that straighten a limb) and flexors (muscles that bend a limb).

Antares or *Alpha Scorpii* brightest star in the constellation Scorpius and the 15th brightest star in the sky. It is a red supergiant several hundred times larger than the Sun and perhaps 10,000 times as luminous, lies about 300 light years away, and fluctuates slightly in brightness.

antenna in zoology, an appendage ('feeler') on the head. Insects, centipedes, and millipedes each have one pair of antennae but there are two pairs in crustaceans, such as shrimps. In insects, the antennae are involved with the senses of smell and touch; they are frequently complex structures with large surface areas that increase the ability to detect scents.

antenna in radio and television, another name for ◊aerial.

anterior in biology, the front of an organism, usually the part that goes forward first when the animal is moving. The anterior end of the nervous system, over the course of evolution, has developed into a brain with associated receptor organs able to detect stimuli including light and chemicals.

anther in a flower, the terminal part of a stamen in which the ◊pollen grains are produced. It is usually borne on a slender stalk or filament, and has two lobes, each containing two chambers, or pollen sacs, within which the pollen is formed.

antheridium organ producing the male gametes, ◊antherozoids, in algae, bryophytes (mosses and liverworts), and pteridophytes (ferns, club mosses, and horsetails). It may be either single-celled, as in most algae, or multicellular, as in bryophytes and pteridophytes.

antherozoid motile (or independently moving) male gamete produced by algae, bryophytes (mosses and liverworts), pteridophytes (ferns, club mosses, and horsetails), and some gymnosperms (notably the cycads). Antherozoids are formed in an antheridium and, after being released, swim by means of one or more ◊flagella, to the female gametes.

antagonistic muscles *The triceps and biceps are antagonistic muscles in the human arm. When the arm is raised the triceps muscle is relaxed and the biceps is contracted. The triceps contracts and the biceps relaxes when the arm is lowered. An insect straightens its legs using antagonistic pairs of muscles.*

Higher plants have nonmotile male gametes contained within ♦pollen grains.

anthracene white, glistening, crystalline, tricyclic, aromatic hydrocarbon with a faint blue fluorescence when pure. Its melting point is about 216°C/421°F and its boiling point 351°C/664°F. It occurs in the high-boiling-point fractions of coal tar, where it was discovered 1832 by the French chemists Auguste Laurent (1808–1853) and Jean Dumas (1800–1884).

anthracite hard, dense, shiny variety of ♦coal, containing over 90% carbon and a low percentage of ash and impurities, which causes it to burn without flame, smoke, or smell. Because of its purity, anthracite gives off relatively little sulphur dioxide when burnt.

anthropometry science dealing with the measurement of the human body, particularly stature, body weight, cranial capacity, and length of limbs, in samples of living populations, as well as the remains of buried and fossilized humans.

antibiotic drug that kills or inhibits the growth of bacteria and fungi. It is derived from living organisms such as fungi or bacteria, which distinguishes it from synthetic antimicrobials.

The earliest antibiotics, the ♦penicillins, came into use from 1941 and were quickly joined by chloramphenicol, the cephalosporins, erythromycins, tetracyclines, and aminoglycosides. A range of broad-spectrum antibiotics, the 4-quinolones, was developed 1989, of which ciprofloxacin was the first. Each class and individual antibiotic acts in a different way and may be effective against either a broad spectrum or a specific type of disease-causing agent.

antibody protein molecule produced in the blood by ◊lymphocytes in response to the presence of foreign or invading substances (◊antigens); such substances include the proteins carried on the surface of infecting microorganisms. Antibody production is only one aspect of ◊immunity in vertebrates.

anticline in geology, a fold in the rocks of the Earth's crust in which the layers or beds bulge upwards to form an arch (seldom preserved intact).

anticoagulant substance that inhibits the formation of blood clots. Common anticoagulants are heparin, produced by the liver and some white blood cells, and derivatives of coumarin. Anticoagulants are used medically in the prevention and treatment of thrombosis and heart attacks. Anticoagulant substances are also produced by blood-feeding animals, such as mosquitoes, leeches, and vampire bats, to keep the victim's blood flowing.

anticyclone area of high atmospheric pressure caused by descending air, which becomes warm and dry. Winds radiate from a calm centre, taking a clockwise direction in the northern hemisphere and an anticlockwise direction in the southern hemisphere. Anticyclones are characterized by clear weather and the absence of rain and violent winds. In summer they bring hot, sunny days and in winter they bring fine, frosty spells, although fog and low cloud are not uncommon in the UK. *Blocking anticyclones*, which prevent the normal air circulation of an area, can cause summer droughts and severe winters.

antidiuretic hormone pituitary hormone that prevents excessive fluid loss. See ◊ADH.

antiferromagnetic material material with a very low magnetic ◊susceptibility that increases with temperature up to a certain temperature, called the ◊Néel temperature. Above the Néel temperature, the material is only weakly attracted to a strong magnet.

antifreeze substance added to a water-cooling system (for example, that of a car) to prevent it freezing in cold weather.

The most common types of antifreeze contain the chemical ethylene ◊glycol, an organic alcohol with a freezing point of about $-15°C/5°F$.

The addition of this chemical depresses the freezing point of water significantly. A solution containing 33.5% by volume of ethylene glycol will not freeze until about $-20°C/-4°F$. A 50% solution will not freeze until $-35°C/-31°F$.

antigen any substance that causes the production of ◊antibodies by the body's immune system. Common antigens include the proteins carried on the surface of bacteria, viruses, and pollen grains. The proteins of incompatible blood groups or tissues also act as antigens, which has to be taken into account in medical procedures such as blood transfusions and organ transplants.

antiknock substance added to petrol to reduce knocking in car engines. It is a mixture of dibromoethane and tetraethyl lead.

antilogarithm or *antilog* the inverse of ◊logarithm, or the number whose logarithm to a given base is a given number. If $y = \log_a x$, then $x = \text{antilog}_a y$.

antimatter in physics, a form of matter in which most of the attributes (such as electrical charge, magnetic moment, and spin) of ◊elementary particles are reversed. Such particles (◊antiparticles) can be created in particle accelerators, such as those at ◊CERN in Geneva, Switzerland, and at ◊Fermilab in the USA.

antimony silver-white, brittle, semimetallic element (a metalloid), symbol Sb (from Latin *stibium*), atomic number 51, relative atomic mass 121.75. It occurs chiefly as the ore stibnite, and is used to make alloys harder; it is also used in photosensitive substances in colour photography, optical electronics, fireproofing, pigment, and medicine. It was employed by the ancient Egyptians in a mixture to protect the eyes from flies.

antinode in physics, the position in a ◊standing wave pattern at which the amplitude of vibration is greatest (compare ◊node). The standing wave of a stretched string vibrating in the fundamental mode has one antinode at its midpoint. A vibrating air column in a pipe has an antinode at the pipe's open end and at the place where the vibration is produced.

antioxidant any substance that prevents deterioration by oxidation in fats, oils, paints, plastics, and rubbers. When used as food ◊additives, antioxidants prevent fats and oils from becoming rancid when exposed to air, and thus extend their shelf life.

antiparticle in nuclear physics, a particle corresponding in mass and properties to a given ◊elementary particle but with the opposite electrical charge, magnetic properties, or coupling to other fundamental forces. For example, an electron carries a negative charge whereas its antiparticle, the positron, carries a positive one. When a particle and its antiparticle collide, they destroy each other, in the process called 'annihilation', their total energy being converted to lighter particles and/or photons. A substance consisting entirely of antiparticles is known as ◊antimatter.

antiseptic any substance that kills or inhibits the growth of microorganisms. The use of antiseptics was pioneered by Joseph Lister (1827–1912). He used carbolic acid (◊phenol), which is a weak antiseptic; antiseptics such as TCP are derived from this.

anus opening at the end of the alimentary canal that allows undigested food and other waste materials to pass out of the body, in the form of faeces. It is found in all types of multicellular animal except the coelenterates (sponges) and the platyhelminths (flatworms), which have a mouth only.

aorta the body's main ◊artery, arising from the left ventricle of the heart in birds and mammals. Carrying freshly oxygenated blood, it arches over the top of the heart and descends through the trunk, finally splitting to form the two iliac arteries. Loss of elasticity in the aorta provides evidence of atherosclerosis, which may lead to heart disease.

a.p. in physics, abbreviation for *atmospheric pressure*.

apastron the point at which an object travelling in an elliptical orbit around a star is at its furthest from the star. The term is usually applied to the position of the minor component of a ◊binary star in relation to the primary. Its opposite is ◊periastron.

apatite common calcium phosphate mineral, $Ca_5(PO_4)_3(F,OH,Cl)$. Apatite has a hexagonal structure and occurs widely in igneous rocks, such as pegmatite, and in contact metamorphic rocks, such as marbles. It is used in the manufacture of fertilizer and as a source of phosphorus. Apatite is the chief constituent of tooth enamel while hydroxyapatite, $Ca_5(PO_4)_3(OH)_2$, is the chief inorganic constituent of bone marrow. Apatite ranks 5 on the ◊Mohs' scale of hardness.

apex the highest point of a triangle, cone, or pyramid – that is, the vertex (corner) opposite a given base.

aphelion the point at which an object, travelling in an elliptical orbit around the Sun, is at its furthest from the Sun.

API abbreviation for ◊*Applications Program Interface*, a standard environment in which computer programs are written.

apogee the point at which an object, travelling in an elliptical orbit around the Earth, is at its furthest from the Earth.

Apollo asteroid member of a group of ◊asteroids whose orbits cross that of the Earth. They are named after the first of their kind, Apollo, discovered 1932 and then lost until 1973. Apollo asteroids are so small and faint that they are difficult to see except when close to Earth (Apollo is about 2 km/1.2 mi across).

Apollo project US space project to land a person on the Moon, achieved 20 July 1969, when Neil Armstrong was the first to set foot there. He was accompanied on the Moon's surface by Col Edwin E Aldrin, Jr; Michael Collins remained in the orbiting command module.

The first Apollo mission carrying a crew, *Apollo 7*, Oct 1968, was a test flight in orbit around the Earth. After three other preparatory flights, *Apollo 11* made the first lunar landing. Five more crewed landings followed, the last 1972. The total cost of the programme was over $24 billion.

aposematic coloration in biology, the technical name for *warning coloration* markings that make a dangerous, poisonous, or foul-tasting animal particularly conspicuous and recognizable to a predator. Examples include the yellow and black stripes of bees and wasps, and the bright red or yellow colours of many poisonous frogs. See also ◊mimicry.

apothecaries' weights obsolete units of mass, formerly used in pharmacy: 20 grains made one scruple; three scruples made one drachm; eight drachms made an apothecary's ounce (oz apoth), and 12 such ounces made an apothecary's pound (lb apoth). There are 7,000 grains in one pound avoirdupois (0.454 kg).

apparent depth depth that a transparent material such as water or glass appears to have when viewed from above. This is less than its real depth because of the ◊refraction that takes place when light passes into a less dense medium. The ratio of the real depth to the apparent depth of a transparent material is equal to its ◊refractive index.

appendix a short, blind-ended tube attached to the ◊caecum. It has no known function in humans, but in herbivores it may be large, containing millions of bacteria that secrete enzymes to digest grass (as no vertebrate can secrete enzymes that will digest cellulose, the main constituent of plant cell walls).

Apple US computer company, manufacturer of the ◊Macintosh range of computers.

Appleton layer band containing ionized gases in the Earth's upper atmosphere, above the ◊E layer (formerly the Kennelly–Heaviside layer). It can act as a reflector of radio signals, although its ionic composition varies with the sunspot cycle. It is named after the English physicist Edward Appleton.

application in computing, a program or job designed for the benefit of the end user, such as a payroll system or a ◊word processor. The term is used to distinguish such programs from those that control the computer or assist the programmer, such as a ◊compiler.

application in mathematics, a curved line that connects a series of points (or 'nodes') in the smoothest possible way. The shape of the curve is governed by a series of complex mathematical formulae. Applications are used in ◊computer graphics and ◊CAD.

applications program program designed to carry out a task for the benefit of the computer user – for example, calculating pay slips or word processing. By contrast, a ◊systems program carries out tasks related to the operation and performance of the computer itself.

Applications Program Interface (API) in computing, standard environment, including tools, protocols, and other routines, in which programs can be written. An API ensures that all applications are consistent with the operating system and have a similar ◊user interface.

appropriate technology simple or small-scale machinery and tools that, because they are cheap and easy to produce and maintain, may be of most use in the developing world; for example, hand ploughs and simple looms. This equipment may be used to supplement local crafts and traditional skills to encourage small-scale industrialization.

aqualung or *scuba* underwater breathing apparatus worn by divers, developed in the early 1940s by French diver Jacques Cousteau. Compressed-air cylinders strapped to the diver's back are regulated by a valve system and by a mouth tube to provide air to the diver at the same pressure as that of the surrounding water (which increases with the depth).

aquamarine blue variety of the mineral ◊beryl. A semiprecious gemstone, it is used in jewellery.

aqua regia a mixture of three parts concentrated hydrochloric acid and one part concentrated nitric acid, which dissolves all metals except silver.

aquarium tank or similar container used for the study and display of living aquatic plants and animals. The same name is used for institutions that exhibit aquatic life. These have been common since Roman times, but the first modern public aquarium was opened in Regent's Park, London, in 1853. A recent development is the oceanarium or seaquarium, a large display of marine life forms.

Aquarius zodiacal constellation a little south of the celestial equator near Pegasus. Aquarius is represented as a man pouring water from a jar. The Sun passes through Aquarius from late Feb to early March. In astrology, the dates for Aquarius are between about 20 Jan and 18 Feb.

aquatic living in water. All life on Earth originated in the early oceans, because the aquatic environment has several advantages for organisms. Dehydration is almost impossible, temperatures usually remain stable, and the density of water provides physical support.

aqueduct any artificial channel or conduit for water, often an elevated structure of stone, wood, or iron built for conducting water across a valley. The Greeks built a tunnel 1,280 m/4,200 ft long near Athens, 2,500 years ago. Many Roman aqueducts are still standing, for example the one at Nîmes in S France, built about AD 18 (which is 48 m/160 ft high).

aqueous humour watery fluid found in the chamber between the cornea and lens of the vertebrate eye. Similar to blood serum in composition, it is constantly renewed.

aqueous solution solution in which the solvent is water.

aquifer any rock formation containing water. The rock of an aquifer must be porous and permeable (full of interconnected holes) so that it can absorb water. Aquifers are an important source of fresh water, for example, for drinking and irrigation, in many arid areas of the world and are exploited by the use of artesian wells.

Aquila constellation on the celestial equator (see ◊celestial sphere). Its brightest star is first-magnitude ◊Altair, flanked by the stars Beta and Gamma Aquilae. It is represented by an eagle.

arc in geometry, a section of a curved line or circle. A circle has three types of arc: a *semicircle*, which is exactly half of the circle; *minor arcs*, which are less than the semicircle;

and *major arcs*, which are greater than the semicircle.

Archaean or *Archaeozoic* the earliest eon of geological time; the first part of the Precambrian, from the formation of Earth up to about 2,500 million years ago. It was a time when no life existed, and with every new discovery of ancient life its upper boundary is being pushed further back.

archaebacteria three groups of bacteria whose DNA differs significantly from that of other bacteria (called the 'eubacteria'). All are strict anaerobes, that is, they are killed by oxygen. This is thought to be a primitive condition and to indicate that the archaebacteria are related to the earliest life forms, which appeared about 4 billion years ago, when there was little oxygen in the Earth's atmosphere. Archaebacteria are found in undersea vents, hot springs, the Dead Sea, and salt pans, and have even adapted to refuse tips.

archaeology study of prehistory and history, based on the examination of physical remains. Principal activities include preliminary field (or site) surveys, excavation (where necessary), and the classification, ◊dating, and interpretation of finds. Since 1958 ◊radiocarbon dating has been used to establish the age of archaeological strata and associated materials. *related disciplines* Useful in archaeological studies are stratigraphy (the study of geological strata), dendrochronology (the establishment of chronological sequences through the study of tree rings), palaeobotany (the study of ancient pollens, seeds, and grains), epigraphy (the study of inscriptions), and numismatics (the study of coins).

archegonium female sex organ found in bryophytes (mosses and liverworts), pteridophytes (ferns, club mosses, and horsetails), and some gymnosperms. It is a multicellular, flask-shaped structure consisting of two parts: the swollen base or venter containing the egg cell, and the long, narrow neck. When the egg cell is mature, the cells of the neck dissolve, allowing the passage of the male gametes, or ◊ant herozoids.

Archimedes' principle in physics, law stating that an object totally or partly submerged in a fluid displaces a volume of fluid that weighs the same as the apparent loss in weight of the object (which, in turn, equals the upwards force, or upthrust, experienced by that object). It was discovered by the Greek mathematician Archimedes.

Archimedes screw one of the earliest kinds of pump, thought to have been invented by Archimedes. It consists of a spiral screw revolving inside a close-fitting cylinder. It is used, for example, to raise water for irrigation.

arc lamp or *arc light* electric light that uses the illumination of an electric arc maintained between two electrodes. The British scientist Humphry Davy developed an arc lamp 1808, and its main use in recent years has been in cinema projectors. The lamp consists of two carbon electrodes, between which a very high voltage is maintained. Electric current arcs (jumps) between the two, creating a brilliant light.

arc minute, arc second units for measuring small angles, used in geometry, surveying, map-making, and astronomy. An arc minute (") is one-sixtieth of a degree, and an arc second (') is one-sixtieth of an arc minute. Small distances in the sky, as between two close stars or the apparent width of a planet's disc, are expressed in minutes and seconds of arc.

Arcturus or *Alpha Boötis* brightest star in the constellation Boötes and the fourth-brightest star in the sky. Arcturus is a red giant about 28 times larger than the Sun and 70 times more luminous, 36 light years away from Earth.

are metric unit of area, equal to 100 square metres (119.6 sq yd); 100 ares make one ◊hectare.

area the size of a surface. It is measured in square units, usually square centimetres (cm^2), square metres (m^2), or square kilometres (km^2). See also ◊surface area.

arena virus any of a group of medium-sized, RNA-containing enveloped viruses that appear grainy under the electron microscope. The group includes the Lassa fever virus.

Argand diagram in mathematics, a method for representing complex numbers by Cartesian coordinates (x, y). Along the x-axis (horizontal axis) are plotted the real numbers, and along the y-axis (vertical axis) the nonreal, or ◊imaginary, numbers.

argon colourless, odourless, nonmetallic, gaseous element, symbol Ar, atomic number 18, relative atomic mass 39.948. It is grouped with the ◊inert gases, since it was long believed not to react with other substances, but observations now indicate that it can be made to combine with boron fluoride to form

area *Areas of some common plane shapes.*

compounds. It constitutes almost 1% of the Earth's atmosphere, and was discovered 1894 by British chemists John Rayleigh (1842–1919) and William Ramsay (1852–1916) after all oxygen and nitrogen had been removed chemically from a sample of air. It is used in electric discharge tubes and argon lasers.

argument in mathematics, a specific value of the independent variable of a ◊function of x. It is also another name for ◊amplitude.

argument in computing, the value on which a ◊function operates. For example, if the argument 16 is operated on by the function 'square root', the answer 4 is produced.

Ariane launch vehicle built in a series by the European Space Agency (first flight 1979). The launch site is at Kourou in French Guiana. Ariane is a three-stage rocket using liquid fuels. Small solid-fuel and liquid-fuel boosters can be attached to its first stage to increase carrying power.

Ariel series of six UK satellites launched by the USA 1962–79, the most significant of which was *Ariel 5*, 1974, which made a pioneering survey of the sky at X-ray wavelengths.

Aries zodiacal constellation in the northern hemisphere between Pisces and Taurus, near Auriga, represented as the legendary ram whose golden fleece was sought by Jason and the Argonauts.

Its most distinctive feature is a curve of three stars of decreasing brightness. The brightest of these is Hamal or Alpha Arietis, 65 light years from Earth.

aril accessory seed cover other than a ◊fruit; it may be fleshy and sometimes brightly coloured, woody, or hairy. In flowering plants (◊angiosperms) it is often derived from the stalk that originally attached the ovule to the ovary wall. Examples of arils include the bright-red, fleshy layer surrounding the yew seed (yews are ◊gymnosperms so they lack true fruits), and the network of hard filaments that partially covers the nutmeg seed and yields the spice known as mace.

arithmetic branch of mathematics concerned with the study of numbers and their properties. The fundamental operations of arithmetic are addition, subtraction, multiplication, and division. Raising to powers (for example, squaring or cubing a number), the extraction of roots (for example, square roots), percentages, fractions, and ratios are developed from these operations.

arithmetic and logic unit (ALU) in a computer, the part of the ◊central processing unit (CPU) that performs the basic arithmetic and logic operations on data.

arithmetic mean the average of a set of n numbers, obtained by adding the numbers and dividing by n. For example, the arithmetic mean of the set of 5 numbers 1, 3, 6, 8, and 12 is $(1 + 3 + 6 + 8 + 12)/5 = 30/5 = 6$.

arithmetic progression or *arithmetic sequence* sequence of numbers or terms that have a common difference between any one term and the next in the sequence. For example, 2, 7, 12, 17, 22, 27, ... is an arithmetic sequence with a common difference of 5.

benzene C_6H_6

pyrimidine $C_4H_4N_2$

a pyridine (nicotinic acid, vitamin B complex) $C_5H_4N \cdot COOH$

— covalent bond
● carbon atom
○ hydrogen atom
Ⓞ oxygen atom
Ⓝ nitrogen atom

pyridine C_5H_5N

imidazole $C_3H_4N_2$

purine $C_5H_4N_4$

aromatic compound *Compounds whose molecules contain the benzene ring, or variations of it, are called aromatic. The term was originally used to distinguish sweet-smelling compounds from others.*

armature in a motor or generator, the wire-wound coil that carries the current and rotates in a magnetic field. (In alternating-current machines, the armature is sometimes stationary.) The pole piece of a permanent magnet or electromagnet and the moving, iron part of a ◊solenoid, especially if the latter acts as a switch, may also be referred to as armatures.

aromatic compound organic chemical compound in which some of the bonding electrons are delocalized (shared among several atoms within the molecule and not localized in the vicinity of the atoms involved in bonding). The commonest aromatic compounds have ring structures, the atoms comprising the ring being either all carbon or containing one or more different atoms (usually nitrogen, sulphur, or oxygen). Typical examples are benzene (C_6H_6) and pyridine (C_6H_5N).

array in computing, a list of values that can all be referred to by a single ◊variable name. Separate values are distinguished by using a *subscript* with each variable name.

array in mathematics, collection of numbers (or letters representing numbers) arranged in rows and columns. A ◊matrix is an array shown inside a pair of brackets; it indicates that the array should be treated as a single entity.

arsenic brittle, greyish-white, semimetallic element (a metalloid), symbol As, atomic number 33, relative atomic mass 74.92. It occurs in many ores and occasionally in its elemental state, and is widely distributed, being present in minute quantities in the soil, the sea, and the human body. In larger quantities, it is poisonous. The chief source of arsenic compounds is as a by-product from metallurgical processes. It is used in making semiconductors, alloys, and solders.

artery vessel that carries blood from the heart to the rest of the body. It is built to withstand considerable pressure, having thick walls which contain smooth muscle fibres. During contraction of the heart muscle, arteries expand in diameter to allow for the sudden increase in pressure that occurs; the resulting ◊pulse or pressure wave can be felt at the wrist. Not all arteries carry oxygenated (oxygen-rich) blood; the pulmonary arteries convey deoxygenated (oxygen-poor) blood from the heart to the lungs.

Arteries are flexible, elastic tubes, consisting of three layers, the middle of which is muscular; its rhythmic contraction aids the pumping of blood around the body. In middle and old age, the walls degenerate and are vulnerable to damage by the build-up of fatty deposits. These reduce elasticity, hardening the arteries and decreasing the internal bore. This condition, known as atherosclerosis, can lead to high blood pressure, loss of circulation, heart disease, and death.

arthropod member of the phylum Arthropoda; an invertebrate animal with jointed legs and a segmented body with a horny or chitinous casing (exoskeleton), which is shed periodically and replaced as the animal grows. Included are arachnids such as spiders and mites, as well as crustaceans, millipedes, centipedes, and insects.

artificial intelligence (AI) branch of science concerned with creating computer programs that can perform actions comparable with those of an intelligent human. Current AI research covers such areas as planning (for robot behaviour), language understanding, pattern recognition, and knowledge representation.

artificial limb device to replace a limb that has been removed by surgery or lost through injury, or one that is malformed because of genetic defects. It is one form of ◊prosthesis.

artificial radioactivity natural and spontaneous radioactivity arising from radioactive isotopes or elements that are formed when elements are bombarded with subatomic particles – protons, neutrons, or electrons – or small nuclei.

artificial selection in biology, selective breeding of individuals that exhibit the particular characteristics that a plant or animal breeder wishes to develop. In plants, desirable features might include resistance to disease, high yield (in crop plants), or attractive appearance. In animal breeding, selection has led to the development of particular breeds of cattle for improved meat production (such as the Aberdeen Angus) or milk production (such as Jerseys).

asbestos any of several related minerals of fibrous structure that offer great heat resistance because of their nonflammability and poor conductivity. Commercial asbestos is generally either made from serpentine ('white' asbestos) or from sodium iron silicate ('blue' asbestos). The fibres are woven together or bound by an inert material. Over time the fibres can work loose and, because they are small enough to float freely in the air or be inhaled, asbestos usage is now strictly controlled; exposure to its dust can cause cancer.

ASCII (acronym for *American standard code for information interchange*) in computing, a coding system in which numbers are assigned to letters, digits, and punctuation symbols. Although computers work in ◊binary number code, ASCII numbers are usually quoted as decimal or ◊hexadecimal numbers. For example, the decimal number 45 (binary 0101101) represents a hyphen, and 65 (binary 1000001) a capital A. The first 32 codes are used for control functions, such as carriage return and backspace.

ascorbic acid $C_6H_8O_6$ or *vitamin C* a relatively simple organic acid found in citrus fruits and vegetables. It is soluble in water and destroyed by prolonged boiling, so soaking or overcooking of vegetables reduces their vitamin C content. Lack of ascorbic acid results in scurvy.

The richest natural source of vitamin C is the Australian billygoat plum, *Terminalia ferdiandiana*. It contains 100 times the concentration found in oranges.

asepsis practice of ensuring that bacteria are excluded from open sites during surgery, wound dressing, blood sampling, and other medical procedures. Aseptic technique is a first line of defence against infection.

asexual reproduction in biology, reproduction that does not involve the manufacture and fusion of sex cells, nor the necessity for two parents. The process carries a clear advantage in that there is no need to search for a mate or to develop complex pollinating mechanisms; every asexual organism can reproduce on its own. Asexual reproduction can therefore lead to a rapid population build-up.

aspartame noncarbohydrate sweetener used in foods under the trade-name Nutrasweet. It is about 200 times as sweet as sugar and, unlike saccharine, has no aftertaste.

asphalt mineral mixture containing semi-solid brown or black ◊bitumen, used in the construction industry. Asphalt is mixed with rock chips to form paving material, and the purer varieties are used for insulating material and for waterproofing masonry. It can be produced artificially by the distillation of ◊petroleum.

aspirin acetylsalicylic acid, a popular pain-relieving drug (◊analgesic) developed in the late 19th century as a household remedy for aches and pains. It relieves pain and reduces inflammation and fever. It is derived from the white willow tree *Salix alba*.

Regular use of aspirin is recommended for people at increased risk of heart attack, thrombosis and some kinds of stroke. However, aspirin may cause stomach bleeding, kidney damage, and hearing defects. It is no longer considered suitable for children under 12 because of a suspected link with a rare disease, Reye's syndrome.

assay in chemistry, the determination of the quantity of a given substance present in a sample. Usually it refers to determining the purity of precious metals.

assembler in computing, a program that translates a program written in an assembly language into a complete ◊machine code program that can be executed by a computer. Each instruction in the assembly language is translated into only one machine-code instruction.

assembly language low-level computer-programming language closely related to a computer's internal codes. It consists chiefly of a set of short sequences of letters (mnemonics), which are translated, by a program called an assembler, into ◊machine code for the computer's ◊central processing unit (CPU) to follow directly. In assembly language, for example, 'JMP' means 'jump' and 'LDA' means 'load accumulator'. Assembly code is used by programmers who need to write very fast or efficient programs.

assimilation in animals, the process by which absorbed food molecules, circulating in the blood, pass into the cells and are used for growth, tissue repair, and other metabolic activities. The actual destiny of each food molecule depends not only on its type, but also on the body requirements at that time.

associative operation in mathematics, an operation in which the outcome is independent of the grouping of the numbers or symbols concerned. For example, multiplication is associative, as $4 \times (3 \times 2) = (4 \times 3) \times 2 = 24$; however, division is not, as $12 \div (4 \div 2) = 6$, but $(12 \div 4) \div 2 = 1.5$. Compare ◊commutative operation and ◊distributive operation.

assortative mating in ◊population genetics, selective mating in a population between individuals that are genetically related or have similar characteristics. If sufficiently consistent, assortative mating can theoretically result in the evolution of new species without geographical isolation (see ◊speciation).

astatine nonmetallic, radioactive element, symbol At, atomic number 85, relative atomic mass 210. It is a member of the ◊halogen group, and is very rare in nature. Astatine is highly unstable, with at least 19 isotopes; the longest lived has a half-life of about eight hours.

asteroid or *minor planet* any of many thousands of small bodies, composed of rock and iron, that orbit the Sun. Most lie in a belt between the orbits of Mars and Jupiter, and are thought to be fragments left over from the formation of the ◊Solar System. About 100,000 may exist, but their total mass is only a few hundredths the mass of the Moon.

They include ◊Ceres (the largest asteroid, 940 km/584 mi in diameter), Vesta (which has a light-coloured surface, and is the brightest as seen from Earth), ◊Eros, and ◊Icarus. Some asteroids are in orbits that bring them close to Earth, and some, such as the ◊Apollo asteroids, even cross Earth's orbit; at least some of these may be remnants of former comets. *Bifurcated asteroids*, first discovered in 1990, are in fact two chunks of rock that touch each other. It may be that at least 10% of asteroids approaching the Earth are bifurcated.

asthenosphere division of the Earth's structure lying beneath the ◊lithosphere, at a depth of approximately 70 km/45 mi to 260 km/160 mi. It is thought to be the soft, partially molten layer of the ◊mantle on which the rigid plates of the Earth's surface move to produce the motions of ◊plate tectonics.

astigmatism aberration occurring in the lens of the eye. It results when the curvature of the lens differs in two perpendicular planes, so that rays in one plane may be in focus while rays in the other are not. With astigmatic eyesight, the vertical and horizontal cannot be in focus at the same time; correction is by the use of a cylindrical lens that reduces the overall focal length of one plane so that both planes are seen in sharp focus.

astrocyte one of several different cell types that form the ◊glia (non-nervous tissue) in the ◊central nervous system.

astrometry measurement of the precise positions of stars, planets, and other bodies in space. Such information is needed for practical purposes including accurate timekeeping, surveying and navigation, and calculating orbits and measuring distances in space. Astrometry is not concerned with the surface features or the physical nature of the body under study.

astronaut person making flights into space; the term *cosmonaut* is used in the West for any astronaut from the former Soviet Union.

astronautics science of space travel. See ◊rocket; ◊satellite; ◊space probe.

astronomical unit unit (symbol AU) equal to the mean distance of the Earth from the Sun: 149,597,870 km/92,955,800 mi. It is used to describe planetary distances. Light travels this distance in approximately 8.3 minutes.

astronomy science of the celestial bodies: the Sun, the Moon, and the planets; the stars and galaxies; and all other objects in the universe. It is concerned with their positions, motions, distances, and physical conditions, and with their origins and evolution. Astronomy is subdivided into fields including astrophysics, celestial mechanics, and cosmology. See also ◊gamma-ray astronomy, ◊infrared astronomy, ◊radio astronomy, ◊ultraviolet astronomy, and ◊X-ray astronomy.

astrophotography use of photography in astronomical research. The first successful photograph of a celestial object was the daguerreotype plate of the Moon taken by John W Draper (1811–1882) of the USA in March 1840. The first photograph of a star, Vega, was taken by US astronomer William C Bond (1789–1859) in 1850. Modern-day astrophotography uses techniques such as ◊charge-coupled devices (CCDs).

astrophysics study of the physical nature of stars, galaxies, and the universe. It began with the development of spectroscopy in the 19th century, which allowed astronomers to analyse the composition of stars from their light. Astrophysicists view the universe as a vast natural laboratory in which they can study matter under conditions of temperature, pressure, and density that are unattainable on Earth.

asymptote in ◊coordinate geometry, a straight line that a curve approaches more and more closely but never reaches. The x and y axes are asymptotes to the graph of $xy =$ constant (a rectangular ◊hyperbola).

atavism (Latin *atavus* 'ancestor') in ◊genetics, the reappearance of a characteristic not apparent in the immediately preceding generations; in psychology, the manifestation of primitive forms of behaviour.

Atlas rocket US rocket, originally designed and built as an intercontinental missile, but subsequently adapted for space use. Atlas rockets launched astronauts in the Mercury series into orbit, as well as numerous other satellites and space probes.

atmosphere mixture of gases that surrounds the Earth, prevented from escaping by the pull of the Earth's gravity. Atmospheric

astronomy: chronology

2300 BC	Chinese astronomers made their earliest observations.
2000	Babylonian priests made their first observational records.
1900	Stonehenge was constructed: first phase.
365	The Chinese observed the satellites of Jupiter with the naked eye.
3rd c.	Aristarchus argued that the Sun is the centre of the Solar System.
2nd c. AD	Ptolemy's complicated Earth-centred system was promulgated, which dominated the astronomy of the Middle Ages.
1543	Copernicus revived the ideas of Aristarchus in *De Revolutionibus*.
1608	Hans Lippershey invented the telescope, which was first used by Galileo 1609.
1609	Johannes Kepler's first two laws of planetary motion were published (the third appeared 1619).
1632	The world's first official observatory was established in Leiden in the Netherlands.
1633	Galileo's theories were condemned by the Inquisition.
1675	The Royal Greenwich Observatory was founded in England.
1687	Isaac Newton's *Principia* was published, including his 'law of universal gravitation'.
1705	Edmond Halley correctly predicted that the comet that had passed the Earth in 1682 would return in 1758; the comet was later to be known by his name.
1781	William Herschel discovered Uranus and recognized stellar systems beyond our Galaxy.
1796	Pierre Laplace elaborated his theory of the origin of the Solar System.
1801	Giuseppe Piazzi discovered the first asteroid, Ceres.
1814	Joseph von Fraunhofer first studied absorption lines in the solar spectrum.
1846	Neptune was identified by Johann Galle, following predictions by John Adams and Urbain Leverrier.
1859	Gustav Kirchhoff explained dark lines in the Sun's spectrum.
1887	The earliest photographic star charts were produced.
1889	Edward Barnard took the first photographs of the Milky Way.
1908	Fragment of comet fell at Tunguska, Siberia.
1920	Arthur Eddington began the study of interstellar matter.
1923	Edwin Hubble proved that the galaxies are systems independent of the Milky Way, and by 1930 had confirmed the concept of an expanding universe.
1930	The planet Pluto was discovered by Clyde Tombaugh at the Lowell Observatory, Arizona, USA.
1931	Karl Jansky founded radio astronomy.
1945	Radar contact with the Moon was established by Z Bay of Hungary and the US Army Signal Corps Laboratory.
1948	The 5-m/200-in Hale reflector telescope was installed at Mount Palomar, California, USA.
1957	The Jodrell Bank telescope dish in England was completed.
1957	The first Sputnik satellite (USSR) opened the age of space observation.
1962	The first X-ray source was discovered in Scorpius.
1963	The first quasar was discovered.
1967	The first pulsar was discovered by Jocelyn Bell and Antony Hewish.
1969	The first crewed Moon landing was made by US astronauts.
1976	A 6-m/240-in reflector telescope was installed at Mount Semirodniki, USSR.
1977	Uranus was discovered to have rings.
1977	The spacecrafts *Voyager 1* and *2* were launched, passing Jupiter and Saturn 1979–1981.
1978	The spacecraft *Pioneer Venus 1* and *2* reached Venus.
1978	A satellite of Pluto, Charon, was discovered by James Christy of the US Naval Observatory.
1986	Halley's comet returned. *Voyager 2* flew past Uranus and discovered six new moons.
1987	Supernova SN1987A flared up, becoming the first supernova to be visible to the naked eye since 1604. The 4.2 m/165 in William Herschel Telescope on La Palma, Canary Islands, and the James Clerk Maxwell Telescope on Mauna Kea, Hawaii, began operation.
1988	The most distant individual star was recorded – a supernova, 5 billion light years away, in the AC118 cluster of galaxies.
1989	*Voyager 2* flew by Neptune and discovered eight moons and three rings.
1990	Hubble Space Telescope was launched into orbit by the US space shuttle.
1991	The space probe *Galileo* flew past the asteroid Gaspra, approaching it to within 26,000 km/16,200 mi.
1992	COBE satellite detected ripples from the Big Bang that mark the first stage in the formation of galaxies.

composition of the atmosphere

gas	symbol	volume (%)	role
nitrogen	N_2	78.08	cycled through human activities and through the action of microorganisms on animal and plant waste
oxygen	O_2	20.94	cycled mainly through the respiration of animals and plants and through the action of photosynthesis
carbon dioxide	CO_2	0.03	cycled through respiration and photosynthesis in exchange reactions with oxygen. It is also a product of burning fossil fuels
argon	Ar	0.093	chemically inert and with only a few industrial uses
neon	Ne	0.0018	as argon
helium	He	0.0005	as argon
krypton	Kr	trace	as argon
xenon	Xe	trace	as argon
ozone	O_3	0.00006	a product of oxygen molecules split into single atoms by the Sun's radiation and unaltered oxygen molecules
hydrogen	H_2	0.00005	unimportant

pressure decreases with height in the atmosphere. In its lowest layer, the atmosphere consists of nitrogen (78%) and oxygen (21%), both in molecular form (two atoms bounded together). The other 1% is largely argon, with very small quantities of other gases, including water vapour and carbon dioxide. The atmosphere plays a major part in the various cycles of nature (the ◊water cycle, ◊carbon cycle, and ◊nitrogen cycle). It is the principal industrial source of nitrogen, oxygen, and argon, which are obtained by fractional distillation of liquid air.

atmosphere or *standard atmosphere* in physics, a unit (symbol atm) of pressure equal to 760 torr, 1013.25 millibars, or 1.01325×10^5 newtons per square metre. The actual pressure exerted by the atmosphere fluctuates around this value, which is assumed to be standard at sea level and 0°C, and is used when dealing with very high pressures.

atmospheric pressure the pressure at any point on the Earth's surface that is due to the weight of the column of air above it; it therefore decreases as altitude increases. At sea level the average pressure is 101 kilopascals (1,013 millibars, 760 mmHg, or 14.7 lb per sq in).

Changes in atmospheric pressure, measured with a barometer, are used in weather forecasting. Areas of relatively high pressure are called ◊anticyclones; areas of low pressure are called ◊depressions.

atom smallest unit of matter that can take part in a chemical reaction, and which cannot be broken down chemically into anything simpler. An atom is made up of protons and neutrons in a central nucleus surrounded by electrons (see ◊atomic structure). The atoms of the various elements differ in atomic number, relative atomic mass, and chemical behaviour. There are 109 different types of atom, corresponding with the 109 known elements as listed in the ◊periodic table of the elements.

atom, electronic structure of the arrangement of electrons around the nucleus of an atom, in distinct energy levels, also called orbitals or shells (see ◊orbital, atomic). These shells can be regarded as a series of concentric spheres, each of which can contain a certain maximum number of electrons; the noble gases have an arrangement in which every shell contains this number (see ◊noble gas structure). The energy levels are usually numbered beginning with the shell nearest to the nucleus. The outermost shell is known as the ◊valency shell as it contains the valence electrons.

The lowest energy level, or innermost shell, can contain no more than two electrons. Outer shells are considered to be stable when they

atom *It is impossible to give a true visual representation of the electronic structure of an atom. Diagrams such as these representing sodium and sulphur atoms should not be regarded as pictures of the atoms, but rather as a 'shorthand' description of their electronic structure.*

ATMOSPHERE

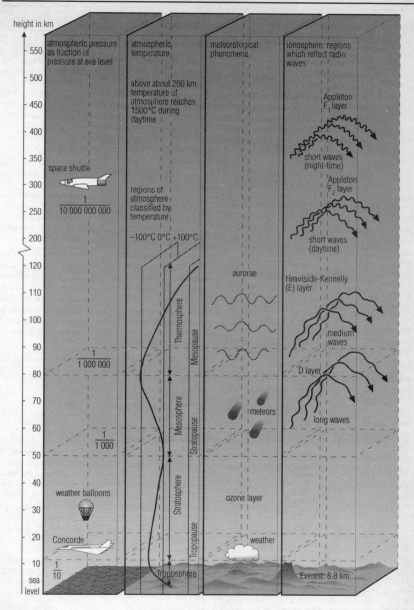

atmosphere All but 1% of the Earth's atmosphere lies in a layer 30 km/19 mi above the ground. At a height of 5,500 m/18,000 ft, air pressure is half that at sea level. The temperature of the atmosphere varies greatly with height; this produces a series of layers, called the troposphere, stratosphere, mesosphere, and thermosphere.

contain eight electrons but additional electrons can sometimes be accommodated provided that the outermost shell has a stable configuration. Electrons in unfilled shells are available to take part in chemical bonding, giving rise to the concept of valency. In ions, the electron shells contain more or fewer electrons than are required for a neutral atom, generating negative or positive charges.

The atomic number of an element indicates the number of electrons in a neutral atom. From this it is possible to deduce its electronic structure. For example, sodium has atomic number 11 ($Z = 11$) and its electronic arrangement (configuration) is two electrons in the first energy level, eight electrons in the second energy level and one electron in the third energy level – generally written as 2.8.1. Similarly for sulphur ($Z = 16$), the electron arrangement will be 2.8.6. The electronic structure dictates whether two elements will combine by ionic or covalent bonding (see ◊bond) or not at all.

atomic absorption spectrometry technique used in archaeology to determine quantitatively the chemical composition of artifactual metals, minerals, and rocks, in order to identify raw material sources, to relate artifacts of the same material, or to trace trade routes. A sample of the material is atomized in a flame, and its light intensity measured. The method is slow and destructive.

atomic clock timekeeping device regulated by various periodic processes occurring in atoms and molecules, such as atomic vibration or the frequency of absorbed or emitted radiation.

atomic energy another name for ◊nuclear energy.

atomic force microscope (AFM) microscope developed in the late 1980s that produces a magnified image using a diamond probe, with a tip so fine that it may consist of a single atom, dragged over the surface of a specimen to 'feel' the contours of the surface. In effect, the tip acts like the stylus of a record player, reading the surface. The tiny up-and-down movements of the probe are converted to an image of the surface by computer and displayed on a screen. The AFM is useful for examination of biological specimens since, unlike the ◊scanning tunnelling microscope, the specimen does not have to be electrically conducting.

atomicity number of atoms of an ◊element that combine together to form a molecule. A molecule of oxygen (O_2) has atomicity 2; sulphur (S_8) has atomicity 8.

atomic mass see ◊relative atomic mass.

atomic mass unit or *dalton unit* (symbol amu or u) unit of mass that is used to measure the relative mass of atoms and molecules. It is equal to one-twelfth of the mass of a carbon-12 atom, which is equivalent to the mass of a proton or 1.66×10^{-27} kg. The ◊relative atomic mass of an atom has no units; thus oxygen-16 has an atomic mass of 16 daltons but a relative atomic mass of 16.

atomic number or *proton number* the number (symbol Z) of protons in the nucleus of an atom. It is equal to the positive charge on the nucleus. In a neutral atom, it is also equal to the number of electrons surrounding the nucleus. The 109 elements are arranged in the ◊periodic table of the elements according to their atomic number. See also ◊nuclear notation.

atomic radiation energy given out by disintegrating atoms during ◊radioactive decay, whether natural or synthesized. The energy may be in the form of fast-moving particles, known as ◊alpha particles and ◊beta particles, or in the form of high-energy electromagnetic waves known as ◊gamma radiation. Overlong exposure to atomic radiation can lead to ◊radiation sickness.

atomic size or *atomic radius* size of an atom expressed as the radius in ◊angstroms or other units of length. The sodium atom has an atomic radius of 1.57 angstroms (1.57×10^{-8} cm). For metals, the size of the atom is always greater than the size of its ion. For non-metals the reverse is true.

atomic structure internal structure of an ◊atom. The core of the atom is the *nucleus*, a dense body only one ten-thousandth the diameter of the atom itself. The simplest nucleus, that of hydrogen, comprises a single stable positively charged particle, the *proton*. Nuclei of other elements contain more protons and additional particles, called *neutrons*, of about the same mass as the proton but with no electrical charge. Each element has its own characteristic nucleus with a unique number of protons, the atomic number. The number of neutrons may vary. Where atoms of a single element have different numbers of neutrons, they are called ◊isotopes. Although some isotopes tend to be unstable and exhibit ◊radioactivity, they all have identical chemical properties.

The nucleus is surrounded by a number of moving *electrons*, each of which has a negative charge equal to the positive charge on a proton, but which weighs only 1/1,839 times as much. In a neutral atom, the nucleus is surrounded by the same number of electrons as it contains protons. According to ◊quantum theory, the position of an electron is uncertain; it may be found at any point. However, it is more likely to be found in some places than others. The region of space in which an electron is most likely to be found is called an orbital (see ◊orbital, atomic). The chemical properties of an element are determined by the ease with which its atoms can gain or lose electrons from its outer orbitals.

High-energy physics research has discovered the existence of subatomic particles (see ◊particle physics) other than the proton, neutron, and electron. More than 300 kinds of particle are now known, and these are classified into several classes according to their mass, electric charge, spin, magnetic moment, and interaction. The ◊elementary particles, which include the electron, are indivisible and may be regarded as the fundamental units of matter; the *hadrons*, such as the proton and neutron, are composite particles made up of either two or three elementary particles called quarks.

Atoms are held together by the electrical forces of attraction between each negative electron and the positive protons within the nucleus. The latter repel one another with enormous forces; a nucleus holds together only because an even stronger force, called the *strong nuclear force*, attracts the protons and neutrons to one another. The strong force acts over a very short range – the protons and neutrons must be in virtual contact with one another (see ◊forces, fundamental). If, therefore, a fragment of a complex nucleus, containing some protons, becomes only slightly loosened from the main group of neutrons and protons, the natural repulsion between the protons will cause this fragment to fly apart from the rest of the nucleus at high speed. It is by such fragmentation of atomic nuclei (nuclear ◊fission) that nuclear energy is released.

atomic time time as given by ◊atomic clocks, which are regulated by natural resonance frequencies of particular atoms, and display a continuous count of seconds.

atomic weight another name for ◊relative atomic mass.

atomizer device that produces a spray of fine droplets of liquid. A vertical tube connected with a horizontal tube dips into a bottle of liquid, and at one end of the horizontal tube is a nozzle, at the other a rubber bulb. When the bulb is squeezed, air rushes over the top of the vertical tube and out through the nozzle. Following ◊Bernoulli's principle, the pressure at the top of the vertical tube is reduced, allowing the liquid to rise. The air stream picks up the liquid, breaks it up into tiny drops, and carries it out of the nozzle as a spray. Scent spray, paint spray guns, and carburettors all use the principle of the atomizer.

ATP (abbreviation for *adenosine triphosphate*) nucleotide molecule found in all cells. It can yield large amounts of energy, and is used to drive the thousands of biological processes needed to sustain life, growth, movement, and reproduction. Green plants use light energy to manufacture ATP as part of the process of ◊photosynthesis. In animals, ATP is formed by the breakdown of glucose molecules, usually obtained from the carbohydrate component of a diet, in a series of reactions termed ◊respiration. It is the driving force behind muscle contraction and the synthesis of complex molecules needed by individual cells.

atrium either of the two upper chambers of the heart. The left atrium receives freshly oxygenated blood from the lungs via the pulmonary vein; the right atrium receives deoxygenated blood from the ◊vena cava. Atrium walls are thin and stretch easily to allow blood into the heart. On contraction, the atria force blood into the thick-walled ventricles, which then give a second, more powerful beat.

auditory canal tube leading from the outer ◊ear opening to the eardrum. It is found only in animals whose eardrums are located inside the skull, principally mammals and birds.

Auriga constellation of the northern hemisphere, represented as a man driving a chariot. Its brightest star is the first-magnitude ◊Capella, about 45 light years from Earth; Epsilon Aurigae is an ◊eclipsing binary star with a period of 27 years, the longest of its kind (last eclipse 1983).

aurora coloured light in the night sky near the Earth's magnetic poles, called *aurora borealis* ('northern lights') in the northern hemisphere and *aurora australis* in the southern hemisphere. Although auroras are usually restricted to the polar skies, fluctuations in the ◊solar wind occasionally cause them to be visible at lower latitudes. An aurora is usually in the form of a luminous arch with its apex towards the magnetic pole followed by arcs, bands, rays, curtains, and coronas, usually green but often showing shades of blue and red, and sometimes yellow or white. Auroras are caused at heights of over 100 km/60 mi by a fast stream of charged particles from solar flares and low-density 'holes' in the Sun's corona. These are guided by the Earth's magnetic field towards the north and south magnetic poles, where they enter the upper atmosphere and bombard the gases in the atmosphere, causing them to emit visible light.

AUSSAT organization formed in 1981 by the federal government of Australia and Telecom Australia to own and operate Australia's domestic satellite system. The first stage, *Aussat 1*, was launched by the US space shuttle *Discovery* in 1985 and the third and final stage was launched from French Guiana, South America, in 1987. The AUSSAT satellite system enables people in remote outback areas of Australia to receive television broadcasts.

autoclave pressurized vessel that uses superheated steam to sterilize materials and equipment such as surgical instruments. It is similar in principle to a pressure cooker.

autoimmunity in medicine, condition where the body's immune responses are mobilized not against 'foreign' matter, such as invading germs, but against the body itself. Diseases considered to be of autoimmune origin include myasthenia gravis, rheumatoid arthritis, and lupus erythematosus.

autolysis in biology, the destruction of a ◊cell after its death by the action of its own ◊enzymes, which break down its structural molecules.

autonomic nervous system in mammals, the part of the nervous system that controls those functions not controlled voluntarily, including the heart rate, activity of the intestines, and the production of sweat. There are two divisions of the autonomic nervous system. The *sympathetic* system responds to stress, when it speeds the heart rate, increases blood pressure and generally prepares the body for action. The *parasympathetic* system is more important when the body is at rest, since it slows the heart rate, decreases blood pressure, and stimulates the digestive system.

autoradiography in biology, a technique for following the movement of molecules within an organism, especially a plant, by labelling with a radioactive isotope that can be traced on photographs. It is used to study ◊photosynthesis, where the pathway of radioactive carbon dioxide can be traced as it moves through the various chemical stages.

autosome any ◊chromosome in the cell other than a sex chromosome. Autosomes are of the same number and kind in both males and females of a given species.

autotroph any living organism that synthesizes organic substances from inorganic molecules by using light or chemical energy. Autotrophs are the *primary producers* in all food chains since the materials they synthesize and store are the energy sources of all other organisms. All green plants and many planktonic organisms are autotrophs, using sunlight to convert carbon dioxide and water into sugars by ◊photosynthesis.

It is estimated that 10% of the energy in autotrophs can pass into the next stage of the ◊food chain, the rest being lost as heat or indigestible matter. See also ◊heterotroph.

autumnal equinox see ◊equinox.

auxin plant ◊hormone that promotes stem and root growth in plants. Auxins influence many aspects of plant growth and development, including cell enlargement, inhibition of development of axillary buds, ◊tropisms, and the initiation of roots. *Synthetic auxins* are used in rooting powders for cuttings, and in some weedkillers, where high auxin concentrations cause such rapid growth that the plants die. They are also used to prevent premature fruitdrop in orchards. The most common naturally occurring auxin is known as indoleacetic acid, or IAA. It is produced in the shoot apex and transported to other parts of the plant.

aviation term used to describe both the science of powered ◊flight and also aerial navigation by means of an aeroplane.

Avogadro's hypothesis in chemistry, the law stating that equal volumes of all gases, when at the same temperature and pressure, have the same numbers of molecules. It was first propounded by Italian physicist Amedeo Avogadro (1776–1856).

Avogadro's number or *Avogadro's constant* the number of carbon atoms in 12 g of the carbon-12 isotope (6.022045×10^{23}). The relative atomic mass of any element, expressed in grams, contains this number of atoms. It is named after Amedeo Avogadro.

avoirdupois system of units of mass based on the pound (0.45 kg), which consists of 16 ounces (each of 16 drams) or 7,000 grains (each equal to 65 mg).

axil upper angle between a leaf (or bract) and the stem from which it grows. Organs developing in the axil, such as shoots and buds, are termed axillary, or lateral.

axilla anatomical term for the armpit.

axiom in mathematics, a statement that is assumed to be true and upon which theorems are proved by using logical deduction; for example, two straight lines cannot enclose a space. The Greek mathematician Euclid used a series of axioms that he considered could not be demonstrated in terms of simpler concepts to prove his geometrical theorems.

axis (plural *axes*) in geometry, one of the reference lines by which a point on a graph may be located. The horizontal axis is usually referred to as the x-axis, and the vertical axis as the y-axis. The term is also used to refer to the imaginary line about which an object may be said to be symmetrical (*axis of symmetry*) – for example, the diagonal of a square – or the line about which an object may revolve (*axis of rotation*).

axon long threadlike extension of a ◊nerve cell that conducts electrochemical impulses away from the cell body towards other nerve cells, or towards an effector organ such as a muscle. Axons terminate in ◊synapses, junctions with other nerve cells, muscles, or glands.

azimuth in astronomy, the angular distance of an object eastwards along the horizon, measured from due north, between the astronomical ◊meridian (the vertical circle passing through the centre of the sky and the north and south points on the horizon) and the vertical circle containing the celestial body whose position is to be measured.

azo dye synthetic dye containing the azo group of two nitrogen atoms (N=N) connecting aromatic ring compounds. Azo dyes are usually red, brown, or yellow, and make up about half the dyes produced. They are manufactured from aromatic ◊amines.

AZT drug used in the treatment of AIDS; see ◊zidovudine.

Babbit metal soft, white metal, an ◊alloy of tin, lead, copper, and antimony, used to reduce friction in bearings, developed by the US inventor Isaac Babbit 1839.

bacillus member of a group of rodlike ◊bacteria that occur everywhere in the soil and air. Some are responsible for diseases such as anthrax or for causing food spoilage.

backcross a breeding technique used to determine the genetic makeup of an individual organism.

background radiation radiation that is always present in the environment. By far the greater proportion (87%) of it is emitted from natural sources. Alpha and beta particles, and gamma radiation are radiated by the traces of radioactive minerals that occur naturally in the environment and even in the human body, and by radioactive gases such as radon and thoron, which are found in soil and may seep upwards into buildings. Radiation from space (◊cosmic radiation) also contributes to the background level.

backing storage in computing, memory outside the ◊central processing unit used to store programs and data that are not in current use. Backing storage must be nonvolatile – that is, its contents must not be lost when the power supply to the computer system is disconnected.

backup in computing, a copy file that is transferred to another medium, usually a ◊floppy disc or tape. The purpose of this is to have available a copy of a file that can be restored in case of a fault in the system or the file itself. Backup files are also created by many applications (with the extension .BAC or .BAK); a version is therefore available of the original file before it was modified by the current application.

backup system in computing, a duplicate computer system that can take over the operation of a main computer system in the event of equipment failure. A large interactive system, such as an airline's ticket-booking system, cannot be out of action for even a few hours without causing considerable disruption. In such cases a complete duplicate computer system may be provided to take over and run the system should the main computer develop a fault or need maintenance.

bacteria (singular *bacterium*) microscopic single-celled organisms lacking a nucleus. Bacteria are widespread, present in soil, air, and water, and as parasites on and in other living things. Some parasitic bacteria cause disease by producing toxins, but others are harmless and may even benefit their hosts. Bacteria usually reproduce by ◊binary fission (dividing into two equal parts), and since this may occur approximately every 20 minutes, a single bacterium is potentially capable of producing 16 million copies of itself in a day. It is thought that 1–10% of the world's bacteria have been identified.

Bacteria can be classified into two broad classes (called Gram positive and negative) according to their reactions to certain stains, or dyes, used in microscopy. The staining technique, called the Gram test after Danish bacteriologist Hans Gram, allows doctors to identify many bacteria quickly.

Bacteria have a large loop of ◊DNA, sometimes called a bacterial chromosome. In addition there are often small, circular pieces of DNA known as ◊plasmids that carry spare genetic information. These plasmids can readily move from one bacterium to another, even though the bacteria may be of different species. In a sense, they are parasites within the bacterial cell, but they survive by coding characteristics that promote the survival of their hosts. For example, some plasmids confer antibiotic resistance on the bacteria they inhabit. The rapid and problematic spread of antibiotic resistance among bacteria is due to plasmids, but they are also useful to humans in ◊genetic engineering.

bacteriology the study of ◊bacteria.

bacteriophage virus that attacks ◊bacteria. Such viruses are now of use in genetic engineering.

Baily's beads bright spots of sunlight seen around the edge of the Moon for a few seconds immediately before and after a total ◊eclipse of the Sun, caused by sunlight shining between mountains at the Moon's edge. Sometimes one bead is much brighter than the others, producing the so-called ***diamond ring*** effect. The effect was described 1836 by the English astronomer Francis Baily (1774–1844), a wealthy stockbroker who retired in 1825 to devote himself to astronomy.

Bakelite first synthetic ♢plastic, created by US chemist Leo Baekeland (1863–1944) in 1909. Bakelite is hard, tough, and heatproof, and is used as an electrical insulator. It is made by the reaction of phenol with formaldehyde, producing a powdery resin that sets solid when heated. Objects are made by subjecting the resin to compression moulding (simultaneous heat and pressure in a mould).

baking powder mixture of ♢bicarbonate of soda, an acidic compound, and a nonreactive filler (usually starch or calcium sulphate), used in baking as a raising agent. It gives a light open texture to cakes and scones, and is used as a substitute for yeast in making soda bread.

balance apparatus for weighing or measuring mass. The various types include the ***beam balance*** consisting of a centrally pivoted lever with pans hanging from each end, and the ***spring balance***, in which the object to be weighed stretches (or compresses) a vertical coil spring fitted with a pointer that indicates the weight on a scale. Kitchen and bathroom scales are balances.

balanced diet diet that includes carbohydrate, protein, fat, vitamins, water, minerals, and fibre (roughage). Although it is agreed that all these substances are needed if a person is to be healthy, argument occurs over how much of each type a person needs.

balance of nature in ecology, the idea that there is an inherent equilibrium in most ♢ecosystems, with plants and animals interacting so as to produce a stable, continuing system of life on Earth. Organisms in the ecosystem are adapted to each other – for example, waste products produced by one species are used by another and resources used by some are replenished by others; the oxygen needed by animals is produced by plants while the waste product of animal respiration, carbon dioxide, is used by plants as a raw material in photosynthesis. The ♢nitrogen cycle, the ♢water cycle, and the control of animal populations by natural predators are other examples. The activities of human beings can, and frequently do, disrupt the balance of nature.

ball-and-socket joint a joint allowing considerable movement in three dimensions, for instance the joint between the pelvis and the femur. To facilitate movement, such joints are rimmed with cartilage and lubricated by synovial fluid. The bones are kept in place by ligaments and moved by muscles.

ballistics study of the motion and impact of projectiles such as bullets, bombs, and missiles. For projectiles from a gun, relevant exterior factors include temperature, barometric pressure, and wind strength; and for nuclear missiles these extend to such factors as the speed at which the Earth turns.

ball valve valve that works by the action of external pressure raising a ball and thereby opening a hole.

bar unit of pressure equal to 10^5 pascals or 10^6 dynes/cm^2, approximately 750 mmHg or 0.987 atm. Its diminutive, the ***millibar*** (one-thousandth of a bar), is commonly used by meteorologists.

barium soft, silver-white, metallic element, symbol Ba, atomic number 56, relative atomic mass 137.33. It is one of the alkaline-earth metals, found in nature as barium carbonate and barium sulphate. As the sulphate it is used in medicine: taken as a suspension (a 'barium meal'), its progress is followed by using X-rays to reveal abnormalities of the alimentary canal. Barium is also used in alloys, pigments, and safety matches and, with strontium, forms the emissive surface in cathode-ray tubes. It was first discovered in barytes or heavy spar.

barium chloride BaCl$_2$ white, crystalline solid, used in aqueous solution to test for the presence of the sulphate ion. When a solution of barium chloride is mixed with a solution of a sulphate, a white precipitate of barium sulphate is produced.

$$Ba^{2+}_{(aq)} + SO_4^{2-}_{(aq)} \rightarrow BaSO_{4(s)}$$

barium nitrate Ba(NO$_3$)$_2$ compound that, like barium chloride, can be used to test for the presence of the sulphate ion as it readily dissolves in water. When sulphate is added, a white precipitate of barium sulphate is formed.

bark protective outer layer on the stems and roots of woody plants, composed mainly of dead cells. To allow for expansion of the stem, the bark is continually added to from within, and the outer surface often becomes cracked or is shed as scales. Trees deposit a variety of chemicals in their bark, including poisons. Many of these chemical substances have economic value because they can be used in the manufacture of drugs. Quinine, derived from the bark of the *Cinchona* tree, is used to fight malarial infections; curare, an anaesthetic used in medicine, comes from the *Strychnus toxifera* tree in the Amazonian rainforest.

Barnard's star second closest star to the Sun, six light years away in the constellation Ophiuchus. It is a faint red dwarf of 10th magnitude, visible only through a telescope. It is named after the US astronomer Edward E Barnard (1857–1923), who discovered 1916 that it has the fastest proper motion of any star, crossing 1 degree of sky every 350 years.

barograph device for recording variations in atmospheric pressure. A pen, governed by the movements of an aneroid ◊barometer, makes a continuous line on a paper strip on a cylinder that rotates over a day or week to create a *barogram*, or permanent record of variations in atmospheric pressure.

barometer instrument that measures atmospheric pressure as an indication of weather. Most often used are the *mercury barometer* and the *aneroid barometer*.

In a mercury barometer a column of mercury in a glass tube, roughly 0.75 m/2.5 ft high (closed at one end, curved upwards at the other), is balanced by the pressure of the atmosphere on the open end; any change in the height of the column reflects a change in pressure. In an aneroid barometer, a shallow cylindrical metal box containing a partial vacuum expands or contracts in response to changes in pressure.

barrel unit of liquid capacity, the value of which depends on the liquid being measured. It is used for petroleum, a barrel of which contains 159 litres/35 imperial gallons; a barrel of alcohol contains 189 litres/41.5 imperial gallons.

barrier island long island of sand, lying offshore and parallel to the coast. Some are over 100 km/60 mi in length.

barrier reef coral reef that lies offshore, separated from the mainland by a shallow lagoon.

baryon in nuclear physics, a heavy subatomic particle made up of three indivisible elementary particles called quarks. The baryons form a subclass of the ◊hadrons and comprise the nucleons (protons and neutrons) and hyperons.

baryte barium sulphate, $BaSO_4$, the most common mineral of barium. It is white or light-coloured, and has a comparatively high density (specific gravity 4.6); the latter property makes

barometer *In the mercury barometer, the weight of the column of mercury is balanced by the pressure of the atmosphere on the lower end. A change in height of the column indicates a change in atmospheric pressure.*

it useful in the production of high-density drilling muds (muds used to cool and lubricate drilling equipment). Baryte occurs mainly in ore veins, where it is often found with calcite and with lead and zinc minerals. It crystallizes in the orthorhombic system and can form tabular crystals or radiating fibrous masses.

basal ganglia several masses of grey matter located deep in the ◊cerebrum, including the caudate nucleus, the globus pallidus and putamen. They are involved with the subconscious regulation of movement. See also ◊brain.

basal metabolic rate (BMR) minimum amount of energy needed by the body to maintain life. It is measured when the subject is awake but resting, and includes the energy required to keep the heart beating, sustain breathing, repair tissues, and keep the brain and nerves functioning. Measuring the subject's consumption of oxygen gives an accurate value for BMR, because oxygen is needed to release energy from food.

basalt commonest volcanic ◊igneous rock, and the principal rock type on the ocean floor; it is basic, that is, it contains relatively little silica: about 50%. It is usually dark grey but can also be green, brown, or black.

base in mathematics, the number of different single-digit symbols used in a particular number system. In our usual (decimal) counting

base

binary (base 2)	octal (base 8)	decimal (base 10)	hexadecimal (base 16)
0	0	0	0
1	1	1	1
10	2	2	2
11	3	3	3
100	4	4	4
101	5	5	5
110	6	6	6
111	7	7	7
1000	10	8	8
1001	11	9	9
1010	12	10	A
1011	13	11	B
1100	14	12	C
1101	15	13	D
1110	16	14	E
1111	17	15	F
10000	20	16	10
11111111	377	255	FF
11111010001	3721	2001	7D1

system of numbers (with symbols 0, 1, 2, 3, 4, 5, 6, 7, 8, 9) the base is 10. In the ◊binary number system, which has only the symbols 1 and 0, the base is two. A base is also a number that, when raised to a particular power (that is, when multiplied by itself a particular number of times as in $10^2 = 10 \times 10 = 100$), has a ◊logarithm equal to the power. For example, the logarithm of 100 to the base ten is 2.

In geometry, the term is used to denote the line or area on which a polygon or solid stands.

base in chemistry, a substance that accepts protons, such as the hydroxide ion (OH⁻) and ammonia (NH₃). Bases react with acids to give a salt. Those that dissolve in water are called ◊alkalis.

Inorganic bases are usually oxides or hydroxides of metals, which react with dilute acids to form a salt and water. A number of carbonates also react with dilute acids, additionally giving off carbon dioxide. Many organic compounds that contain nitrogen are bases.

base pair in biochemistry, the linkage of two base (purine or pyrimidine) molecules in ◊DNA. They are found in nucleotides, and form the basis of the genetic code.

One base lies on one strand of the DNA double helix, and one on the other, so that the base pairs link the two strands like the rungs of a ladder. In DNA, there are four bases: adenine and guanine (purines) and cytosine and thymine (pyrimidines). Adenine always pairs with thymine, and cytosine with guanine.

BASIC (acronym for *beginner's all-purpose symbolic instruction code*) high-level computer-programming language, developed 1964, originally designed to take advantage of ◊multi-user systems (which can be used by many people at the same time). The language is relatively easy to learn and is popular among microcomputer users.

basicity number of replaceable hydrogen atoms in an acid. Nitric acid (HNO₃) is monobasic, sulphuric acid (H₂SO₄) is dibasic, and phosphoric acid (H₃PO₄) is tribasic.

basic oxide compound formed by a metal and oxygen, containing the O^{2-} ion. If, like sodium oxide (Na₂O), it is soluble, it forms the metal hydroxide when added to water.

$$Na_2O_{(s)} + H_2O_{(l)} \rightarrow 2NaOH_{(aq)}$$

All basic oxides react with acids to form a salt and water.

$$CaO + 2HNO_3 \rightarrow Ca(NO_3)_2 + H_2O$$

basic–oxygen process most widely used method of steelmaking, involving the blasting of oxygen at supersonic speed into molten pig iron.

basidiocarp spore-bearing body, or 'fruiting body', of all basidiomycete fungi (see ◊fungus), except the rusts and smuts. An example is the edible mushroom *Agaricus brunnescens*. Other types include globular basidiocarps (puffballs)

basic–oxygen process *The basic–oxygen process is the primary method used to produce steel. Oxygen is blown at high pressure through molten pig iron and scrap steel in a converter lined with basic refractory materials. The impurities, principally carbon, quickly burn out, producing steel.*

or flat ones that project from tree trunks (brackets). They are made up of a mass of tightly packed, intermeshed ◊hyphae.

bathyscaph or *bathyscaphe* or *bathyscape* deep-sea diving apparatus used for exploration at great depths in the ocean. In 1960, Jacques Piccard and Don Walsh took the bathyscaph *Trieste* to a depth of 10,917 m/35,820 ft in the Challenger Deep in the Mariana Trench off the island of Guam in the Pacific Ocean.

bathysphere watertight steel sphere used for observation in deep-sea diving. It is lowered into the ocean by cable to a maximum depth of 900 m/3,000 ft.

battery any energy-storage device allowing release of electricity on demand. It is made up of one or more electrical ◊cells. Primary-cell batteries are disposable; secondary-cell batteries, or ◊accumulators, are rechargeable. Primary-cell batteries are an extremely uneconomical form of energy, since they produce only 2% of the power used in their manufacture.

The common *dry cell* is a primary-cell battery based on the ◊Leclanché cell and consists of a central carbon electrode immersed in a paste of manganese dioxide and ammonium chloride as the electrolyte. The zinc casing forms the other electrode. It is dangerous to try to recharge a primary-cell battery.

The lead–acid *car battery* is a secondary-cell battery. The car's generator continually recharges the battery. It consists of sets of lead (positive) and lead peroxide (negative) plates in an electrolyte of sulphuric acid (◊battery acid).

The introduction of rechargeable nickel–cadmium batteries has revolutionized portable electronic newsgathering (sound recording, video) and information processing (computing). These batteries offer a stable, short-term source of power free of noise and electrical hazards.

battery acid ◊sulphuric acid of approximately 70% concentration used in lead–acid cells (as found in car batteries). The chemical reaction within the battery that generates electricity also changes the acid's composition.

baud in computing, unit that measures the speed of data transmission. One baud represents a transmission rate of one bit per second.

baud in engineering, a unit of electrical signalling speed equal to one pulse per second, measuring the rate at which signals are sent between electronic devices such as telegraphs and computers; 300 baud is about 300 words a minute.

battery *The common dry cell relies on chemical changes occurring between the electrodes – the central carbon rod and the outer zinc casing – and the ammonium chloride electrolyte to produce electricity. The mixture of carbon and manganese is used to increase the life of the cell.*

bauxite principal ore of ◊aluminium, consisting of a mixture of hydrated aluminium oxides and hydroxides, generally contaminated with compounds of iron, which give it a red colour. To produce aluminium the ore is processed into a white powder (alumina), which is then smelted by passing a large electric current through it. Chief producers of bauxite are Australia, Guinea, Jamaica, Russia, Kazakhstan, Surinam, and Brazil.

Bayesian statistics form of statistics that uses the knowledge of prior probability together with the probability of actual data to determine posterior probabilities, using Bayes' theorem.

Bayes' theorem in statistics, a theorem relating the ◊probability of particular events taking place to the probability that events conditional upon them have occurred.

B cell or *B ◊lymphocyte* immune cell that produces antibodies. Each B cell produces just one type of ◊antibody, specific to a single ◊antigen. Lymphocytes are related to ◊T cells.

Bear, Great and Little common names (and translations of the Latin) for the constellations ◊Ursa Major and ◊Ursa Minor respectively.

Beaufort scale system of recording wind velocity, devised by Francis Beaufort 1806. It is a numerical scale ranging from 0 to 17, calm being indicated by 0 and a hurricane by 12; 13–17 indicate degrees of hurricane force.

Beaufort scale

number and description	features	air speed	
		mi per hr	m per sec
0 calm	smoke rises vertically; water smooth	less than 1	less than 0.3
1 light air	smoke shows wind direction; water ruffled	1–3	0.3–1.5
2 slight breeze	leaves rustle; wind felt on face	4–7	1.6–3.3
3 gentle breeze	loose paper blows around	8–12	3.4–5.4
4 moderate breeze	branches sway	13–18	5.5–7.9
5 fresh breeze	small trees sway, leaves blown off	19–24	8.0–10.7
6 strong breeze	whistling in telephone wires; sea spray from waves	25–31	10.8–13.8
7 moderate gale	large trees sway	32–38	13.9–17.1
8 fresh gale	twigs break from trees	39–46	17.2–20.7
9 strong gale	branches break from trees	47–54	20.8–24.4
10 whole gale	trees uprooted, weak buildings collapse	55–63	24.5–28.4
11 storm	widespread damage	64–72	28.5–32.6
12 hurricane	widespread structural damage	73–82	above 32.7
13		83–92	
14		93–103	
15		104–114	
16		115–125	
17		126–136	

becquerel SI unit (symbol Bq) of ◊radioactivity, equal to one radioactive disintegration (change in the nucleus of an atom when a particle or ray is given off) per second. It is named after French physicist Henri Becquerel.

bel unit of sound measurement equal to ten ◊decibels. It is named after Scottish scientist Alexander Graham Bell.

bentonite type of clay, consisting mainly of montmorillonite and resembling ◊fuller's earth, which swells when wet. It is used in papermaking, moulding sands, drilling muds for oil wells, and as a decolorant in food processing.

benzaldehyde C_6H_5CHO colourless liquid with the characteristic odour of almonds. It is used as a solvent and in the making of perfumes and dyes. It occurs in certain leaves, such as the cherry, laurel, and peach, and in a combined form in certain nuts and kernels. It can be extracted from such natural sources, but is usually made from ◊toluene.

benzene C_6H_6 clear liquid hydrocarbon of characteristic odour, occurring in coal tar. It is used as a solvent and in the synthesis of many chemicals.

The benzene molecule consists of a ring of six carbon atoms, all of which are in a single plane, and it is one of the simplest ◊cyclic compounds. Benzene is the simplest of a class of compounds collectively known as *aromatic compounds*. Some are considered carcinogenic (cancer-inducing).

benzoic acid C_6H_5COOH white crystalline solid, sparingly soluble in water, that is used as a preservative for certain foods and as an antiseptic. It is obtained chemically by the direct oxidation of benzaldehyde and occurs in certain natural resins, some essential oils, and as hippuric acid.

benzpyrene one of a number of organic compounds associated with a particular polycyclic ring structure. Benzpyrenes are present in coal tar at low levels and are considered carcinogenic (cancer-inducing). Traces of benzpyrenes are present in wood smoke, and this has given rise to some concern about the safety of naturally smoked foods.

beriberi nutritional disorder occurring mostly in the tropics and resulting from a deficiency of vitamin B_1 (◊thiamine). The disease

benzene *The molecule of benzene consists of six carbon atoms arranged in a ring, with six hydrogen atoms attached. The benzene ring structure is found in many naturally occurring organic compounds.*

takes two forms: in one oedema (waterlogging of the tissues) occurs; in the other there is severe emaciation. There is nerve degeneration in both forms and many victims succumb to heart failure.

berkelium synthesized, radioactive, metallic element of the actinide series, symbol Bk, atomic number 97, relative atomic mass 247.

It was first produced 1949 by Glenn Seaborg and his team, at the University of California at Berkeley, USA, after which it is named.

Bernoulli's principle law stating that the speed of a fluid varies inversely with pressure, an increase in speed producing a decrease in pressure (such as a drop in hydraulic pressure as the fluid speeds up flowing through a constriction in a pipe) and vice versa. The principle also explains the pressure differences on each surface of an aerofoil, which gives lift to the wing of an aircraft. The principle was named after Swiss mathematician and physicist Daniel Bernoulli.

berry fleshy, many-seeded ◊fruit that does not split open to release the seeds. The outer layer of tissue, the exocarp, forms an outer skin that is often brightly coloured to attract birds to eat the fruit and thus disperse the seeds. Examples of berries are the tomato and the grape.

beryl mineral, beryllium aluminium silicate, $Be_3Al_2Si_6O_{18}$, which forms crystals chiefly in granite. It is the chief ore of beryllium. Two of its gem forms are aquamarine (light-blue crystals) and emerald (dark-green crystals).

beryllium hard, light-weight, silver-white, metallic element, symbol Be, atomic number 4, relative atomic mass 9.012. It is one of the ◊alkaline-earth metals, with chemical properties similar to those of magnesium; in nature it is found only in combination with other elements. It is used to make sturdy, light alloys and to control the speed of neutrons in nuclear reactors. Beryllium oxide was discovered in 1798 by French chemist Louis-Nicolas Vauquelin (1763–1829), but the element was not isolated until 1828, by Friedrich Wöhler and Antoine-Alexandre-Brutus Bussy independently.

Bessemer process the first cheap method of making ◊steel, invented by Henry Bessemer in England 1856. It has since been superseded by more efficient steel-making processes, such as the ◊basic–oxygen process. In the Bessemer process compressed air is blown into the bottom of a converter, a furnace shaped like a cement mixer, containing molten pig iron. The excess carbon in the iron burns out, other impurities form a slag, and the furnace is emptied by tilting.

beta decay the disintegration of the nucleus of an atom to produce a beta particle, or high-speed electron, and an electron-antineutrino. During beta decay, a neutron in the nucleus changes into a proton, thereby increasing the atomic number by one while the mass number stays the same. The mass lost in the change is converted into kinetic (movement) energy of the beta particle.

Beta decay is caused by the weak nuclear force, one of the fundamental ◊forces of nature operating inside the nucleus.

beta particle electron ejected with great velocity from a radioactive atom that is undergoing spontaneous disintegration. Beta particles do not exist in the nucleus but are created on disintegration, beta decay, when a neutron converts to a proton to emit an electron.

Beta particles are more penetrating than ◊alpha particles, but less so than ◊gamma radiation; they can travel several metres in air, but are stopped by 2–3 mm of aluminium. They are less strongly ionizing than alpha particles and, like cathode rays, are easily deflected by magnetic and electric fields.

Betelgeuse or *Alpha Orionis* red supergiant star in the constellation of Orion and the tenth brightest star in the sky, although its brightness varies. It is is 1,100 million km/700 million mi across, about 800 times larger than the Sun, roughly the same size as the orbit of Mars. It is over 10,000 times as luminous as the Sun, and lies 650 light years from Earth.

Bezier curve curved line invented by Pierre Bézier that connects a series of points (or 'nodes') in the smoothest possible way. The shape of the curve is governed by a series of complex mathematical formulae. They are used in ◊computer graphics and ◊CAD.

bicarbonate common name for ◊hydrogencarbonate

bicarbonate indicator pH indicator sensitive enough to show a colour change as the concentration of the gas carbon dioxide increases. The indicator is used in photosynthesis and respiration experiments to find out whether carbon dioxide is being liberated. The initial red colour changes to yellow as the pH becomes more acidic.

bicarbonate of soda or *baking soda* (technical name *sodium hydrogencarbonate*) $NaHCO_3$ white crystalline solid that neutralizes acids and is used in medicine to treat acid indigestion. It is also used in baking powders and effervescent drinks.

bicuspid valve or *mitral valve* in the left side of the ◊heart, a flap of tissue that prevents blood flowing back into the atrium when the ventricle contracts.

biennial plant plant that completes its life cycle in two years. During the first year it grows vegetatively and the surplus food produced is stored in its ◊perennating organ, usually the root. In the following year these food reserves are used for the production of leaves, flowers, and seeds, after which the plant dies. Many root vegetables are biennials, including the carrot *Daucus carota* and parsnip *Pastinaca sativa*. Some garden plants that are grown as biennials are actually perennials, for example, the wallflower *Cheiranthus cheiri*.

Big Bang in astronomy, the hypothetical 'explosive' event that marked the origin of the universe as we know it. At the time of the Big Bang, the entire universe was squeezed into a hot, superdense state. The Big Bang explosion threw this compacted material outwards, producing the expanding universe (see ◊red shift). The cause of the Big Bang is unknown; observations of the current rate of expansion of the universe suggest that it took place about 10–20 billion years ago. The Big Bang theory began modern ◊cosmology.

According to a modified version of the Big Bang, called the *inflationary theory*, the universe underwent a rapid period of expansion shortly after the Big Bang, which accounts for its current large size and uniform nature. The inflationary theory is supported by the most recent observations of the ◊cosmic background radiation. Scientists have calculated that one 10^{-36} second (equivalent to one million-million-million-million-million-millionth of a second) after the Big Bang, the universe was the size of a pea, and the temperature was 10 billion million million million°C (18 billion million million million°F). One second after the Big Bang, the temperature was about 10 billion°C (18 billion°F).

bile brownish alkaline fluid produced by the liver. Bile is stored in the gall bladder and is intermittently released into the duodenum (small intestine) to aid digestion. Bile consists of bile salts, bile pigments, cholesterol, and lecithin. *Bile salts* assist in the breakdown and absorption of fats; *bile pigments* are the breakdown products of old red blood cells that are passed into the gut to be eliminated with the faeces.

bimetallic strip strip made from two metals each having a different coefficient of ◊thermal expansion; it therefore bends when subjected to a change in temperature. Such strips are used widely for temperature measurement and control.

binary fission in biology, a form of ◊asexual reproduction, whereby a single-celled organism, such as the amoeba, divides into two smaller 'daughter' cells. It can also occur in a few simple multicellular organisms, such as sea anemones, producing two smaller sea anemones of equal size.

binary number system system of numbers to ◊base two, using combinations of the digits 1 and 0. Codes based on binary numbers are used to represent instructions and data in all modern digital computers, the values of the binary digits (contracted to 'bits') being stored or transmitted as, for example, open/closed switches, magnetized/unmagnetized discs and tapes, and high/low voltages in circuits.

binary search in computing, a rapid technique used to find any particular record in a list of records held in sequential order. The computer is programmed to compare the record sought with the record in the middle of the ordered list. This being done, the computer discards the half of the list in which the record does not appear, thereby reducing the number of records left to search by half. This process of selecting the middle record and discarding the unwanted half of the list is repeated until the required record is found.

binary star pair of stars moving in orbit around their common centre of mass. Observations show that most stars are binary, or even multiple – for example, the nearest star system to the Sun, ◊Alpha Centauri.

One of the stars in the binary system Epsilon Aurigae may be the largest star known. Its diameter is 2,800 times that of the Sun. A *spectroscopic binary* is a binary in which two stars are so close together that they cannot be seen separately, but their separate light spectra can be distinguished by a spectroscope. Another type is the ◊eclipsing binary.

binding energy in physics, the amount of energy needed to break the nucleus of an atom into the neutrons and protons of which it is made.

binoculars optical instrument for viewing an object in magnification with both eyes; for example, field glasses and opera glasses. Binoculars consist of two telescopes containing lenses and prisms, which produce a stereoscopic effect as well as magnifying the image.

Use of prisms has the effect of 'folding' the light path, allowing for a compact design.

binomial in mathematics, an expression consisting of two terms, such as $a + b$ or $a\,b$.

binomial system of nomenclature in biology, the system in which all organisms are identified by a two-part Latinized name. Devised by the biologist Linnaeus (1707–1778), it is also known as the Linnaean system. The first name is capitalized and identifies the ♢genus; the second identifies the ♢species within that genus.

biochemistry science concerned with the chemistry of living organisms: the structure and reactions of proteins (such as enzymes), nucleic acids, carbohydrates, and lipids.

Its study has led to an increased understanding of life processes, such as those by which organisms synthesize essential chemicals from food materials, store and generate energy, and pass on their characteristics through their genetic material. A great deal of medical research is concerned with the ways in which these processes are disrupted. Biochemistry also has applications in agriculture and in the food industry (for instance, in the use of enzymes).

biodegradable capable of being broken down by living organisms, principally bacteria and fungi. In biodegradable substances, such as food and sewage, the natural processes of decay lead to compaction and liquefaction, and to the release of nutrients that are then recycled by the ecosystem.

biodiversity (contraction of *biological diversity*) measure of the variety of the Earth's animal, plant, and microbial species; of genetic differences within species; and of the ecosystems that support those species. Its maintenance is important for ecological stability and as a resource for research into, for example, new drugs and crops. Research suggests that biodiversity is far greater than previously realized, especially among smaller organisms – for instance, it is thought that there are 30–40 million insects, of which only a million have so far been identified. Today, the most significant threat to biodiversity comes from the destruction of rainforests and other habitats in the southern hemisphere. In the 20th century, the destruction of habitats is believed to have resulted in the most severe and rapid loss of diversity in the history of the planet.

bioengineering the application of engineering to biology and medicine. Common applications include the design and use of artificial limbs, joints, and organs, including hip joints and heart valves.

biofeedback modification or control of a biological system by its results or effects. For example, a change in the position or ♢trophic level of one species affects all levels above it.

Many biological systems are controlled by negative feedback. When enough of the hormone thyroxine has been released into the blood, the hormone adjusts its own level by 'switching off' the gland that produces it. In ecology, as the numbers in a species rise, the food supply available to each individual is reduced. This acts to reduce the population to a sustainable level.

biofuel any solid, liquid, or gaseous fuel produced from organic (once living) matter, either directly from plants or indirectly from industrial, commercial, domestic, or agricultural wastes. There are three main methods for the development of biofuels: the burning of dry organic wastes (such as household refuse, industrial and agricultural wastes, straw, wood, and peat); the fermentation of wet wastes (such as animal dung) in the absence of oxygen to produce biogas (containing up to 60% methane), or the fermentation of sugar cane or corn to produce alcohol and esters; and energy forestry (producing fast-growing wood for fuel).

biogenesis biological term coined 1870 by English scientist Thomas Henry Huxley to express the hypothesis that living matter always arises out of other similar forms of living matter. It superseded the opposite idea of ♢spontaneous generation or abiogenesis (that is, that living things may arise out of non-living matter).

biogeography study of how and why plants and animals are distributed around the world, in the past as well as in the present; more specifically, a theory describing the geographical distribution of ♢species developed by Robert MacArthur and US zoologist Edward O Wilson. The theory argues that for many species, ecological specializations mean that suitable habitats are patchy in their occurrence. Thus for a dragonfly, ponds in which to breed are separated by large tracts of land, and for edelweiss adapted to alpine peaks the deep valleys between cannot be colonized.

biological clock regular internal rhythm of activity, produced by unknown mechanisms,

biochemistry: chronology

c. 1830	Johannes Müller discovered proteins.
1833	Anselme Payen and J F Persoz first isolated an enzyme.
1862	Haemoglobin was first crystallized.
1869	The genetic material DNA (deoxyribonucleic acid) was discovered by Friedrich Mieschler.
1899	Emil Fischer postulated the 'lock-and-key' hypothesis to explain the specificity of enzyme action.
1913	Leonor Michaelis and M L Menten developed a mathematical equation describing the rate of enzyme-catalysed reactions.
1915	The hormone thyroxine was first isolated from thyroid gland tissue.
1920	The chromosome theory of heredity was postulated by Thomas H Morgan; growth hormone was discovered by Herbert McLean Evans and J A Long.
1921	Insulin was first isolated from the pancreas by Frederick Banting and Charles Best.
1926	Insulin was obtained in pure crystalline form.
1927	Thyroxine was first synthesized.
1928	Alexander Fleming discovered penicillin.
1931	Paul Karrer deduced the structure of retinol (vitamin A); vitamin D compounds were obtained in crystalline form by Adolf Windaus and Askew, independently of each other.
1932	Charles Glen King isolated ascorbic acid (vitamin C).
1933	Tadeus Reichstein synthesized ascorbic acid.
1935	Richard Kuhn and Karrer established the structure of riboflavin (vitamin B_2).
1936	Robert Williams established the structure of thiamine (vitamin B_1); biotin was isolated by Kogl and Tonnis.
1937	Niacin was isolated and identified by Conrad Arnold Elvehjem.
1938	Pyridoxine (vitamin B_6) was isolated in pure crystalline form.
1939	The structure of pyridoxine was determined by Kuhn.
1940	Hans Krebs proposed the citric acid (Krebs) cycle; Hickman isolated retinol in pure crystalline form; Williams established the structure of pantothenic acid; biotin was identified by Albert Szent-Györgyi, Vincent Du Vigneaud, and co-workers.
1941	Penicillin was isolated and characterized by Howard Florey and Ernst Chain.
1943	The role of DNA in genetic inheritance was first demonstrated by Oswald Avery, Colin MacLeod, and Maclyn McCarty.
1950	The basic components of DNA were established by Erwin Chargaff; the alpha-helical structure of proteins was established by Linus Pauling and R B Corey.
1953	James Watson and Francis Crick determined the molecular structure of DNA.
1956	Mahlon Hoagland and Paul Zamecnick discovered transfer RNA (ribonucleic acid); mechanisms for the biosynthesis of RNA and DNA were discovered by Arthur Kornberg and Severo Ochoa.
1957	Interferon was discovered by Alick Isaacs and Jean Lindemann.
1958	The structure of RNA was determined.
1960	Messenger RNA was discovered by Sydney Brenner and François Jacob.
1961	Marshall Nirenberg and Ochoa determined the chemical nature of the genetic code.
1965	Insulin was first synthesized.
1966	The immobilization of enzymes was achieved by Chibata.
1968	Brain hormones were discovered by Roger Guillemin and Andrew Schally.
1975	J Hughes and Hans Kosterlitz discovered encephalins.
1976	Guillemin discovered endorphins.
1977	J Baxter determined the genetic code for human growth hormone.
1978	Human insulin was first produced by genetic engineering.
1979	The biosynthetic production of human growth hormone was announced by Howard Goodman and J Baxter of the University of California, and by D V Goeddel and Seeburg of Genentech.
1982	Louis Chedid and Michael Sela developed the first synthesized vaccine.
1983	The first commercially available product of genetic engineering (Humulin) was launched.
1985	Alec Jeffreys devised genetic fingerprinting.
1993	UK researchers introduced a healthy version of the gene for cystic fibrosis into the lungs of mice with induced cystic fibrosis, restoring normal function.

and not dependent on external time signals. Such clocks are known to exist in almost all animals, and also in many plants, fungi, and unicellular organisms. In higher organisms, there appears to be a series of clocks of graded importance. For example, although body temperature and activity cycles in human beings are normally 'set' to 24 hours, the two cycles may vary independently, showing that two clock mechanisms are involved.

biological control control of pests such as insects and fungi through biological means, rather than the use of chemicals. This can include breeding resistant crop strains; inducing sterility in the pest; infecting the pest species with disease organisms; or introducing the pest's natural predator. Biological control tends to be naturally self-regulating, but as ecosystems are so complex, it is difficult to predict all the consequences of introducing a biological controlling agent.

biological oxygen demand (BOD) the amount of dissolved oxygen taken up by microorganisms in a sample of water. Since these microorganisms live by decomposing organic matter, and the amount of oxygen used is proportional to their number and metabolic rate, BOD can be used as a measure of the extent to which the water is polluted with organic compounds.

biological shield shield around a nuclear reactor that is intended to protect personnel from the effects of ◊radiation. It usually consists of a thick wall of steel and concrete.

biology science of life. Biology includes all the life sciences – for example, anatomy and physiology (the study of the structure of living things), cytology (the study of cells), zoology (the study of animals) and botany (the study of plants), ecology (the study of habitats and the interaction of living species), animal behaviour, embryology, and taxonomy, and plant breeding. Increasingly this century biologists have concentrated on molecular structures: biochemistry, biophysics, and genetics (the study of inheritance and variation). The pinnacle of this development during the 1990s is the international ◊Human Genome Project, which will attempt to map the entire genetic code contained in the 23 pairs of human chromosomes.

bioluminescence production of light by living organisms. It is a feature of many deep-sea fishes, crustaceans, and other marine animals. On land, bioluminescence is seen in some nocturnal insects such as glow-worms and fireflies, and in certain bacteria and fungi. Light is usually produced by the oxidation of luciferin, a reaction catalysed by the ◊enzyme luciferase. This reaction is unique, being the only known biological oxidation that does not produce heat. Animal luminescence is involved in communication, camouflage, or the luring of prey, but its function in other organisms is unclear.

biomass the total mass of living organisms present in a given area. It may be specified for a particular species (such as earthworm biomass) or for a general category (such as herbivore biomass). Estimates also exist for the entire global plant biomass. Measurements of biomass can be used to study interactions between organisms, the stability of those interactions, and variations in population numbers.

biome broad natural assemblage of plants and animals shaped by common patterns of vegetation and climate. Examples include the tundra biome and the desert biome.

biomechanics study of natural structures to improve those produced by humans. For example, mother-of-pearl is structurally superior to glass fibre, and deer antlers have outstanding durability because they are composed of microscopic fibres. Such natural structures may form the basis of high-tech composites.

biometry literally, the measurement of living things, but generally used to mean the application of mathematics to biology. The term is now largely obsolete, since mathematical or statistical work is an integral part of most biological disciplines.

bionics design and development of electronic or mechanical artificial systems that imitate those of living things. The bionic arm, for example, is an artificial limb (◊prosthesis) that uses electronics to amplify minute electrical signals generated in body muscles to work electric motors, which operate the joints of the fingers and wrist.

biophysics application of physical laws to the properties of living organisms. Examples include using the principles of ◊mechanics to calculate the strength of bones and muscles, and ◊thermodynamics to study plant and animal energetics.

biorhythm rhythmic change, mediated by ◊hormones, in the physical state and activity patterns of certain plants and animals that have seasonal activities. Examples include winter hibernation, spring flowering or breeding, and periodic migration. The hormonal changes themselves are often a response to changes in day length (◊photoperiodism); they signal the time of year to the animal or plant. Other biorhythms are innate and continue even if external stimuli such as day length are removed. These include a 24-hour or ◊circadian rhythm, a 28-day or circalunar rhythm (corresponding to the phases of the Moon), and even a year-long rhythm in some organisms.

BIOS (acronym for *basic input/output system*) in computing, the part of the ◊operating system that handles input and output. The term is also used to describe the programs stored in ◊ROM (and called ROM BIOS), which are automatically run when a computer is switched on allowing it to ◊boot. BIOS is unaffected by upgrades to the operating system stored on disc.

biosensor device based on microelectronic circuits that can directly measure medically significant variables for the purpose of diagnosis or monitoring treatment. One such device measures the blood sugar level of diabetics using a single drop of blood, and shows the result on a liquid crystal display within a few minutes.

biosphere the narrow zone that supports life on our planet. It is limited to the waters of the Earth, a fraction of its crust, and the lower regions of the atmosphere.

biosynthesis synthesis of organic chemicals from simple inorganic ones by living cells – for example, the conversion of carbon dioxide and water to glucose by plants during ◊photosynthesis. Other biosynthetic reactions produce cell constituents including proteins and fats.

biotechnology industrial use of living organisms to manufacture food, drugs, or other products. The brewing and baking industries have long relied on the yeast microorganism for ◊fermentation purposes, while the dairy industry employs a range of bacteria and fungi to convert milk into cheeses and yoghurts. ◊Enzymes, whether extracted from cells or produced artificially, are central to most biotechnological applications.

Recent advances include ◊genetic engineering, in which single-celled organisms with modified ◊DNA are used to produce insulin and other drugs. In 1993 two-thirds of biotechnology companies were concentrating on human health developments, whilst only 1 in 10 were concerned with applications for food and agriculture.

biotic factor organic variable affecting an ecosystem for example, the changing population of elephants and its effect on the African savanna.

biotin or *vitamin H* vitamin of the B complex, found in many different kinds of food; egg yolk, liver, legumes, and yeast contain large amounts. Biotin is essential to the metabolism of fats. Its absence from the diet may lead to dermatitis.

biotite dark mica, $K(Mg, Fe)_3 Al\, Si_3 O_{10}(OH, F)_2$, a common silicate mineral. It is brown to black with shiny surfaces, and like all micas, it splits into very thin flakes along its one perfect cleavage. Biotite is a mineral of igneous rocks such as granites, and metamorphic rocks such as schists and gneisses.

bird backboned animal of the class Aves, the biggest group of land vertebrates, characterized by warm blood, feathers, wings, breathing through lungs, and egg-laying by the female. There are nearly 8,500 species of birds.

Birds are bipedal, with the front limb modified to form a wing and retaining only three digits. The heart has four chambers, and the body is maintained at a high temperature (about 41°C/106°F). Most birds fly, but some groups (such as ostriches) are flightless, and others include flightless members. Many communicate by sounds, or by visual displays, in connection with which many species are brightly coloured. Birds have highly developed patterns of instinctive behaviour. Hearing and eyesight are well developed, but the sense of smell is usually poor. Typically the eggs are brooded in a nest and, on hatching, the young receive a period of parental care. The study of birds is called ◊ornithology.

birth act of producing live young from within the body of female animals. Both viviparous and ovoviviparous animals give birth to young. In viviparous animals, embryos obtain nourishment from the mother via a ◊placenta or other means.

In ovoviviparous animals, fertilized eggs develop and hatch in the oviduct of the mother and gain little or no nourishment from maternal tissues. See also ◊pregnancy.

bisector a line that bisects an angle or another line (known as a *perpendicular bisector* when it bisects at right angles).

bismuth hard, brittle, pinkish-white, metallic element, symbol Bi, atomic number 83, relative atomic mass 208.98. It has the highest atomic number of all the stable elements (the elements from atomic number 84 up are radioactive). Bismuth occurs in ores and occasionally as a free metal (◊native metal). It is a poor conductor of heat and electricity, and is used in alloys of low melting point and in medical compounds to soothe gastric ulcers. The name comes from the Latin *besemutum*, from the earlier German *Wismut*.

bistable circuit or *flip-flop* simple electronic circuit that remains in one of two stable states until it receives a pulse (logic 1 signal) through one of its inputs, upon which it switches, or 'flips', over to the other state. Because it is a two-state device, it can be used to store binary digits and is widely used in the ◊integrated circuits used to build computers.

bisulphate another term for ◊hydrogen sulphate.

bit (contraction of *bi*nary digi*t*) in computing, a single binary digit, either 0 or 1. A bit is the smallest unit of data stored in a computer; all other data must be coded into a pattern of individual bits. A ◊byte represents sufficient computer memory to store a single ◊character of data, and usually contains eight bits. For example, in the ◊ASCII code system used by most microcomputers the capital letter A would be stored in a single byte of memory as the bit pattern 01000001.

bit map in computing, a pattern of ◊bits used to describe the organization of data. Bit maps are used to store typefaces or graphic images (bit-mapped or ◊raster graphics), with 1 representing black (or a colour) and 0 white.

bitumen impure mixture of hydrocarbons, including such deposits as petroleum, asphalt, and natural gas, although sometimes the term is restricted to a soft kind of pitch resembling asphalt.

bivalent in biology, a name given to the pair of homologous chromosomes during reduction division (◊meiosis). In chemistry, the term is sometimes used to describe an element or group with a ◊valency of two, although the term 'divalent' is more common.

black body in physics, a hypothetical object that completely absorbs all thermal (heat) radiation striking it. It is also a perfect emitter of thermal radiation.

black earth exceedingly fertile soil that covers a belt of land in NE North America, Europe, and Asia.

black hole object in space whose gravity is so great that nothing can escape from it, not even light. Thought to form when massive stars shrink at the ends of their lives, a black hole sucks in more matter, including other stars, from the space around it. Matter that falls into a black hole is squeezed to infinite density at the centre of the hole. Black holes can be detected because gas falling towards them becomes so hot that it emits X-rays.

Massive black holes containing the mass of millions of stars are thought to lie at the centres of ◊quasars. Microscopic black holes may have been formed in the chaotic conditions of the ◊Big Bang.

bladder hollow elastic-walled organ which stores the urine produced in the kidneys. It is present in the ◊urinary systems of some fishes, most amphibians, some reptiles, and all mammals. Urine enters the bladder through two ureters, one leading from each kidney, and leaves it through the urethra.

blast furnace smelting furnace used to extract metals from their ores, chiefly pig iron from iron ore. The temperature is raised by the injection of an air blast.

In the extraction of iron the ingredients of the furnace are iron ore, coke (carbon), and limestone. The coke is the fuel and provides the carbon monoxide for the reduction of the iron ore; the limestone acts as a flux, removing impurities.

blastocyst in mammals, the hollow ball of cells which is an early stage in the development of the ◊embryo, roughly equivalent to the blastula of other animal groups.

blastomere in biology, a cell formed in the first stages of embryonic development, after the splitting of the fertilized ovum, but before the formation of the ◊blastula or blastocyst.

blast furnace *The raw materials are fed into the top of a blast furnace. Hot high-pressure air is blown in at the sides of the furnace. As iron forms, it melts and runs to the bottom of the furnace; molten slag floats on its surface.*

blastula early stage in the development of a fertilized egg, when the egg changes from a solid mass of cells (the morula) to a hollow ball of cells (the blastula), containing a fluid-filled cavity (the blastocoel). See also ◊embryology.

bleaching decolorization of coloured materials. The two main types of bleaching agent are the *oxidizing bleaches*, which bring about the ◊oxidation of pigments and include the ultraviolet rays in sunshine, hydrogen peroxide, and chlorine in household bleaches, and the *reducing bleaches*, which bring about ◊reduction and include sulphur dioxide.

blight any of a number of plant diseases caused mainly by parasitic species of ◊fungus, which produce a whitish appearance on leaf and stem surfaces; for example, potato blight *Phytophthora infestans*. General damage caused by aphids or pollution is sometimes known as blight.

blind spot area where the optic nerve and blood vessels pass through the retina of the ◊eye. No visual image can be formed as there are no light-sensitive cells in this part of the retina. Thus the organism is blind to objects that fall in this part of the visual field.

BL Lacertae object starlike object that forms the centre of a distant galaxy, with a prodigious energy output. BL Lac objects, as they are called, seem to be related to ◊quasars and are thought to be the brilliant nuclei of elliptical galaxies. They are so named because the first to be discovered lies in the small constellation Lacerta.

blood fluid circulating in the arteries, veins, and capillaries of vertebrate animals; the term also refers to the corresponding fluid in those invertebrates that possess a closed ◊circulatory system. Blood carries nutrients and oxygen to each body cell and removes waste products, such as carbon dioxide. It is also important in the immune response and, in many animals, in the distribution of heat throughout the body.

In humans blood makes up 5% of the body weight, occupying a volume of 5.5 l/10 pt in the average adult. It is composed of a colourless, transparent liquid called *plasma*, in which are suspended microscopic cells of three main varieties.

Red cells (erythrocytes) form nearly half the volume of the blood, with about six million red cells in every millilitre of an adult's blood. Their red colour is caused by ◊haemoglobin. *White cells* (leucocytes) are of various kinds. Some (phagocytes) ingest invading bacteria and so protect the body from disease; these also help to repair injured tissues. Others (lymphocytes) produce antibodies, which help provide immunity. Blood *platelets* (thrombocytes) assist in the clotting of blood.

blood clotting complex series of events (known as the blood clotting cascade) that prevents excessive bleeding after injury. It is triggered by ◊vitamin K. The result is the formation of a meshwork of protein fibres (fibrin) and trapped blood cells over the cut blood vessels.

When platelets (cell fragments) in the bloodstream come into contact with a damaged blood vessel, they and the vessel wall itself release the enzyme *thrombokinase*, which brings about the conversion of the inactive enzyme *prothrombin* into the active *thrombin*. Thrombin in turn catalyses the conversion of the soluble protein *fibrinogen*, present in blood plasma, to the insoluble *fibrin*. This fibrous protein forms a net over the wound that traps red blood cells and seals the wound; the resulting jellylike clot hardens on exposure to air to form a scab. Calcium, vitamin K, and a variety of enzymes called factors are also necessary for efficient blood clotting. ◊Haemophilia is one of several diseases in which the clotting mechanism is impaired.

blood group any of the types into which blood is classified according to the presence or otherwise of certain ◊antigens on the surface of its red cells. Red blood cells of one individual may carry molecules on their surface that act as antigens in another individual whose red blood cells lack these molecules. The two main antigens are designated A and B. These give rise to four blood groups: having A only (A), having B only (B), having both (AB), and having neither (O). Each of these groups may or may not contain the ◊rhesus factor. Correct typing of blood groups is vital in transfusion, since incompatible types of donor and recipient blood will result in coagulation, with possible death of the recipient.

blood pressure pressure, or tension, of the blood against the inner walls of blood vessels, especially the arteries, due to the muscular pumping activity of the heart. Abnormally high blood pressure (hypertension) may be associated with various conditions or arise with no obvious cause; abnormally low blood pressure (hypotension) occurs in shock and after excessive fluid or blood loss from any cause.

blood vessel tube that conducts blood either away from or towards the heart in multicellular animals. Freshly oxygenated blood is carried in the arteries – major vessels which give way to the arterioles (small arteries) and finally capillaries; deoxygenated blood is returned to the heart by way of capillaries, then venules (small veins) and veins.

bloom whitish powdery or waxlike coating over the surface of certain fruits that easily rubs off when handled. It often contains ◊yeasts that live on the sugars in the fruit. The term bloom is also used to describe a rapid increase in number of certain species of algae found in lakes, ponds, and oceans.

blubber thick layer of ◊fat under the skin of marine mammals, which provides an energy store and an effective insulating layer, preventing the loss of body heat to the surrounding water. Blubber has been used (when boiled down) in engineering, food processing, cosmetics, and printing, but all of these products can now be produced synthetically.

blue-green algae or *cyanobacteria* single-celled, primitive organisms that resemble bacteria in their internal cell organization, sometimes joined together in colonies or filaments. Blue-green algae are among the oldest known living organisms and, with bacteria, belong to the kingdom Monera; remains have been found in rocks up to 3.5 billion years old. They are widely distributed in aquatic habitats, on the damp surfaces of rocks and trees, and in the soil.

blue shift in astronomy, a manifestation of the ◊Doppler effect in which an object appears bluer when it is moving towards the observer or the observer is moving towards it (blue light is of a higher frequency than other colours in the spectrum). The blue shift is the opposite of the ◊red shift.

BMR abbreviation for ◊*basal metabolic rate*.

Bode's law numerical sequence that gives the approximate distances, in astronomical units (distance between Earth and Sun = one astronomical unit), of the planets from the Sun by adding 4 to each term of the series 0, 3, 6, 12, 24, ... and then dividing by 10. Bode's law predicted the existence of a planet between ◊Mars and ◊Jupiter, which led to the discovery of the asteroids.

The 'law' breaks down for ◊Neptune and ◊Pluto.

body temperature intensity of heat deep in the tissues (core temperature) as measured on a thermometer. In the healthy adult it is normally 37°C. Any significant departure from this norm is potentially serious (although fever is a necessary response to infection). Hyperthermia (temperature above 41°C) and ◊hypothermia (below 3°C) are life-threatening.

Bohr model model of the atom conceived by Danish physicist Neils Bohr 1913. It assumes that the following rules govern the behaviour of electrons: (1) electrons revolve in orbits of specific radius around the nucleus without emitting radiation; (2) within each orbit, each electron has a fixed amount of energy; electrons in orbits farther away from the nucleus have greater energies; (3) an electron may 'jump' from one orbit of high energy to another of lower energy causing the energy difference to be emitted as a ◊photon of electromagnetic radiation such as light. The Bohr model has been superseded by wave mechanics (see ◊quantum theory).

boiler any vessel that converts water into steam. Boilers are used in conventional power stations to generate steam to feed steam ◊turbines, which drive the electricity generators. They are also used in steamships, which are propelled by steam turbines, and in steam locomotives. Every boiler has a furnace in which fuel (coal, oil, or gas) is burned to produce hot gases, and a system of tubes in which heat is transferred from the gases to the water.

boiling process of changing a liquid into its vapour, by heating it at the maximum possible temperature for that liquid (see ◊boiling point) at atmospheric pressure.

boiling point for any given liquid, the temperature at which the application of heat raises the temperature of the liquid no further, but converts it into vapour.

bolometer sensitive ◊thermometer that measures the energy of radiation by registering the change in electrical resistance of a fine wire when it is exposed to heat or light.

The US astronomer Samuel Langley devised it 1880 for measuring radiation from stars.

Boltzmann constant in physics, the constant (symbol k) that relates the kinetic energy (energy of motion) of a gas atom or molecule to temperature. Its value is 1.380662×10^{-23} joules per Kelvin. It is equal to the gas constant R, divided by ◊Avogadro's number.

bolus mouthful of chewed food mixed with saliva, ready for swallowing. Most vertebrates swallow food immediately, but grazing mammals chew their food a great deal, allowing a mechanical and chemical breakdown to begin.

bond in chemistry, the result of the forces of attraction that hold together atoms of an element or elements to form a molecule. The principal types of bonding are ◊ionic, ◊covalent, ◊metallic, and ◊intermolecular (such as hydrogen bonding).

The type of bond formed depends on the elements concerned and their electronic structure. In an ionic or electrovalent bond, common in inorganic compounds, the combining atoms gain or lose electrons to become ions; for example, sodium (Na) loses an electron to form a sodium ion (Na+) while chlorine (Cl) gains an electron to form a chloride ion (Cl-) in the ionic bond of sodium chloride (NaCl).

In a covalent bond, the atomic orbitals of two atoms overlap to form a molecular orbital containing two electrons, which are thus effectively shared between the two atoms. Covalent bonds are common in organic compounds, such as the four carbon-hydrogen bonds in methane (CH_4). In a dative covalent or coordinate bond, one of the combining atoms supplies both of the valence electrons in the bond.

A metallic bond joins metals in a crystal lattice; the atoms occupy lattice positions as positive ions, and valence electrons are shared between all the ions in an 'electron gas'.

In a hydrogen bond, a hydrogen atom joined to an electronegative atom, such as nitrogen or oxygen, becomes partially positively charged, and is weakly attracted to another electronegative atom on a neighbouring molecule.

The strongest known noncovalent bond is the ***superbond*** that is formed between avidin, a protein found in egg white, and the growth factor ◊biotin. It is almost impossible to separate the two molecules once bonded. The superbond is used in a number of biomedical research applications.

bone hard connective tissue comprising the ◊skeleton of most vertebrate animals. Bone is composed of a network of collagen fibres impregnated with mineral salts (largely calcium phosphate and calcium carbonate), a combination that gives it great density and strength, comparable in some cases with that of reinforced concrete. Enclosed within this solid matrix are bone cells, blood vessels, and nerves. The interior of the long bones of the limbs consists of a spongy matrix filled with a soft marrow that produces blood cells.

There are two types of bone: those that develop by replacing ◊cartilage and those that form directly from connective tissue. The

bone *Bone is a network of fibrous material impregnated with mineral salts and as strong as reinforced concrete. The upper end of the thighbone or femur is made up of spongy bone, which has a fine lacework structure designed to transmit the weight of the body. The shaft of the femur consists of hard compact bone designed to resist bending. Fine channels carrying blood vessels, nerves, and lymphatics interweave even the densest bone.*

latter, which includes the bones of the cranium, are usually platelike in shape and form in the skin of the developing embryo. Humans have about 206 distinct bones in the skeleton, of which the smallest are the three ossicles in the middle ear.

bone marrow substance found inside the cavity of bones. In early life it produces red blood cells but later on lipids (fat) accumulate and its colour changes from red to yellow.

Boolean algebra set of algebraic rules, named after mathematician George Boole, in which TRUE and FALSE are equated to 0 and 1. Boolean algebra includes a series of operators (AND, OR, NOT, NAND (NOT AND), NOR, and XOR (exclusive OR)), which can be used to manipulate TRUE and FALSE values (see ◊truth table). It is the basis of computer logic because the truth values can be directly associated with ◊bits.

booster first-stage rocket of a space-launching vehicle, or an additional rocket strapped to the main rocket to assist takeoff.

boot or *bootstrap* in computing, the process of starting up a computer. Most computers have a small, built-in boot program that starts automatically when the computer is switched on – its only task is to load a slightly larger program, usually from a disc, which in turn loads the main ◊operating system.

In microcomputers the operating system is often held in the permanent ◊ROM memory and the boot program simply triggers its operation.

Boötes constellation of the northern hemisphere represented by a herdsman driving a bear (Ursa Major) around the pole. Its brightest star is ◊Arcturus (or Alpha Boötis) about 36 light years from Earth.

borax hydrous sodium borate, $Na_2B_4O_7.10H_2O$, found as soft, whitish crystals or encrustations on the shores of hot springs and in the dry beds of salt lakes in arid regions, where it occurs with other borates, halite, and ◊gypsum. It is used in bleaches and washing powders.

boric acid or *boracic acid* H_3BO_3, acid formed by the combination of hydrogen and oxygen with nonmetallic boron. It is a weak antiseptic and is used in the manufacture of glass and enamels. It is also an efficient insecticide against ants and cockroaches.

boron nonmetallic element, symbol B, atomic number 5, relative atomic mass 10.811. In nature it is found only in compounds, as with sodium and oxygen in borax. It exists in two allotropic forms (see ◊allotropy): brown amorphous powder and very hard, brilliant crystals. Its compounds are used in the preparation of boric acid, water softeners, soaps, enamels, glass, and pottery glazes. In alloys it is used to harden steel. Because it absorbs slow neutrons, it is used to make boron carbide control rods for nuclear reactors. It is a necessary trace element in the human diet. The element was named by Humphry Davy, who isolated it 1808, from *bor*ax + -*on*.

boson in physics, an elementary particle whose spin can only take values that are whole numbers or zero. Bosons may be classified as ◊gauge bosons (carriers of the four fundamental forces) or ◊mesons. All elementary particles are either bosons or ◊fermions.

botany the study of plants. It is subdivided into a number of specialized studies, such as the identification and classification of plants (taxonomy), their external formation (plant morphology), their internal arrangement (plant anatomy), their microscopic examination (plant histology), their functioning and life history (plant physiology), and their distribution over the Earth's surface in relation to their surroundings (plant ecology). Palaeobotany concerns the study of fossil plants, while economic botany deals with the utility of plants. Horticulture, agriculture, and forestry are branches of botany.

Bourdon gauge instrument for measuring pressure, patented by French watchmaker Eugène Bourdon 1849. The gauge contains a C-shaped tube, closed at one end. When the pressure inside the tube increases, the tube uncurls slightly causing a small movement at its closed end. A system of levers and gears magnifies this movement and turns a pointer, which indicates the pressure on a circular scale. Bourdon gauges are often fitted to cylinders of compressed gas used in industry and hospitals.

Bowman's capsule in the vertebrate kidney, a cup-shaped structure enclosing the glomerulus, which is the initial site of filtration of the blood leading to urine formation.

There are approximately a million of these capsules in a human kidney, each containing a tight knot of capillaries and leading into a kidney tubule or nephron where unwanted fluid and waste molecules are filtered from the blood to be excreted in the form of urine.

Boyle's law law stating that the volume of a given mass of gas at a constant temperature is

brain *The structure of the human brain. At the back of the skull lies the cerebellum, which coordinates reflex actions that control muscular activity. The medulla controls respiration, heartbeat, and blood pressure. The hypothalamus is concerned with instinctive drives and emotions. The thalamus relays signals to and from various parts of the brain. The pituitary gland controls the body's hormones. Distinct areas of the large convoluted cerebral hemispheres that fill most of the skull are linked to sensations, such as hearing and sight, and voluntary activities, such as movement.*

inversely proportional to its pressure. For example, if the pressure of a gas doubles, its volume will be reduced by a half, and vice versa. The law was discovered in 1662 by Irish physicist and chemist Robert Boyle. See also ◊gas laws.

bps abbreviation for *bits per second*, a measure used in specifying data transmission rates.

brachiopod or *lamp shell* any member of the phylum Brachiopoda, marine invertebrates with two shells, resembling but totally unrelated to bivalves.

There are about 300 living species; they were much more numerous in past geological ages. They are suspension feeders, ingesting minute food particles from water. A single internal organ, the lophophore, handles feeding, aspiration, and excretion.

bract leaflike structure in whose ◊axil a flower or inflorescence develops. Bracts are generally green and smaller than the true leaves. However, in some plants they may be brightly coloured and conspicuous, taking over the role of attracting pollinating insects to the flowers, whose own petals are small; examples include poinsettia *Euphorbia pulcherrima* and bougainvillea.

brain in higher animals, a mass of interconnected ◊nerve cells forming the anterior part of the ◊central nervous system, whose activities it coordinates and controls. In ◊vertebrates, the brain is contained by the skull. An enlarged portion of the upper spinal cord, the *medulla oblongata*, contains centres for the control of respiration, heartbeat rate and strength, and blood pressure. Overlying this is the *cerebellum*, which is concerned with coordinating complex muscular processes such as maintaining posture and moving limbs. The cerebral hemispheres (*cerebrum*) are paired outgrowths of the front end of the forebrain, in early vertebrates mainly concerned with the senses, but in higher vertebrates greatly developed and involved in the integration of all sensory input and motor output, and in thought, emotions, memory, and behaviour.

In vertebrates, many of the nerve fibres from the two sides of the body cross over as they enter the brain, so that the left cerebral hemisphere is associated with the right side of the body and vice versa. In humans, a certain asymmetry develops in the two halves of the cerebrum. In right-handed people, the left hemisphere seems to play a greater role in controlling verbal and some mathematical skills, whereas the right hemisphere is more involved in spatial perception. In general, however, skills and abilities are not closely localized. In the brain, nerve impulses are passed across ◊synapses by neurotransmitters, in the same way as in other parts of the nervous system.

In mammals the cerebrum is the largest part of the brain, carrying the *cerebral cortex*. This consists of a thick surface layer of cell bodies (grey matter), below which fibre tracts (white matter) connect various parts of the cortex to each other and to other points in the central nervous system. As cerebral complexity grows, the surface of the brain becomes convoluted into deep folds. In higher mammals, there are large unassigned areas of the brain that seem to be connected with intelligence, personality, and higher mental faculties. Language is controlled in two special regions usually in the left side of the brain: *Broca's area* governs the ability to talk, and *Wernicke's area* is responsible for the comprehension of spoken and written words. In 1990, scientists at Johns Hopkins University, Baltimore, succeeded in culturing human brain cells.

brainstem region where the top of the spinal cord merges with the undersurface of the brain, consisting largely of the medulla oblongata and midbrain.

The oldest part of the brain in evolutionary terms, the brainstem is the body's life-support centre, containing regulatory mechanisms for vital functions such as breathing, heart rate, and blood pressure. It is also involved in controlling the level of consciousness by acting as a relay station for nerve connections to and from the higher centres of the brain.

brake *Two common braking systems: the disc brake (top) and the drum brake (bottom). In the disc brake, increased hydraulic pressure of the brake fluid in the pistons forces the brake pads against the steel disc attached to the wheel. A self-adjusting mechanism balances the force on each pad. In the drum brake, increased pressure of the brake fluid within the slave cylinder forces the brake pad against the brake drum attached to the wheel.*

brake device used to slow down or stop the movement of a moving body or vehicle. The mechanically applied calliper brake used on bicycles uses a scissor action to press hard rubber blocks against the wheel rim. The main braking system of a car works hydraulically: when the driver depresses the brake pedal, liquid pressure forces pistons to apply brakes on each wheel.

brass metal ◊alloy of copper and zinc, with not more than 5% or 6% of other metals. The zinc content ranges from 20% to 45%, and the colour of brass varies accordingly from coppery to whitish yellow. Brasses are characterized by the ease with which they may be shaped and machined; they are strong and ductile, resist many forms of corrosion, and are used for electrical fittings, ammunition cases, screws, household fittings, and ornaments.

brazing method of joining two metals by melting an ◊alloy into the joint. It is similar to soldering but takes place at a much higher temperature. Copper and silver alloys are widely used for brazing, at temperatures up to about 900°C/1,650°F.

breast one of a pair of organs on the chest of the human female, also known as a ◊mammary gland. Each of the two breasts contains milk-producing cells and a network of tubes or ducts that lead to openings in the nipple.

Breathalyzer trademark for an instrument for on-the-spot checking by police of the amount of alcohol consumed by a suspect driver. The driver breathes into a plastic bag connected to a tube containing a chemical (such as a diluted solution of potassium dichromate in 50% sulphuric acid) that changes colour in the presence of alcohol. Another method is to use a gas chromatograph, again from a breath sample.

breathing in terrestrial animals, the muscular movements whereby air is taken into the lungs and then expelled, a form of ◊gas exchange. Breathing is sometimes referred to as external respiration, for true respiration is a cellular (internal) process.

Lungs are specialized for gas exchange but are not themselves muscular, consisting of spongy material. In order for oxygen to be passed to the blood and carbon dioxide removed, air is drawn into the lungs (inhaled) by the contraction of the diaphragm and

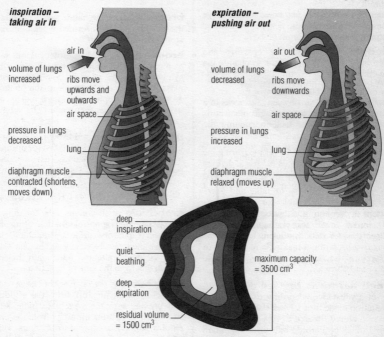

subdivisions of lung air

breathing *To breathe in, the diaphragm muscle contracts. This allows the lungs to increase in volume, and air flows into the lungs. To breathe out, the diaphragm muscle is relaxed and moves up, forcing air from the lungs. Deep inspiration increases the volume of air breathed in, but there is always some air left in the lungs, known as the residual volume.*

intercostal muscles; relaxation of these muscles enables air to be breathed out (exhaled). The rate of breathing is controlled by the brain. High levels of activity lead to a greater demand for oxygen and an increased rate of breathing.

breathing rate the number of times a minute the lungs inhale and exhale. The rate increases during exercise because the muscles require an increased supply of oxygen and nutrients. At the same time very active muscles produce a greater volume of carbon dioxide, a waste gas that must be removed by the lungs via the blood.

breccia coarse clastic ◊sedimentary rock, made up of broken fragments (clasts) of pre-existing rocks. It is similar to ◊conglomerate but the fragments in breccia are jagged in shape.

breed recognizable group of domestic animals, within a species, with distinctive characteristics that have been produced by ◊artificial selection.

breeder reactor or *fast breeder* alternative names for ◊fast reactor, a type of nuclear reactor.

breeding in biology, the crossing and selection of animals and plants to change the characteristics of an existing ◊breed or ◊cultivar (variety), or to produce a new one.

breeding in nuclear physics, a process in a reactor in which more fissionable material is produced than is consumed in running the reactor.

brewster unit (symbol B) for measuring the reaction of optical materials to stress, defined in terms of the slowing down of light passing through the material when it is stretched or compressed.

bridge in computing, a device that connects two similar local area networks (LANs). Bridges transfer data in packets between the two networks, without making any changes or interpreting the data in any way. See also ◊router and ◊brouter.

Brinell hardness test test of the hardness of a substance according to the area of indentation made by a 10 mm/0.4 in hardened steel or sintered (heated to very high temperatures) tungsten carbide ball under standard loading conditions in a test machine. The resulting Brinell number is equal to the load (kg) divided by the surface area (mm^2) and is named after its inventor, Swedish metallurgist Johann Brinell.

British Standards Institute (BSI) UK national standards body. Although government funded, the institute is independent. The BSI interprets international technical standards for the UK, and also sets its own.

British thermal unit imperial unit (symbol Btu) of heat, now replaced in the SI system by the ◊joule (one British thermal unit is approximately 1,055 joules). Burning one cubic foot of natural gas releases about 1,000 Btu of heat.

brittle material material that breaks suddenly under stress at a point just beyond its elastic limit (see ◊elasticity). Brittle materials may also break suddenly when given a sharp knock. Pottery, glass, and cast iron are examples of brittle materials. Compare ◊ductile material.

broad-leaved tree another name for a tree belonging to the ◊angiosperms, such as ash, beech, oak, maple, or birch. The leaves are generally broad and flat, in contrast to the needlelike leaves of most ◊conifers. See also ◊deciduous tree.

bromide salt of the halide series containing the Br$^-$ ion, which is formed when a bromine atom gains an electron.

bromine dark, reddish-brown, nonmetallic element, a volatile liquid at room temperature, symbol Br, atomic number 35, relative atomic mass 79.904. It is a member of the ◊halogen group, has an unpleasant odour, and is very irritating to mucous membranes. Its salts are known as bromides.

bromine water red-brown solution of bromine in water. It is used to test for unsaturation in organic compounds, as such compounds decolorize the solution as they react with the bromine.

$$CH_2=CH_2 + Br_2 \rightarrow CH_2BrCH_2Br$$

bronchiole small-bore air tube found in the vertebrate lung responsible for delivering air to the main respiratory surfaces. Bronchioles lead off from the larger bronchus and branch extensively before terminating in the many thousand alveoli that form the bulk of lung tissue.

bronchus one of a pair of large tubes (bronchii) branching off from the windpipe and passing into the vertebrate lung. Apart from their size, bronchii differ from the bronchioles in possessing cartilaginous rings, which give rigidity and prevent collapse during breathing movements.

bronze alloy of copper and tin, yellow or brown in colour. It is harder than pure copper, more suitable for ◊casting, and also resists

◊corrosion. Bronze may contain as much as 25% tin, together with small amounts of other metals, mainly lead.

brouter device for connecting computer networks that incorporates the facilities of both a ◊bridge and a ◊router. Brouters usually offer routing over a limited number of ◊protocols, operating by routing where possible and bridging the remaining protocols.

brown dwarf hypothetical object less massive than a star, but heavier than a planet. Brown dwarfs would not have enough mass to ignite nuclear reactions at their centres, but would shine by heat released during their contraction from a gas cloud. Because of the difficulty of detection, no brown dwarfs have been spotted with certainty, but some astronomers believe that vast numbers of them may exist throughout the Galaxy.

Brownian movement the continuous random motion of particles in a fluid medium (gas or liquid) as they are subjected to impact from the molecules of the medium. The phenomenon was explained by German physicist Albert Einstein in 1905 but was probably observed as long ago as 1827 by the Scottish botanist Robert Brown. It provides evidence for the ◊kinetic theory of matter.

brown ring test in analytical chemistry, a test for the presence of ◊nitrates.

Iron(II) sulphate is added to an aqueous solution containing the test substance. Concentrated sulphuric acid is then carefully poured down the inside wall of the test tube so that it forms a distinct layer at the bottom. The formation of a brown colour at the boundary between the two layers indicates the presence of nitrate.

brush in certain electric motors, one of a pair of contacts that pass electric current into and out of the rotating coils by means of a device known as a ◊commutator. The brushes, which are often replaceable, are usually made of a soft, carbon material to reduce wear of the copper contacts on the rotating commutator.

bryophyte member of the Bryophyta, a division of the plant kingdom containing three classes: the Hepaticae (◊liverwort), Musci (◊moss), and Anthocerotae (◊hornwort). Bryophytes are generally small, low-growing, terrestrial plants with no vascular (water-conducting) system as in higher plants. Their life cycle shows a marked ◊alternation of generations. Bryophytes chiefly occur in damp habitats and require water for the dispersal of the male gametes (◊antherozoids).

BSI abbreviation for ◊*British Standards Institute*.

Btu symbol for ◊*British thermal unit*.

bubble chamber in physics, a device for observing the nature and movement of atomic particles, and their interaction with radiations. It is a vessel filled with a superheated liquid through which ionizing particles move and collide. The paths of these particles are shown by strings of bubbles, which can be photographed and studied. By using a pressurized liquid medium instead of a gas, it overcomes drawbacks inherent in the earlier ◊cloud chamber. It was invented by US physicist Donald Glaser 1952. See ◊particle detector.

bubble-jet printer in computing, an ◊ink-jet printer in which the ink is heated to boiling point so that it forms a bubble at the end of a nozzle. When the bubble bursts, the ink is transferred to the paper.

bubble memory in computing, a memory device based on the creation of small 'bubbles' on a magnetic surface. Bubble memories typically store up to 4 megabits (4 million ◊bits) of information. They are not sensitive to shock and vibration, unlike other memory devices such as disc drives, yet, like magnetic discs, they are nonvolatile and do not lose their information when the computer is switched off.

bubble sort in computing, a technique for ◊sorting data. Adjacent items are continually exchanged until the data are in sequence.

buckminsterfullerene form of carbon, made up of molecules (buckyballs) consisting of 60 carbon atoms arranged in 12 pentagons and 20 hexagons to form a perfect sphere. It was named after the US architect and engineer Richard Buckminster Fuller because of its structural similarity to the geodesic dome that he designed. See ◊fullerene.

buckyballs popular name for molecules of ◊buckminsterfullerene.

bud undeveloped shoot usually enclosed by protective scales; inside is a very short stem and numerous undeveloped leaves, or flower parts, or both. Terminal buds are found at the tips of shoots, while axillary buds develop in the ◊axils of the leaves, often remaining dormant unless the terminal bud is removed or damaged. Adventitious buds may be produced anywhere on the plant, their formation sometimes stimulated by an injury, such as that caused by pruning.

budding type of ◊asexual reproduction in which an outgrowth develops from a cell to form a new individual. Most yeasts reproduce in this way.

buffer in computing, a part of the ◊memory used to store data temporarily while it is waiting to be used. For example, a program might store data in a printer buffer until the printer is ready to print it.

buffer mixture of chemical compounds chosen to maintain a steady ◊pH. The commonest buffers consist of a mixture of a weak organic acid and one of its salts or a mixture of acid salts of phosphoric acid. The addition of either an acid or a base causes a shift in the ◊chemical equilibrium, thus keeping the pH constant.

bulb underground bud with fleshy leaves containing a reserve food supply and with roots growing from its base. Bulbs function in vegetative reproduction and are characteristic of many monocotyledonous plants such as the daffodil, snowdrop, and onion. Bulbs are grown on a commercial scale in temperate countries, such as England and the Netherlands.

bulbil small bulb that develops above ground from a bud. Bulbils may be formed on the stem from axillary buds, as in members of the saxifrage family, or in the place of flowers, as seen in many species of onion *Allium*. They drop off the parent plant and develop into new individuals, providing a means of ◊vegetative reproduction and dispersal.

bulletin board in computing, a centre for the electronic storage of messages, usually accessed over the telephone network via ◊electronic mail and a ◊modem. Bulletin boards are usually dedicated to specific interest groups, and may carry messages, notices, and programs.

Bunsen burner gas burner used in laboratories, consisting of a vertical metal tube through which a fine jet of fuel gas is directed. Air is drawn in through airholes near the base of the tube and the mixture is ignited and burns at the tube's upper opening. Named after the German chemist Robert von Bunsen.

buoyancy lifting effect of a fluid on a body wholly or partly immersed in it. This was studied by Archimedes in the 3rd century BC.

bur or *burr* in botany, a type of 'false fruit' or ◊pseudocarp, surrounded by numerous hooks; for instance, that of burdock *Arctium*, where the hooks are formed from bracts surrounding the flowerhead. Burs catch in the feathers or fur of passing animals, and thus may be dispersed over considerable distances.

burette in chemistry, a piece of apparatus, used in ◊titration, for the controlled delivery of measured variable quantities of a liquid.

burning common name for ◊combustion.

bus in computing, the electrical pathway through which a computer processor communicates with some of its parts and/or peripherals. Physically, a bus is a set of parallel tracks that can carry digital signals; it may take the form of copper tracks laid down on the computer's ◊printed circuit boards (PCBs), or of an external cable or connection.

bushel dry or liquid measure equal to eight gallons or four pecks (2,219.36 cu in/36.37 litres) in the UK; some US states have different standards according to the goods measured.

butadiene or *buta-1,3-diene* $CH_2{:}CH\ CH{:}CH_2$ inflammable gas derived from petroleum, used in making synthetic rubber and resins.

butane C_4H_{10} one of two gaseous alkanes (paraffin hydrocarbons) having the same formula but differing in structure. Normal butane is derived from natural gas; isobutane is a by-product of petroleum manufacture. Liquefied under pressure, it is used as a fuel for industrial and domestic purposes (for example, in portable cookers).

butene C_4H_8 fourth member of the ◊alkene series of hydrocarbons. It is an unsaturated compound, containing one double bond.

by-product substance formed incidentally during the manufacture of some other substance; for example, slag is a by-product of the production of iron in a ◊blast furnace. For industrial processes to be economical, by-products must be recycled or used in other ways as far as possible; in this example, slag is used for making roads.

byte sufficient computer memory to store a single character of data. The character is stored in the byte of memory as a pattern of ◊bits (binary digits), using a code such as ◊ASCII. A byte usually contains eight bits – for example, the capital letter F can be stored as the bit pattern 01000110.

C high-level general-purpose computer-programming language popular on minicomputers and microcomputers. Developed in the early 1970s from an earlier language called BCPL, C was first used as the language of the operating system ◊Unix, though it has since become widespread beyond Unix. It is useful for writing fast and efficient systems programs, such as operating systems (which control the operations of the computer).

°C symbol for degrees ◊Celsius, sometimes called centigrade.

C++ in computing, a high-level programming language used in ◊object-oriented programming.

cable unit of length, used on ships, originally the length of a ship's anchor cable or 120 fathoms (219 m/720 ft), but now taken as one-tenth of a ◊nautical mile (185.3 m/608 ft).

cable television distribution of broadcast signals through cable relay systems. Narrowband systems were originally used to deliver services to areas with poor regular reception; systems with wider bands, using coaxial and fibreoptic cable, are increasingly used for distribution and development of home-based interactive services.

cache memory in computing, a reserved area of the ◊immediate access memory used to increase the running speed of a computer program.

CAD (acronym for *computer-aided design*) the use of computers in creating and editing design drawings. CAD also allows such things as automatic testing of designs and multiple or animated three-dimensional views of designs. CAD systems are widely used in architecture, electronics, and engineering, for example in the motor-vehicle industry, where cars designed with the assistance of computers are now commonplace.

A related development is ◊CAM (computer-assisted manufacturing).

cadmium soft, silver-white, ductile, and malleable metallic element, symbol Cd, atomic number 48, relative atomic mass 112.40. Cadmium occurs in nature as a sulphide or carbonate in zinc ores. It is a toxic metal that, because of industrial dumping, has become an environmental pollutant. It is used in batteries, electroplating, and as a constituent of alloys used for bearings with low coefficients of friction; it is also a constituent of an alloy with a very low melting point.

caecum in the ◊digestive system of animals, a blind-ending tube branching off from the first part of the large intestine, terminating in the appendix. It has no function in humans but is used for the digestion of cellulose by some grass-eating mammals.

caesium soft, silvery-white, ductile metallic element, symbol Cs, atomic number 55, relative atomic mass 132.905. It is one of the ◊alkali metals, and is the most electropositive of all the elements. In air it ignites spontaneously, and it reacts vigorously with water. It is used in the manufacture of photocells. The name comes from the blueness of its spectral line.

caffeine ◊alkaloid organic substance found in tea, coffee, and kola nuts; it stimulates the heart and central nervous system. When isolated, it is a bitter crystalline compound, $C_8H_{10}N_4O_2$. Too much caffeine (more than six average cups of tea or coffee a day) can be detrimental to health.

caisson hollow cylindrical or boxlike structure, usually of reinforced ◊concrete, sunk into a riverbed to form the foundations of a bridge.

cal symbol for ◊*calorie*.

CAL (acronym for *computer-assisted learning*) the use of computers in education and training: the computer displays instructional material to a student and asks questions about the information given; the student's answers determine the sequence of the lessons.

calamine $ZnCO_3$ zinc carbonate, an ore of zinc. The term also refers to a pink powder made of a mixture of zinc oxide and iron(II) oxide used in lotions and ointments as an astringent for treating, for example, sunburn, eczema, measles rash, and insect bites and stings.

calcination ◊oxidation of metals by burning in air.

calcite colourless, white, or light-coloured common rock-forming mineral, calcium carbonate, $CaCO_3$. It is the main constituent of ◊limestone and marble and forms many types of invertebrate shell.

calcium soft, silvery-white metallic element, symbol Ca, atomic number 20, relative atomic mass 40.08. It is one of the ◊alkaline-earth metals. It is the fifth most abundant element (the third most abundant metal) in the Earth's crust. It is found mainly as its carbonate $CaCO_3$, which occurs in a fairly pure condition as chalk and limestone (see ◊calcite). Calcium is an essential component of bones, teeth, shells, milk, and leaves, and it forms 1.5% of the human body by mass.

calcium carbonate $CaCO_3$ white solid, found in nature as limestone, marble, and chalk. It is a valuable resource, used in the making of iron, steel, cement, glass, slaked lime, bleaching powder, sodium carbonate and bicarbonate, and many other industrially useful substances.

calcium hydrogencarbonate $Ca(HCO_3)_2$ substance found in ◊hard water, formed when rainwater passes over limestone rock.

$$CaCO_{3\,(s)} + CO_{2\,(g)} + H_2O_{(l)} \rightarrow Ca(HCO_3)_{2\,(aq)}$$

When this water is boiled it reforms calcium carbonate, removing the hardness; this type of hardness is therefore known as temporary hardness.

calcium hydrogenphosphate $Ca(H_2PO_4)_2$ substance made by heating calcium phosphate with 70% sulphuric acid. It is more soluble in water than calcium phosphate, and is used as a fertilizer.

calcium hydroxide $Ca(OH)_2$ or *slaked lime* white solid, slightly soluble in water. A solution of calcium hydroxide is called ◊limewater and is used in the laboratory to test for the presence of carbon dioxide. It is manufactured industrially by adding water to calcium oxide (quicklime) in a strongly exothermic reaction.

$$CaO + H_2O \rightarrow Ca(OH)_2$$

It is used to reduce soil acidity and as a cheap alkali in many industrial processes.

calcium nitrate $Ca(NO_3)_2$ white, crystalline compound that is very soluble in water. The solid decomposes on heating to form oxygen, brown nitrogen(IV) oxide gas and the white solid calcium oxide.

$$2Ca(NO_3)_2 \rightarrow 2CaO + 4NO_2 + O_2$$

calcium oxide or *quicklime* CaO white solid compound, formed by heating ◊calcium carbonate.

$$CaCO_3 \rightarrow CaO + CO_2$$

When water is added it forms calcium hydroxide (slaked lime) in an ◊exothermic reaction.

$$CaO + H_2O \rightarrow Ca(OH)_2$$

It is a typical basic oxide, turning litmus blue.

calcium phosphate $Ca_3(PO_4)_2$ or *calcium orthophosphate* white solid, the main constituent of animal bones. It occurs naturally as the mineral apatite and in rock phosphate, and is used in the preparation of phosphate fertilizers.

calcium sulphate $CaSO_4$ white, solid compound, found in nature as gypsum and anhydrite. It dissolves slightly in water to form ◊hard water; this hardness is not removed by boiling, and hence is sometimes called permanent hardness.

calcium superphosphate common name for ◊calcium hydrogenphosphate.

calculus branch of mathematics that permits the manipulation of continuously varying quantities, used in practical problems involving such matters as changing speeds, problems of flight, varying stresses in the framework of a bridge, and alternating current theory. *Integral calculus* deals with the method of summation or adding together the effects of continuously varying quantities. *Differential calculus* deals in a similar way with rates of change. Many of its applications arose from the study of the gradients of the tangents to curves.

calibration the preparation of a usable scale on a measuring instrument. A mercury ◊thermometer, for example, can be calibrated with a Celsius scale by noting the heights of the mercury column at two standard temperatures the freezing point (0°C) and boiling point (100°C) of water – and dividing the distance between them into 100 equal parts and continuing these divisions above and below.

californium synthesized, radioactive, metallic element of the actinide series, symbol Cf, atomic number 98, relative atomic mass 251. It is produced in very small quantities and used in nuclear reactors as a neutron source. The longest-lived isotope, Cf-251, has a half-life of 800 years.

callipers measuring instrument used, for example, to measure the internal and external diameters of pipes. Some callipers are made like a pair of compasses, having two legs, often curved, pivoting about a screw at one end. The ends of the legs are placed in contact with the object to be measured, and the gap between the ends is then measured against a rule. The slide calliper looks like an adjustable spanner, and carries a scale for direct measuring, usually with a ◊vernier scale for accuracy.

Callisto second largest moon of Jupiter, 4,800 km/3,000 mi in diameter, orbiting every 16.7 days at a distance of 1.9 million km/1.2 million mi from the planet. Its surface is covered with large craters.

callus in botany, a tissue that forms at a damaged plant surface. Composed of large, thin-walled ◊parenchyma cells, it grows over and around the wound, eventually covering the exposed area.

In animals, a callus is a thickened pad of skin, formed where there is repeated rubbing against a hard surface. In humans, calluses often develop on the hands and feet of those involved in heavy manual work.

calomel Hg_2Cl_2 (technical name ***mercury(I) chloride***) white, heavy powder formerly used as a laxative, now used as a pesticide and fungicide.

calorie c.g.s. unit of heat, now replaced by the ◊joule (one calorie is approximately 4.2 joules). It is the heat required to raise the temperature of one gram of water by 1°C. In dietetics, the Calorie or kilocalorie is equal to 1,000 calories.

calorific value the amount of heat generated by a given mass of fuel when it is completely burned. It is measured in joules per kilogram. Calorific values are measured experimentally with a bomb calorimeter.

calorimeter instrument used in physics to measure heat. A simple calorimeter consists of a heavy copper vessel that is polished (to reduce heat losses by radiation) and covered with insulating material (to reduce losses by convection and conduction).

calyptra in mosses and liverworts, a layer of cells that encloses and protects the young ◊sporophyte (spore capsule), forming a sheathlike hood around the capsule. The term is also used to describe the root cap, a layer of ◊parenchyma cells covering the end of a root that gives protection to the root tip as it grows through the soil. This is constantly being worn away and replaced by new cells from a special ◊meristem, the calyptrogen.

calyx collective term for the ◊sepals of a flower, forming the outermost whorl of the ◊perianth. It surrounds the other flower parts and protects them while in bud. In some flowers, for example, the campions *Silene*, the sepals are fused along their sides, forming a tubular calyx.

cam part of a machine that converts circular motion to linear motion or vice versa. The ***edge cam*** in a car engine is in the form of a rounded projection on a shaft, the camshaft. When the camshaft turns, the cams press against linkages (plungers or followers) that open the valves in the cylinders.

CAM (acronym for ***computer-aided manufacturing***) the use of computers to control production processes; in particular, the control of machine tools and ◊robots in factories. In some factories, the whole design and production system has been automated by linking ◊CAD (computer-aided design) to CAM.

cambium in botany, a layer of actively dividing cells (lateral ◊meristem), found within stems and roots, that gives rise to ◊secondary growth in perennial plants, causing an increase in girth. There are two main types of cambium: ***vascular cambium***, which gives rise to secondary ◊xylem and ◊phloem tissues, and ***cork cambium*** (or phellogen), which gives rise to secondary cortex and cork tissues (see ◊bark).

Cambrian period of geological time 570–510 million years ago; the first period of the Palaeozoic era. All invertebrate animal life appeared, and marine algae were widespread. The earliest fossils with hard shells, such as trilobites, date from this period.

camcorder another name for a ◊video camera.

camera obscura darkened box with a tiny hole for projecting the inverted image of the scene outside on to a screen inside.

camouflage colours or structures that allow an animal to blend with its surroundings to avoid detection by other animals. Camouflage can take the form of matching the background colour, of countershading (darker on top, lighter below, to counteract natural shadows), or of irregular patterns that break up the outline of the animal's body. More elaborate camouflage involves closely resembling a feature of the natural environment, as with the stick insect; this is closely akin to ◊mimicry.

camphor $C_{10}H_{16}O$ volatile, aromatic ◊ketone substance obtained from the camphor tree *Cinnamomum camphora*. It is distilled from chips of the wood, and is used in insect repellents and medicinal inhalants and liniments, and in the manufacture of celluloid.

Campylobacter genus of bacteria that cause serious outbreaks of gastroenteritis. They grow best at 43°C, and so are well suited to the digestive tract of birds. Poultry is therefore the most likely source of a *Campylobacter* outbreak, although the bacteria can also be transmitted via beef or milk. *Campylobacter* can survive in water for up to 15 days, so may be present in drinking water if supplies are contaminated by sewage or reservoirs are polluted by seagulls.

cancer group of diseases characterized by abnormal proliferation of cells.

Cancer (malignant) cells are usually degenerate, capable only of reproducing themselves (tumour formation). Malignant cells tend to spread from their site of origin by travelling through the bloodstream or lymphatic system.

There are more than 100 types of cancer. Some, like lung or bowel cancer, are common; others are rare. The likely causes remain unexplained. Triggering agents (◊carcinogens) include chemicals such as those found in cigarette smoke, other forms of smoke, asbestos dust, exhaust fumes, and many industrial chemicals. Some viruses can also trigger the cancerous growth of cells (see ◊oncogenes), as can X-rays and radioactivity. Dietary factors are important in some cancers; for example, lack of fibre in the diet may predispose people to bowel cancer and a diet high in animal fats and low in fresh vegetables and fruit increases the risk of breast cancer. Psychological ◊stress may increase the risk of cancer, more so if the person concerned is not able to control the source of the stress. In some families there is a genetic tendency towards a particular type of cancer. Cancer is one of the leading causes of death in the industrialized world.

Cancer faintest of the zodiacal constellations (its brightest stars are fourth magnitude). It lies in the northern hemisphere, between Leo and Gemini, and is represented as a crab. Cancer's most distinctive feature is the star cluster Praesepe, popularly known as the Beehive. The Sun passes through the constellation during late July and early Aug. In astrology, the dates for Cancer are between about 22 June and 22 July.

candela SI unit (symbol cd) of luminous intensity, which replaced the old units of candle and standard candle. It measures the brightness of a light itself rather than the amount of light falling on an object, which is called *illuminance* and measured in ◊lux.

candle vertical cylinder of wax (such as tallow or paraffin wax) with a central wick of string. A flame applied to the end of the wick melts the wax, thereby producing a luminous flame. The wick is treated with a substance such as alum so that it carbonizes but does not rapidly burn out.

Canes Venatici constellation of the northern hemisphere near Ursa Major, identified with the hunting dogs of ◊Boötes, the herder. Its stars are faint, and it contains the Whirlpool galaxy (M51), the first spiral galaxy to be recognized.

canine in mammalian carnivores, any of the long, often pointed teeth found at the front of the mouth between the incisors and premolars. Canine teeth are used for catching prey, for killing, and for tearing flesh. They are absent in herbivores such as rabbits and sheep, and are much reduced in humans.

Canis Major brilliant constellation of the southern hemisphere, identified with one of the two dogs following at the heel of Orion. Its main star, ◊Sirius, is the brightest star in the sky.

Canis Minor small constellation along the celestial equator, identified with the second of the two dogs of Orion. Its brightest star is Procyon.

Canopus or *Alpha Carinae* second brightest star in the sky (after Sirius), lying in the constellation Carina. It is a yellow-white supergiant about 120 light years from Earth, and thousands of times more luminous than the Sun.

cantilever beam or structure that is fixed at one end only, though it may be supported at some point along its length; for example, a diving board. The cantilever principle, widely used in construction engineering, eliminates the need for a second main support at the free end of the beam, allowing for more elegant structures and reducing the amount of materials required. Many large-span bridges have been built on the cantilever principle.

capacitance, electrical property of a capacitor that determines how much charge can be stored in it for a given potential difference between its terminals. It is equal to the ratio of the electrical charge stored to the potential difference. It is measured in ◊farads.

capacitor or *condenser* device for storing electric charge, used in electronic circuits; it consists of two or more metal plates separated by an insulating layer called a dielectric.

Its *capacitance* is the ratio of the charge stored on either plate to the potential difference between the plates. The SI unit of capacitance is the farad, but most capacitors have much smaller capacitances, and the microfarad (a millionth of a farad) is the commonly used practical unit.

Capella or *Alpha Aurigae* brightest star in the constellation Auriga and the sixth brightest star in the sky. It consists of a pair of yellow giant stars 41 light years from Earth, orbiting each other every 104 days.

capillarity spontaneous movement of liquids up or down narrow tubes, or capillaries. The movement is due to unbalanced molecular attraction at the boundary between the liquid and the tube. If liquid molecules near the boundary are more strongly attracted to molecules in the material of the tube than to other nearby liquid molecules, the liquid will rise in the tube. If liquid molecules are less attracted to the material of the tube than to other liquid molecules, the liquid will fall.

capillary narrowest blood vessel in vertebrates, 0.008–0.02 mm in diameter, barely wider than a red blood cell. Capillaries are distributed as *beds*, complex networks connecting arteries and veins. Capillary walls are extremely thin, consisting of a single layer of cells, and so nutrients, dissolved gases, and waste products can easily pass through them. This makes the capillaries the main area of exchange between the fluid (◊lymph) bathing body tissues and the blood.

capillary in physics, a very narrow, thick-walled tube, usually made of glass, such as in a thermometer. Properties of fluids, such as surface tension and viscosity, can be studied using capillary tubes.

capitulum in botany, a flattened or rounded head (inflorescence) of numerous, small, stalkless flowers. The capitulum is surrounded by a circlet of petal-like bracts and has the appearance of a large, single flower. It is characteristic of plants belonging to the daisy family (Compositae) such as the daisy *Bellis perennis* and the garden marigold *Calendula officinalis*; but is also seen in parts of other families, such as scabious *Knautia* and teasels *Dipsacus*. The individual flowers are known as ◊florets.

Capricorn alternative term for Capricornus.

Capricornus zodiacal constellation in the southern hemisphere next to Sagittarius. It is represented as a fish-tailed goat, and its brightest stars are third magnitude. The Sun passes through it late Jan to mid-Feb. In astrology, the dates for Capricornus (popularly known as Capricorn) are between about 22 Dec and 19 Jan.

capsule in botany, a dry, usually many-seeded fruit formed from an ovary composed of two or more fused ◊carpels, which splits open to release the seeds. The same term is used for the spore-containing structure of mosses and liverworts; this is borne at the top of a long stalk or seta.

caramel complex mixture of substances produced by heating sugars, without charring, until they turn brown. Caramel is used as colouring and flavouring in foods. Its production in the manufacture of sugar confection gives rise to a toffeelike sweet of the same name.

carat unit for measuring the mass of precious stones; it is equal to 0.2 g/0.00705 oz, and is part of the troy system of weights. It is also the unit of purity in gold (US karat). Pure gold is 24-carat; 22-carat (the purest used in jewellery) is 22 parts gold and two parts alloy (to give greater strength).

carbide compound of carbon and one other chemical element, usually a metal, silicon, or boron.

Calcium carbide (CaC_2) can be used as the starting material for many basic organic chemical syntheses, by the addition of water and generation of ethyne (acetylene). Some metallic carbides are used in engineering because of their extreme hardness and strength. Tungsten carbide is an essential ingredient of carbide tools and high-speed tools.

carbohydrate chemical compound composed of carbon, hydrogen, and oxygen, with the basic formula $C_m(H_2O)_n$, and related compounds with the same basic structure but modified ◊functional groups. As sugar and starch, carbohydrates form a major energy-providing part of the human diet.

The simplest carbohydrates are sugars (*monosaccharides*, such as glucose and fructose, and *disaccharides*, such as sucrose), which are soluble compounds, some with a sweet taste. When these basic sugar units are joined together in long chains or branching structures they form *polysaccharides*, such as starch and glycogen, which often serve as food stores in living organisms. Even more complex

carbohydrate *A molecule of the polysaccharide glycogen (animal starch) is formed from linked glucose ($C_6H_{12}O_6$) molecules. A typical glycogen molecule has 100–1,000 glucose units.*

carbohydrates are known, including ◊chitin, which is found in the cell walls of fungi and the hard outer skeletons of insects, and ◊cellulose, which makes up the cell walls of plants. Carbohydrates form the chief foodstuffs of herbivorous animals.

carbolic acid common name for the aromatic compound ◊phenol.

carbon nonmetallic element, symbol C, atomic number 6, relative atomic mass 12.011. It occurs on its own as diamond, graphite, and as fullerenes (the allotropes), as compounds in carbonaceous rocks such as chalk and limestone, as carbon dioxide in the atmosphere, as hydrocarbons in petroleum, coal, and natural gas, and as a constituent of all organic substances.

In its amorphous form, it is familiar as coal, charcoal, and soot. The atoms of carbon can link with one another in rings or chains, giving rise to innumerable complex compounds. Of the inorganic carbon compounds, the chief ones are *carbon dioxide*, a colourless gas formed when carbon is burned in an adequate supply of air; and *carbon monoxide* (CO), formed when carbon is oxidized in a limited supply of air. *Carbon disulphide* (CS_2) is a dense liquid with a sweetish odour. Another group of compounds is the *carbon halides*, including ◊carbon tetrachloride (tetrachloromethane, CCl_4). When added to steel, carbon forms a wide range of alloys with useful properties. In pure form, it is used as a moderator in nuclear reactors; as colloidal graphite it is a good lubricant and, when deposited on a surface in a vacuum, obviates photoelectric and secondary emission of electrons. Carbon is used as a fuel in the form of coal or coke. The radioactive isotope carbon-14 (half-life 5,730 years) is used as a tracer in biological research. Analysis of interstellar dust has led to the discovery of ◊buckminsterfullerenes, carbon molecules containing 60 carbon atoms.

carbonate CO_3^{2-} ion formed when carbon dioxide dissolves in water; any salt formed by this ion and another chemical element, usually a metal.

carbonated water water in which carbon dioxide is dissolved under pressure. It forms the basis of many fizzy soft drinks such as soda water and lemonade.

carbonation in earth science, a form of chemical ◊weathering caused by rainwater that has absorbed carbon dioxide from the atmosphere and formed a weak carbonic acid. The slightly acidic rainwater is then capable of dissolving certain minerals in rocks. ◊Limestone is particularly vulnerable to this form of weathering.

carbon cycle sequence by which ◊carbon circulates and is recycled through the natural world. The carbon element from carbon dioxide, released into the atmosphere by living things as a result of ◊respiration, is taken up by plants during ◊photosynthesis and converted into carbohydrates; the oxygen component is released back into the atmosphere. The simplest link in the carbon cycle occurs when an animal eats a plant and carbon is transferred from, say, a leaf cell to the animal body. Today, the carbon cycle is in danger of being disrupted by the increased consumption and burning of fossil fuels, and the burning of large tracts of tropical forests, as a result of which levels of carbon dioxide are building up in the atmosphere and probably contributing to the ◊greenhouse effect.

carbon cycle *The carbon cycle is necessary for the continuation of life. Since there is only a limited amount of carbon in the Earth and its atmosphere, carbon must be continuously recycled if life is to continue. Other chemicals necessary for life – nitrogen, sulphur, and phosphorus, for example – also circulate in natural cycles.*

carbon dating alternative name for ◊radiocarbon dating.

carbon dioxide CO_2 colourless, odourless gas, slightly soluble in water and denser than air. It is formed by the complete oxidation of carbon.

It is produced by living things during the processes of respiration and the decay of organic matter, and plays a vital role in the carbon cycle. It is used as a coolant in its solid form (known as 'dry ice'), and in the chemical industry. Its increasing density contributes to the ◊greenhouse effect and ◊global warming.

carbon fibre fine, black, silky filament of pure carbon produced by heat treatment from a special grade of Courtelle acrylic fibre, used for reinforcing plastics. The resulting composite is very stiff and, weight for weight, has four times the strength of high-tensile steel. It is used in the aerospace industry, cars, and electrical and sports equipment.

carbonic acid H_2CO_3 weak, dibasic acid formed by dissolving carbon dioxide in water.

$$H_2O + CO_2 \leftrightarrow H_2CO_3$$

It forms two series of salts: ◊carbonates and ◊hydrogencarbonates. Fizzy drinks are made by dissolving carbon dioxide in water under pressure; soda water is a solution of carbonic acid.

Carboniferous period of geological time 363–290 million years ago, the fifth period of the Palaeozoic era. In the USA it is divided into two periods: the Mississippian (lower) and the Pennsylvanian (upper).

Typical of the lower-Carboniferous rocks are shallow-water ◊limestones, while upper-Carboniferous rocks have ◊delta deposits with ◊coal (hence the name). Amphibians were abundant, and reptiles evolved during this period.

carbon monoxide CO colourless, odourless gas formed when carbon is oxidized in a limited supply of air. It is a poisonous constituent of car exhaust fumes, forming a stable compound with haemoglobin in the blood, thus preventing the haemoglobin from transporting oxygen to the body tissues.

In industry, carbon monoxide is used as a reducing agent in metallurgical processes – for example, in the extraction of iron in ◊blast furnaces – and is a constituent of cheap fuels such as water gas. It burns in air with a luminous blue flame to form carbon dioxide.

carbon tetrachloride former name for ◊tetrachloromethane.

Carborundum trademark for a very hard, black abrasive, consisting of silicon carbide (SiC), an artificial compound of carbon and silicon. It is harder than ◊corundum but not as hard as ◊diamond.

carboxyl group –COOH in organic chemistry, the acidic functional group that determines the properties of fatty acids (carboxylic acids) and amino acids.

carboxylic acid organic acid containing the carboxyl group (–COOH) attached to another group (R), which can be hydrogen (giving methanoic acid, HCOOH) or a larger molecule (up to 24 carbon atoms). When R is a straight-chain alkyl group (such as CH_3 or CH_3CH_2), the acid is known as a ◊fatty acid.

carburation mixing of a gas, such as air, with a volatile hydrocarbon fuel, such as petrol, kerosene, or fuel oil, in order to form an explosive mixture. The process, which increases the amount of potential heat energy released during combustion, is used in internal-combustion engines. In most petrol engines the liquid fuel is atomized and mixed with air by means of a device called a *carburettor*.

carcinogen any agent that increases the chance of a cell becoming cancerous, including various chemical compounds, some viruses, X-rays, and other forms of ionizing radiation. The term is often used more narrowly to mean chemical carcinogens only.

carcinoma malignant ◊tumour arising from the skin, the glandular tissues, or the mucous membranes that line the gut and lungs.

cardiac pertaining to the ◊heart.

cardinal number in mathematics, one of the series of numbers 0, 1, 2, 3, 4, ... Cardinal numbers relate to quantity, whereas ordinal numbers (first, second, third, fourth, ...) relate to order.

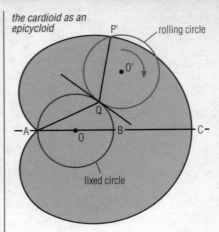

cardioid *The cardioid is the curve formed when one circle rolls around the edge of another circle of the same size. It is named after its heart shape.*

cardioid heart-shaped curve traced out by a point on the circumference of a circle, resulting from the circle rolling around the edge of another circle of the same diameter.

Carina constellation of the southern hemisphere, represented as a ship's keel. Its brightest star is Canopus; it also contains Eta Carinae, a massive and highly luminous star embedded in a gas cloud, perhaps 8,000 light years away. It has varied unpredictably in the past; some astronomers think it is likely to explode as a supernova within 10,000 years.

carnassial tooth one of a powerful scissorlike pair of molars, found in all mammalian carnivores except seals. Carnassials are formed from an upper premolar and lower molar, and are shaped to produce a sharp cutting surface. Carnivores such as dogs transfer meat to the back of the mouth, where the carnassials slice up the food ready for swallowing.

carnelian semiprecious gemstone variety of ◊chalcedony consisting of quartz (silica) with iron impurities, which give it a translucent red colour. It is found mainly in Brazil, India, and Japan.

carnivore mammal of the order Carnivora. Although its name describes the flesh-eating ancestry of the order, it includes pandas, which are herbivorous, and civet cats that eat fruit.

Carnot cycle series of changes in the physical condition of a gas in a reversible heat engine, necessarily in the following order: (1) isothermal expansion (without change of temperature), (2) adiabatic expansion (without change of heat content), (3) isothermal compression, and (4) adiabatic compression.

carnotite potassium uranium vanadate, $K_2(UO_2)_2(VO_4)_2.3H_2O$, a radioactive ore of vanadium and uranium with traces of radium.

A yellow powdery mineral, it is mined chiefly in the Colorado Plateau, USA; Radium Hill, Australia; and Shaba, Zaire.

carotene naturally occurring pigment of the ◊carotenoid group. Carotenes produce the orange, yellow, and red colours of carrots, tomatoes, oranges, and crustaceans.

carotenoid any of a group of yellow, orange, red, or brown pigments found in many living organisms, particularly in the ◊chloroplasts of plants. There are two main types, the *carotenes* and the *xanthophylls*. Both types are long-chain lipids (◊fats).

carotid artery one of a pair of major blood vessels, one on each side of the neck, supplying blood to the head.

carpel female reproductive unit in flowering plants (◊angiosperms). It usually comprises an ◊ovary containing one or more ovules, the stalk or style, and a ◊stigma at its top which receives the pollen. A flower may have one or more carpels, and they may be separate or fused together. Collectively the carpels of a flower are known as the ◊gynoecium.

carrying capacity in ecology, the maximum number of animals of a given species that a particular area can support. When the carrying capacity is exceeded, there is insufficient food (or other resources) for the members of the population. The population may then be reduced by emigration, reproductive failure, or death through starvation.

Cartesian coordinates in ◊coordinate geometry, components used to define the position of a point by its perpendicular distance from a set of two or more axes, or reference lines. For a two-dimensional area defined by two axes at right angles (a horizontal x-axis and a vertical y-axis), the coordinates of a point are given by its perpendicular distances from the y-axis and x-axis, written in the form (x,y). For example, a point P that lies three units from the y-axis and four units from the x-axis has Cartesian coordinates (3,4) (see ◊abscissa and ◊ordinate). In three-dimensional coordinate geometry, points are located with reference to a third, z-axis, mutually at right angles to the $\times$ and y axes.

cartilage flexible bluish-white ◊connective tissue made up of the protein collagen. In cartilaginous fish it forms the skeleton; in other vertebrates it forms the greater part of the embryonic skeleton, and is replaced by ◊bone in the course of development, except in areas of wear such as bone endings, and the discs between the backbones. It also forms structural tissue in the larynx, nose, and external ear of mammals.

caryopsis dry, one-seeded ◊fruit in which the wall of the seed becomes fused to the carpel wall during its development. It is a type of ◊achene, and therefore develops from one ovary and does not split open to release the seed. Caryopses are typical of members of the grass family (Gramineae), including the cereals.

casein main protein of milk, from which it can be separated by the action of acid, the enzyme rennin, or bacteria (souring); it is also the main protein in cheese. Casein is used as a protein supplement in the treatment of malnutrition. It is used commercially in cosmetics, glues, and as a sizing for coating paper.

Cassini joint space probe of the US agency NASA and the European Space Agency to the planet Saturn. *Cassini* is scheduled to be launched 1997 and to go into orbit around Saturn 2004, dropping off a sub-probe, *Huygens*, to land on Saturn's largest moon, Titan.

Cassiopeia prominent constellation of the northern hemisphere, named after the mother of Andromeda. It has a distinctive W-shape, and contains one of the most powerful radio sources in the sky, Cassiopeia A, the remains of a ◊supernova (star explosion).

cassiterite or *tinstone* chief ore of tin, consisting of reddish-brown to black stannic oxide (SnO_2), usually found in granite rocks. When fresh it has a bright ('adamantine') lustre. It was formerly extensively mined in Cornwall, England; today Malaysia is the world's main supplier. Other sources of cassiterite are Africa, Indonesia, and South America.

casting process of producing solid objects by pouring molten material into a shaped mould and allowing it to cool. Casting is used to shape such materials as glass and plastics, as well as metals and alloys.

cast iron cheap but invaluable constructional material, most commonly used for car engine blocks. Cast iron is partly refined pig (crude) ◊iron, which is very fluid when molten and highly suitable for shaping by casting; it contains too many impurities (for example, carbon) to be readily shaped in any other way. Solid cast iron is heavy and can absorb great shock but is very brittle.

Castor or *Alpha Geminorum* second brightest star in the constellation Gemini and the 23rd brightest star in the sky. Along with ◊Pollux, it forms a prominent pair at the eastern end of Gemini.

catabolism in biology, the destructive part of ◊metabolism where living tissue is changed into energy and waste products.

It is the opposite of ◊anabolism. It occurs continuously in the body, but is accelerated during many disease processes, such as fever, and in starvation.

catalyst substance that alters the speed of, or makes possible, a chemical or biochemical reaction but remains unchanged at the end of the reaction. ◊Enzymes are natural biochemical catalysts. In practice most catalysts are used to speed up reactions.

catalytic converter device fitted to the exhaust system of a motor vehicle in order to reduce toxic emissions from the engine. It converts harmful exhaust products to relatively harmless ones by passing the exhaust gases over a mixture of catalysts coated on a metal or ceramic honeycomb (a structure that increases the surface area and therefore the amount of active catalyst with which the exhaust gases will come into contact). *Oxidation catalysts* (small amounts of precious palladium and platinum metals) convert hydrocarbons (unburnt fuel) and carbon monoxide into carbon dioxide and water, but do not affect nitrogen oxide emissions. *Three-way catalysts* (platinum and rhodium metals) convert nitrogen oxide gases into nitrogen and oxygen.

catastrophe theory mathematical theory developed by René Thom in 1972, in which he showed that the growth of an organism proceeds by a series of gradual changes that are triggered by, and in turn trigger, large-scale changes or 'catastrophic' jumps. It also has applications in engineering – for example, the gradual strain on the structure of a bridge that can eventually result in a sudden collapse – and has been extended to economic and psychological events.

catecholamine chemical that functions as a ◊neurotransmitter or a ◊hormone. Dopamine, adrenaline (epinephrine), and noradrenaline (norepinephrine) are catecholamines.

catenary curve taken up by a flexible cable suspended between two points, under gravity; for example, the curve of overhead suspension cables that hold the conductor wire of an electric railway or tramway.

cathode in chemistry, the negative electrode of an electrolytic ◊cell, towards which positive particles (cations), usually in solution, are attracted. See ◊electrolysis.

cathode in electronics, the part of an electronic device in which electrons are generated. In a thermionic valve, electrons are produced by the heating effect of an applied current; in a photocell, they are produced by the interaction of light and a semiconducting material. The cathode is kept at a negative potential relative to the device's other electrodes (anodes) in order to ensure that the liberated electrons stream away from the cathode and towards the anodes.

cathode-ray oscilloscope (CRO) instrument used to measure electrical potentials or voltages that vary over time and to display the waveforms of electrical oscillations or signals. Readings are displayed graphically on the screen of a ◊cathode-ray tube.

cathode-ray tube vacuum tube in which a beam of electrons is produced and focused onto a fluorescent screen. It is an essential component of television receivers, computer visual display units, and oscilloscopes.

cation ◊ion carrying a positive charge. During electrolysis, cations in the electrolyte move to the cathode (negative electrode).

catkin in flowering plants (◊angiosperms), a pendulous inflorescence, bearing numerous small, usually unisexual flowers. The tiny flowers are stalkless and the petals and sepals are usually absent or much reduced in size. Many types of trees bear catkins, including willows, poplars, and birches. Most plants with catkins are wind-pollinated, so the male catkins produce large quantities of pollen. Some ◊gymnosperms also have catkin-like structures that produce pollen, for example, the swamp cypress *Taxodium*.

CAT scan or *CT scan* (acronym for *computerized axial tomography scan*) sophisticated method of X-ray imaging. Quick and noninvasive, CAT scanning is used in medicine as an aid to diagnosis, helping to pinpoint problem areas without the need for exploratory surgery. It is also used in archaeology to investigate mummies.

The CAT scanner passes a narrow fan of X-rays through successive slices of the suspect body part. These slices are picked up by crystal detectors in a scintillator and converted electronically into cross-sectional images displayed on a viewing screen. Gradually, using views taken from various angles, a three-dimensional picture of the organ or tissue can be built up and irregularities analysed.

cat's eyes reflective studs used to mark the limits of traffic lanes, invented by Percy Shaw (1890–1976) in England, as a road safety device in 1934.

A cat's eye stud has two pairs of reflective prisms (the eyes) set in a rubber pad, which reflect the light of a vehicle's headlamps back to the driver. When a vehicle goes over a stud, it moves down inside an outer rubber case; the surfaces of the prisms brush against the rubber and are thereby cleaned.

cauda tail, or taillike appendage; part of the *cauda equina*, a bundle of nerves at the bottom of the spinal cord in vertebrates.

caustic soda former name for ◊sodium hydroxide (NaOH).

cavitation formation of partial vacuums in fluids at high velocities, produced by propellers or other machine parts in hydraulic engines, in accordance with ◊Bernoulli's principle. When these vacuums collapse, pitting, vibration, and noise can occur in the metal parts in contact with the fluids.

cc symbol for *cubic centimetre*.

CD abbreviation for ◊*compact disc*.

CD-I (abbreviation for *compact disc-interactive*) compact disc developed by Philips for storing a combination of video, audio, text, and pictures. It is intended principally for the consumer market to be used in systems using a combination of computer and television. An alternative format is ◊DVI (digital video interactive).

CD-R (abbreviation for *compact disc-recordable*) compact disc on which data can be overwritten. The disc combines magnetic and optical technology: during the writing process, a laser melts the surface of the disc, thereby allowing the magnetic elements of the surface layer to be realigned.

CD-ROM (abbreviation for *compact-disc read-only memory*) computer storage device developed from the technology of the audio ◊compact disc. It consists of a plastic-coated metal disc, on which binary digital information is etched in the form of microscopic pits. This can then be read optically by passing a light beam over the disc. CD-ROMs typically hold about 550 ◊megabytes of data, and are used in distributing large amounts of text and graphics, such as encyclopedias, catalogues, and technical manuals.

CD-ROM drive in computing, a disc drive for reading CD-ROM discs. The vast majority of CD-ROM drives conform to the Yellow Book standard, defined by Philips and Sony. Because of this, all drives are essentially interchangeable. CD-ROM drives are available either as stand-alone or built-in units with a variety of interfaces (connections) and access times.

CD-ROM XA (CD-ROM extended architecture) in computing, a set of standards for storing multimedia information on CD-ROM. Developed by Philips, Sony, and Microsoft, it is a partial development of the ◊CD-I standard. It interleaves data (as in CD-I) so that blocks of audio data are sandwiched between blocks of text, graphics, or video. This allows parallel streams of data to be handled, so that information can be seen and heard simultaneously. Unlike CD-I, CD-ROM XA discs are playable on any ◊CD-ROM drive with an appropriate interface.

CDTV (abbreviation for *C*ommodore *D*ynamic *T*otal *V*ision) multimedia computer system developed by Commodore. It is designed for the home and is a direct rival to ◊CD-I. It consists of a box about the size of a home video recorder, containing an Amiga 500 computer and a ◊CD-ROM drive. CDTV discs can store a combination of text, pictures, sound, and video. Like CD-I, CDTV plugs into a TV set and stereo system. CDTV cannot play CD-ROM or CD-I discs.

celestial mechanics the branch of astronomy that deals with the calculation of the orbits of celestial bodies, their gravitational attractions (such as those that produce the Earth's tides), and also the orbits of artificial satellites and space probes. It is based on the laws of motion and gravity laid down by Isaac Newton (1642–1727).

celestial sphere imaginary sphere surrounding the Earth, on which the celestial bodies seem to lie. The positions of bodies such as stars, planets, and galaxies are specified by their coordinates on the celestial sphere. The equivalents of latitude and longitude on the eelestial sphere are called ◊declination and ◊right ascension (which is

cell *Cells are the building units of living things. Most of them are tiny, having a diameter of less than 0.03 mm/0.001 in. Animal and plant cells have similar internal structures and composition, but plant cells are more rigid in structure because of the cellulose cell wall. Animal cells lack chloroplasts, the photosynthesizing structures of plant cells.*

measured in hours from 0 to 24). The *celestial poles* lie directly above the Earth's poles, and the *celestial equator* lies over the Earth's equator. The celestial sphere appears to rotate once around the Earth each day, actually a result of the rotation of the Earth on its axis.

celestine or *celestite* mineral consisting of strontium sulphate, $SrSO_4$, occurring as white or light blue crystals. It is the principal source of strontium.

cell in biology, a discrete, membrane-bound portion of living matter, the smallest unit capable of an independent existence. All living organisms consist of one or more cells, with the exception of ◊viruses. Bacteria, protozoa, and many other microorganisms consist of single cells, whereas a human is made up of billions of cells. Essential features of a cell are the membrane, which encloses it and restricts the flow of substances in and out; the jellylike material within, the ◊cytoplasm; the ◊ribosomes, which carry out protein synthesis; and the ◊DNA, which forms the hereditary material.

cell differentiation in developing embryos, the process by which cells acquire their specialization, such as heart cells, muscle cells, skin cells, and brain cells. The seven-day-old human pre-embryo consists of thousands of individual cells, each of which is destined to take up a particular role in the body.

Research has shown that the eventual function of a cell, in for example, a chicken embryo, is determined by the cell's position. The embryo can be mapped into areas corresponding with the spinal cord, the wings, the legs, and many other tissues. If the embryo is relatively young, a cell transplanted from one area to another will develop according to its new position. As the embryo develops the cells lose their flexibility and become unable to change their destiny.

cell division the process by which a cell divides, either ◊meiosis, associated with sexual reproduction, or ◊mitosis, associated with growth, cell replacement, or repair. Both forms involve the duplication of DNA and the splitting of the nucleus.

cell, electrical or *voltaic cell* or *galvanic cell* device in which chemical energy is converted into electrical energy; the popular name *the basic principle of a chemical cell*

a simple cell

cell, electrical *The basic principle of an electrical cell: rods of different metals immersed in an electrolyte generate a voltage. The now obsolete Daniel cell was the first reliable source of electricity, producing a constant voltage for a considerable time.*

is '◊battery', but this actually refers to a collection of cells in one unit. The reactive chemicals of a *primary cell* cannot be replenished, whereas *secondary cells* – such as storage batteries – are rechargeable: their chemical reactions can be reversed and the original condition restored by applying an electric current. It is dangerous to attempt to recharge a primary cell.

Each cell contains two conducting ◊electrodes immersed in an ◊electrolyte, in a container. A spontaneous chemical reaction within the cell generates a negative charge (excess of electrons) on one electrode, and a positive charge (deficiency of electrons) on the other. The accumulation of these equal but opposite charges prevents the reaction from continuing unless an outer connection (external circuit) is made between the electrodes allowing the charges to dissipate. When this occurs, electrons escape from the cell's negative terminal and are replaced at the positive, causing a current to flow. After prolonged use, the cell will become flat (cease to supply current).

cell, electrolytic device to which electrical energy is applied in order to bring about a chemical reaction; see ◊electrolysis.

cell membrane or *plasma membrane* thin layer of protein and fat surrounding cells that controls substances passing between the cytoplasm and the intercellular space. The cell membrane is semipermeable, allowing some substances to pass through and some not.

cellophane transparent wrapping film made from wood ◊cellulose, widely used for packaging, first produced by Swiss chemist Jacques Edwin Brandenberger 1908.

cellphone contraction of ◊cellular phone.

cell sap dilute fluid found in the large central vacuole of many plant cells. It is made up of water, amino acids, glucose, and salts. The sap has many functions, including storage of useful materials, and provides mechanical support for non-woody plants.

cellular phone or *cellphone* mobile radio telephone, one of a network connected to the telephone system by a computer-controlled communication system. Service areas are divided into small 'cells', about 5 km/3 mi across, each with a separate low-power transmitter.

celluloid transparent or translucent, highly flammable, plastic material (a ◊thermoplastic) made from cellulose nitrate and camphor. It was once used for toilet articles, novelties, and photographic film, but has now been replaced by the nonflammable substance ◊cellulose acetate.

cellulose complex ◊carbohydrate composed of long chains of glucose units. It is the principal constituent of the cell wall of higher plants, and a vital ingredient in the diet of many ◊herbivores. Molecules of cellulose are organized into long, unbranched microfibrils that give support to the cell wall. No mammal produces cellulase, the enzyme necessary for digesting cellulose; mammals such as rabbits and cows are only able to digest grass because the bacteria present in their gut can manufacture it.

cellulose acetate or *cellulose ethanoate* chemical (an ◊ester) made by the action of acetic acid (ethanoic acid) on cellulose. It is used in making transparent film, especially photographic film; unlike its predecessor, celluloid, it is not flammable.

cellulose nitrate or *nitrocellulose* series of esters of cellulose with up to three nitrate (NO_3) groups per monosaccharide unit. It is made by the action of concentrated nitric acid on cellulose (for example, cotton waste) in the presence of concentrated sulphuric acid. Fully nitrated cellulose (gun cotton) is explosive, but esters with fewer nitrate groups were once used in making lacquers, rayon, and plastics, such as coloured and photographic film, until replaced by the nonflammable cellulose acetate. ◊Celluloid is a form of cellulose nitrate.

cell wall in plants, the tough outer surface of the cell. It is constructed from a mesh of ◊cellulose and is very strong and relatively inelastic. Most living cells are turgid (swollen with water; see ◊turgor) and develop an internal hydrostatic pressure (wall pressure) that acts against the cellulose wall. The result of this turgor pressure is to give the cell, and therefore the plant, rigidity. Plants that are not woody are particularly reliant on this form of support.

Celsius scale of temperature, previously called centigrade, in which the range from freezing to boiling of water is divided into 100 degrees, freezing point being 0 degrees and boiling point 100 degrees, measured in °C.

cement any bonding agent used to unite particles in a single mass or to cause one surface to adhere to another. *Portland cement* is a powder obtained from burning together a mixture of lime (or chalk) and clay, and when mixed with water and sand or gravel, turns into mortar or concrete. In geology, a chemically precipitated material such as carbonate that occupies the interstices of clastic rocks.

Cenozoic or *Caenozoic* era of geological time that began 65 million years ago and is still in process. It is divided into the Tertiary and Quaternary periods. The Cenozoic marks the emergence of mammals as a dominant group, including humans, and the formation of the mountain chains of the Himalayas and the Alps.

Centaurus large bright constellation of the southern hemisphere, represented as a centaur. Its brightest star, ◊Alpha Centauri, is a triple star and contains the closest star to the Sun, Proxima Centauri. Omega Centauri, the largest and brightest globular cluster of stars in the sky, is 16,000 light years away.

centigrade former name for the ◊Celsius temperature scale.

central nervous system *The central nervous system (CNS) with its associated nerves. The CNS controls and integrates body functions. In humans and other vertebrates it consists of a brain and a spinal cord, which are linked to the body's muscles and organs by means of the peripheral nervous system.*

central dogma in genetics and evolution, the fundamental belief that ◊genes can affect the nature of the physical body, but that changes in the body (◊acquired character, for example, through use or accident) cannot be translated into changes in the genes.

central heating system of heating from a central source, typically of a house, larger building, or group of buildings, as opposed to heating each room individually. Steam heat and hot-water heat are the most common systems in use. Water is heated in a furnace burning oil, gas, or solid fuel, and, as steam or hot water, is then pumped through radiators in each room. The level of temperature can be selected by adjusting a ◊thermostat on the burner or in a room.

central nervous system (CNS) the brain and spinal cord, as distinct from other components of the ◊nervous system. The CNS integrates all nervous function.

In invertebrates it consists of a paired ventral nerve cord with concentrations of nerve-cell bodies, known as ◊*ganglia* in each segment, and a small brain in the head. Some simple invertebrates, such as sponges and jellyfishes, have no CNS but a simple network of nerve cells called a ***nerve net***.

central processing unit (CPU) main component of a computer, the part that executes individual program instructions and controls the operation of other parts. It is sometimes called the central processor or, when contained on a single integrated circuit, a microprocessor.

centre of gravity the point in an object about which its weight is evenly balanced. In a uniform gravitational field, this is the same as the centre of mass.

centre of mass or *centre of gravity* point in or near an object from which its total weight appears to originate and can be assumed to act. A symmetrical homogeneous object such as a sphere or cube has its centre of mass at its physical centre; a hollow object (such as a cup) may have its centre of mass in space inside the hollow.

For an object to be in stable equilibrium, a perpendicular line down through its centre of mass must run within the boundaries of its base; if tilted until this line falls outside the base, the object becomes unstable and topples over.

centrifugal force useful concept in physics, based on an apparent (but not real) force. It may be regarded as a force that acts radially outward from a spinning or orbiting object, thus balancing the ◊centripetal force (which is real). For an object of mass m moving with a velocity v in a circle of radius r, the centrifugal force F equals mv^2/r (outward).

centrifuge apparatus that rotates at high speeds, causing substances inside it to be thrown outward. One use is for separating mixtures of substances of different densities.

centriole structure found in the ◊cells of animals that plays a role in the processes of ◊meiosis and ◊mitosis (cell division).

centripetal force force that acts radially inward on an object moving in a curved path. For example, with a weight whirled in a circle

central processing unit *The central processing unit is the 'brain' of a computer; it is here that all the computer's work is done. The arithmetic and logic unit (ALU) does the arithmetic, using the registers to store intermediate results, supervised by the control unit. Input and output circuits connect the ALU to external memory, input, and output devices.*

at the end of a length of string, the centripetal force is the tension in the string. For an object of mass m moving with a velocity v in a circle of radius r, the centripetal force F equals mv^2/r (inward). The reaction to this force is the ◊centrifugal force.

centromere part of the ◊chromosome where there are no ◊genes. Under the microscope, it usually appears as a constriction in the strand of the chromosome, and is the point at which the spindle fibres are attached during ◊meiosis and ◊mitosis (cell division).

centrosome cell body that contains the ◊centrioles. During cell division the centrosomes organize the microtubules to form the spindle that divides the chromosomes into daughter cells. Centrosomes were first described 1887, independently by German biologist Theodor Boveri (1862–1915) and Edouard van Beneden.

cephalopod any predatory marine mollusc of the class Cephalopoda, with the mouth and head surrounded by tentacles. Cephalopods are the most intelligent, the fastest-moving, and the largest of all animals without backbones, and there are remarkable luminescent forms which swim or drift at great depths. They have the most highly developed nervous and sensory systems of all invertebrates, the eye in some closely paralleling that found in vertebrates. Examples include octopus, squid, and cuttlefish. Shells are rudimentary or absent in most cephalopods.

Cepheid variable yellow supergiant star that varies regularly in brightness every few days or weeks as a result of pulsations. The time that a Cepheid variable takes to pulsate is directly related to its average brightness; the longer the pulsation period, the brighter the star.

This relationship, the *period luminosity law* allows astronomers to use Cepheid variables as 'standard candles' to measure distances in our Galaxy and to nearby galaxies.

Cepheus constellation of the north polar region, named after King Cepheus of Greek mythology, husband of Cassiopeia and father of Andromeda. It contains the Garnet Star (Mu Cephei), a red supergiant of variable brightness that is one of the reddest-coloured stars known, and Delta Cephei, prototype of the ◊Cepheid variables.

cereal grass grown for its edible, nutrient-rich, starchy seeds. The term refers primarily to wheat, oats, rye, and barley, but may also refer to maize (corn), millet, and rice. Cereals contain about 75% complex carbohydrates and 10% protein, plus fats and fibre (roughage). They store well. If all the world's cereal crop were consumed as whole-grain products directly by humans, everyone could obtain adequate protein and carbohydrate; however, a large proportion of cereal production in affluent nations is used as animal feed to boost the production of meat, dairy products, and eggs.

cerebellum part of the brain of ◊vertebrate animals which controls muscle tone, movement, balance, and coordination. It is relatively small in lower animals such as newts and lizards, but large in birds since flight demands precise coordination. The human cerebellum is also well developed, because of the need for balance when walking or running, and for finely coordinated hand movements.

cerebral pertaining to the brain, especially the part known as the ◊cerebrum, concerned with higher brain functions.

cerebral cortex or *grey matter* fissured outer layer of the ◊cerebrum. Some 3 mm thick in the adult, it is the most sophisticated part of the brain, responsible for all higher functions and for initiating voluntary movement. Anatomists divide it into four lobes, named after the skull plates beneath which they lie: frontal, parietal, temporal and occipital.

cerebral hemisphere one of the two halves of the ◊cerebrum.

cerebrospinal fluid (CSF) clear fluid that buffers the brain and spinal cord. It is present in the subarachnoid space, the four ventricles of the brain and the central canal of the spinal cord. It is secreted in clusters of blood vessels in the ventricles known as choroid plexuses, and is eventually absorbed into the bloodstream. Its normal contents include glucose, salts, proteins and some white cells. Small amounts of CSF may be withdrawn by lumbar puncture for diagnostic tests.

cerebrum part of the vertebrate ◊brain, formed from the two paired cerebral hemispheres, separated by a central fissure. In birds and mammals it is the largest and most developed part of the brain. It is covered with an infolded layer of grey matter, the cerebral cortex, which integrates brain functions. The cerebrum coordinates all voluntary activity.

Ceres the largest asteroid, 940 km/584 mi in diameter, and the first to be discovered (by Italian astronomer Giuseppe Piazzi 1801). Ceres orbits the Sun every 4.6 years at an average distance of 414 million km/257 million mi. Its mass is about one-seventieth of that of the Moon.

cerium malleable and ductile, grey, metallic element, symbol Ce, atomic number 58, relative atomic mass 140.12. It is the most abundant member of the lanthanide series, and is used in alloys, electronic components, nuclear fuels, and lighter flints. It was discovered 1804 by the Swedish chemists Jöns Berzelius and Wilhelm Hisinger (1766–1852), and, independently, by Martin Klaproth. The element was named after the then recently discovered asteroid Ceres.

cermet (acronym of *ceramics and matal*) bonded material containing ceramics and metal, widely used in jet engines and nuclear reactors. Cermets behave much like metals but have the great heat resistance of ceramics. Tungsten carbide, molybdenum boride, and aluminium oxide are among the ceramics used; iron, cobalt, nickel, and chromium are among the metals.

CERN nuclear research organization founded 1954 as a cooperative enterprise among European governments. It has laboratories at Meyrin, near Geneva, Switzerland. It was originally known as the *Conseil Européen pour la Recherche Nucléaire* but subsequently renamed *Organisation Européenne pour la Recherche Nucléaire*, although still familiarly known as CERN. It houses the world's largest particle ◊accelerator, the ◊Large Electron Positron Collider (LEP), with which notable advances have been made in ◊particle physics.

cervix abbreviation for *cervix uteri*, the neck of the womb; see ◊uterus.

Cetus constellation straddling the celestial equator (see ◊celestial sphere), represented as a sea monster. Cetus contains the long-period variable star ◊Mira, and Tau Ceti, one of the nearest stars visible with the naked eye.

c.g.s. system system of units based on the centimetre, gram, and second, as units of length, mass, and time respectively. It has been replaced for scientific work by the ◊SI units to avoid inconsistencies in definition of the thermal calorie and electrical quantities.

chain reaction in chemistry, a succession of reactions, usually involving ◊free radicals, where the products of one stage are the reactants of the next. A chain reaction is characterized by the continual generation of reactive substances.

A chain reaction comprises three separate stages: *initiation* – the initial generation of reactive species; *propagation* – reactions that involve reactive species and generate similar or different reactive species; and *termination* – reactions that involve the reactive species but produce only stable, nonreactive substances. Chain reactions may occur slowly (for example, the oxidation of edible oils) or accelerate as the number of reactive species increases, ultimately resulting in explosion.

chain reaction in nuclear physics, a fission reaction that is maintained because neutrons released by the splitting of some atomic nuclei themselves go on to split others, releasing even more neutrons. Such a reaction can be controlled (as in a nuclear reactor) by using moderators to absorb excess neutrons. Uncontrolled, a chain reaction produces a nuclear explosion (as in an atom bomb).

chalaza glutinous mass of transparent albumen supporting the yolk inside birds' eggs. The chalaza is formed as the egg slowly passes down the oviduct, when it also acquires its coiled structure.

chalcedony form of quartz, SiO_2, in which the crystals are so fine-grained that they are impossible to distinguish with a microscope (cryptocrystalline). Agate, onyx, and carnelian are ◊gem varieties of chalcedony.

chalcopyrite copper iron sulphide, $CuFeS_2$, the most common ore of copper. It is brassy yellow in colour and may have an iridescent surface tarnish. It occurs in many different types of mineral vein, in rocks ranging from basalt to limestone.

chalk soft, fine-grained, whitish rock composed of calcium carbonate, $CaCO_3$, extensively quarried for use in cement, lime, and mortar, and in the manufacture of cosmetics and toothpaste. *Blackboard chalk* in fact consists of ◊gypsum (calcium sulphate, $CaSO_4.2H_2O$).

Chalk is composed chiefly of coccolithophores, unicellular lime-secreting algae, and so is primarily of plant origin. It is formed from deposits of deep-sea sediments called oozes.

Challenger orbiter used in the US ◊space shuttle programme which on 28 Jan 1986 exploded on takeoff, killing all seven crew members.

change of state in science, a change in the physical state (solid, liquid, or gas) of a material. For instance, melting, boiling, evaporation, and their opposites, solidification and condensation, are changes of state. The former set of changes are brought about by heating or decreased pressure; the latter by cooling or increased pressure.

These changes involve the absorption or release of heat energy, called ◊latent heat, even though the temperature of the material does not change during the transition between states. See also ◊states of matter. In the unusual change of state called *sublimation*, a solid changes directly to a gas without passing through the liquid state. For example, solid carbon dioxide (dry ice) sublimes to carbon dioxide gas.

chaos theory or *chaology* branch of mathematics used to determine unpredictable systems – for example, ◊turbulence, and the weather.

Within chaos theory, order and chaos are viewed as two aspects of the same process, the evolution of a system in time. Chaotic systems are difficult to predict because any initial error of measurement, however small, will grow and distort the prediction. This is known as the *butterfly effect*: a butterfly flaps its wings and a month later the resulting air disturbance causes a hurricane thousands of miles away.

The geometry of chaos can be explored using ◊topology and ◊fractals.

characteristic in mathematics, the integral (whole-number) part of a ◊logarithm. The fractional part is the ◊mantissa.

charcoal black, porous form of ◊carbon, produced by heating wood or other organic materials in the absence of air. It is used as a fuel in the smelting of metals such as copper and zinc, and by artists for making black line drawings. *Activated charcoal* has been powdered and dried so that it presents a much increased surface area for adsorption; it is used for filtering and purifying liquids and gases – for example, in drinking-water filters and gas masks.

charge see ◊electric charge.

charge-coupled device (CCD) device for forming images electronically, using a layer of silicon that releases electrons when struck by incoming light. The electrons are stored in ◊pixels and read off into a computer at the end of the exposure. CCDs have now almost entirely replaced photographic film for applications such as astrophotography where extreme sensitivity to light is essential.

charged particle beam high-energy beam of electrons or protons. Such beams are being developed as weapons.

Charles's law law stating that the volume of a given mass of gas at constant pressure is directly proportional to its absolute temperature (temperature in kelvin). It was discovered by French physicist Jacques Charles 1787, and independently by French chemist Joseph Gay-Lussac 1802.

The gas increases by 1/273 of its volume at 0°C for each °C rise of temperature. This means that the coefficient of expansion of all gases is the same. The law is only approximately true and the coefficient of expansion is generally taken as 0.003663 per °C.

charm in physics, a property possessed by one type of ◊quark (very small particles found inside protons and neutrons), called the charm quark. The effects of charm are only seen in experiments with particle ◊accelerators. See ◊elementary particles.

chelate chemical compound whose molecules consist of one or more metal atoms or charged ions joined to chains of organic residues by coordinate (or dative covalent) chemical ◊bonds.

The parent organic compound is known as a *chelating agent* – for example, EDTA (ethylene-diaminetetraacetic acid), used in chemical analysis. Chelates are used in analytical chemistry, in agriculture and horticulture as carriers of essential trace metals, in water softening, and in the treatment of thalassaemia by removing excess iron, which may build up to toxic levels in the body. Metalloproteins (natural chelates) may influence the performance of enzymes or provide a mechanism for the storage of iron in the spleen and plasma of the human body.

chemical change change that occurs when two or more substances (reactants) interact with each other, resulting in the production of different substances (products) with different chemical compositions. A simple example of chemical change is the burning of carbon in oxygen to produce carbon dioxide.

chemical element alternative name for ◊element.

chemical equation method of indicating the reactants and products of a chemical reaction by using chemical symbols and formulae. A chemical equation gives two basic pieces of information: (1) the reactants (on the left-hand side) and products (right-hand side); and (2) the reacting proportions (stoichiometry) – that is, how many units of each reactant and product are involved. The equation must balance; that is, the total number of atoms of a particular element on the left-hand side must be the

chemical equilibrium condition in which the products of a reversible chemical reaction are formed at the same rate at which they decompose back into the reactants, so that the concentration of each reactant and product remains constant.

chemical family collection of elements that have very similar chemical and physical properties. In the ◊periodic table of the elements such collections are to be found in the vertical columns (groups). The groups that contain the most markedly similar elements are group I, the ◊alkali metals; group II, the ◊alkaline-earth metals; group VII, the ◊halogens; and group 0, the noble or ◊inert gases.

chemiluminescence the emission of light from a substance as a result of a chemical reaction (rather than raising its temperature). See ◊luminescence.

chemisorption the attachment, by chemical means, of a single layer of molecules, atoms, or ions of gas to the surface of a solid or, less frequently, a liquid. It is the basis of catalysis (see ◊catalyst) and is of great industrial importance.

chemistry science concerned with the composition of matter (gas, liquid, or solid) and of the changes that take place in it under certain conditions.

Organic chemistry is the branch of chemistry that deals with carbon compounds. *Inorganic chemistry* deals with the description, properties, reactions, and preparation of all the elements and their compounds, with the exception of carbon compounds. *Physical chemistry* is concerned with the quantitative explanation of chemical phenomena and reactions, and the measurement of data required for such explanations. This branch studies in particular the movement of molecules and the effects of temperature and pressure, often with regard to gases and liquids.

chemosynthesis method of making ◊protoplasm (contents of a cell) using the energy from chemical reactions, in contrast to the use of light energy employed for the same purpose in ◊photosynthesis. The process is used by certain bacteria, which can synthesize organic compounds from carbon dioxide and water using the energy from special methods of ◊respiration.

chemotherapy any medical treatment with chemicals. It usually refers to treatment of cancer with cytotoxic and other drugs. The term was coined by the German bacteriologist Paul Ehrlich for the use of synthetic chemicals against infectious diseases.

chemotropism movement by part of a plant in response to a chemical stimulus. The response by the plant is termed 'positive' if the growth is towards the stimulus or 'negative' if the growth is away from the stimulus.

Cherenkov radiation in physics, radiation emitted as a bluish light when charged atomic particles pass through water or other transparent media at a greater speed than that of light. It is named after the Russian physicist Pavel Cherenkov (1904–1990).

Chernobyl town in central Ukraine; site of a nuclear power station. In April 1986 two huge explosions destroyed a central reactor, breaching the 1,000 tonne roof. In the immediate vicinity of Chernobyl, 31 people died and 135,000 were permanently evacuated. It has been estimated that there will be an additional 20–40,000 deaths from cancer in the next 60 years.

childbirth the expulsion of a baby from its mother's body following ◊pregnancy. In a broader sense, it is the period of time involving labour and delivery of the baby.

chimera in biology, an organism composed of tissues that are genetically different. Chimeras can develop naturally if a ◊mutation occurs in a cell of a developing embryo, but are more commonly produced artificially by implanting cells from one organism into the embryo of another.

china clay clay mineral formed by the decomposition of ◊feldspars. The alteration of aluminium silicates results in the formation of *kaolinite*, $Al_2Si_2O_5(OH)_4$, from which *kaolin*, or white china clay, is derived.

chip or *silicon chip* another name for an ◊*integrated circuit*, a complete electronic circuit on a slice of silicon (or other semiconductor) crystal only a few millimetres square.

Chiron unusual Solar-System object orbiting between Saturn and Uranus, discovered 1977 by US astronomer Charles T Kowal (1940–). Initially classified as an asteroid, it is now believed to be a giant cometary nucleus about 200 km/120 mi across, composed of ice with a dark crust of carbon dust.

chitin complex long-chain compound, or ◊polymer; a nitrogenous derivative of glucose. Chitin is widely found in invertebrates. It forms the ◊exoskeleton of insects and other

chemistry: chronology

c. 3000 BC	Egyptians were producing bronze – an alloy of copper and tin.
c. 450 BC	Greek philosopher Empedocles proposed that all substances are made up of a combination of four elements – Earth, air, fire, and water – an idea that was developed by Plato and Aristotle and persisted for over 2,000 years.
c. 400 BC	Greek philosopher Democritus theorized that matter consists ultimately of tiny, indivisible particles, *atomos*.
AD 1	Gold, silver, copper, lead, iron, tin, and mercury were known.
200	The techniques of solution, filtration, and distillation were known.
7th–17th centuries	Chemistry was dominated by alchemy, the attempt to transform nonprecious metals such as lead and copper into gold. Though misguided, it led to the discovery of many new chemicals and techniques, such as sublimation and distillation.
12th century	Alcohol was first distilled in Europe.
1242	Gunpowder introduced to Europe from the Far East.
1620	Scientific method of reasoning expounded by Francis Bacon in his *Novum Organum*.
1650	Leyden University in the Netherlands set up the first chemistry laboratory.
1661	Robert Boyle defined an element as any substance that cannot be broken down into still simpler substances and asserted that matter is composed of 'corpuscles' (atoms) of various sorts and sizes, capable of arranging themselves into groups, each of which constitutes a chemical substance.
1662	Boyle described the inverse relationship between the volume and pressure of a fixed mass of gas (Boyle's law).
1697	Georg Stahl proposed the erroneous theory that substances burn because they are rich in a certain substance, called phlogiston.
1755	Joseph Black discovered carbon dioxide.
1774	Joseph Priestley discovered oxygen, which he called 'dephlogisticated air'. Antoine Lavoisier demonstrated his law of conservation of mass.
1777	Lavoisier showed air to be made up of a mixture of gases, and showed that one of these – oxygen – is the substance necessary for combustion (burning) and rusting to take place.
1781	Henry Cavendish showed water to be a compound.
1792	Alessandra Volta demonstrated the electrochemical series.
1807	Humphry Davy passed electric current through molten compounds (the process of electrolysis) in order to isolate elements, such as potassium, that had never been separated by chemical means. Jöns Berzelius proposed that chemicals produced by living creatures should be termed 'organic'.
1808	John Dalton published his atomic theory, which states that every element consists of similar indivisible particles – called atoms – which differ from the atoms of other elements in their mass; he also drew up a list of relative atomic masses. Joseph Gay-Lussac announced that the volumes of gases that combine chemically with one another are in simple ratios.
1811	Publication of Amedeo Avogadro's hypothesis on the relation between the volume and number of molecules of a gas, and its temperature and pressure.
1813–14	Berzelius devised the chemical symbols and formulae still used to represent elements and compounds.
1828	Franz Wöhler converted ammonium cyanate into urea – the first synthesis of an organic compound from an inorganic substance.
1832–33	Michael Faraday expounded the laws of electrolysis, and adopted the term 'ion' for the particles believed to be responsible for carrying current.
1846	Thomas Graham expounded his law of diffusion.
1853	Robert Bunsen invented the Bunsen burner.
1858	Stanislao Cannizzaro differentiated between atomic and molecular weights (masses).
1861	Organic chemistry was defined by German chemist Friedrich Kekulé as the chemistry of carbon compounds.
1864	John Newlands devised the first periodic table of the elements.
1869	Dmitri Mendeleyev expounded his periodic table of the elements (based on atomic mass), leaving gaps for elements that had not yet been discovered.
1874	Jacobus van't Hoff suggested that the four bonds of carbon are arranged tetrahedrally, and that carbon compounds can therefore be three-dimensional and asymmetric.
1884	Swedish chemist Svante Arrhenius suggested that electrolytes (solutions or molten compounds that conduct electricity) dissociate into ions, atoms or groups of atoms that carry a positive or negative charge.
1894	William Ramsey and Lord Rayleigh discovered the first inert gases, argon.

1897	The electron was discovered by J J Thomson.
1901	Mikhail Tsvet invented paper chromatography as a means of separating pigments.
1909	Sören Sörensen devised the pH scale of acidity.
1912	Max von Laue showed crystals to be composed of regular, repeating arrays of atoms by studying the patterns in which they diffract X-rays.
1913–14	Henry Moseley equated the atomic number of an element with the positive charge on its nuclei, and drew up the periodic table, based on atomic number, that is used today.
1916	Gilbert Newton Lewis explained covalent bonding between atoms as a sharing of electrons.
1927	Nevil Sidgwick published his theory of valency, based on the numbers of electrons in the outer shells of the reacting atoms.
1930	Electrophoresis, which separates particles in suspension in an electric field, was invented by Arne Tiselius.
1932	Deuterium (heavy hydrogen), an isotope of hydrogen, was discovered by Harold Urey.
1940	Edwin McMillan and Philip Abelson showed that new elements with a higher atomic number than uranium can be formed by bombarding uranium with neutrons, and synthesized the first transuranic element, neptunium.
1942	Plutonium was first synthesized by Glenn T Seaborg and Edwin McMillan.
1950	Derek Barton deduced that some properties of organic compounds are affected by the orientation of their functional groups (the study of which became known as conformational analysis).
1954	Einsteinium and fermium were synthesized.
1955	Ilya Prigogine described the thermodynamics of irreversible processes (the transformations of energy that take place in, for example, many reactions within living cells).
1962	Neil Bartlett prepared the first compound of an inert gas, xenon hexafluoroplatinate; it was previously believed that inert gases could not take part in a chemical reaction.
1965	Robert B Woodward synthesized complex organic compounds.
1981	Quantum mechanics applied to predict course of chemical reactions by US chemist Roald Hoffmann and Kenichi Fukui of Japan.
1982	Element 109, unnilennium, synthesized.
1985	Fullerenes, a new class of carbon solids made up of closed cages of carbon atoms, were discovered by Harold Kroto and David Walton at the University of Sussex, England.
1987	US chemists Donald Cram and Charles Pederson, and Jean-Marie Lehn of France created artificial molecules that mimic the vital chemical reactions of life processes.
1990	Jean-Marie Lehn, Ulrich Koert, and Margaret Harding reported the synthesis of a new class of compounds, called nucleohelicates, that mimic the double helical structure of DNA, turned inside out.
1993	US chemists at the University of California and the Scripps Institute synthesized rapamycin, one of a group of complex, naturally occurring antibiotics and immunosuppressants that are being tested as anticancer agents.

arthropods. It combines with protein to form a covering that can be hard and tough, as in beetles, or soft and flexible, as in caterpillars and other insect larvae. It is insoluble in water and resistant to acids, alkalis, and many organic solvents. In crustaceans such as crabs, it is impregnated with calcium carbonate for extra strength.

Chitin also occurs in some ◊protozoans and coelenterates (such as certain jellyfishes), in the jaws of annelid worms, and as the cell-wall polymer of fungi. Its uses include coating apples (still fresh after six months), coating seeds, and dressing wounds. In 1993 chemists at North Carolina State University found that it can be used to filter pollutants from industrial waste water.

chlamydia viruslike bacteria which live parasitically in animal cells, and cause disease in humans and birds. Chlamydiae are thought to be descendants of bacteria that have lost certain metabolic processes. In humans, a strain of chlamydia causes trachoma, a disease found mainly in the tropics (a leading cause of blindness); venereally transmitted chlamydiae cause genital and urinary infections.

chloral or ***trichloroethanal*** CCl_3CHO oily, colourless liquid with a characteristic pungent smell, produced by the action of chlorine on ethanol. It is soluble in water and its compound chloral hydrate is a powerful sleep-inducing agent.

chlorate any salt derived from an acid containing both chlorine and oxygen and possessing the negative ion ClO^-, ClO_2^{2-}, ClO_3^-, or ClO_4^{4-}. Common chlorates are those of sodium, potassium, and barium. Certain chlorates are used in weedkillers.

chloride Cl⁻ negative ion formed when hydrogen chloride dissolves in water, and any salt containing this ion, commonly formed by the action of hydrochloric acid (HCl) on various metals or by direct combination of a metal and chlorine. Sodium chloride (NaCl) is common table salt.

chlorinated solvent any liquid organic compound that contains chlorine atoms, often two or more. These compounds are very effective solvents for fats and greases, but many also have toxic properties. They include trichloromethane (chloroform, $CHCl_3$), tetrachloromethane (carbon tetrachloride, CCl_4), and trichloroethene ($CH_2ClCHCl_2$).

chlorination the treatment of water with chlorine in order to disinfect it; also, any chemical reaction in which a chlorine atom is introduced into a chemical compound.

chlorine greenish-yellow, gaseous, non-metallic element with a pungent odour, symbol Cl, atomic number 17, relative atomic mass 35.453. It is a member of the ◊halogen group and is widely distributed, in combination with the ◊alkali metals, as chlorates or chlorides.

chlorofluorocarbon (CFC) synthetic chemical that is odourless, nontoxic, nonflammable, and chemically inert. The first CFC was synthesized 1892, but no use was found for it until the 1920s. Since then their stability and apparently harmless properties have made CFCs popular in ◊aerosols, as refrigerants in refrigerators and air conditioners, and in the manufacture of foam packaging. They are partly responsible for the destruction of the ◊ozone layer. In June 1990 representatives of 93 nations, including the UK and the USA, agreed to phase out production of CFCs and various other ozone-depleting chemicals by the end of the 20th century.

chloroform (technical name *trichloromethane*) CCl_3 clear, colourless, toxic, carcinogenic liquid with a characteristic pungent, sickly sweet smell and taste, formerly used as an anaesthetic (now superseded by less harmful substances). It is used as a solvent and in the synthesis of organic chemical compounds.

chlorophyll green pigment present in most plants; it is responsible for the absorption of light energy during ◊photosynthesis.

The pigment absorbs the red and blue-violet parts of sunlight but reflects the green, thus giving plants their characteristic colour.

Chlorophyll is found within chloroplasts, present in large numbers in leaves. Cyanobacteria (blue-green algae) and other photo-synthetic bacteria also have chlorophyll, though of a slightly different type. Chlorophyll is similar in structure to ◊haemoglobin, but with magnesium instead of iron as the reactive part of the molecule.

chloroplast *Green chlorophyll molecules on the membranes of the vesicle stacks capture light energy to produce food by photosynthesis.*

chloroplast structure (◊organelle) within a plant cell containing the green pigment chlorophyll. Chloroplasts occur in most cells of the green plant that are exposed to light, often in large numbers. Typically, they are flattened and disclike, with a double membrane enclosing the stroma, a gel-like matrix. Within the stroma are stacks of fluid-containing cavities, or vesicles, where ◊photosynthesis occurs.

chlorosis abnormal condition of green plants in which the stems and leaves turn pale green or yellow. The yellowing is due to a reduction in the levels of the green chlorophyll pigments. It may be caused by a deficiency in essential elements (such as magnesium, iron, or manganese), a lack of light, genetic factors, or viral infection.

choke coil in physics, a coil employed to limit or suppress alternating current without stopping direct current, particularly the type used as a 'starter' in the circuit of fluorescent lighting.

cholecalciferol or *vitamin D* fat-soluble chemical important in the uptake of calcium and phosphorous for bones. It is found in liver, fish oils and margarine. It can be produced in the skin, provided that the skin is adequately exposed to sunlight. Lack of vitamin D leads to rickets and other bone diseases.

cholesterol white, crystalline ◊sterol found throughout the body, especially in fats, blood, nerve tissue, and bile; it is also provided in the diet by foods such as eggs, meat, and butter. A high level of cholesterol in the blood is thought to contribute to atherosclerosis (hardening of the arteries).

chord in geometry, a straight line joining any two points on a curve. The chord that passes through the centre of a circle (its longest chord) is the diameter. The longest and shortest chords of an ellipse (a regular oval) are called the major and minor axes respectively.

chordate animal belonging to the phylum Chordata, which includes vertebrates, sea squirts, amphioxi, and others. All these animals, at some stage of their lives, have a supporting rod of tissue (notochord or backbone) running down their bodies.

chorion outermost of the three membranes enclosing the embryo of reptiles, birds, and mammals; the ◊amnion is the innermost membrane.

choroid layer found at the rear of the ◊eye beyond the retina. By absorbing light that has already passed through the retina, it stops back-reflection and so prevents blurred vision.

chromatography technique for separating or analysing a mixture of gases, liquids, or dissolved substances. This is brought about by means of two immiscible substances, one of which (*the mobile phase*) transports the sample mixture through the other (*the stationary phase*). The mobile phase may be a gas or a liquid; the stationary phase may be a liquid or a solid, and may be in a column, on paper, or in a thin layer on a glass or plastic support. The components of the mixture are absorbed or impeded by the stationary phase to different extents and therefore become separated. The technique is used for both qualitative and quantitive analyses in biology and chemistry.

In *paper chromatography*, the mixture separates because the components have differing solubilities in the solvent flowing through the paper and in the chemically bound water of the paper.

In *thin-layer chromatography*, a wafer-thin layer of adsorbent medium on a glass plate replaces the filter paper. The mixture separates because of the differing solubilities of the components in the solvent flowing up the solid layer, and their differing tendencies to stick to the solid (adsorption). The same principles apply in *column chromatography*.

In *gas–liquid chromatography*, a gaseous mixture is passed into a long, coiled tube (enclosed in an oven) filled with an inert powder coated in a liquid. A carrier gas flows through the tube. As the mixture proceeds along the tube it separates as the components dissolve in the liquid to differing extents or stay as a gas. A detector locates the different components as they emerge from the tube. The technique is very powerful, allowing tiny quantities of substances (fractions of parts per million) to be separated and analysed.

chromite $FeCr_2O_3$, iron chromium oxide, the main chromium ore. It is one of the ◊spinel group of minerals, and crystallizes in dark-coloured octahedra of the cubic system. Chromite is usually found in association with ultrabasic and basic rocks; in Cyprus, for example, it occurs with ◊serpentine, and in South Africa it forms continuous layers in a layered intrusion.

chromium hard, brittle, grey-white, metallic element, symbol Cr, atomic number 24, relative atomic mass 51.996. It takes a high polish, has a high melting point, and is very resistant to corrosion. It is used in chromium electroplating, in the manufacture of stainless steel and other alloys, and as a catalyst. Its compounds are used for tanning leather and for ◊alums. In human nutrition it is a vital trace element. In nature, it occurs chiefly as chrome iron ore or chromite ($FeCr_2O_4$). Kazakhstan, Zimbabwe, and Brazil are sources.

chromium ore essentially chromite, $FeCr_2O_3$, from which chromium is extracted. South Africa and Zimbabwe are major producers.

chromosome structure in a cell nucleus that carries the ◊genes. Each chromosome consists of one very long strand of DNA, coiled and folded to produce a compact body. The point on a chromosome where a particular gene occurs is known as its locus. Most higher organisms have two copies of each chromosome (they are ◊diploid) but some have only one (they are ◊haploid). There are 46 chromosomes in a normal human cell. See also ◊mitosis and ◊meiosis.

chromosphere layer of mostly hydrogen gas about 10,000 km/6,000 mi deep above the visible surface of the Sun (the photosphere). It appears pinkish red during ◊eclipses of the Sun.

chronometer instrument for measuring time precisely, originally used at sea. It is designed to remain accurate through all conditions of temperature and pressure. The first accurate marine chronometer, capable of an accuracy of half a minute a year, was made 1761 by John Harrison in England.

chrysotile mineral in the ◊olivine group.

chyme general term for the stomach contents. Chyme resembles a thick creamy fluid and is made up of partly digested food, hydrochloric acid, and a range of enzymes.

cilia (singular *cilium*) small hairlike organs on the surface of some cells, particularly the cells lining the upper respiratory tract. Their wavelike movements waft particles of dust and debris towards the exterior. Some single-celled organisms move by means of cilia. In multicellular animals, they keep lubricated surfaces clear of debris. They also move food in the digestive tracts of some invertebrates.

ciliary muscle ring of muscle surrounding and controlling the lens inside the vertebrate eye, used in ◊accommodation (focusing). Suspensory ligaments, resembling spokes of a wheel, connect the lens to the ciliary muscle and pull the lens into a flatter shape when the muscle relaxes. When the muscle is relaxed the lens has its longest ◊focal length and focuses rays from distant objects. On contraction, the lens returns to its normal spherical state; the lens therefore has a shorter focal length and focuses images of near objects.

cinnabar mercuric sulphide, HgS, the only commercially useful ore of mercury. It is deposited in veins and impregnations near recent volcanic rocks and hot springs. The mineral itself is used as a red pigment, commonly known as *vermilion*. Cinnabar is found in the USA (California), Spain (Almadén), Peru, Italy, and Slovenia.

circadian rhythm metabolic rhythm found in most organisms, which generally coincides with the 24-hour day. Its most obvious manifestation is the regular cycle of sleeping and waking, but body temperature and the concentration of ◊hormones that influence mood and behaviour also vary over the day. In humans, alteration of habits (such as rapid air travel round the world) may result in the circadian rhythm being out of phase with actual activity patterns, causing malaise until it has had time to adjust.

circle perfectly round shape, the path of a point that moves so as to keep a constant distance from a fixed point (the centre). Each circle has a *radius* (the distance from any point on the circle to the centre), a *circumference* (the boundary of the circle), *diameters* (straight lines crossing the circle through the centre), *chords* (lines joining two points on the circumference), *tangents* (lines that touch the circumference at one point only), *sectors* (regions inside the circle between two radii), and *segments* (regions between a chord and the circumference).

circuit in physics or electrical engineering, an arrangement of electrical components through which a current can flow. There are two basic circuits, series and parallel. In a ◊series circuit, the components are connected end to end so that the current flows through all components one after the other. In a ◊parallel circuit, components are connected side by side so that part of the current passes through each component. A circuit diagram shows in graphical form how components are connected together, using standard symbols for the components.

circuit breaker switching device designed to protect an electric circuit from excessive current. It has the same action as a ◊fuse, and

circle Technical terms used in the geometry of the circle; the area of a circle can be seen to equal πr^2 by dividing the circle into segments which form a rectangle.

many houses now have a circuit breaker between the incoming mains supply and the domestic circuits. Circuit breakers usually work by means of ◊solenoids. Those at electricity-generating stations have to be specially designed to prevent dangerous arcing (the release of luminous discharge) when the high-voltage supply is switched off. They may use an air blast or oil immersion to quench the arc.

circuit diagram simplified drawing of an electric circuit. The circuit's components are represented by internationally recognized symbols, and the connecting wires by straight lines. A dot indicates where wires join.

circulatory system system of vessels in an animal's body that transports essential substances (blood or other circulatory fluid) to and from the different parts of the body. Except for simple animals such as sponges and coelenterates (jellyfishes, sea anemones, corals), all animals have a circulatory system.

In fishes, blood passes once around the body before returning to a two-chambered heart (single circulation). In birds and mammals, blood passes to the lungs and back to the heart before circulating around the remainder of the body (double circulation). In all vertebrates, blood flows in one direction. Valves in the heart, large arteries, and veins prevent backflow, and the muscular walls of the arteries assist in pushing the blood around the body.

Although most animals have a heart or hearts to pump the blood, normal body movements circulate the fluid in some small invertebrates. In the *open system*, found in snails and other molluscs, the blood (more correctly called ◊haemolymph) passes from the arteries into a body cavity (haemocoel), and from here is gradually returned to the heart, via the gills, by other blood vessels. Insects and other arthropods have an open system with a heart. In the *closed system* of earthworms, blood flows directly from the main artery to the main vein, via smaller lateral vessels in each body segment. Vertebrates, too, have a closed system with a network of tiny ◊capillaries carrying the blood from arteries to veins. See ◊human body for illustration.

circumference in geometry, the curved line that encloses a curved plane figure, for example a ◊circle or an ellipse. Its length varies according to the nature of the curve, and may be ascertained by the appropriate formula. The circumference of a circle is πd or $2\pi r$, where d is the diameter of the circle, r is its radius, and π is the constant pi, approximately equal to 3.1416.

CISC (acronym for *complex instruction set computer*) in computing, a microprocessor (processor on a single chip) that can carry out a large number of ◊machine code instructions – for example, the Intel 80386. The term was introduced to distinguish them from the more rapid ◊RISC (reduced instruction set computer) processors, which handle only a smaller set of instructions.

cistron in genetics, the segment of ◊DNA that is required to synthesize a complete polypeptide chain. It is the molecular equivalent of a ◊gene.

CITES (abbreviation for *Convention on International Trade in Endangered Species*) international agreement under the auspices of the IUCN with the aim of regulating trade in ◊endangered species of animals and plants. The agreement came into force 1975 and by 1991 had been signed by 110 states. It prohibits any trade in a category of 8,000 highly endangered species and controls trade in a further 30,000 species.

citric acid $HOOCCH_2C(OH)(COOH)CH_2COOH$ organic acid widely distributed in the plant kingdom; it is found in high concentrations in citrus fruits and has a sharp, sour taste. At one time it was commercially prepared from concentrated lemon juice, but now the main source is the fermentation of sugar with certain moulds.

civil engineering branch of engineering concerned with the construction of roads, bridges, aqueducts, waterworks, tunnels, canals, irrigation works, and harbours.

cladistics method of biological ◊classification (taxonomy) that uses a formal step-by-step procedure for objectively assessing the extent to which organisms share particular characters, and for assigning them to taxonomic groups (for example, ◊species, ◊genus, family). Taxonomic groups are termed *clades*.

cladode in botany, a flattened stem that is leaflike in appearance and function. It is an adaptation to dry conditions because a stem contains fewer ◊stomata than a leaf, and water loss is thus minimized. The true leaves in such plants are usually reduced to spines or small scales. Examples of plants with cladodes are butcher's-broom *Ruscus aculeatus*, asparagus, and certain cacti. Cladodes may bear flowers or fruit on their surface, and this distinguishes them from leaves.

CLASSIFICATION

classification in biology, the arrangement of organisms into a hierarchy of groups on the basis of their similarities in biochemical, anatomical, or physiological characters. The basic grouping is a ◊species, several of which may constitute a ◊genus, which in turn are grouped into families, and so on up through orders, classes, phyla (in plants, sometimes called divisions), to kingdoms.

classify in mathematics, to put into separate classes, or ◊sets, which may be uniquely defined.

clathrate compound formed when the small molecules of one substance fill in the holes in the structural lattice of another, solid, substance – for example, sulphur dioxide molecules in ice crystals. Clathrates are therefore intermediate between mixtures and true compounds (which are held together by ◊ionic or covalent chemical bonds).

clathration in chemistry, a method of removing water from an aqueous solution, and therefore increasing the solution's concentration, by trapping it in a matrix with inert gases such as freons.

clausius in engineering, a unit of ◊entropy (the loss of energy as heat in any physical process). It is defined as the ratio of energy to temperature above absolute zero.

clavicle the collar bone of many vertebrates. In humans it is vulnerable to fracture; falls involving a sudden force on the arm may result in very high stresses passing into the chest region by way of the clavicle and other bones.

claw hard, hooked, pointed outgrowth of the digits of mammals, birds, and most reptiles. Claws are composed of the protein keratin, and grow continuously from a bundle of cells in the lower skin layer. Hooves and nails are modified structures with the same origin as claws.

clay very fine-grained ◊sedimentary deposit that has undergone a greater or lesser degree of consolidation. When moistened it is plastic, and it hardens on heating, which renders it impermeable. It may be white, grey, red, yellow, blue, or black, depending on its composition. Clay minerals consist largely of hydrous silicates of aluminium and magnesium together with iron, potassium, sodium, and organic substances. The crystals of clay minerals have a layered structure, capable of holding water, and are responsible for its plastic properties. According to international classification, in mechanical analysis of soil, clay has a grain size of less than 0.002 mm/0.00008 in.

clay mineral one of a group of hydrous silicate minerals that form most of the fine-grained particles in clays. Clay minerals are normally formed by weathering or alteration of other silicates. Virtually all have sheet silicate structures similar to the ◊micas. They exhibit the following useful properties: loss of water on heating, swelling and shrinking in different conditions, cation exchange with other media, and plasticity when wet. Examples are kaolinite, illite, and montmorillonite.

cleavage in mineralogy, the tendency of a mineral to split along defined, parallel planes related to its internal structure. It is a useful distinguishing feature in mineral identification. Cleavage occurs where bonding between atoms is weakest, and cleavages may be perfect, good, or poor, depending on the bond strengths; a given mineral may possess one, two, three, or more orientations along which it will cleave.

cleistogamy production of flowers that never fully open and that are automatically self-fertilized. Cleistogamous flowers are often formed late in the year, after the production of normal flowers, or during a period of cold weather, as seen in several species of violet *Viola*.

client-server architecture in computing, a system in which the mechanics of looking after data are separated from the programs that use the data. For example, the 'server' might be a central database, typically located on a large computer that is reserved for this purpose. The 'client' would be an ordinary program that requests data from the server as needed.

climate the average weather conditions of a particular place over a long period of time, usually 30 years. Several different climate zones may be identified. The main factors determining the variations of climate over the surface of the Earth are: (1) the effect of latitude and the tilt of the Earth's axis; (2) the large-scale movement of wind and ocean currents; and (3) the temperature difference between land and sea. Recent research indicates that human activity may influence world climate (see ◊greenhouse effect).

climate model computer simulation, based on physical and mathematical data, of the entire climatic system of the Earth. It is used by researchers to study such topics as the possible long-term disruptive effects of the greenhouse gases, or of variations in the amount of radiation given off by the Sun.

climatology study of climate, its global variations and causes.

climax community assemblage of plants and animals that is relatively stable in its environment. It is brought about by ecological ♦succession, and represents the point at which succession ceases to occur.

climax vegetation the plants in a climax community.

clinometer hand-held surveying instrument for measuring angles of slope.

clo unit of thermal insulation of clothing. Standard clothes have an insulation of about 1 clo; the warmest clothing is about 4 clo per 2.5 cm/1 in of thickness. See also ♦tog.

cloaca the common posterior chamber of most vertebrates into which the digestive, urinary, and reproductive tracts all enter; a cloaca is found in most reptiles, birds, and amphibians; many fishes; and, to a reduced degree, marsupial mammals. The cloaca forms a chamber in which products can be stored before being voided from the body via a muscular opening, the cloacal aperture. Placental mammals have a separate digestive opening (the anus) and urinogenital opening.

clone an exact replica. In genetics, any one of a group of genetically identical cells or organisms. An identical ♦twin is a clone; so, too, are bacteria living in the same colony. The term clone has also been adopted by computer technology to describe a (nonexistent) device that mimics an actual one to enable certain software programs to run correctly.

closed-circuit television (CCTV) localized television system in which programmes are sent over relatively short distances, the camera, receiver, and controls being linked by cable. Closed-circuit TV systems are used in department stores and large offices as a means of internal security, monitoring people's movements.

cloud water vapour condensed into minute water particles that float in masses in the atmosphere. Clouds, like fogs or mists, which occur at lower levels, are formed by the cooling of air containing water vapour, which generally condenses around tiny dust particles.

cloud chamber apparatus for tracking ionized particles. It consists of a vessel fitted with a piston and filled with air or other gas, saturated with water vapour. When the volume of the vessel is suddenly expanded by moving the piston outwards, the vapour cools and a cloud of tiny droplets forms on any nuclei, dust, or ions present. As fast-moving ionizing particles collide with the air or gas molecules, they show as visible tracks.

clubroot disease affecting cabbages, turnips, and allied plants of the Cruciferae family. It is caused by a ♦slime mould, *Plasmodiophora brassicae*. This attacks the roots of the plant, which send out knotty outgrowths. Eventually the whole plant decays.

clusec unit for measuring the power of a vacuum pump.

clutch any device for disconnecting rotating shafts, used especially in a car's transmission system. In a car with a manual gearbox, the driver depresses the clutch when changing gear, thus disconnecting the engine from the gearbox.

cm symbol for *centimetre*.

CMOS (abbreviation for ♦*complementary metal-oxide semiconductor*) family of integrated circuits (chips) widely used in building electronic systems.

CNS abbreviation for ♦central nervous system.

coal black or blackish mineral substance formed from the compaction of ancient plant matter in tropical swamp conditions. It is used as a fuel and in the chemical industry. Coal is classified according to the proportion of carbon it contains. The main types are ♦*anthracite* (shiny, with about 90% carbon), *bituminous coal* (shiny and dull patches, about 75% carbon), and *lignite* (woody, grading into peat, about 50% carbon). Coal burning is one of the main causes of ♦acid rain.

coal gas gas produced when coal is destructively distilled or heated out of contact with the air. Its main constituents are methane, hydrogen, and carbon monoxide. Coal gas has been superseded by ♦natural gas for domestic purposes.

coal tar black oily material resulting from the destructive distillation of bituminous coal.

coaxial cable electric cable that consists of a solid or stranded central conductor insulated from and surrounded by a solid or braided conducting tube or sheath. It can transmit the high-frequency signals used in television, telephone, and other telecommunications transmissions.

cobalt hard, lustrous, grey, metallic element, symbol Co, atomic number 27, relative atomic mass 58.933. It is found in various ores and occasionally as a free metal, sometimes in metallic meteorite fragments. It is used in the preparation of magnetic, wear-resistant, and high-strength alloys; its compounds are used in inks, paints, and varnishes.

cobalt-60 radioactive (half-life 5.3 years) isotope produced by neutron radiation of cobalt in heavy-water reactors, used in large amounts for gamma rays in cancer therapy, industrial radiography, and research, substituting for the much more costly radium.

cobalt chloride $CoCl_2$ compound that exists in two forms: the hydrated salt ($CoCl_2.6H_2O$), which is pink, and the anhydrous salt, which is blue. The anhydrous form is used as an indicator because it turns pink if water is present. When the hydrated salt is gently heated the blue anhydrous salt is reformed.

cobalt ore cobalt is extracted from a number of minerals, the main ones being *smaltite*, $(CoNi)As_3$; *linnaeite*, Co_3S_4; *cobaltite*, CoAsS; and *glaucodot*, (CoFe)AsS.

COBOL (acronym for ***common business-oriented language***) high-level computer-programming language, designed in the late 1950s for commercial data-processing problems; it has become the major language in this field. COBOL features powerful facilities for file handling and business arithmetic. Program instructions written in this language make extensive use of words and look very much like English sentences. This makes COBOL one of the easiest languages to learn and understand.

cocaine alkaloid $C_{17}H_{21}NO_4$ extracted from the leaves of the coca tree. It has limited medical application, mainly as a local anaesthetic agent that is readily absorbed by mucous membranes (lining tissues) of the nose and throat. It is both toxic and addictive. Its use as a stimulant is illegal. ◊Crack is a derivative of cocaine.

coccus (plural *cocci*) member of a group of globular bacteria, some of which are harmful to humans. The cocci contain the subgroups ***streptococci***, where the bacteria associate in straight chains, and ***staphylococci***, where the bacteria associate in branched chains.

cochlea part of the inner ◊ear. It is equipped with approximately 10,000 hair cells, which move in response to sound waves and thus stimulate nerve cells to send messages to the brain. In this way they turn vibrations of the air into electrical signals.

cocktail effect the effect of two toxic, or potentially toxic, chemicals when taken together rather than separately. Such effects are known to occur with some mixtures of drugs, with the active ingredient of one making the body more sensitive to the other. This sometimes occurs because both drugs require the same ◊enzyme to break them down. Chemicals such as pesticides and food additives are only ever tested singly, not in combination with other chemicals that may be consumed at the same time, so no allowance is made for cocktail effects.

COD abbreviation for ◊*chemical oxygen demand*, a measure of water and effluent quality.

codominance in genetics, the failure of a pair of alleles, controlling a particular characteristic, to show the classic recessive–dominant relationship. Instead, aspects of both alleles may show in the phenotype.

codon in genetics, a triplet of bases (see ◊base pair) in a molecule of DNA or RNA that directs the placement of a particular amino acid during the process of protein (polypeptide) synthesis. There are 64 codons in the ◊genetic code.

coefficient the number part in front of an algebraic term, signifying multiplication. For example, in the expression $4x^2 + 2xy - x$, the coefficient of x^2 is 4 (because $4x^2$ means $4 \times x^2$), that of xy is 2, and that of $\times$ is -1 (because $-1 \times x = -x$).

coefficient of relationship the probability that any two individuals share a given gene by virtue of being descended from a common ancestor. In sexual reproduction of diploid species, an individual shares half its genes with each parent, with its offspring, and (on average) with each sibling; but only a quarter (on average) with its grandchildren or its siblings' offspring; an eighth with its great-grandchildren, and so on.

coelacanth lobe-finned fish *Latimeria chalumnae* up to 2 m/6 ft long. It has bone and muscle at the base of the fins, and is distantly related to the freshwater lobefins, which were the ancestors of all land animals with backbones. Coelacanths live in deep water (200 m/ 650 ft) around the Comoros Islands, off the coast of Madagascar. They were believed to be extinct until one was caught in 1938.

coelenterate any freshwater or marine organism of the phylum Coelenterata, having a body wall composed of two layers of cells. They also possess stinging cells. Examples are jellyfish, hydra, and coral.

coelom in all but the simplest animals, the fluid-filled cavity that separates the body wall from the gut and associated organs, and allows the gut muscles to contract independently of the rest of the body.

coevolution evolution of those structures and behaviours within a species that can best be understood in relation to another species. For example, insects and flowering plants have evolved together: insects have produced mouthparts suitable for collecting pollen or drinking nectar, and plants have developed chemicals and flowers that will attract insects to them.

coherence in physics, property of two or more waves of a beam of light or other electromagnetic radiation having the same frequency and the same ◊phase, or a constant phase difference.

cohesion in physics, a phenomenon in which interaction between two surfaces of the same material in contact makes them cling together (with two different materials the similar phenomenon is called adhesion). According to kinetic theory, cohesion is caused by attraction between particles at the atomic or molecular level. ◊Surface tension, which causes liquids to form spherical droplets, is caused by cohesion.

coke clean, light fuel produced when coal is strongly heated in an airtight oven. Coke contains 90% carbon and makes a useful domestic and industrial fuel (used, for example in the iron and steel industries and in the production of town gas).

cold-blooded of animals, dependent on the surrounding temperature; see ◊*poikilothermy*.

cold fusion in nuclear physics, the fusion of atomic nuclei at room temperature. Were cold fusion to become possible it would provide a limitless, cheap, and pollution-free source of energy, and it has therefore been the subject of research around the world. In 1989, Martin Fleischmann (1927–) and Stanley Pons (1943–) of the University of Utah, USA, claimed that they had achieved cold fusion in the laboratory, but their results could not be substantiated.

cold-working method of shaping metal at or near atmospheric temperature.

coleoptile the protective sheath that surrounds the young shoot tip of a grass during its passage through the soil to the surface. Although of relatively simple structure, most coleoptiles are very sensitive to light, ensuring that seedlings grow upwards.

collagen protein that is the main constituent of ◊connective tissue. Collagen is present in skin, cartilage, tendons, and ligaments. Bones are made up of collagen, with the mineral calcium phosphate providing increased rigidity.

collenchyma plant tissue composed of relatively elongated cells with thickened cell walls, in particular at the corners where adjacent cells meet. It is a supporting and strengthening tissue found in nonwoody plants, mainly in the stems and leaves.

colligative property property that depends on the concentration of particles in a solution. Such properties include osmotic pressure (see ◊osmosis), elevation of ◊boiling point, depression of ◊freezing point, and lowering of ◊vapour pressure.

collimator (1) small telescope attached to a larger optical instrument to fix its line of sight; (2) optical device for producing a nondivergent beam of light; (3) any device for limiting the size and angle of spread of a beam of radiation or particles.

collision theory theory that explains how chemical reactions take place and why rates of reaction alter. For a reaction to occur the reactant particles must collide. Only a certain fraction of the total collisions cause chemical change; these are called *fruitful collisions*. The fruitful collisions have sufficient energy (activation energy) at the moment of impact to break the existing bonds and form new bonds, resulting in the products of the reaction. Increasing the concentration of the reactants and raising the temperature bring about more collisions and therefore more fruitful collisions, increasing the rate of reaction.

colloid substance composed of extremely small particles of one material (the dispersed phase) evenly and stably distributed in another material (the continuous phase). The size of the dispersed particles (1–1,000 nanometres across) is less than that of particles in suspension but greater than that of molecules in true solution. Colloids involving gases include *aerosols* (dispersions of liquid or solid particles in a gas, as in fog or smoke) and *foams* (dispersions of gases in liquids).

Those involving liquids include *emulsions* (in which both the dispersed and the continuous

a fruitful collision

an unfruitful collision

collision theory *In a fruitful collision, an unstable activated complex is formed which then breaks apart to form new compounds. In an unfruitful collision, the intermediate complex reforms the original compounds when it dissociates.*

phases are liquids) and *sols* (solid particles dispersed in a liquid). Sols in which both phases contribute to a molecular three-dimensional network have a jellylike form and are known as *gels*; gelatin, starch 'solution', and silica gel are common examples.

colon in anatomy, the main part of the large intestine, between the caecum and rectum. Water and mineral salts are absorbed from undigested food in the colon, and the residue passes as faeces towards the rectum.

colonization in ecology, the spread of species into a new habitat, such as a freshly cleared field, a new motorway verge, or a recently flooded valley. The first species to move in are called *pioneers*, and may establish conditions that allow other animals and plants to move in (for example, by improving the condition of the soil or by providing shade). Over time a range of species arrives and the habitat matures; early colonizers will probably be replaced, so that the variety of animal and plant life present changes. This is known as ◊succession.

colour quality or wavelength of light emitted or reflected from an object. Visible white light consists of electromagnetic radiation of various wavelengths, and if a beam is refracted through a prism, it can be spread out into a spectrum, in which the various colours correspond to different wavelengths. From long to short wavelengths (from about 700 to 400 nanometres) the colours are red, orange, yellow, green, blue, indigo, and violet.

The light entering our eyes is either reflected from the objects we see, or emitted by hot or luminous objects. Sources of light have a characteristic ◊spectrum or range of wavelengths. Hot solid objects emit light with a broad range of wavelengths, the maximum intensity being at a wavelength which depends on the temperature. The hotter the object, the shorter the wavelengths emitted, as described by ◊Wien's law. When an object is illuminated by white light, some of the wavelengths are absorbed and some are reflected to the eye of an observer. The object appears coloured because of the mixture of wavelengths in the reflected light. For instance, a red object absorbs all wavelengths falling on it except those in the red end of the spectrum. In the light sensitive lining of our eyeball (the ◊retina) cells called cones are responsible for colour vision.

colouring food ◊additive used to alter or improve the colour of processed foods. Colourings include artificial colours, such as tartrazine and amaranth, which are made from petrochemicals, and the 'natural' colours such as chlorophyll, caramel, and carotene. Some of the natural colours are actually synthetic copies of the naturally occurring substances, and some of these, notably the synthetically produced caramels, may be injurious to health.

columbium (Cb) former name for the chemical element ◊niobium. The name is still used occasionally in metallurgy.

coma in astronomy, the hazy cloud of gas and dust that surrounds the nucleus of a ◊comet.

coma in optics, one of the geometrical aberrations of a lens, whereby skew rays from a pointed object make a comet-shaped spot on the image plane instead of meeting at a point.

combined cycle generation system of electricity generation that makes use of both a gas turbine and a steam turbine. Combined cycle plants are more efficient than conventional generating plants, with an efficiency of energy conversion of around 40% (compared with under 38% for conventional plants).

combined heat and power generation (CHP generation) simultaneous production of electricity and useful heat in a power station. The heat is often in the form of hot water or steam, which can be used for local district heating or in industry. The electricity output from a CHP plant is lower than from a conventional station, but the overall efficiency of energy conversion is higher. A typical CHP plant may convert 80% of the original fuel energy into a mix of electricity and useful heat, whereas a conventional power station rarely even manages a 40% conversion rate.

combustion burning, defined in chemical terms as the rapid combination of a substance with oxygen, accompanied by the evolution of heat and usually light. A slow-burning candle flame and the explosion of a mixture of petrol vapour and air are extreme examples of combustion.

comet small, icy body orbiting the Sun, usually on a highly elliptical path. A comet consists of a central nucleus a few kilometres across, and has been likened to a dirty snowball because it consists mostly of ice mixed with dust. As the comet approaches the Sun the nucleus heats up, releasing gas and dust which form a tenuous coma, up to 100,000 km/ 60,000 mi wide, around the nucleus. Gas and dust stream away from the coma to form one or more tails, which may extend for millions of kilometres.

Comets are believed to have been formed at the birth of the Solar System. Billions of them may reside in a halo (the ♦*Oort cloud*) beyond Pluto. The gravitational effect of passing stars pushes some towards the Sun, when they eventually become visible from Earth. Most comets swing around the Sun and return to distant space, never to be seen again for thousands or millions of years, although some, called *periodic comets*, have their orbits altered by the gravitational pull of the planets so that they reappear every 200 years or less. Of the 800 or so comets whose orbits have been calculated, about 160 are periodic. The brightest is ♦Halley's comet.

command language in computing, a set of commands and the rules governing their use, by which users control a program. For example, an ♦operating system may have commands such as SAVE and DELETE, or a payroll program may have commands for adding and amending staff records.

command line interface (CLI) in computing, a character-based interface in which a prompt is displayed on the screen at which the user types a command, followed by carriage return, at which point the command, if valid, is executed.

commensalism in biology, a relationship between two ♦species whereby one (the commensal) benefits from the association, whereas the other neither benefits nor suffers. For example, certain species of millipede and silverfish inhabit the nests of army ants and live by scavenging on the refuse of their hosts, but without affecting the ants.

common logarithm another name for a ♦logarithm to the base ten.

communication in biology, the signalling of information by one organism to another, usually with the intention of altering the recipient's behaviour. Signals used in communication may be *visual* (such as the human smile or the display of colourful plumage in birds), *auditory* (for example, the whines or barks of a dog), *olfactory* (such as the odours released by the scent glands of a deer), *electrical* (as in the pulses emitted by electric fish), or *tactile* (for example, the nuzzling of male and female elephants).

communications satellite relay station in space for sending telephone, television, telex, and other messages around the world. Messages are sent to and from the satellites via ground stations. Most communications satellites are in ♦geostationary orbit, appearing to hang fixed over one point on the Earth's surface.

The first satellite to carry TV signals across the Atlantic Ocean was *Telstar* in July 1962. The world is now linked by a system of communications satellites called Intelsat.

community in ecology, an assemblage of plants, animals, and other organisms living within a circumscribed area. Communities are usually named by reference to a dominant feature such as characteristic plant species (for example, beech-wood community), or a prominent physical feature (for example, a freshwater-pond community).

compact disc *The compact disc is a digital storage device; music is recorded as a series of etched pits representing numbers in digital code. During playing, a laser scans the pits and the pattern of reflected light reveals the numbers representing the sound recorded. The optical signal is converted to electrical form by a photocell and sent to the amplifiers and loudspeakers.*

commutative operation in mathematics, an operation that is independent of the order of the numbers or symbols concerned. For example, addition is commutative: the result of adding 4 + 2 is the same as that of adding 2 + 4; subtraction is not as 4 − 2 = 2, but 2 − 4 = −2. Compare ◊associative operation and ◊distributive operation.

commutator device in a DC (direct-current) electric motor that reverses the current flowing in the armature coils as the armature rotates.

A DC generator, or ◊dynamo, uses a commutator to convert the AC (alternating current) generated in the armature coils into DC. A commutator consists of opposite pairs of conductors insulated from one another, and contact to an external circuit is provided by carbon or metal brushes.

compact disc or *CD* disc for storing digital information, about 12 cm/4.5 in across, mainly used for music, when it can have over an hour's playing time. The compact disc is made of aluminium with a transparent plastic coating; the metal disc underneath is etched by a ◊laser beam with microscopic pits that carry a digital code representing the sounds. During playback, a laser beam reads the code and produces signals that are changed into near-exact replicas of the original sounds.

compass any instrument for finding direction. The most commonly used is a magnetic compass, consisting of a thin piece of magnetic material with the north-seeking pole indicated, free to rotate on a pivot and mounted on a compass card on which the points of the compass are marked. When the compass is properly adjusted and used, the north-seeking pole will point to

compensation point in biology, the point at which there is just enough light for a plant to survive. At this point all the food produced by ◊photosynthesis is used up by ◊respiration. For aquatic plants, the compensation point is the depth of water at which there is just enough light to sustain life (deeper water = less light = less photosynthesis).

competition in ecology, the interaction between two or more organisms, or groups of organisms (for example, species), that use a common resource which is in short supply. Competition invariably results in a reduction in the numbers of one or both competitors, and in ◊evolution contributes both to the decline of certain species and to the evolution of ◊adaptations.

compiler computer program that translates programs written in a ◊high-level language into machine code (the form in which they can be run by the computer). The compiler translates each high-level instruction into several machine-code instructions – in a process called *compilation* – and produces a complete independent program that can be run by the computer as often as required, without the original source program being present.

complement in mathematics, the set of the elements within the universal set that are not contained in the designated set. For example, if the universal set is the set of all positive whole numbers and the designated set S is the set of all even numbers, then the complement of S (denoted S') is the set of all odd numbers.

complementary metal-oxide semiconductor (CMOS) in electronics, a particular way of manufacturing integrated circuits (chips). The main advantage of CMOS chips is their low power requirement and heat dissipation, which enables them to be used in electronic watches and portable microcomputers. However, CMOS circuits are expensive to manufacture and have lower operating speeds than have circuits of the ◊transistor–transistor logic (TTL) family.

complementary number in number theory, the number obtained by subtracting a number from its base. For example, the complement of 7 in numbers to base 10 is 3. Complementary numbers are necessary in computing, as the only mathematical operation of which digital computers (including pocket calculators) are directly capable is addition. Two numbers can be subtracted by adding one number to the complement of the other; two numbers can be divided by using successive subtraction (which, using complements, becomes successive addition); and multiplication can be performed by using successive addition.

complementation in genetics, the interaction that can occur between two different mutant alleles of a gene in a ◊diploid organism, to make up for each other's deficiencies and allow the organism to function normally.

complex number in mathematics, a number written in the form $a + ib$, where a and b are ◊real numbers and i is the square root of -1 (that is, $i^2 = -1$); i used to be known as the 'imaginary' part of the complex number. Some equations in algebra, such as those of the form $x^2 + 5 = 0$, cannot be solved without recourse to complex numbers, because the real numbers do not include square roots of negative numbers.

component in mechanics, one of two or more forces (or other ◊vector quantities) acting on a body whose combined effect is represented by a single ◊resultant force. For purposes of calculation, a force may be resolved (see ◊resolution of forces) into components acting at right angles to each other.

Compositae daisy family, comprising dicotyledonous flowering plants characterized by flowers borne in composite heads (see ◊capitulum). It is the largest family of flowering plants, the majority being herbaceous. Birds seem to favour the family for use in nest 'decoration', possibly because many species either repel or kill insects. Species include the daisy and dandelion; food plants such as the artichoke, lettuce, and safflower; and the garden varieties of chrysanthemum, dahlia, and zinnia.

composite in industry, any purpose-designed engineering material created by combining single materials with complementary properties into a composite form. Most composites have a structure in which one component consists of discrete elements, such as fibres, dispersed in a continuous matrix. For example, lengths of asbestos, glass, or carbon steel, or 'whiskers' (specially grown crystals a few millimetres long) of substances such as silicon carbide may be dispersed in plastics, concrete, or steel.

compound chemical substance made up of two or more ◊elements bonded together, so that they cannot be separated by physical means. Compounds are held together by ionic or covalent bonds.

computing: chronology

1614	John Napier invented logarithms.
1615	William Oughtred invented the slide rule.
1623	Wilhelm Schickard (1592–1635) invented the mechanical calculating machine.
1645	Blaise Pascal produced a calculator.
1672–74	Gottfried Leibniz built his first calculator, the Stepped Reckoner.
1801	Joseph-Marie Jacquard developed an automatic loom controlled by punch cards.
1820	The first mass-produced calculator, the Arithometer, was developed by Charles Thomas de Colmar (1785–1870).
1822	Charles Babbage completed his first model for the difference engine.
1830s	Babbage created the first design for the analytical engine.
1890	Herman Hollerith developed the punched-card ruler for the US census.
1936	Alan Turing published the mathematical theory of computing.
1938	Konrad Zuse constructed the first binary calculator, using Boolean algebra.
1939	US mathematician and physicist J V Atanasoff (1903–) became the first to use electronic means for mechanizing arithmetical operations.
1943	The Colossus electronic code-breaker was developed at Bletchley Park, England. The Harvard University Mark I or Automatic Sequence Controlled Calculator (partly financed by IBM) became the first program-controlled calculator.
1946	ENIAC (acronym for electronic numerator, integrator, analyser, and computer), the first general purpose, fully electronic digital computer, was completed at the University of Pennsylvania, USA.
1948	Manchester University (England) Mark I, the first stored-program computer, was completed. William Shockley of Bell Laboratories invented the transistor.
1951	Launch of Ferranti Mark I, the first commercially produced computer. Whirlwind, the first real-time computer, was built for the US air-defence system. Grace Murray Hopper of Remington Rand invented the compiler computer program.
1952	EDVAC (acronym for electronic discrete variable computer) was completed at the Institute for Advanced Study, Princeton, USA (by John Von Neumann and others).
1953	Magnetic core memory was developed.
1958	The first integrated circuit was constructed.
1963	The first minicomputer was built by Digital Equipment (DEC). The first electronic calculator was built by Bell Punch Company.
1964	Launch of IBM System/360, the first compatible family of computers. John Kemeny and Thomas Kurtz of Dartmouth College invented BASIC (Beginner's All-purpose Symbolic Instruction Code), a computer language similar to FORTRAN.
1965	The first supercomputer, the Control Data CD6600, was developed.
1971	The first microprocessor, the Intel 4004, was announced.
1974	CLIP–4, the first computer with a parallel architecture, was developed by John Backus at IBM.
1975	Altair 8800, the first personal computer (PC), or microcomputer, was launched.
1981	The Xerox Star system, the first WIMP system (acronym for windows, icons, menus, and pointing devices), was developed. IBM launched the IBM PC.
1984	Apple launched the Macintosh computer.
1985	The Inmos T414 transputer, the first 'off-the-shelf' microprocessor for building parallel computers, was announced.
1988	The first optical microprocessor, which uses light instead of electricity, was developed.
1989	Wafer-scale silicon memory chips, able to store 200 million characters, were launched.
1990	Microsoft released Windows 3, a popular windowing environment for PCs.
1992	Philips launched the CD-I (Compact-Disc Interactive) player, based on CD audio technology, to provide interactive multimedia programs for the home user.
1993	Intel launched the Pentium chip containing 3.1 million transistors and capable of 100 MIPs (millions of instructions per second). The Personal Digital Assistant (PDA), which recognizes user's handwriting, went on sale.

compressor machine that compresses a gas, usually air, commonly used to power pneumatic tools, such as road drills, paint sprayers, and dentist's drills.

CompuServe large (US-based) public on-line data service. It is widely used for ◊electronic mail and ◊bulletin boards.

computer programmable electronic device that processes data and performs calculations and other symbol-manipulation tasks. There are three types: the ◊*digital computer*, which manipulates information coded as binary numbers (see ◊binary number system); the ◊*analogue computer*, which works with continuously varying quantities; and the ***hybrid computer***, which has characteristics of both analogue and digital computers.

computer-aided design use of computers to create and modify design drawings; see ◊CAD.

computer-aided manufacturing use of computers to regulate production processes in industry; see ◊CAM.

computer-assisted learning use of computers in education and training; see ◊CAL.

computer generation any of the five broad groups into which computers may be classified: ***first generation*** the earliest computers, developed in the 1940s and 1950s, made from valves and wire circuits; ***second generation*** from the early 1960s, based on transistors and printed circuits; ***third generation*** from the late 1960s, using integrated circuits and often sold as families of computers, such as the IBM 360 series; ***fourth generation*** using ◊microprocessors, large-scale integration (LSI), and sophisticated programming languages, still in use in the 1990s; and ***fifth generation*** based on parallel processing and very large-scale integration, currently under development.

computer graphics use of computers to display and manipulate information in pictorial form. Input may be achieved by scanning an image, by drawing with a mouse or stylus on a graphics tablet, or by drawing directly on the screen with a light pen.

computerized axial tomography medical technique, usually known as ◊CAT scan, for noninvasive investigation of disease or injury.

computer program coded instructions for a computer; see ◊program.

computer simulation representation of a real-life situation in a computer program. For example, the program might simulate the flow of customers arriving at a bank. The user can alter variables, such as the number of cashiers on duty, and see the effect.

computer terminal the device whereby the operator communicates with the computer; see ◊terminal.

concave of a surface, curving inwards, or away from the eye. For example, a bowl appears concave when viewed from above. In geometry, a concave polygon is one that has an interior angle greater than 180°. Concave is the opposite of ◊convex.

concave lens lens that possesses at least one surface that curves inwards. It is a ◊converging lens, spreading out those light rays that have been refracted through it. A concave lens is thinner at its centre than at its edges.

concave mirror curved or spherical mirror that reflects light from its inner surface. It may be either circular or parabolic in section. A concave mirror converges light rays to form a reduced, inverted, real image in front, or an enlarged, upright, virtual image seemingly behind it, depending on how close the object is to the mirror.

concentration in chemistry, the amount of a substance (◊solute) present in a specified amount of a solution. Either amount may be specified as a mass or a volume (liquids only). Common units used are ◊moles per cubic decimetre, grams per cubic decimetre, grams per 100 cubic centimetres, and grams per 100 grams.

concentration gradient change in the concentration of a substance from one area to another. Particles, such as sugar molecules, in a fluid move over time so that they become evenly distributed throughout the fluid. They seem to move from an area of high concentration to an area of low concentration; that is, they diffuse along the concentration gradient (see ◊diffusion).

conceptacle flask-shaped cavities found in the swollen tips of certain brown seaweeds, notably the wracks, *Fucus*. The gametes are formed within them and released into the water via a small pore in the conceptacle, known as an ostiole.

Concorde the only supersonic airliner, which cruises at Mach 2, or twice the speed of sound, about 2,170 kph/1,350 mph. Concorde, the result of Anglo–French cooperation, made its first flight 1969 and entered commercial service seven years later. It is 62 m/202 ft long and has a wing span of nearly 26 m/84 ft.

concrete building material composed of cement, stone, sand, and water. It has been used since Egyptian and Roman times. Since the late 19th century, it has been increasingly employed as an economical alternative to materials such as brick and wood, and has been combined with steel to increase its tension capacity.

Reinforced concrete is strengthened by combining concrete with another material, such as steel rods or glass fibres. The addition of carbon fibres to concrete increases its conductivity. The electrical resistance of the concrete changes with increased stress or fracture, so this 'smart concrete' can be used as an early indicator of structural damage.

concurrent lines two or more lines passing through a single point; for example, the diameters of a circle are all concurrent at the centre of the circle.

condensation in organic chemistry, a reaction in which two organic compounds combine to form a larger molecule, accompanied by the removal of a smaller molecule (usually water). This is also known as an addition–elimination reaction. Polyamides (such as nylon) and polyesters (such as Terylene) are made by condensation ◊polymerization.

condensation conversion of a vapour to a liquid. This is frequently achieved by letting the vapour come into contact with a cold surface. It is an essential step in ◊distillation processes, and is the process by which water vapour turns into fine water droplets to form ◊cloud.

condensation number in physics, the ratio of the number of molecules condensing on a surface to the total number of molecules touching that surface.

condensation polymerization ◊polymerization reaction in which one or more monomers, with more than one reactive functional group, combine to form a polymer with the elimination of water or another small molecule.

condenser in optics, a ◊lens or combination of lenses with a short focal length used for concentrating a light source onto a small area, as used in a slide projector or microscope substage lighting unit. A condenser can also be made using a concave mirror.

condenser laboratory apparatus used to condense vapours back to liquid so that the liquid can be recovered. It is used in ◊distillation and in reactions where the liquid mixture can be kept boiling without the loss of solvent.

condenser in electronic circuits, another name for a ◊capacitor.

conditioning in psychology, two major principles of behaviour modification.

In *classical conditioning*, described by Ivan Pavlov, a new stimulus can evoke an automatic response by being repeatedly associated with a stimulus that naturally provokes that response. For example, the sound of a bell repeatedly associated with food will eventually trigger salivation, even if sounded without food being presented. In *operant conditioning*, described by US psychologists Edward Lee Thorndike (1874–1949) and B F Skinner, the frequency of a voluntary response can be increased by following it with a reinforcer or reward.

condom or *sheath* barrier contraceptive, made of rubber, which fits over an erect penis and holds in the sperm produced by ejaculation. It is an effective means of preventing pregnancy if used carefully, preferably with a ◊spermicide. A condom with spermicide is 97% effective; one without spermicide is 85% effective as a contraceptive. Condoms also give protection against sexually transmitted diseases, including AIDS.

conductance ability of a material to carry an electrical current, usually given the symbol G. For a direct current, it is the reciprocal of ◊resistance: a conductor of resistance R has a conductance of $1/R$. For an alternating current, conductance is the resistance R divided by the ◊impedance Z: $G = R/Z$. Conductance was formerly expressed in reciprocal ohms (or mhos); the SI unit is the siemens (S).

conduction, electrical flow of charged particles through a material that gives rise to electric current. Conduction in metals involves the flow of negatively charged free ◊electrons. Conduction in gases and some liquids involves the flow of ◊ions that carry positive charges in one direction and negative charges in the other.

conduction, heat flow of heat energy through a material without the movement of any part of the material itself (compare ◊conduction, electrical). Heat energy is present in all materials in the form of the kinetic energy of their vibrating molecules, and may be conducted from one molecule to the next in the form of this mechanical vibration. In the case of metals, which are particularly good conductors of heat, the free electrons within the material carry heat around very quickly.

conductor any material that conducts heat or electricity (as opposed to an insulator, or nonconductor). A good conductor has a high electrical or heat conductivity, and is generally a substance rich in free electrons such as a metal. A poor conductor (such as the non-metals glass and porcelain) has few free electrons. Carbon is exceptional in being nonmetallic and yet (in some of its forms) a relatively good conductor of heat and electricity. Substances such as silicon and germanium, with intermediate conductivities that are improved by heat, light, or voltage, are known as ◊semiconductors.

cone in botany, the reproductive structure of the conifers and cycads; also known as a

volume = $\frac{1}{3}\pi r^2 h$

surface area = $\pi r(l+r)$

cone *The volume and surface area of a cone are given by formulae involving a few simple dimensions.*

◊strobilus. It consists of a central axis surrounded by numerous, overlapping, scalelike, modified leaves (sporophylls) that bear the reproductive organs. Usually there are separate male and female cones, the former bearing pollen sacs containing pollen grains, and the larger female cones bearing the ovules that contain the ova or egg cells. The pollen is carried from male to female cones by the wind (◊anemophily). The seeds develop within the female cone and are released as the scales open in dry atmospheric conditions, which favour seed dispersal.

cone in geometry, a solid or surface consisting of the set of all straight lines passing through a fixed point (the vertex) and the points of a circle or ellipse whose plane does not contain the vertex.

configuration in chemistry, the arrangement of atoms in a molecule or of electrons in atomic orbitals.

configuration in computing, the way in which a system, whether it be ◊hardware and/or ◊software, is set up. A minimum configuration is often referred to for a particular application, and this will usually include a specification of processor, disc and memory size, and peripherals required.

congenital disease in medicine, a disease that is present at birth. It is not necessarily

conic section

conglomerate in mineralogy, coarse clastic ◊sedimentary rock, composed of rounded fragments (clasts) of pre-existing rocks cemented in a finer matrix, usually sand.

conic section curve obtained when a conical surface is intersected by a plane. If the intersecting plane cuts both extensions of the cone, it yields a ◊hyperbola; if it is parallel to the side of the cone, it produces a ◊parabola. Other intersecting planes produce ◊circles or ◊ellipses.

conidium (plural *conidia*) asexual spore formed by some fungi at the tip of a specialized ◊hypha or conidiophore. The conidiophores grow erect, and cells from their ends round off and separate into conidia, often forming long chains. Conidia easily become detached and are dispersed by air movements.

conifer a tree or shrub of the order Coniferales, in the gymnosperm or naked-seedbearing group of plants. They are often pyramidal in form, with leaves that are either scaled or made up of needles; most are evergreen. Conifers include pines, spruces, firs, yews, junipers, monkey puzzles, and larches.

The reproductive organs are the male and female cones, and pollen is distributed by the wind. The seeds develop in the female cones. The processes of maturation, fertilization, and seed ripening may extend over several years. Most conifers grow quickly and can tolerate poor soil, steep slopes, and short growing seasons.

conjugation in biology, the bacterial equivalent of sexual reproduction. A fragment of the ◊DNA from one bacterium is passed along a thin tube, the pilus, into another bacterium.

conjugation in organic chemistry, the alternation of double (or triple) and single carbon–carbon bonds in a molecule – for example, in penta-1,3-diene, $H_2C=CH-CH=CH-CH_3$. Conjugation imparts additional stability as the double bonds are less reactive than isolated double bonds.

conjunction in astronomy, the alignment of two celestial bodies as seen from Earth. A ◊superior planet (or other object) is in conjunction when it lies behind the Sun. An ◊inferior planet (or other object) comes to *inferior conjunction* when it passes between the Earth and the Sun; it is at *superior conjunction* when it passes behind the Sun. *Planetary conjunction* takes place when a planet is closely aligned with another celestial object, such as the Moon, a star, or another planet.

conjunctiva membrane covering the front of the vertebrate ◊eye. It is continuous with the epidermis of the eyelids, and lies on the surface of the cornea.

connective tissue in animals, tissue made up of a noncellular substance, the ◊extracellular matrix, in which some cells are embedded. Skin, bones, tendons, cartilage, and adipose tissue (fat) are the main connective tissues. There are also small amounts of connective tissue in organs such as the brain and liver, where they maintain shape and structure.

conservation in the life sciences, action taken to protect and preserve the natural world, usually from pollution, overexploitation, and other harmful features of human activity. The late 1980s saw a great increase in public concern for the environment, with membership of conservation groups, such as Friends of the Earth, the Sierra Club, and the Nature Conservancy rising sharply. Globally the most important issues include the depletion of atmospheric ozone by the action of chlorofluorocarbons (CFCs), the build-up of carbon dioxide in the atmosphere (thought to contribute to an intensification of the ◊greenhouse effect), and ◊deforestation.

conservation of energy in chemistry, the principle that states that in a chemical reaction, the total amount of energy in the system remains unchanged.

conservation of mass in chemistry, the principle that states that in a chemical reaction the sum of all the masses of the substances involved in the reaction (reactants) is equal to the sum of all of the masses of the substances produced by the reaction (products) – that is, no matter is gained or lost.

conservation of momentum in mechanics, a law that states that total ◊momentum is conserved (remains constant) in all collisions, providing no external resultant force acts on the colliding bodies. The principle may be expressed as an equation used to solve numerical problems: total momentum before collision = total momentum after collision.

console in computing, a combination of keyboard and screen (also described as a terminal). For a multiuser system, such as ◊Unix, there is only one system console from which the system can be administered, while there may be many user terminals.

constellations

constellation	popular name	constellation	popular name
Andromeda	–	Lacerta	Lizard
Antlia	Airpump	Leo	Lion
Apus	Bird of Paradise	Leo Minor	Little Lion
Aquarius	Water-bearer	Lepus	Hare
Aquila	Eagle	Libra	Balance
Ara	Altar	Lupus	Wolf
Aries	Ram	Lynx	–
Auriga	Charioteer	Lyra	Lyre
Boötes	Herdsman	Mensa	Table
Caelum	Chisel	Microscopium	Microscope
Camelopadalis	Giraffe	Monoceros	Unicorn
Cancer	Crab	Musca	Southern Fly
Canes Venatici	Hunting Dogs	Norma	Rule
Canis Major	Great Dog	Octans	Octant
Canis Minor	Little Dog	Ophiuchus	Serpent-bearer
Capricornus	Sea-goat	Orion	–
Carina	Keel	Pavo	Peacock
Cassiopeia	–	Pegasus	Flying Horse
Centaurus	Centaur	Perseus	–
Cepheus	–	Phoenix	Phoenix
Cetus	Whale	Pictor	Painter
Chamaeleon	Chameleon	Pisces	Fishes
Circinus	Compasses	Piscis Austrinus	Southern Fish
Columba	Dove	Puppis	Poop
Coma Berenices	Bernice's Hair	Pyxis	Compass
Corona Australis	Southern Crown	Reticulum	Net
Corona Borealis	Northern Crown	Sagitta	Arrow
Corvus	Crow	Sagittarius	Archer
Crater	Cup	Scorpius	Scorpion
Crux	Southern Cross	Sculptor	–
Cygnus	Swan	Scutum	Shield
Delphinus	Dolphin	Serpens	Serpent
Dorado	Goldfish	Sextans	Sextant
Draco	Dragon	Taurus	Bull
Equuleus	Foal	Telescopium	Telescope
Eridanus	River	Triangulum	Triangle
Fornax	Furnace	Triangulum Australe	Southern Triangle
Gemini	Twins	Tuscana	Toucan
Grus	Crane	Ursa Major	Great Bear
Hercules	–	Ursa Minor	Little Bear
Horologium	Clock	Vela	Sails
Hydra	Watersnake	Virgo	Virgin
Hydrus	Little Snake	Volans	Flying Fish
Indus	Indian	Vulpecula	Fox

constant in mathematics, a fixed quantity or one that does not change its value in relation to ◊variables. For example, in the algebraic expression $y^2 = 5x - 3$, the numbers 3 and 5 are constants. In physics, certain quantities are regarded as universal constants, such as the speed of light in a vacuum.

constantan or *eureka* high-resistance alloy of approximately 40% nickel and 60% copper with a very low coefficient of ◊thermal expansion (measure of expansion on heating). It is used in electrical resistors.

constant composition, law of in chemistry, the law that states that the proportions of the amounts of the elements in a pure compound are always the same and are independent of the method by which the compound was produced.

constellation one of the 88 areas into which the sky is divided for the purposes of identifying and naming celestial objects. The first constellations were simple, arbitrary patterns of stars in which early civilizations visualized gods, sacred beasts, and mythical heroes.

contact force force or push produced when two objects are pressed together and their surface atoms try to keep them apart. Contact forces always come in pairs – for example, the downwards force exerted on a floor by the sole of a person's foot is matched by an equal upwards force exerted by the floor on that sole.

contact process the main industrial method of manufacturing the chemical ◊sulphuric acid. Sulphur dioxide (produced by burning sulphur) and air are passed over a hot (450°C) ◊catalyst of vanadium(V) oxide. The sulphur trioxide produced is absorbed in concentrated sulphuric acid to make fuming sulphuric acid (oleum), which is then diluted with water to give concentrated sulphuric acid (98%). Unreacted gases are recycled.

continental drift in geology, the theory that, about 250–200 million years ago, the Earth consisted of a single large continent (◊Pangaea), which subsequently broke apart to form the continents known today. The theory was proposed 1912 by German meteorologist Alfred Wegener, but such vast continental movements could not be satisfactorily explained until the study of ◊plate tectonics in the 1960s.

continuous data data that can take any of an infinite number of values between whole numbers and so may not be measured completely accurately. This type of data contrasts with ◊discrete data, in which the variable can only take one of a finite set of values. For example, the sizes of apples on a tree form continuous data, whereas the numbers of apples form discrete data.

continuous variation the slight difference of an individual character, such as height, across a sample of the population. Although there are very tall and very short humans, there are also many people with an intermediate height. The same applies to weight. Continuous variation can result from the genetic make-up of a population, or from environmental influences, or from a combination of the two.

contraceptive any drug, device, or technique that acts to prevent pregnancy. The contraceptive pill (the ◊Pill) contains female hormones that interfere with egg production or the first stage of pregnancy. The 'morning-after' pill can be taken up to 72 hours after unprotected intercourse. Barrier contraceptives include ◊condoms (sheaths) and ◊diaphragms, also called caps or Dutch caps; they prevent the sperm entering the cervix (neck of the womb).

Upper Carboniferous period

Eocene

Lower Quaternary

continental drift *The drifting continents. The continents are slowly shifting their positions, driven by fluid motion beneath the Earth's crust. Over 200 million years ago, there was a single large continent called Pangaea. By 200 million years ago, the continents had started to move apart. By 50 million years ago, the continents were approaching their present positions.*

◊Intrauterine devices, also known as IUDs or coils, cause a slight inflammation of the lining of the womb; this prevents the fertilized egg from becoming implanted. See also ◊Norplant.

contractile root in botany, a thickened root at the base of a corm, bulb, or other organ that helps position it at an appropriate level in the ground. Contractile roots are found, for example, on the corms of plants of the genus *Crocus*. After they have become anchored in the soil, the upper portion contracts, pulling the plant deeper into the ground.

contractile vacuole tiny organelle found in many single-celled fresh-water organisms. It slowly fills with water, and then contracts, expelling the water from the cell.

control in biology, the process by which a tissue, an organism, a population, or an ecosystem maintains itself in a balanced, stable state. Blood sugar must be kept at a stable level if the brain is to function properly, and this steady-state is maintained by an interaction between the liver, the hormone insulin, and a detector system in the pancreas.

control experiment essential part of a scientifically valid experiment, designed to show that the factor being tested is actually responsible for the effect observed. In the control experiment all factors, apart from the one under test, are exactly the same as in the test experiments, and all the same measurements are carried out. In drug trials, a placebo (a harmless substance) is given alongside the substance being tested in order to compare effects.

control total in computing, a ◊validation check in which an arithmetic total of a specific field from a group of records is calculated. This total is input together with the data to which it refers. The program recalculates the control total and compares it with the one entered to ensure that no entry errors have been made.

control unit the component of the ◊central processing unit that decodes, synchronizes, and executes program instructions.

convection heat energy transfer that involves the movement of a fluid (gas or liquid). According to ◊kinetic theory, molecules of fluid in contact with the source of heat expand and tend to rise within the bulk of the fluid. Less energetic, cooler molecules sink to take their place, setting up convection currents. This is the principle of natural convection in many domestic hot-water systems and space heaters.

conventional current direction in which an electric current is considered to flow in a circuit. By convention, the direction is that in which positive-charge carriers would flow – from the positive terminal of a cell to its negative terminal. In circuit diagrams, the arrows shown on symbols for components such as diodes and transistors point in the direction of conventional current flow.

convergence in mathematics, the property of a series of numbers in which the difference between consecutive terms gradually decreases. The sum of a converging series approaches a limit as the number of terms tends to ◊infinity.

convergent evolution the independent evolution of similar structures in species (or other taxonomic groups) that are not closely related, as a result of living in a similar way. Thus, birds and bats have wings, not because they are descended from a common winged ancestor, but because their respective ancestors independently evolved flight.

converging lens lens that converges or brings to a focus those light rays that have been refracted by it. It is a ◊convex lens, with at least one surface that curves outwards, and is thicker towards the centre than at the edge. Converging lenses are used to form real images in many ◊optical instruments, such as cameras and projectors. A converging lens that forms a virtual, magnified image may be used as a ◊magnifying glass or to correct ◊long-sightedness.

convertiplane ◊vertical takeoff and landing craft (VTOL) with rotors on its wings that spin horizontally for takeoff, but tilt to spin in a vertical plane for forward flight.

convex of a surface, curving outwards, or towards the eye. For example, the outer surface of a ball appears convex. In geometry, the term is used to describe any polygon possessing no interior angle greater than 180°. Convex is the opposite of ◊concave.

convex lens lens that possesses at least one surface that curves outwards. It is a ◊converging lens, bringing rays of light to a focus. A convex lens is thicker at its centre than at its edges.

convex mirror curved or spherical mirror that reflects light from its outer surface. It diverges reflected light rays to form a reduced, upright, virtual image. Convex mirrors give a wide field of view and are therefore particularly suitable for surveillance purposes in shops.

coordinate in geometry, a number that defines the position of a point relative to a point or axis (reference line). ◊Cartesian coordinates define a point by its perpendicular distances from two or more axes drawn through a fixed point mutually at right angles to each other. ◊Polar coordinates define a point in a plane by its distance from a fixed point and direction from a fixed line.

coordinate geometry or *analytical geometry* system of geometry in which points, lines, shapes, and surfaces are represented by algebraic expressions. In plane (two-dimensional) coordinate geometry, the plane is usually defined by two axes at right angles to each other, the horizontal x-axis and the vertical y-axis, meeting at O, the origin. A point on the plane can be represented by a pair of ◊Cartesian coordinates, which define its position in terms of its distance along the x-axis and along the y-axis from O. These distances are respectively the x and y coordinates of the point.

Lines are represented as equations; for example, $y = 2x + 1$ gives a straight line, and $y = 3x^2 + 2x$ gives a ◊parabola (a curve). The graphs of varying equations can be drawn by plotting the coordinates of points that satisfy their equations, and joining up the points. One of the advantages of coordinate geometry is that geometrical solutions can be obtained without drawing but by manipulating algebraic expressions. For example, the coordinates of the point of intersection of two straight lines can be determined by finding the unique values of x and y that satisfy both of the equations for the lines, that is, by solving them as a pair of ◊simultaneous equations. The curves studied in simple coordinate geometry are the ◊conic sections (circle, ellipse, parabola, and hyperbola), each of which has a characteristic equation.

copepod ◊crustacean of the subclass Copepoda, mainly microscopic and found in plankton.

copper orange-pink, very malleable and ductile, metallic element, symbol Cu (from Latin *cuprum*), atomic number 29, relative atomic mass 63.546. It is used for its durability, pliability, high thermal and electrical conductivity, and resistance to corrosion.

copper(II) carbonate $CuCO_3$ green solid that readily decomposes to form black copper(II) oxide on heating.

$$CuCO_3 \rightarrow CuO + CO_2$$

It dissolves in dilute acids to give blue solutions with effervescence caused by the giving off of carbon dioxide.

$$CuCO_3 + H_2SO_4 \rightarrow CuSO_4 + CO_2 + H_2O$$

copper ore any mineral from which copper is extracted, including native copper, Cu; chalcocite, Cu_2S; chalcopyrite, $CuFeS_2$; bornite, Cu_5FeS_4; azurite, $Cu_3(CO_3)_2(OH)_2$; malachite, $Cu_2CO_3(OH)_2$; and chrysocolla, $CuSiO_3.2H_2O$.

copper(II) oxide CuO black solid that is readily reduced to copper by carbon, carbon monoxide, or hydrogen if heated with any of these.

$$CuO + C \rightarrow Cu + CO$$
$$CuO + CO \rightarrow Cu + CO_2$$
$$CuO + H_2 \rightarrow Cu + H_2O$$

It is usually made in the laboratory by heating copper(II) carbonate, nitrate, or hydroxide.

$$2Cu(NO_3)_2 \rightarrow 2CuO + 4NO_2 + O_2$$

Copper(II) oxide is a typical basic oxide, dissolving readily in most dilute acids.

copper(II) sulphate $CuSO_4$ substance usually found as a blue, crystalline, hydrated salt $CuSO_4.5H_2O$ (also called blue vitriol). It is made from the action of dilute sulphuric acid on copper(II) oxide, hydroxide, or carbonate.

$$CuO + H_2SO_4 + 4H_2O \rightarrow CuSO_4.5H_2O$$

When the hydrated salt is heated gently it loses its water of crystallization and the blue crystals turn to a white powder. The reverse reaction is used as a chemical test for water.

$$CuSO_4.5H_2O \rightleftharpoons CuSO_4 + 5H_2O$$

coprocessor in computing, an additional ◊processor that works with the main ◊central processing unit to carry out a specific function. The two most common coprocessors are the ***mathematical coprocessor***, used to speed up calculations, and the ***graphic coprocessor***, used to improve the handling of graphics.

copulation act of mating in animals with internal ◊fertilization. Male mammals have a ◊penis or other organ that is used to introduce spermatozoa into the reproductive tract of the female. Most birds transfer sperm by pressing their cloacas (the openings of their reproductive tracts) together.

coral marine invertebrate of the class Anthozoa in the phylum Cnidaria, which also includes sea anemones and jellyfish. It has a skeleton of lime (calcium carbonate) extracted from the surrounding water. Corals exist in warm seas, at moderate depths with sufficient light. Their accumulated skeletons make up large coral reefs and atolls.

cordierite silicate mineral, $(Mg,Fe)_2Al_4Si_5O_{18}$, blue to purplish in colour. It is characteristic of metamorphic rocks formed from clay sediments under conditions of low pressure but moderate temperature; it is the mineral that forms the spots in spotted slate and spotted hornfels.

core in earth science, the innermost part of the Earth. It is divided into an inner core, the upper boundary of which is 1,700 km/1,060 mi from the centre, and an outer core, 1,820 km/1,130 mi thick. Both parts are thought to consist of iron-nickel alloy, with the inner core being solid and the outer core being semisolid. The temperature may be 3,000°C/5,400°F.

core in archaeology, a solid cylinder of sediment or soil collected with a coring device and used to evaluate the geological context and stratigraphy of archaeological material or to obtain palaeobotanical samples. Core can also mean the tool used to extract a core sample from the ground, or a stone blank from which flakes or blades are removed.

Coriolis effect the effect of the Earth's rotation on the atmosphere and on all objects on the Earth's surface. In the northern hemisphere it causes moving objects and currents to be deflected to the right; in the southern hemisphere it causes deflection to the left. The effect is named after its discoverer, French mathematician Gaspard Coriolis (1792–1843).

cork light, waterproof outer layers of the bark of the stems and roots of almost all trees and shrubs. The cork oak *Quercus suber*, a native of S Europe and N Africa, is cultivated in Spain and Portugal; the exceptionally thick outer layers of its bark provide the cork that is used commercially.

corm short, swollen, underground plant stem, surrounded by protective scale leaves, as seen in the genus *Crocus*. It stores food, provides a means of ◊vegetative reproduction, and acts as a ◊perennating organ.

cornea transparent front section of the vertebrate ◊eye. The cornea is curved and behaves as a fixed lens, so that light entering the eye is partly focused before it reaches the lens.

cornified layer the upper layer of the skin where the cells have died, their cytoplasm being replaced by keratin, a fibrous protein also found in nails and hair. Cornification gives the skin its protective waterproof quality.

corolla collective name for the petals of a flower. In some plants the petal margins are partly or completely fused to form a ***corolla tube***, for example in bindweed *Convolvulus arvensis*.

corona faint halo of hot (about 2,000,000°C/ 3,600,000°F) and tenuous gas around the Sun, which boils from the surface. It is visible at solar ◊eclipses or through a ***coronagraph***, an instrument that blocks light from the Sun's brilliant disc.

Gas flows away from the corona to form the ◊solar wind.

Corona Australis or ***Southern Crown*** constellation of the southern hemisphere, located near the constellation Sagittarius.

Corona Borealis or ***Northern Crown*** constellation of the northern hemisphere, between Hercules and Boötes, traditionally identified with the jewelled crown of Ariadne that was cast into the sky by Bacchus (in Greek mythology). Its brightest star is Alphecca (or Gemma), which is 78 light years from Earth.

corpuscular theory hypothesis about the nature of light championed by Isaac Newton, who postulated that it consists of a stream of particles or corpuscles. The theory was superseded at the beginning of the 19th century by English physicist Thomas Young's wave theory. ◊Quantum theory and wave mechanics embody both concepts.

corpus luteum glandular tissue formed in the mammalian ◊ovary after ovulation from the Graafian follicle, a group of cells associated with bringing the egg to maturity. It secretes the hormone progesterone in anticipation of pregnancy.

After the release of an egg the follicle enlarges under the action of luteinizing hormone, released from the pituitary. The corpus luteum secretes the hormone progesterone, which maintains the uterus wall ready for pregnancy. If pregnancy occurs, the corpus luteum continues to secrete progesterone until the fourth month; if pregnancy does not occur the corpus luteum breaks down.

corrasion the grinding away of solid rock surfaces by particles carried by water, ice, and wind. It is generally held to be the most significant form of ◊erosion. As the eroding particles are carried along they become eroded themselves due to the process of attrition.

correlation the degree of relationship between two sets of information. If one set of data increases at the same time as the other, the relationship is said to be positive or direct. If one set of data increases as the other decreases, the relationship is negative or inverse. Correlation can be shown by plotting a best-fit line on a scatter diagram.

corrosion the eating away and eventual destruction of metals and alloys by chemical attack. The rusting of ordinary iron and steel is the most common form of corrosion. Rusting takes place in moist air, when the iron combines with oxygen and water to form a brown-orange deposit of ◊rust (hydrated iron oxide). The rate of corrosion is increased where the atmosphere is polluted with sulphur dioxide. Salty road and air conditions accelerate the rusting of car bodies.

corrosion in earth science, an alternative name for ◊solution, the process by which water dissolves rocks such as limestone.

cortex in biology, the outer part of a structure such as the brain, kidney, or adrenal gland. In

cosine The cotangent of angle β is equal to the ratio of the length of the adjacent side to the length of the hypotenuse (the longest side, opposite to the right angle).

botany the cortex includes non-specialized cells lying just beneath the surface cells of the root and stem.

corticosteroid any of several steroid hormones secreted by the cortex of the ♦adrenal glands; also synthetic forms with similar properties. Corticosteroids have anti-inflammatory and ♦immunosuppressive effects and may be used to treat a number of conditions, including rheumatoid arthritis, severe allergies, asthma, some skin diseases, and some cancers. Side effects can be serious, and therapy must be withdrawn very gradually.

corundum native aluminium oxide, Al_2O_3, the hardest naturally occurring mineral known apart from diamond (corundum rates 9 on the Mohs' scale of hardness); lack of ♦cleavage also increases its durability. Its crystals are barrel-shaped prisms of the trigonal system. Varieties of gem-quality corundum are *ruby* (red) and *sapphire* (any colour other than red, usually blue). Poorer-quality and synthetic corundum is used in industry, for example as an ♦abrasive.

cosecant in trigonometry, a ♦function of an angle in a right-angled triangle found by dividing the length of the hypotenuse (the longest side) by the length of the side opposite the angle. Thus the cosecant of an angle A, usually shortened to cosec A, is always greater than (or equal to) 1. It is the reciprocal of the sine of the angle, that is, cosec $A = 1/\sin A$.

cosine in trigonometry, a ♦function of an angle in a right-angled triangle found by dividing the length of the side adjacent to the angle by the length of the hypotenuse (the longest side). It is usually shortened to *cos*.

cosine rule in trigonometry, a rule that relates the sides and angles of triangles. The rule has the formula:

$$a^2 = b^2 + c^2 - 2bc \cos A$$

where a, b, and c are the sides of the triangle, and A is the angle opposite a.

cosmic background radiation or *3° radiation* electromagnetic radiation left over from the original formation of the universe in the Big Bang around 15 billion years ago. It corresponds to an overall background temperature of 3K (–270°C/–454°F), or 3°C above absolute zero. In 1992 the Cosmic Background Explorer satellite, COBE, detected slight 'ripples' in the strength of the background radiation that are believed to mark the first stage in the formation of galaxies.

cosmic radiation streams of high-energy particles from outer space, consisting of protons, alpha particles, and light nuclei, which collide with atomic nuclei in the Earth's atmosphere, and produce secondary nuclear particles (chiefly ♦mesons, such as pions and muons) that shower the Earth.

Those of low energy seem to be galactic in origin, and those of high energy of extragalactic origin. The galactic particles may come from ♦supernova explosions or ♦pulsars. At higher energies, other sources are necessary, possibly the giants jets of gas which are emitted from some galaxies.

cosmid fragment of ♦DNA from the human genome inserted into a bacterial cell. The bacterium replicates the fragment along with its own DNA. In this way the fragments are copied for a gene library. Cosmids are characteristically 40,000 base pairs in length. The

the cosine rule states that
$a^2 = b^2 + c^2 - 2bc \cos A$

cosine rule

most commonly used bacterium is *Escherichia coli*. A ◊yeast artificial chromosome works in the same way.

cosmogony study of the origin and evolution of cosmic objects, especially the Solar System.

cosmological principle in astronomy, a hypothesis that any observer anywhere in the ◊universe has the same view that we have; that is, that the universe is not expanding from any centre but all galaxies are moving away from one another.

cosmology study of the structure of the universe. Modern cosmology began in the 1920s with the discovery that the universe is expanding, which suggested that it began in an explosion, the ◊Big Bang. An alternative, now discarded view, the ◊steady-state theory, claimed that the universe has no origin, but is expanding because new matter is being continually created.

cosmonaut term used in the West for any astronaut from the former Soviet Union.

Cosmos name used from the early 1960s for nearly all Soviet artificial satellites. Over 2,200 Cosmos satellites had been launched by the end of 1992.

cotangent in trigonometry, a ◊function of an angle in a right-angled triangle found by dividing the length of the side adjacent to the angle by the length of the side opposite it. It is usually written as cotan, or cot and it is the reciprocal of the tangent of the angle, so that cot $A = 1/\tan A$, where A is the angle in question.

cotyledon structure in the embryo of a seed plant that may form a 'leaf' after germination and is commonly known as a seed leaf. The number of cotyledons present in an embryo is an important character in the classification of flowering plants (◊angiosperms).

Monocotyledons (such as grasses, palms, and lilies) have a single cotyledon, whereas dicotyledons (the majority of plant species) have two. In seeds that also contain ◊endosperm (nutritive tissue), the cotyledons are thin, but where they are the primary food-storing tissue, as in peas and beans, they may be quite large. After germination the cotyledons either remain below ground (hypogeal) or, more commonly, spread out above soil level (epigeal) and become the first green leaves. In gymnosperms there may be up to a dozen cotyledons within each seed.

coulomb SI unit (symbol C) of electrical charge. One coulomb is the quantity of electricity conveyed by a current of one ◊ampere in one second.

count rate in physics, the number of particles emitted per unit time by a radioactive source. It is measured by a counter, such as a ◊Geiger counter, or ◊ratemeter.

couple in mechanics, a pair of forces acting on an object that are equal in magnitude and opposite in direction, but do not act along the same straight line. The two forces produce a turning effect or moment that tends to rotate the object; however, no single resultant (unbalanced) force is produced and so the object is not moved from one position to another.

courtship behaviour exhibited by animals as a prelude to mating. The behaviour patterns vary considerably from one species to another, but are often ritualized forms of behaviour not obviously related to courtship or mating (for example, courtship feeding in birds).

covalent bond chemical ◊bond produced when two atoms share one or more pairs of electrons (usually each atom contributes an electron). The bond is often represented by a single line drawn between the two atoms. Covalently bonded substances include hydrogen (H_2), water (H_2O), and most organic substances.

CPU in computing, abbreviation for ◊*central processing unit*.

Crab nebula cloud of gas 6,000 light years from Earth, in the constellation Taurus. It is the remains of a star that exploded as a ◊supernova (observed as a brilliant point of light on Earth 1054). At its centre is a ◊pulsar that flashes 30 times a second. The name comes from its crab-like shape.

cracking reaction in which a large ◊alkane molecule is broken down by heat into a smaller alkane and a small ◊alkene molecule. The reaction is carried out at a high temperature (600°C or higher) and often in the presence of a catalyst. Cracking is a commonly used process in the ◊petrochemical industry.

It is the main method of preparation of alkenes and is also used to manufacture petrol from the higher-boiling-point ◊fractions that are obtained by fractional ◊distillation (fractionation) of crude oil.

crane in engineering, a machine for raising, lowering, or placing in position heavy loads. The three main types are the jib crane, the overhead travelling crane, and the tower crane. Most cranes have the machinery mounted on a revolving turntable. This may be mounted on

two hydrogen atoms

or H ˣ H, H–H
a molecule of hydrogen sharing an electron pair

two hydrogen atoms and one oxygen atom

or H ˣ O ˣ H, H–O–H
a molecule of water showing the two covalent bonds

covalent bond *The formation of a covalent bond between two hydrogen atoms to form a hydrogen molecule (H_2), and between two hydrogen atoms and an oxygen atom to form a molecule of water (H_2O). The sharing means that each atom has a more stable arrangement of electrons (its outer electron shells are full).*

trucks or be self-propelled, often being fitted with caterpillar tracks.

cranium the dome-shaped area of the vertebrate skull that protects the brain. It consists of eight bony plates fused together by sutures (immovable joints). Fossil remains of the human cranium have aided the development of theories concerning human evolution.

crank handle bent at right angles and connected to the shaft of a machine; it is used to transmit motion or convert reciprocating (back-and-forwards or up-and-down) movement into rotary movement, or vice versa.

crankshaft essential component of piston engines that converts the up-and-down (reciprocating) motion of the pistons into useful rotary motion. The car crankshaft carries a number of cranks. The pistons are connected to the cranks by connecting rods and bearings; when the pistons move up and down, the connecting rods force the offset crank pins to describe a circle, thereby rotating the crankshaft.

creationism theory concerned with the origins of matter and life, claiming, as does the Bible in Genesis, that the world and humanity were created by a supernatural Creator, not more than 6,000 years ago. It was developed in response to Darwin's theory of ◊evolution; it is not recognized by most scientists as having a factual basis.

creep in civil and mechanical engineering, the property of a solid, typically a metal, under continuous stress that causes it to deform below its ◊yield point (the point at which any elastic solid normally stretches without any increase in load or stress). Lead, tin, and zinc, for example, exhibit creep at ordinary temperatures, as seen in the movement of the lead sheeting on the roofs of old buildings.

creosote black, oily liquid derived from coal tar, used as a wood preservative. Medicinal creosote, which is transparent and oily, is derived from wood tar.

crescent curved shape of the Moon when it appears less than half-illuminated.

Cretaceous period of geological time 146–65 million years ago. It is the last period of the Mesozoic era, during which angiosperm (seed-bearing) plants evolved, and dinosaurs reached a peak before their almost complete extinction at the end of the period. Chalk is a typical rock type of the second half of the period.

crith unit of mass used for weighing gases. One crith is the mass of one litre of hydrogen gas (H_2) at standard temperature and pressure.

critical angle in optics, for a ray of light passing from a denser to a less dense medium (such as from glass to air), the smallest angle of incidence at which the emergent ray grazes the surface of the denser medium – at an angle of refraction of 90°.

When the angle of incidence is less than the critical angle, the ray passes out (is refracted) into the less dense medium; when the angle of incidence is greater than the critical angle, the ray is reflected back into the denser medium.

critical mass in nuclear physics, the minimum mass of fissile material that can undergo a continuous ◊chain reaction. Below this mass, too many ◊neutrons escape from the surface for a chain reaction to carry on; above the critical mass, the reaction may accelerate into a nuclear explosion.

critical reaction in a nuclear reactor, a self-sustaining chain reaction in which the number of neutrons being released by the fission of uranium-235 nuclei and the number of neutrons being absorbed by uranium-238 nuclei and by control rods are balanced. If balance is not achieved the reaction will either slow down and cease to generate enough power, or will build up and go out of control, as in a nuclear explosion. Control rods are used to adjust the rate of reaction and maintain balance.

critical temperature temperature above which a particular gas cannot be converted into a liquid by pressure alone. It is also the temperature at which a magnetic material loses its magnetism (the Curie temperature or point).

CRO abbreviation for ◊*cathode-ray oscilloscope*.

crop in birds, the thin-walled enlargement of the digestive tract between the oesophagus and stomach. It is an effective storage organ especially in seed-eating birds; a pigeon's crop can hold about 500 cereal grains. Digestion begins in the crop, by the moisturizing of food. A crop also occurs in insects and annelid worms.

crossing over in biology, a process that occurs during ◊meiosis. While the chromosomes are lying alongside each other in pairs, each partner may twist around the other and exchange corresponding chromosomal segments. It is a form of genetic ◊recombination, which increases variation and thus provides the raw material of evolution.

critical angle *The critical angle is the angle at which light from within a transparent medium just grazes the surface of the medium.*

cross-linking in chemistry, lateral linking between two or more long-chain molecules in a ◊polymer. Cross-linking gives the polymer a higher melting point and makes it harder. Examples of cross-linked polymers include Bakelite and vulcanized rubber.

CRT abbreviation for ◊cathode-ray tube.

crude oil the unrefined form of ◊petroleum.

crumple zone region at the front and rear of a motor vehicle that is designed to crumple gradually during a collision, so reducing the risk of serious injury to passengers. The progressive crumpling absorbs the kinetic energy of the vehicle more gradually than a rigid structure would, thereby diminishing the forces of deceleration acting on the vehicle and on the people inside.

crust the outermost part of the structure of Earth, consisting of two distinct parts, the oceanic crust and the continental crust. The *oceanic* crust is on average about 10 km/6.2 mi thick and consists mostly of basaltic types of rock. By contrast, the *continental* crust is largely made of granite and is more complex in its structure. It varies in thickness from about 40 km/25 mi to 70 km/45 mi being deeper beneath mountain ranges. Because of the movements of ◊plate tectonics, the oceanic crust is in no place older than about 200 million years. However, parts of the continental crust are over 3 billion years old.

crustacean one of the class of arthropods that includes crabs, lobsters, shrimps, woodlice, and barnacles. The external skeleton is made of protein and chitin hardened with lime. Each segment bears a pair of appendages that may be modified as sensory feelers (antennae), as mouthparts, or as swimming, walking, or grasping structures.

Crux constellation of the southern hemisphere, popularly known as the Southern Cross, the smallest of the 88 constellations. Its brightest stars are Alpha Crucis (or Acrux), a ◊double star about 400 light years from Earth, and Beta Crucis (or Mimosa). Near Beta Crucis lies a glittering star cluster known as the Jewel Box. The constellation also contains the Coalsack, a dark cloud of dust silhouetted against the bright starry background of the Milky Way.

cryogenics science of very low temperatures (approaching ◊absolute zero), including the production of very low temperatures and the exploitation of special properties associated with them, such as the disappearance of electrical resistance (◊superconductivity).

crystal system

crystal system	minimum symmetry	possible shape	mineral examples
cubic	4 threefold axes	cube, octahedron, dodecahedron	diamond, garnet, pyrite
tetragonal	1 fourfold axis	square-based prism	zircon
orthorhombic	3 twofold axes or mirror planes	matchbox shape	baryte
monoclinic	1 twofold axis or mirror plane	matchbox distorted in one plane	gypsum
triclinic	no axes or mirror planes	matchbox distorted in three planes	plagioclase feldspar
trigonal	1 threefold axis	triangular prism, rhombohedron	calcite, quartz
hexagonal	1 sixfold axis	hexagonal prism	beryl

cryolite rare granular crystalline mineral (sodium aluminium fluoride), Na_3AlF_6, used in the electrolytic reduction of ◊bauxite to aluminium. It is chiefly found in Greenland.

cryptogam obsolete name applied to the lower plants. It included the algae, liverworts, mosses, and ferns (plus the fungi and bacteria in very early schemes of classification). In such classifications seed plants were known as ◊phanerogams.

cryptography science of creating and reading codes; for example, those produced by the Enigma coding machine used by the Germans in World War II and those used in commerce by banks encoding electronic fund-transfer messages, business firms sending computer-conveyed memos between headquarters, and in the growing field of electronic mail. No method of encrypting is completely unbreakable, but decoding can be made extremely complex and time consuming.

cryptosporidium waterborne parasite which causes disease in humans and other animals. It has been found in drinking water in the UK and USA, causing diarrhoea, abdominal cramps, vomiting, and fever, and can be fatal in people with damaged immune systems, such as AIDS sufferers or those with leukaemia.

crystal substance with an orderly three-dimensional arrangement of its atoms or molecules, thereby creating an external surface of clearly defined smooth faces having characteristic angles between them. Examples are table salt and quartz.

Each geometrical form, many of which may be combined in one crystal, consists of two or more faces – for example, dome, prism, and pyramid. A mineral can often be identified by the shape of its crystals and the system of crystallization determined. A single crystal can vary in size from a submicroscopic particle to a mass some 30 m/100 ft in length. Crystals fall into seven crystal systems or groups, classified on the basis of the relationship of three or four imaginary axes that intersect at the centre of any perfect, undistorted crystal.

crystallization formation of crystals from a liquid, gas, or solution.

crystallography the scientific study of crystals. In 1912 it was found that the shape and size of the repeating atomic patterns (unit cells) in a crystal could be determined by passing X-rays through a sample. This method, known as ◊X-ray diffraction, opened up an entirely new way of 'seeing' atoms. It has been found that many substances have a unit cell that exhibits all the symmetry of the whole crystal; in table salt (sodium chloride, NaCl), for instance, the unit cell is an exact cube.

crystal system all known crystalline substances crystallize in one of the seven crystal systems defined by symmetry. The elements of symmetry used for this purpose are: (1) planes of *mirror symmetry*, across which a mirror image is seen, and (2) axes of *rotational symmetry*, about which, in a 360° rotation of the crystal, equivalent faces are seen twice, three, four, or six times. To be assigned to a particular crystal system, a mineral must possess a certain minimum symmetry, but it may also possess additional symmetry elements. Since crystal symmetry is related to internal structure, a given mineral will always crystallize in the same system, although the crystals may not always grow into precisely the same shape. In cases where two minerals have the same chemical composition (for example graphite and diamond, or quartz and cristobalite), they will generally have different crystal systems because their internal structures are different.

CT-scan another version of ◊CAT scan.

CT scanner or *CAT scanner* medical device used to obtain detailed X-ray pictures of the inside of a patient's body.

cu abbreviation for *cubic* (measure).

cube in geometry, a regular solid figure whose faces are all squares. It has six equal-area faces and 12 equal-length edges.

cube to multiply a number by itself and then by itself again. For example, 5 cubed = 5^3 = 5 × 5 × 5 = 125. The term also refers to a number formed by cubing; for example, 1, 8, 27, 64 are the first four cubes.

cubic centimetre (or metre) the metric measure of volume, corresponding to the volume of a cube whose edges are all 1 cm (or 1 metre) in length.

cubic decimetre metric measure (symbol dm^3) of volume corresponding to the volume of a cube whose edges are all 1 dm (10 cm) long; it is equivalent to a capacity of one litre.

cubic equation any equation in which the largest power of × is x^3. For example, $x^3 + 3x^2y + 4y^2 = 0$ is a cubic equation.

cubit earliest known unit of length, which originated between 2800 and 2300 BC. It is approximately 50.5 cm/20.6 in long, which is about the length of the human forearm measured from the tip of the middle finger to the elbow.

cultivar variety of a plant developed by horticultural or agricultural techniques. The term derives from *culti*vated *var*iety.

culture in biology, the growing of living cells and tissues in laboratory conditions.

cumulative frequency in statistics, the total frequency of a given value up to and including a certain point in a set of data. It is used to draw the cumulative frequency curve, the ogive.

cuprite Cu_2O ore (copper(I) oxide), found in crystalline form or in earthy masses. It is red to black in colour, and is often called ruby copper.

cupronickel copper alloy (75% copper and 25% nickel), used in hardware products and for coinage.

curie former unit (symbol Ci) of radioactivity, equal to 37×10^9 ◊becquerels. One gram of radium has a radioactivity of about one curie. It was named after French physicist Pierre Curie.

Curie temperature the temperature above which a magnetic material cannot be strongly magnetized. Above the Curie temperature, the energy of the atoms is too great for them to join together to form the small areas of magnetized material, or ◊domains, which combine to produce the strength of the overall magnetization.

curium synthesized, radioactive, metallic element of the actinide series, symbol Cm, atomic number 96, relative atomic mass 247. It is produced by bombarding plutonium or americium with neutrons. Its longest-lived isotope has a half-life of 1.7×10^7 years.

current, electric see ◊electric current.

cuticle the horny noncellular surface layer of many invertebrates such as insects; in botany, the waxy surface layer on those parts of plants that are exposed to the air, continuous except for ◊stomata and ◊lenticels. All types are secreted by the cells of the ◊epidermis. A cuticle reduces water loss and, in arthropods, acts as an ◊exoskeleton.

cutting technique of vegetative propagation involving taking a section of root, stem, or leaf and treating it so that it develops into a new plant.

cwt symbol for ◊*hundredweight,* a unit of weight equal to 112 pounds (50.802 kg); 100 lb (45.36 kg) in the USA.

cyanamide process process used in the manufacture of calcium cyanamide ($CaCN_2$), a colourless crystalline powder used as a fertilizer under the tradename Nitrolime. Calcium carbide is reacted with nitrogen in an electric furnace.

$$CaC_2 + N_2 \rightarrow CaCN_2 + C$$

The calcium cyanamide reacts with water in the soil to form the ammonium ion and calcium carbonate. The ammonium is then oxidized to nitrate, which is taken up by plants. Calcium cyanamide can also be converted commercially into ammonia.

cyanide CN^- ion derived from hydrogen cyanide (HCN), and any salt containing this ion (produced when hydrogen cyanide is neutralized by alkalis), such as potassium cyanide (KCN). The principal cyanides are potassium, sodium, calcium, mercury, gold, and copper. Certain cyanides are poisons.

cyanobacteria (singular *cyanobacterium*) alternative name for ◊blue-green algae.

cyanocobalamin chemical name for ◊vitamin B_{12}, which is normally produced by microorganisms in the gut. The richest sources are liver, fish and eggs. It is essential to the replacement of cells, the maintenance of the myelin sheath which insulates nerve fibres, and the efficient use of folic acid, another vitamin in the B complex. Deficiency can result in pernicious anaemia (defective production of red blood cells), and possible degeneration of the nervous system.

cybernetics science concerned with how systems organize, regulate, and reproduce themselves, and also how they evolve and learn. In the laboratory, inanimate objects are created that behave like living systems. Applications range from the creation of electronic artificial limbs to the running of the fully automated factory where decision-making machines operate up to managerial level.

cyberspace the imaginary, interactive 'worlds' created by computers; often used interchangeably with virtual world or ◊virtual reality. The invention of the word 'cyberspace' is generally credited to US science-fiction writer William Gibson (1948–) in his first novel *Neuromancer* 1984.

cyclamate derivative of cyclohexysulphamic acid, formerly used as an artificial ◊sweetener, 30 times sweeter than sugar. It was first synthesized 1937.

cycle in physics, a sequence of changes that moves a system away from, and then back to, its original state. An example is a vibration that moves a particle first in one direction and then in the opposite direction, with the particle returning to its original position at the end of the vibration.

cyclic compound any of a group of organic chemicals that have rings of atoms in their molecules, giving them a closed-chain structure.

They may be alicyclic (cyclopentane), aromatic (benzene), or heterocyclic (pyridine). *Alicyclic compounds* (*ali*phatic *cyclic*) have localized bonding: all the electrons are confined to their own particular bonds, in contrast to *aromatic compounds*, where certain electrons have free movement between different bonds in the ring. Alicyclic compounds have chemical properties similar to their straight-chain counterparts; aromatic compounds, because of their special structure, undergo entirely different chemical reactions. *Heterocyclic compounds* have a ring of carbon atoms with one or more carbons replaced by another element, usually nitrogen, oxygen, or sulphur. They may be aliphatic or aromatic in nature.

cyclone alternative name for a ◊depression, an area of low atmospheric pressure. A severe cyclone that forms in the tropics is called a tropical cyclone or hurricane.

cyclotron circular type of particle ◊accelerator.

Cygnus large prominent constellation of the northern hemisphere, named after its shape (Latin 'swan'). Its brightest star is first-magnitude ◊Deneb.

cylinder in geometry, a tubular solid figure with a circular base. In everyday use, the term applies to a *right cylinder*, the curved surface of which is at right angles to the base.

The volume V of a cylinder is given by the formula $V = \pi r^2 h$, where r is the radius of the base and h is the height of the cylinder. Its total surface area A has the formula $A = 2\pi r(h + r)$, where $2\pi rh$ is the curved surface area, and $2\pi r^2$ is the area of both circular ends.

cytochrome protein responsible for part of the process of ◊respiration by which food molecules are broken down in ◊aerobic organisms. Cytochromes are part of the electron transport chain, which uses energized electrons to reduce molecular oxygen (O_2) to oxygen ions (O^{2-}). These combine with hydrogen ions (H^+) to form water (H_2O), the end product of aerobic respiration. As electrons are passed from one cytochrome to another, energy is released and used to make ◊ATP.

cytokinin ◊plant hormone that stimulates cell division. Cytokinins affect several different aspects of plant growth and development, but only if ◊auxin is also present. They may delay the process of senescence, or ageing, break the dormancy of certain seeds and buds, and induce flowering.

cytology the study of ◊cells and their functions. Major advances have been made possible in this field by the development of ◊electron microscopes.

cytoplasm the part of the cell outside the ◊nucleus. Strictly speaking, this includes all the ◊organelles (mitochondria, chloroplasts, and so on), but often cytoplasm refers to the jellylike matter in which the organelles are embedded (correctly termed the cytosol).

cytoskeleton in a living cell, a matrix of protein filaments and tubules that occurs within the cytosol (the liquid part of the cytoplasm). It gives the cell a definite shape, transports vital substances around the cell, and may also be involved in cell movement.

DAC abbreviation for ◊*digital-to-analogue converter*.

daisywheel printing head in a computer printer or typewriter that consists of a small plastic or metal disc made up of many spokes (like the petals of a daisy). At the end of each spoke is a character in relief. The daisywheel is rotated until the spoke bearing the required character is facing an inked ribbon, then a hammer strikes the spoke against the ribbon, leaving the impression of the character on the paper beneath.

dam structure built to hold back water in order to prevent flooding, to provide water for irrigation and storage, and to provide hydro-electric power. The biggest dams are of the earth-and-rock-fill type, also called ***embankment dams***. Early dams in Britain, built before about 1800, had a core made from puddled clay (clay which has been mixed with water to make it impermeable). Such dams are generally built on broad valley sites. Deep, narrow gorges dictate a ***concrete dam***, where the strength of reinforced concrete can withstand the water pressures involved.

damper any device that deadens or lessens vibrations or oscillations; for example, one used to check vibrations in the strings of a piano. The term is also used for the movable plate in the flue of a stove or furnace for controlling the draught.

darcy c.g.s. unit (symbol D) of permeability, used mainly in geology to describe the permeability of rock (for example, to oil, gas, or water).

dark matter hypothetical matter that, according to current theories of ◊cosmology, makes up 90–99% of the mass of the universe but so far remains undetected. Dark matter, if shown to exist, would explain many currently unexplained gravitational effects in the movement of galaxies.

Astronomers 1993 identified part of the dark matter in the form of stray planets and ◊brown dwarfs, objects known as MACHOs (massive astrophysical compact halo objects).

DAT abbreviation for ◊*digital audio tape*.

data facts, figures, and symbols, especially as stored in computers. The term is often used to mean raw, unprocessed facts, as distinct from information, to which a meaning or interpretation has been applied.

database in computing, a structured collection of data, which may be manipulated to select and sort desired items of information. For example, an accounting system might be built around a database containing details of customers and suppliers. In larger computers, the database makes data available to the various programs that need it, without the need for those programs to be aware of how the data are stored. The term is also sometimes used for simple record-keeping systems, such as mailing lists, in which there are facilities for searching, sorting, and producing records.

data compression in computing, techniques for reducing the amount of storage needed for a given amount of data. They include word tokenization (in which frequently used words are stored as shorter codes), variable bit lengths (in which common characters are represented by fewer ◊bits than less common ones), and run-length encoding (in which a repeated value is stored once along with a count).

dating science of determining the age of geological structures, rocks, and fossils, and placing them in the context of geological time. The techniques are of two types: relative dating and absolute dating. ***Relative dating*** can be carried out by identifying fossils of creatures that lived only at certain times (marker fossils), and by looking at the physical relationships of rocks to other rocks of a known age.

Absolute dating is achieved by measuring how much of a rock's radioactive elements have changed since the rock was formed, using the process of ◊radiometric dating.

day time taken for the Earth to rotate once on its axis. The ***solar day*** is the time that the Earth takes to rotate once relative to the Sun. It is divided into 24 hours, and is the basis of our civil day. The ***sidereal day*** is the time that the Earth takes to rotate once relative to the stars. It is 3 minutes 56 seconds shorter than the solar day, because the Sun's position against the background of stars as seen from Earth changes as the Earth orbits it.

dBASE family of microcomputer programs used for manipulating large quantities of data; also, a related ◊fourth-generation language. The first version, dBASE II, appeared in 1981; it has since become the basis for a recognized standard for database applications, known as Xbase.

DC in physics, abbreviation for ◊*direct current* (electricity).

DCC abbreviation for ◊*digital compact cassette*.

DDT abbreviation for ***dichloro-diphenyl-trichloroethane*** ($ClC_6H_5)_2CHCHCl_2$) insecticide discovered 1939 by Swiss chemist Paul Müller. It is useful in the control of insects that spread malaria, but resistant strains develop. DDT is highly toxic and persists in the environment and in living tissue. Its use is now banned in most countries, but it continues to be used on food plants in Latin America.

deamination removal of the amino group ($-NH_2$) from an unwanted ◊amino acid. This is the nitrogen-containing part, and it is converted into ammonia, uric acid, or urea (depending on the type of animal) to be excreted in the urine. In vertebrates, deamination occurs in the ◊liver.

death cessation of all life functions, so that the molecules and structures associated with living things become disorganized and indistinguishable from similar molecules found in non-living things. Living organisms expend large amounts of energy preventing their complex molecules from breaking up; cellular repair and replacement are vital processes in multicellular organisms. At death this energy is no longer available, and the processes of disorganization become inevitable.

In medicine, death used to be pronounced with the permanent cessation of heartbeat. The advent of life-support equipment has made this point sometimes difficult to determine, and in controversial cases a person is now pronounced dead when the brain ceases to control the vital functions even if breathing and heartbeat are maintained (artificially).

decay, radioactive see ◊radioactive decay.

Decca navigation system radio-based aid to marine navigation, available in many parts of the world. The system consists of a master radio transmitter and two or three secondary transmitters situated within 100–200 km/60–120 mi from the master. The signals from the transmitters are detected by the ship's navigation receiver, and slight differences (phase differences) between the master and secondary transmitter signals indicate the position of the receiver.

decibel unit (symbol dB) of measure used originally to compare sound intensities and subsequently electrical or electronic power outputs; now also used to compare voltages. An increase of 10 dB is equivalent to a 10-fold increase in intensity or power, and a 20-fold increase in voltage. A whisper has an intensity of 20 dB; 140 dB (a jet aircraft taking off nearby) is the threshold of pain.

deciduous of trees and shrubs, that shed their leaves at the end of the growing season or during a dry season to reduce ◊transpiration, the loss of water by evaporation.

Most deciduous trees belong to the ◊angiosperms, plants in which the seeds are enclosed within an ovary, and the term 'deciduous tree' is sometimes used to mean 'angiosperm tree', despite the fact that many angiosperms are evergreen, especially in the tropics, and a few ◊gymnosperms, plants in which the seeds are exposed, are deciduous (for example, larches). The term ***broad-leaved*** is now preferred to 'deciduous' for this reason.

decimal fraction a ◊fraction in which the denominator is any higher power of 10. Thus $3/_{10}$, $51/_{100}$, and $23/_{1,000}$ are decimal fractions and are normally expressed as 0.3, 0.51, 0.023. The use of decimals greatly simplifies addition and multiplication of fractions, though not all fractions can be expressed exactly as decimal fractions.

decimal number system or *denary number system* the most commonly used number system, to the base ten. Decimal numbers do not necessarily contain a decimal point; 563, 5.63, and −563 are all decimal numbers. Other systems are mainly used in computing and include the ◊binary number system, ◊octal number system, and ◊hexadecimal number system.

declination in astronomy, the coordinate on the ◊celestial sphere (imaginary sphere surrounding the Earth) that corresponds to latitude on the Earth's surface. Declination runs from 0° at the celestial equator to 90° at the north and south celestial poles.

decoder in computing, an electronic circuit used to select one of several possible data pathways. Decoders are, for example, used to direct data to individual memory locations within a computer's immediate access memory.

decomposer in biology, any organism that breaks down dead matter. Decomposers play a vital role in the ⟡ecosystem by freeing important chemical substances, such as nitrogen compounds, locked up in dead organisms or excrement. They feed on some of the released organic matter, but leave the rest to filter back into the soil or pass in gas form into the atmosphere. The principal decomposers are bacteria and fungi, but earthworms and many other invertebrates are often included in this group. The ⟡nitrogen cycle relies on the actions of decomposers.

decomposition process whereby a chemical compound is reduced to its component substances. In biology, it is the destruction of dead organisms either by chemical reduction or by the action of decomposers.

decontamination factor in radiological protection, a measure of the effectiveness of a decontamination process. It is the ratio of the original contamination to the remaining radiation after decontamination: 1,000 and above is excellent; 10 and below is poor.

decrepitation unusual features that accompany the thermal decomposition of some crystals, such as lead(II) nitrate. When these are heated, they spit and crackle and may jump out of the test tube before they decompose.

dedicated computer computer built into another device for the purpose of controlling or supplying information to it. Its use has increased dramatically since the advent of the ⟡microprocessor: washing machines, digital watches, cars, and video recorders all now have their own processors.

deficiency disease any disease arising from the lack of an essential nutrient in the diet. For example, ⟡beriberi is due to a deficiency of vitamin B_1 (thiamine); ⟡rickets is due to a lack of vitamin D. See ⟡nutrition.

deforestation destruction of forest for timber, fuel, charcoal burning, and clearing for agriculture and extractive industries, such as mining, without planting new trees to replace those lost (reafforestation) or working on a cycle that allows the natural forest to regenerate. Deforestation causes fertile soil to be blown away or washed into rivers, leading to ⟡soil erosion, drought, flooding, and loss of wildlife. It may also increase the carbon dioxide content of the atmosphere and intensify the ⟡greenhouse effect, because there are fewer trees absorbing carbon dioxide from the air for photosynthesis.

degaussing neutralization of the magnetic field around a body by encircling it with a conductor through which a current is maintained. Ships were degaussed in World War II to prevent them from detonating magnetic mines.

degree in mathematics, a unit (symbol °) of measurement of an angle or arc. A circle or complete rotation is divided into 360°. A degree may be subdivided into 60 minutes (symbol ″), and each minute may be subdivided in turn into 60 seconds (symbol ′).

Temperature is also measured in degrees, which are divided on a decimal scale. See also ⟡Celsius, and ⟡Fahrenheit.

dehydration in chemistry, the removal of water from a substance to give a product with a new chemical formula; it is not the same as ⟡drying.

There are two types of dehydration. For substances such as hydrated copper sulphate ($CuSO_4.5H_2O$) that contain ⟡water of crystallization, dehydration means removing this water to leave the anhydrous substance. This may be achieved by heating, and is reversible.

Some substances, such as ethanol, contain the elements of water (hydrogen and oxygen) joined in a different form. *Dehydrating agents* such as concentrated sulphuric acid will remove these elements in the ratio 2:1.

Deimos one of the two moons of Mars. It is irregularly shaped, 15 × 12 × 11 km/9 × 7.5 × 7 mi, orbits at a height of 24,000 km/15,000 mi every 1.26 days, and is not as heavily cratered as the other moon, Phobos. Deimos was discovered 1877 by US astronomer Asaph Hall (1829–1907), and is thought to be an asteroid captured by Mars's gravity.

deliquescence phenomenon of a substance absorbing so much moisture from the air that it ultimately dissolves in it to form a solution.

Delta rocket US rocket used to launch many scientific and communications satellites since 1960, based on the Thor ballistic missile. Several increasingly powerful versions produced as satellites became larger and heavier. Solid-fuel boosters were attached to the first stage to increase lifting power.

delta wing aircraft wing shaped like the Greek letter *delta* Δ. Its design enables an aircraft to pass through the ⟡sound barrier with little effect. The supersonic airliner ⟡Concorde and the US ⟡space shuttle have delta wings.

deltoid thick muscle swathing the shoulder joint from the collar bone (⇨clavicle) to its attachment part way down the upper arm bone (⇨humerus).

demodulation in radio, the technique of separating a transmitted audio frequency signal from its modulated radio carrier wave. At the transmitter the audio frequency signal (representing speech or music, for example) may be made to modulate the amplitude (AM broadcasting) or frequency (FM broadcasting) of a continuously transmitted radio-frequency carrier wave. At the receiver, the signal from the aerial is demodulated to extract the required speech or sound component. In early radio systems, this process was called detection. See ⇨modulation.

denaturation irreversible changes occurring in the structure of proteins such as enzymes, usually caused by changes in pH or temperature, by radiation or chemical treatments. An example is the heating of egg albumen resulting in solid egg white.

dendrite part of a ⇨nerve cell or neuron. The dendrites are slender filaments projecting from the cell body. They receive incoming messages from many other nerve cells and pass them on to the cell body.

If the combined effect of these messages is strong enough, the cell body will send an electrical impulse along the axon (the threadlike extension of a nerve cell). The tip of the axon passes its message to the dendrites of other nerve cells.

dendrochronology analysis of the ⇨annual rings of trees to date past events. Samples of wood are obtained by means of a narrow metal tube that is driven into a tree to remove a core extending from the bark to the centre. Samples taken from timbers at an archaeological site can be compared with a master core on file for that region or by taking cores from old, living trees; the year when they were felled can be determined by locating the point where the rings of the two samples correspond and counting back from the present.

Deneb or *Alpha Cygni* brightest star in the constellation Cygnus, and the 19th brightest star in the sky. It is one of the greatest supergiant stars known, with a true luminosity of about 60,000 times that of the Sun. Deneb is about 1,800 light years from Earth.

denitrification process occurring naturally in soil, where bacteria break down ⇨nitrates to give nitrogen gas, which returns to the atmosphere.

dental formula $i\frac{0}{3}$ $c\frac{0}{0}$ $pm\frac{3}{3}$ $m\frac{3}{3}$

dental formula $i\frac{3}{3}$ $c\frac{1}{1}$ $pm\frac{4}{4}$ $m\frac{2}{3}$

dentition *The dentition and dental formulae of a typical herbivore (sheep) and carnivore (dog). The dog has long pointed canines for puncturing and gripping its prey (they are also bared in threat), and has modified premolars and molars (carnassials) for shearing flesh. In the rabbit, by contrast, there is a wide gap, or diastema, between the incisors and premolars; no canines exist.*

density measure of the compactness of a substance; it is equal to its mass per unit volume and is measured in kg per cubic metre/lb per cubic foot. Density is a ⇨scalar quantity. The density D of a mass m occupying a volume V is given by the formula:

$$D = m/V$$

⇨Relative density is the ratio of the density of a substance to that of water at 4°C.

dental formula way of showing what an animal's teeth are like. The dental formula consists of eight numbers separated by a line into two rows. The four above the line represent the teeth in one side of the upper jaw, starting at the front. If this reads 2 1 2 3 (as for humans) it means two incisors, one canine, two premolars, and three molars (see ⇨tooth). The numbers below the line represent the lower jaw. The total number of teeth can be calculated by adding up all the numbers and multiplying by two.

densities of some common substances

	density in kg m^{-3}
solids	
balsa wood	200
oak	700
butter	900
ice	920
ebony	120
sand (dry)	1,600
concrete	2,400
aluminium	2,700
steel	7,800
copper	8,900
lead	11,000
uranium	19,000
liquids	
water	1,000
petrol, paraffin	800
olive oil	900
milk	1,030
sea water	1,030
glycerine	1,300
Dead Sea brine	1,800
gases at standard temperature and pressure (0°C and 1 atm)	
air	1.3
hydrogen	0.09
helium	0.18
methane	0.72
nitrogen	1.25
oxygen	1.43
carbon dioxide	1.98
propane	2.02
butane	2.65

dentition type and number of teeth in a species. Different kinds of teeth have different functions; a grass-eating animal will have large molars for grinding its food, whereas a meat-eater will need powerful canines for catching and killing its prey. The teeth that are less useful may be reduced in size or missing altogether. An animal's dentition is represented diagrammatically by a ◊dental formula.

deoxyribonucleic acid full name of ◊DNA.

depolarizer oxidizing agent used in drycell batteries that converts hydrogen released at the negative electrode into water. This prevents the build-up of gas, which would otherwise impair the efficiency of the battery. ◊Manganese(IV) oxide is used for this purpose.

depression or *cyclone* or *low* in meteorology, a region of low atmospheric pressure. A depression forms as warm, moist air from the tropics mixes with cold, dry polar air, producing warm and cold boundaries (◊fronts) and unstable weather – low cloud and drizzle, showers, or fierce storms. The warm air, being less dense, rises above the cold air to produce the area of low pressure on the ground. Air spirals in towards the centre of the depression in an anticlockwise direction in the northern hemisphere, clockwise in the southern hemisphere, generating winds up to gale force. Depressions tend to travel eastwards and can remain active for several days.

derivative or *differential coefficient* in mathematics, the limit of the gradient of a chord linking two points on a curve as the distance between the points tends to zero; for a function with a single variable, $y = f'(x)$, it is denoted by $f'(x)$, $Df(x)$, or dy/dx, and is equal to the gradient of the curve.

depression A typical depression showing low pressure at the centre.

derrick simple lifting machine consisting of a pole carrying a block and tackle. Derricks are commonly used on ships that carry freight. In the oil industry the tower used for hoisting the drill pipes is known as a derrick.

desiccator airtight vessel, traditionally made of glass, in which materials may be stored either to dry them or to prevent them, once dried, from reabsorbing moisture.

The base of the desiccator is a chamber in which is placed a substance with a strong affinity for water (such as calcium chloride or silica gel), which removes water vapour from the desiccator atmosphere and from substances placed in it.

desktop in computing, a graphical representation of file systems, in which applications and files are represented by pictures (icons), which can be triggered by a single or double click with a ◊mouse button. Such a ◊graphical user interface can be compared with the ◊command line interface, which is character-based.

desktop publishing (DTP) use of microcomputers for small-scale typesetting and page makeup. DTP systems are capable of producing camera-ready pages (pages ready for photographing and printing), made up of text and graphics, with text set in different typefaces and sizes. The page can be previewed on the screen before final printing on a laser printer.

detergent surface-active cleansing agent. The common detergents are made from ◊fats (hydrocarbons) and sulphuric acid, and their long-chain molecules have a type of structure similar to that of ◊soap molecules: a salt group at one end attached to a long hydrocarbon 'tail'. They have the advantage over soap in that they do not produce scum by forming insoluble salts with the calcium and magnesium ions present in hard water.

To remove dirt, which is generally attached to materials by means of oil or grease, the hydrocarbon 'tails' (soluble in oil or grease) penetrate the oil or grease drops, while the 'heads' (soluble in water but insoluble in grease) remain in the water and, being salts, become ionized. Consequently the oil drops become negatively charged and tend to repel one another; thus they remain in suspension and are washed away with the dirt.

There is evidence from Italy 1993 that phosphate-free detergents may cause pollution. Since 1989 the Adriatic has been plagued by the growth of large mucilaginous mats producing foul-smelling brown foam possibly caused by the ◊zeolites (crystalline minerals) and polycarboxylic acids (PCAs) that are used as substitutes for phosphates. The PCAs coat the zeolites and clay particles in the water keeping them afloat. Microorganisms and organic molecules also bind with the PCAs producing noxious foam.

determinant in mathematics, an array of elements written as a square, and denoted by two vertical lines enclosing the array. For a 2×2 matrix, the determinant is given by the difference between the products of the diagonal terms. Determinants are used to solve sets of ◊simultaneous equations by matrix methods.

detonator or *blasting cap* or *percussion cap* small explosive charge used to trigger off a main charge of high explosive. The relatively unstable compounds mercury fulminate and lead acid are often used in detonators, being set off by a lighted fuse or, more commonly, an electric current.

detritus in biology, the organic debris produced during the ◊decomposition of animals and plants.

deuterium naturally occurring heavy isotope of hydrogen, mass number 2 (one proton and one neutron), discovered by Harold Urey 1932. It is sometimes given the symbol D. In nature, about one in every 6,500 hydrogen atoms is deuterium. Combined with oxygen, it produces 'heavy water' (D_2O), used in the nuclear industry.

deuteron nucleus of an atom of deuterium (heavy hydrogen). It consists of one proton and one neutron, and is used in the bombardment of chemical elements to synthesize other elements.

Devarda's alloy mixture of aluminium and zinc in powder form used in the test for nitrate (NO_3^-). Sodium hydroxide solution is added to the solid or solution, followed by the Devarda's alloy, and the mixture is heated. The release of ammonia gas indicates the presence of a nitrate.

development in biology, the process whereby a living thing transforms itself from a single cell into a vastly complicated multicellular organism, with structures, such as limbs, and functions, such as respiration, all able to work correctly in relation to each other. Most of the details of this process remain unknown, although some of the central features are becoming understood.

Devonian period of geological time 408–360 million years ago, the fourth period of the Palaeozoic era. Many desert sandstones from North America and Europe date from this time. The first land plants flourished in the Devonian period, corals were abundant in the seas, amphibians evolved from air-breathing fish, and insects developed on land.

dew point temperature at which the air becomes saturated with water vapour. At temperatures below the dew point, the water vapour condenses out of the air as droplets. If the droplets are large they become deposited on the ground as dew; if small they remain in suspension in the air and form mist or fog.

diabase alternative name for ◊dolerite (a form of basalt that contains very little silica), especially dolerite that has metamorphosed.

diabetes disease *diabetes mellitus* in which a disorder of the islets of Langerhans in the ◊pancreas prevents the body producing the hormone ◊insulin, so that sugars cannot be used properly. Treatment is by strict dietary control and oral or injected insulin, depending on the type of diabetes.

There are two forms of diabetes: Type 1, or insulin-dependent diabetes, which usually begins in childhood (early onset) and is an autoimmune condition; and Type 2, or noninsulin-dependent diabetes, which occurs in later life (late onset).

Insulin-secreting implants were trialled successfully in dogs 1993 using islet cells enclosed in gel capsules. The capsules were small enough to be injected by needle into the peritoneal cavity, and functioned for six months.

Much rarer, *diabetes insipidus* is due to a deficiency of a hormone secreted by the ◊pituitary gland to regulate the body's water balance. It is controlled by hormone therapy.

dialysis Technique for removing waste products from the blood in chronic or acute kidney failure. There are two main methods, haemodialysis and peritoneal dialysis.

diamagnetic of a material weakly repelled by a magnet. All substances are diamagnetic but the behaviour is often masked by stronger forms of magnetism such as ◊paramagnetic or ◊ferromagnetic behaviour.

diameter straight line joining two points on the circumference of a circle that passes through the centre of that circle. It divides a circle into two equal halves.

diamond generally colourless, transparent mineral, the hard crystalline form of carbon. It is regarded as a precious gemstone, and is the hardest substance known (10 on the ◊Mohs' scale). Diamond crystallizes in the cubic system as octahedral crystals, some with curved faces and striations. They have a high refractive index of 2.42. Industrial diamonds, which may be natural or synthetic, are used for cutting, grinding, and polishing.

Diamonds may be found as alluvial diamonds on or close to the Earth's surface in riverbeds or dried watercourses; on the sea bottom (off SW Africa); or, more commonly, in diamond-bearing volcanic pipes composed of 'blue ground', ◊kimberlite or lamproite, where the original matrix has penetrated the Earth's crust from great depths.

diamorphine technical term for ◊heroin.

DIANE (acronym from *direct information access network for Europe*) collection of information suppliers, or 'hosts', for the European computer network.

diapause period of suspended development that occurs in some species of insects, characterized by greatly reduced metabolism. Periods of diapause are often timed to coincide with the winter months, and improve the insect's chances of surviving adverse conditions.

diaphragm thin muscular sheet separating the thorax from the abdomen in mammals. It is attached by way of the ribs at either side and the breastbone and backbone. Arching upwards against the heart and lungs the diaphragm is important in the mechanics of breathing. It contracts at each inhalation, moving downwards to increase the volume of the chest cavity, and relaxes at exhalation.

diaphragm or *cap* or *Dutch cap* barrier ◊contraceptive that is passed into the vagina to fit over the cervix (neck of the uterus), preventing sperm from entering the uterus. For a cap to be effective, a ◊spermicide must be used and the diaphragm left in place for 6–8 hours after intercourse. This method is 97% effective if practised correctly.

diatom microscopic alga of the division Bacillariophyta found in all parts of the world. Diatoms consist of single cells, sometimes grouped in colonies.

The cell wall is made up of two overlapping valves known as *frustules*, which are usually impregnated with silica, and which fit together like the lid and body of a pillbox. Diatomaceous earths (diatomite) are made up of the valves of fossil diatoms, and are used in the manufacture of dynamite and in the rubber and plastics industries.

diatomic molecule molecule composed of two atoms joined together. In the case of an element such as oxygen (O_2), the atoms are identical.

dibasic acid acid containing two replaceable hydrogen atoms, such as sulphuric acid (H_2SO_4). The acid can form two series of salts, the normal salt (sulphate, SO_4^{2-}) and the acid salt (hydrogensulphate HSO_4^-).

dichloro-diphenyl-trichloroethane full name of the insecticide ◊DDT.

dichotomous key method of identifying an organism. The investigator is presented with pairs of statements, for example 'flower has less than five stamens' and 'flower has five or more stamens'. By successively eliminating statements the field naturalist moves closer to a positive identification. Dichotomous keys assume a good knowledge of the subject under investigation.

dicotyledon major subdivision of the ◊angiosperms, containing the great majority of flowering plants. Dicotyledons are characterized by the presence of two seed leaves, or ◊cotyledons, in the embryo, which is usually surrounded by an ◊endosperm. They generally have broad leaves with netlike veins.

Dicotyledons may be small plants such as daisies and buttercups, shrubs, or trees such as oak and beech. The other subdivision of the angiosperms is the ◊monocotyledons.

diecasting form of ◊casting in which molten metal is injected into permanent metal moulds or dies.

dielectric substance (an insulator such as ceramic, rubber, or glass) capable of supporting electric stress. The dielectric constant, or relative permittivity, of a substance is the ratio of the capacitance of a capacitor with the substance as dielectric to that of a similar capacitor in which the dielectric is replaced by a vacuum.

diesel engine ◊internal-combustion engine that burns a lightweight fuel oil. The diesel engine operates by compressing air until it becomes sufficiently hot to ignite the fuel. It is a piston-in-cylinder engine, like the ◊petrol engine, but only air (rather than an air-and-fuel mixture) is taken into the cylinder on the first piston stroke (down). The piston moves up and compresses the air until it is at a very high temperature. The fuel oil is then injected into the hot air, where it burns, driving the piston down on its power stroke. For this reason the engine is called a compression-ignition engine.

diesel oil lightweight fuel oil used in diesel engines. Like petrol, it is a petroleum product. When used in vehicle engines, it is also known as *derv* (*diesel-engine road vehicle*).

diet the range of foods eaten by an animal, also a particular selection of food, or the overall intake and selection of food for a particular person or people. The basic components of a diet are a group of chemicals: proteins, carbohydrates, fats, vitamins, minerals, and water. Different animals require these substances in different proportions, but the necessity of finding and processing an appropriate diet is a very basic drive in animal evolution. For instance, all guts are adapted for digesting and absorbing food, but different guts have adapted to cope with particular diets.

For humans, an adequate diet is one that fulfills the body's nutritional requirements and gives an energy intake proportional to the person's activity level (the average daily requirement is 2,400 calories for men, less for women, more for active children). In the Third World and in famine or poverty areas some 450 million people in the world subsist on fewer than 1,500 calories per day, whereas in the developed countries the average daily intake is 3,300 calories.

differential arrangement of gears in the final drive of a vehicle's transmission system that allows the driving wheels to turn at different speeds when cornering. The differential consists of sets of bevel gears and pinions within a cage attached to the crown wheel. When cornering, the bevel pinions rotate to allow the outer wheel to turn faster than the inner.

differential calculus branch of ◊calculus involving applications such as the determination of maximum and minimum points and rates of change.

differentiation in embryology, the process by which cells become increasingly different and specialized, giving rise to more complex structures that have particular functions in the adult organism. For instance, embryonic cells may develop into nerve, muscle, or bone cells.

differentiation in mathematics, a procedure for determining the ◊derivative or gradient of the tangent to a curve f(x) at any point x.

diffraction the slight spreading of a light beam into a pattern of light and dark bands when it passes through a narrow slit or past the edge of an obstruction. A *diffraction grating* is a plate of glass or metal ruled with close, equidistant parallel lines used for separating a wave train such as a beam of incident light into its component frequencies (white light results in a spectrum).

The regular spacing of atoms in crystals are used to diffract X-rays, and in this way the structure of many substances has been elucidated, including recently that of proteins (see ◊X-ray diffraction). Sound waves can also be diffracted by a suitable array of solid objects.

diffusion spontaneous and random movement of molecules or particles in a fluid (gas or liquid) from a region in which they are at a high concentration to a region of lower concentration, until a uniform concentration is achieved throughout. No mechanical mixing or stirring is involved. For instance, if a drop of ink is added to water, its molecules will diffuse until their colour becomes evenly distributed throughout.

In biological systems, diffusion plays an essential role in the transport, over short distances, of molecules such as nutrients, respiratory gases, and neurotransmitters. It provides the means by which small molecules pass into and out of individual cells and microorganisms, such as amoebae, that possess no circulatory system.

digestion process whereby food eaten by an animal is broken down mechanically, and chemically by ◊enzymes, mostly in the ◊stomach and ◊intestines, to make the nutrients available for absorption and cell metabolism.

In some single-celled organisms, such as amoebae, a food particle is engulfed by the cell and digested in a vacuole within the cell.

digestive system mouth, stomach, intestine, and associated glands of animals, which are responsible for digesting food. The food is broken down by physical and chemical means in the ◊stomach; digestion is completed, and most nutrients are absorbed in the small intestine; what remains is stored and concentrated into faeces in the large intestine. In birds, additional digestive organs are the ◊crop and ◊gizzard.

In smaller, simpler animals such as jellyfishes, the digestive system is simply a cavity (coelenteron or enteric cavity) with a 'mouth' into which food is taken; the digestible portion is dissolved and absorbed in this cavity, and the remains are ejected back through the mouth.

digit in mathematics, any of the numbers 0 to 9 in the decimal system. Different bases have different ranges of digits. For example the hexadecimal system has digits 0 to 9 and A to F whereas the binary system has two digits 0 and 1.

digital in electronics and computing, a term meaning 'coded as numbers'. A digital system uses two-state, either on/off or high/low voltage pulses, to encode, receive, and transmit information. A *digital display* shows discrete values as numbers (as opposed to an analogue signal, such as the continuous sweep of a pointer on a dial).

digital audio tape (DAT) digitally recorded audio tape produced in cassettes that can carry two hours of sound on each side and are about half the size of standard cassettes. DAT players/recorders were developed 1987 but not marketed in the UK until 1989. Prerecorded cassettes are copy-protected. The first DAT for computer data was introduced 1988.

differentiation *A mathematical procedure for determining the gradient, or slope, of the tangent to any curve f(x) at any point x. Part of a branch of mathematics called differential calculus, it is used to solve problems involving continuously varying quantities (such as the change in velocity or altitude of a rocket), to find the rates at which these variations occur and to obtain maximum and minimum values for the quantities.*

digital compact cassette (DCC) digitally recorded audio cassette that is roughly the same size as a standard cassette. It cannot be played on a normal tape recorder, though standard tapes can be played on a DCC machine; this is known as 'backwards compatibility'. The playing time is 90 minutes.

digital computer computing device that operates on a two-state system, using symbols that are internally coded as binary numbers

(numbers made up of combinations of the digits 0 and 1); see ◊computer.

digital data transmission in computing, a way of sending data by converting all signals (whether pictures, sounds, or words) into numeric (normally binary) codes before transmission, then reconverting them on receipt. This virtually eliminates any distortion or degradation of the signal during transmission, storage, or processing.

digital recording technique whereby the pressure of sound waves is sampled more than 30,000 times a second and the values converted by computer into precise numerical values. These are recorded and, during playback, are reconverted to sound waves.

digital sampling electronic process used in ◊telecommunications for transforming a constantly varying (analogue) signal into one composed of discrete units, a digital signal. In the creation of recorded music, sampling enables the composer, producer, or remix engineer to borrow discrete vocal or instrumental parts from other recorded work (it is also possible to sample live sound).

digital-to-analogue converter electronic circuit that converts a digital signal into an ◊analogue (continuously varying) signal. Such a circuit is used to convert the digital output from a computer into the analogue voltage required to produce sound from a conventional loudspeaker.

digital video interactive powerful compression system used for storing video images on computer; see ◊DVI.

digitizer in computing, a device that converts an analogue video signal into a digital format so that video images can be input, stored, displayed, and manipulated by a computer. The term is sometimes used to refer to a ◊graphics tablet.

dihybrid inheritance in genetics, a pattern of inheritance observed when two characteristics are studied in succeeding generations. The first experiments of this type, as well as in ◊monohybrid inheritance, were carried out by Austrian biologist Gregor Mendel (1822–1884) using pea plants.

dilution process of reducing the concentration of a solution by the addition of a solvent.

dimension in science, any directly measurable physical quantity such as mass (M), length (L), and time (T), and the derived units obtainable by multiplication or division from such quantities.

For example, acceleration (the rate of change of velocity) has dimensions (LT^{-2}), and is expressed in such units as km s^{-2}. A quantity that is a ratio, such as relative density or humidity, is dimensionless.

dimethyl sulphoxide (DMSO) $(CH_3)_2SO_4$ colourless liquid used as an industrial solvent and an antifreeze. It is obtained as a by-product of the processing of wood to paper.

DIN (abbreviation for *Deutsches Institut für Normung*) German national standards body, which has set internationally accepted standards for (among other things) paper sizes and electrical connectors.

dinitrogen oxide alternative name for ◊nitrous oxide, or 'laughing gas', one of the nitrogen oxides.

diode combination of a cold anode and a heated cathode (or the semiconductor equivalent, which incorporates a p–n junction). Either device allows the passage of direct current in one direction only, and so is commonly used in a ◊rectifier to convert alternating current (AC) to direct current (DC).

dioecious of plants with male and female flowers borne on separate individuals of the same species. Dioecism occurs, for example, in the willows *Salix*. It is a way of avoiding self-fertilization.

dioptre optical unit in which the power of a ◊lens is expressed as the reciprocal of its focal length in metres. The usual convention is that convergent lenses are positive and divergent lenses negative. Short-sighted people need lenses of power about –0.66 dioptre; a typical value for long sight is about +1.5 dioptre.

diorite igneous rock intermediate in composition; the coarse-grained plutonic equivalent of ◊andesite.

dioxin any of a family of over 200 organic chemicals, all of which are heterocyclic hydrocarbons (see ◊cyclic compounds).

The term is commonly applied, however, to only one member of the family, 2,3,7,8-tetrachlorodibenzo-*p*-dioxin (2,3,7,8-TCDD), a highly toxic chemical that occurs, for example, as an impurity in the defoliant Agent Orange, used in the Vietnam War, and sometimes in the weedkiller 2,4,5-T. It has been associated with a disfiguring skin complaint (chloracne), birth defects, miscarriages, and cancer.

dip the angle and direction in which a bed of rock is inclined to the horizontal. Rocks that are dipping have usually been affected by folding.

diploblastic in biology, having a body wall composed of two layers. The outer layer is the *ectoderm*, the inner layer is the *endoderm*. This pattern of development is shown by ◊coelenterates.

diploid having two sets of ◊chromosomes in each cell. In sexually reproducing species, one set is derived from each parent, the ◊gametes, or sex cells, of each parent being ◊haploid (having only one set of chromosomes) due to ◊meiosis (reduction cell division).

dip, magnetic angle at a particular point on the Earth's surface between the direction of the Earth's magnetic field and the horizontal. It is measured using a *dip circle*, which has a magnetized needle suspended so that it can turn freely in the vertical plane of the magnetic field. In the northern hemisphere the needle dips below the horizontal, pointing along the line of the magnetic field towards its north pole. At the magnetic north and south poles, the needle dips vertically and the angle of dip is 90°. See also ◊angle of declination.

dipole the uneven distribution of magnetic or electrical characteristics within a molecule or substance so that it behaves as though it possesses two equal but opposite poles or charges, a finite distance apart.

The uneven distribution of electrons within a molecule composed of atoms of different ◊electronegativities may result in an apparent concentration of electrons towards one end of the molecule and a deficiency towards the other, so that it forms a dipole consisting of apparently separated but equal positive and negative charges. The product of one charge and the distance between them is the *dipole moment*. A bar magnet behaves as though its magnetism were concentrated in separate north and south magnetic poles because of the uneven distribution of its magnetic field.

dipole, magnetic see ◊magnetic dipole.

direct access or *random access* type of ◊file access. A direct-access file contains records that can be accessed directly by the computer because each record has its own address on the storage disc.

direct combination method of making a simple salt by heating its two constituent elements together. For example, iron(II) sulphide can be made by heating iron and sulphur, and aluminium chloride by passing chlorine over hot aluminium.

$$Fe + S \rightarrow FeS$$
$$2Al + 3Cl_2 \rightarrow 2AlCl_3$$

direct current (DC) electric current that flows in one direction, and does not reverse its flow as ◊alternating current does. The electricity produced by a battery is direct current.

directory in computing, a list of file names, together with information that enables a computer to retrieve those files from ◊backing storage. The computer operating system will usually store and update a directory on the backing storage to which it refers. So, for example, on each ◊disc used by a computer a directory file will be created listing the disc's contents.

disaccharide ◊sugar made up of two monosaccharides or simple sugars, such as glucose or fructose. Sucrose, $C_{12}H_{22}O_{11}$, or table sugar, is a disaccharide.

disc or *disk* in computing, a common medium for storing large volumes of data (an alternative is ◊magnetic tape). A *magnetic disc* is rotated at high speed in a disc-drive unit as a read/write (playback or record) head passes over its surfaces to record or read the magnetic variations that encode the data. Recently, *optical discs*, such as ◊CD-ROM (compact-disc read-only memory) and ◊WORM (write once, read many times), have been used to store computer data. Data are recorded on the disc surface as etched microscopic pits and are read by a laser-scanning device. Optical discs have an enormous capacity – about 550 megabytes

surface of a magnetic disc

disc formatting *The surface of a magnetic disc. Floppy discs typically have 135 tracks per inch. The first track normally contains the disc directory which lists the files held on the disc and the sectors in which they are stored.*

(million ◊bytes) on a compact disc, and thousands of megabytes on a full-size optical disc.

disc drive mechanical device that reads data from and writes data to a magnetic ◊disc.

disc formatting in computing, preparing a blank magnetic disc so that data can be stored on it. Data are recorded on a disc's surface on circular tracks, each of which is divided into a number of sectors. In formatting a disc, the computer's operating system adds control information such as track and sector numbers, which enables the data stored to be accessed correctly by the disc-drive unit.

discharge tube device in which a gas conducting an electric current emits visible light. It is usually a glass tube from which virtually all the air has been removed (so that it 'contains' a near vacuum), with electrodes at each end. When a high-voltage current is passed between the electrodes, the few remaining gas atoms in the tube (or some deliberately introduced ones) ionize and emit coloured light as they conduct the current along the tube. The light originates as electrons change energy levels in the ionized atoms.

discrete data data that can take only whole-number or fractional values. The opposite is ◊continuous data, which can take all in-between values. Examples of discrete data include frequency and population data. However, measurements of time and other dimensions can give rise to continuous data.

disease any condition that impairs the normal state of an organism, and usually alters the functioning of one or more of its organs or systems. A disease is usually characterized by a set of specific symptoms and signs, although these may not always be apparent to the sufferer. Diseases may be inborn (see ◊congenital disease) or acquired through infection, injury, or other cause. Many diseases have unknown causes.

disinfectant agent that kills, or prevents the growth of, bacteria and other microorganisms. Chemical disinfectants include carbolic acid (phenol, used by Joseph Lister (1827–1912) in surgery in the 1870s), ethanol, methanol, chlorine, and iodine.

dispersal phase of reproduction during which gametes, eggs, seeds, or offspring move away from the parents into other areas. The result is that overcrowding is avoided and parents do not find themselves in competition with their own offspring. The mechanisms are various, including a reliance on wind or water currents and, in the case of animals, locomotion. The ability of a species to spread widely through an area and to colonize new habitats has survival value in evolution.

displacement activity in animal behaviour, an action that is performed out of its normal context, while the animal is in a state of stress, frustration, or uncertainty. Birds, for example, often peck at grass when uncertain whether to attack or flee from an opponent; similarly, humans scratch their heads when nervous.

displacement reaction chemical reaction in which a less reactive element is replaced in a compound by a more reactive one.

For example, the addition of powdered zinc to a solution of copper(II) sulphate displaces copper metal, which can be detected by its characteristic colour. See also ◊electrochemical series.

dissection cutting apart of bodies to study their organization, or tissues to gain access to a site in surgery. Postmortem dissection was considered a sin in the Middle Ages. In the UK before 1832, hanged murderers were the only legal source of bodies, supplemented by graverobbing. The Anatomy Act 1832 authorized the use of deceased institutionalized paupers unclaimed by next of kin, and by the 1940s bequests of bodies had been introduced.

dissociation in chemistry, the process whereby a single compound splits into two or more smaller products, which may be capable of recombining to form the reactant.

distance ratio in a machine, the distance moved by the input force, or effort, divided by the distance moved by the output force, or load. The ratio indicates the movement magnification achieved, and is equivalent to the machine's ◊velocity ratio.

distance-time graph graph used to describe the motion of a body by illustrating the relationship between the distance that it travels and the time taken. Plotting distance (on the vertical axis) against time (on the horizontal axis) produces a graph the gradient of which is the body's speed. If the gradient is constant (the graph is a straight line), the body has uniform or constant speed; if the gradient varies (the graph is curved), then so does the speed, and the body may be said to be accelerating or decelerating.

distillation technique used to purify liquids or to separate mixtures of liquids possessing different boiling points. *Simple distillation* is

distillation. Laboratory apparatus for simple distillation. Other forms of distillation include steam distillation, in which steam is passed into the mixture being distilled, and vacuum distillation, in which air is removed from above the mixture to be distilled.

produced with a ◊word ◊CAD package. An ◊OCR document might be ing data that is a paper document contain-computer using a document reader.

documentation in computing, the written information associated with a computer program or ◊applications package. Documentation is usually divided into two categories: program documentation and user documentation.

DOLERITE 126

DNA How the DNA ... DNA molecule consists of t... wrapped around each other in a ... helix. The main strands consist of alterna... sugar (S) and phosphate (P) groups, and attached to each sugar is a nitrogenous base – adenine (A), cytosine (C), guanine (G), or thymine (T). The sequence of bases carries the genetic code which specifies the characteristics of offspring. The strands are held together by weak bonds between the bases, cytosine to guanine, and adenine to thymine. The weak bonds allow the strands to split apart, allowing new bases to attach, forming another double strand.

of limestone rock where the calcite content is replaced by the mineral dolomite. Dolomite rock may be white, grey, brown, or reddish in colour, commonly crystalline. It is used as a building material. The region of the Alps known as the Dolomites is a fine example of dolomite formation.

domain small area in a magnetic material that behaves like a tiny magnet. The magnetism of the material is due to the movement of electrons in the atoms of the domain. In an unmagnetized sample of material, the domains point in random directions, or form closed loops, so that there is no overall magnetization of the sample. In a magnetized sample, the domains are aligned so that their magnetic effects combine to produce a strong overall magnetism.

dominance in genetics, the masking of one allele (an alternative form of a gene) by ...ther allele. For example, if a ◊heterozygous ...ne allele for blue eyes and one for ...her eye colour will be ...s is described as ... eyes as

how a cell divides

1 original double helix
2 forms ladder
3 unzips
4 ...aned zip teeth new bases join onto
5

distributed processing computer processing that uses more than one computer to run an application. ◊Local area networks, ◊client-server architecture, and ◊parallel processing involve distributed processing.

distributive operation in mathematics, an operation, such as multiplication, that bears a relationship to another operation, such as addition, such that $a \times (b + c) = (a \times b) + (a \times c)$. For example, $3 \times (2 + 4) = (3 \times 2) + (3 \times 4) = 18$. Multiplication may be said to be distributive over addition. Addition is not, however, distributive over multiplication because $3 + (2 \times 4) \leftarrow (3 + 2) \times (3 + 4)$.

distributor device in the ignition system of a piston engine that distributes pulses of high-voltage electricity to the ◊spark plugs in the cylinders. The electricity is passed to the plug leads by the tip of a rotor arm, driven by the engine camshaft, and current is fed to the rotor arm from the ignition coil. The distributor also opens and closes to interrupt the battery current to the coil, thus triggering the high-voltage pulses. With electronic ignition it is absent.

DNA (abbreviation for *deoxyribonucleic acid*) complex giant molecule that contains, in chemically coded form, all the information needed to build, control, and maintain a living organism. DNA is a ladderlike double-stranded ◊nucleic acid that forms the basis of genetic inheritance in all organisms, except for a few viruses that have only ◊RNA. In organisms other than bacteria it is organized into ◊chromosomes and contained in the cell nucleus.

The oldest sequenced DNA to be found is that of a nemonychid weevil trapped in amber from the Cretaceous period, and estimated as being 120–135 million years old.

DNA fingerprinting or *DNA profiling* another name for ◊genetic fingerprinting.

document in computing, data associated with a particular application. For example, a *text document* might be produced by a ◊word processor and a ◊draw

the vapours of the piece of apparatus y are cooled and con- oduced (the distillate) is e non-volatile solutes (now in so... ... ain in the distillation vessel to be dis... ... as impurities or recovered as required. Mixtures of liquids (such as ◊petroleum or aqueous ethanol) are separated by *fractional distillation*, or *fractionation*. When the mixture is boiled, the vapours of its most volatile component rise into a vertical ◊fractionating column where they condense to liquid form. However, as this liquid runs back down the column it is reheated to boiling point by the hot rising vapours of the next-most-volatile component and so its vapours ascend the column once more. This boiling–condensing process occurs repeatedly inside the column, eventually bringing about a temperature gradient along its length. The vapours of the more volatile components therefore reach the top of the column and enter the condenser for collection before those of the less volatile components. In the fractional distillation of petroleum, groups of compounds (fractions) possessing similar relative molecular masses and boiling points are tapped off at different points on the column.

Doppler effect change in the observed frequency (or wavelength) of waves due to relative motion between the wave source and the observer. The Doppler effect is responsible for the perceived change in pitch of a siren as it approaches and then recedes, and for the ◊red shift of light from distant stars. It is named after the Austrian physicist Christian Doppler (1803–1853).

Dorado constellation of the southern hemisphere, represented as a gold fish. It is easy to locate, since the Large ◊Magellanic Cloud marks its southern border.

dormancy in botany, a phase of reduced physiological activity exhibited by certain buds, seeds, and spores. Dormancy can help a plant to survive unfavourable conditions, as in annual plants that pass the cold winter season as dormant seeds, and plants that form dormant buds.

dorsal in vertebrates, the surface of the animal closest to the backbone. For most vertebrates and invertebrates this is the upper surface, or the surface furthest from the ground. For bipedal primates such as humans, where the dorsal surface faces backwards, then the word is 'back'.

DOS (acronym for *disc operating system*) computer ◊operating system specifically designed for use with disc storage; also used as an alternative name for a particular operating system, ◊MS-DOS.

dot matrix printer computer printer that produces each character individually by printing a pattern, or matrix, of very small dots. The printing head consists of a vertical line or block of either 9 or 24 printing pins. As the printing head is moved from side to side across the paper, the pins are pushed forwards selectively to strike an inked ribbon and build up the dot pattern for each character on the paper beneath.

double bond two covalent bonds between adjacent atoms, as in the ◊alkenes (–C=C–) and ◊ketones (–C=O–).

double decomposition reaction between two chemical substances (usually ◊salts in solution) that results in the exchange of a constituent from each compound to create two different compounds.

double star two stars that appear close together. Most double stars attract each other due to gravity, and orbit each other, forming a genuine ◊binary star, but other double stars are at different distances from Earth, and lie in the same line of sight only by chance. Through a telescope both types of double star look the same.

Down's syndrome condition caused by a chromosomal abnormality (the presence of an extra copy of chromosome 21), which in humans produces mental retardation; a flattened face; coarse, straight hair; and a fold of skin at the inner edge of the eye (hence the former name 'mongolism'). The condition can be detected by prenatal testing.

dpi abbreviation for *dots per inch*, a measure of the ◊resolution of images produced by computer screens and printers.

Draco in astronomy, a large but faint constellation, representing a dragon coiled around the north celestial pole. The star Alpha Draconis (Thuban) was the pole star 4,800 years ago.

drag resistance to motion a body experiences when passing through a fluid – gas or liquid. The aerodynamic drag aircraft experience when travelling through the air represents a great waste of power, so they must be carefully shaped, or streamlined, to reduce drag to a minimum. Cars benefit from ◊streamlining, and aerodynamic drag is used to slow down spacecraft returning from space. Boats travelling through water experience hydrodynamic drag on their hulls, and the fastest vessels are ◊hydrofoils, whose hulls lift out of the water while cruising.

DRAM (acronym for *dynamic random-access memory*) computer memory device in the form of a silicon chip commonly used to provide the ◊immediate-access memory of microcomputers. DRAM loses its contents unless they are read and rewritten every 2 milliseconds or so. This process is called *refreshing* the memory. DRAM is slower but cheaper than ◊SRAM, an alternative form of silicon-chip memory.

driver in computing, a program that controls a peripheral device. Every device connected to the computer needs a driver program. The driver ensures that communication between the computer and the device is successful.

drug any of a range of substances, natural or synthetic, administered to humans and animals as therapeutic agents: to diagnose, prevent or treat disease or to assist recovery from injury. Traditionally many drugs were obtained from plants or animals; some minerals also had medicinal value. Today, increasing numbers of drugs are synthesized in the laboratory.

drupe fleshy ◊fruit containing one or more seeds which are surrounded by a hard, protective layer – for example cherry, almond, and plum. The wall of the fruit (◊pericarp) is differentiated into the outer skin (exocarp), the fleshy layer of tissues (mesocarp), and the hard layer surrounding the seed (endocarp).

dry ice solid carbon dioxide (CO_2), used as a refrigerant. At temperatures above $-79°C$/$-110.2°F$, it sublimes (turns into vapour without passing through a liquid stage) to gaseous carbon dioxide.

drying removal of liquid water from a substance without altering its chemical composition (unlike ◊dehydration). Drying agents include deliquescent substances such as calcium chloride (see ◊deliquescence), concentrated acids such as sulphuric and nitric acid, and silica gel.

DTP abbreviation for ◊*desktop publishing*.

ductile material material that can sustain large deformations beyond its elastic limit (see ◊elasticity) without fracture. Metals are very ductile, and may be pulled out into wires, or hammered or rolled into thin sheets without breaking. Compare ◊brittle material.

ductless gland alternative name for an ◊endocrine gland.

duodenum in vertebrates, a short length of ◊alimentary canal found between the stomach and the small intestine. Its role is in digesting carbohydrates, fats, and proteins. The smaller molecules formed are then absorbed, either by the duodenum or the ileum.

duralumin lightweight aluminium ◊alloy widely used in aircraft construction, containing copper, magnesium, and manganese.

DVI (abbreviation for *digital video interactive*) in computing, a powerful compression and decompression system for digital video and audio. DVI enables 72 minutes of full-screen, full-motion video and its audio track to be stored on a CD-ROM. Originally developed by the US firm RCA, DVI is now owned by Intel and has active support from IBM and Microsoft. It can be used on the hard disc of a PC as well as on a CD-ROM.

dye substance that, applied in solution to fabrics, imparts a colour resistant to washing. *Direct dyes* combine with the material of the fabric, yielding a coloured compound; *indirect dyes* require the presence of another substance (a mordant), with which the fabric must first be treated; *vat dyes* are colourless soluble substances that on exposure to air yield an insoluble coloured compound.

dynamics or *kinetics* in mechanics, the mathematical and physical study of the behaviour of bodies under the action of forces that produce changes of motion in them.

dynamite explosive consisting of a mixture of nitroglycerine and diatomaceous earth (diatomite, an absorbent, chalklike material). It was first devised by Alfred Nobel.

dynamo simple generator, or machine for transforming mechanical energy into electrical energy. A dynamo in basic form consists of a powerful field magnet between the poles of which a suitable conductor, usually in the form of a coil (armature), is rotated. The mechanical energy of rotation is thus converted into an electric current in the armature.

dyne c.g.s. unit (symbol dyn) of force. 10^5 dynes make one newton. The dyne is defined as the force that will accelerate a mass of one gram by one centimetre per second per second.

dysprosium silver-white, metallic element of the ◊lanthanide series, symbol Dy, atomic number 66, relative atomic mass 162.50. It is among the most magnetic of all known substances and has a great capacity to absorb neutrons.

ear organ of hearing in animals. It responds to the vibrations that constitute sound, and these are translated into nerve signals and passed to the brain. A mammal's ear consists of three parts: outer ear, middle ear, and inner ear. The ***outer ear*** is a funnel that collects sound, directing it down a tube to the ***ear drum*** (tympanic membrane), which separates the outer and ***middle ear***s. Sounds vibrate this membrane, the mechanical movement of which is transferred to a smaller membrane leading to the ***inner ear*** by three small bones, the auditory ossicles. Vibrations of the inner ear membrane move fluid contained in the snail-shaped cochlea, which vibrates hair cells that stimulate the auditory nerve connected to the brain. Three fluid-filled canals of the inner ear detect changes of position; this mechanism, with other sensory inputs, is responsible for the sense of balance.

Earth third planet from the Sun. It is almost spherical, flattened slightly at the poles. About 70% of the surface (including the north and south polar icecaps) is covered with water. The Earth is surrounded by a life-supporting atmosphere and is the only planet on which life is known to exist, and is composed of three concentric layers: the ♢core, the ♢mantle, and the ♢crust.

earth electrical connection between an appliance and the ground. In the event of a fault in an electrical appliance, for example, involving connection between the live part of the circuit and the outer casing, the current flows to earth, causing no harm to the user.

ear The structure of the ear. The three bones of the middle ear – hammer, anvil, and stirrup – vibrate in unison and magnify sounds about 20 times. The spiral-shaped cochlea is the organ of hearing. As sound waves pass down the spiral tube, they vibrate fine hairs lining the tube, which activate the auditory nerve connected to the brain. The semicircular canals are the organs of balance, detecting movements of the head.

Earth Inside the Earth. The surface of the Earth is a thin crust about 6 km/4 mi thick under the sea and 40 km/25 mi thick under the continents. Under the crust lies the mantle, about 2,900 km/1,800 mi thick and with a temperature of 1,500–3,000°C/2,700–5,400°F. The outer core is liquid iron and nickel at 4,000°C/7,200°F. The inner core is probably solid iron and nickel at about 5,000°C/9,000°F.

earth science scientific study of the planet Earth as a whole, a synthesis of several traditional subjects such as ◊geology, ◊meteorology, oceanography, ◊geophysics, ◊geochemistry, and ◊palaeontology.

Earth Summit (official name *United Nations Conference on Environment and Development*) international meeting in Rio de Janeiro, Brazil, June 1992, which drew up measures toward world environmental protection. Treaties were made to combat global warming and protect wildlife ('biodiversity') (the latter was not signed by the USA till 1994).

EBCDIC (abbreviation for *extended binary-coded decimal interchange code*) in computing, a code used for storing and communicating alphabetic and numeric characters. It is an 8-bit code, capable of holding 256 different characters, although only 85 of these are defined in the standard version. It is still used in many mainframe computers, but almost all mini- and microcomputers now use ◊ASCII code.

ecdysis periodic shedding of the ◊exoskeleton by insects and other arthropods to allow growth. Prior to shedding, a new soft and expandable layer is first laid down underneath

earth science: chronology

- **1735** English lawyer George Hadley described the circulation of the atmosphere as large-scale convection currents centred on the equator.
- **1736** A French expedition to Lapland proved the Earth to be flattened at the poles.
- **1743** Christopher Packe produced the first geological map, of S England.
- **1744** The first map produced on modern surveying principles was produced by Cæsar-François Cassini in France.
- **1745** In Russia, Mikhail Vasilievich Lomonosov published a catalogue of over 3,000 minerals.
- **1760** Lomonosov explained the formation of icebergs. John Mitchell proposed that earthquakes are produced when one layer of rock rubs against another.
- **1766** The fossilized bones of a huge animal (later called *Mosasaurus*) were found in a quarry near the river Meuse, the Netherlands.
- **1776** James Keir suggested that some rocks, such as those making up the Giant's Causeway in Ireland, may have formed as molten material that cooled and then crystallized.
- **1779** French naturalist Comte George de Buffon speculated that the Earth may be much older than the 6,000 years suggested by the Bible.
- **1785** Scottish geologist James Hutton proposed the theory of uniformitarianism: all geological features are the result of processes that are at work today, acting over long periods of time.
- **1786** German–Swiss Johann von Carpentier described the European ice age.
- **1793** Jean Baptiste Lamarck argued that fossils are the remains of once-living animals and plants.
- **1794** William Smith produced the first large-scale geological maps of England.
- **1795** In France, Georges Cuvier identified the fossil bones discovered in the Netherlands in 1766 as being those of a reptile, now extinct.
- **1804** French physicists Jean Biot and Joseph Gay-Lussac studied the atmosphere from a hot-air balloon.
- **1809** The first geological survey of the eastern USA was produced by William Maclure.
- **1815** In England, William Smith showed how rock strata (layers) can be identified on the basis of the fossils found in them.
- **1822** Mary Ann Mantell discovered on the English coast the first fossil to be recognized as that of a dinosaur (an iguanodon). In Germany, Friedrich Mohs introduced a scale for specifying mineral hardness.
- **1825** Cuvier proposed his theory of catastrophes as the cause of the extinction of large groups of animals.
- **1830** Scottish geologist Charles Lyell published the first volume of *The Principles of Geology*, which described the Earth as being several hundred million years old.
- **1839** In the USA, Louis Agassiz described the motion and laying down of glaciers, confirming the reality of the ice ages.
- **1842** English palaeontologist Richard Owen coined the name 'dinosaur' for the reptiles, now extinct, that lived about 175 million years ago.
- **1846** Irish physicist William Thomson (Lord Kelvin) estimated, using the temperature of the Earth, that the Earth is 100 million years old.
- **1850** US naval officer Matthew Fontaine Maury mapped the Atlantic Ocean, noting that it is deeper near its edges than at the centre.
- **1852** Edward Sabine in Ireland showed a link between sunspot activity and changes in the Earth's magnetic field.
- **1853** James Coffin described the three major wind bands that girdle each hemisphere.
- **1854** English astronomer George Airy calculated the mass of the Earth by measuring gravity at the top and bottom of a coal mine.
- **1859** Edwin Drake drilled the world's first oil well at Titusville, Pennsylvania, USA.
- **1872** The beginning of the world's first major oceanographic expedition, the four-year voyage of the *Challenger*.
- **1882** Scottish physicist Balfour Stewart postulated the existence of the ionosphere (the ionized layer of the outer atmosphere) to account for differences in the Earth's magnetic field.
- **1884** German meteorologist Vladimir Köppen introduced a classification of the world's temperature zones.
- **1890** English geologist Arthur Holmes used radioactivity to date rocks, establishing the Earth to be 4.6 billion years old.
- **1895** In the USA, Jeanette Picard launched the first balloon to be used for stratospheric research.
- **1896** Swedish chemist Svante Arrhenius discovered a link between the amount of carbon dioxide in the atmosphere and the global temperature.
- **1897** Norwegian-US meteorologist Jacob Bjerknes and his father Vilhelm developed the mathematical theory of weather forecasting.

earth science: continued

1902	British physicist Oliver Heaviside and US engineer Arthur Edwin Kennelly predicted the existence of an electrified layer in the atmosphere that reflects radio waves. In France, Léon Teisserenc discovered layers of different temperatures in the atmosphere, which he called the troposphere and stratosphere.
1906	Richard Dixon Oldham proved the Earth to have a molten core by studying seismic waves.
1909	Yugoslav physicist Andrija Mohorovičč discovered a discontinuity in the Earth's crust, about 30 km/18 mi below the surface, that forms the boundary between the crust and the mantle.
1912	In Germany, Alfred Wegener proposed the theory of continental drift and the existence of a supercontinent, Pangaea, in the distant past.
1913	French physicist Charles Fabry discovered the ozone layer in the upper atmosphere.
1914	German-US geologist Beno Gutenberg discovered the discontinuity that marks the boundary between the Earth's mantle and the outer core.
1922	British meteorologist Lewis Fry Richardson developed a method of numerical weather forecasting.
1925	A German expedition discovered the Mid-Atlantic ridge by means of sonar. Edward Appleton discovered a layer of the atmosphere that reflects radio waves; it was later named after him.
1929	By studying the magnetism of rocks, Japanese geologist Motonori Matuyama showed that the Earth's magnetic field reverses direction from time to time.
1935	US seismologist Charles Francis Richter established a scale for measuring the magnitude of earthquakes.
1936	Danish seismologist Inge Lehmann postulated the existence of a solid inner core of the Earth from the study of seismic waves.
1939	In Germany, Walter Maurice Elsasser proposed that eddy currents in the molten iron core cause the Earth's magnetism.
1950	Hungarian-US mathematician John Von Neumann made the first 24-hour weather forecast by computer.
1956	US geologists Bruce Charles Heezen and Maurice Ewing discovered a global network of oceanic ridges and rifts that divide the Earth's surface into plates.
1958	Using rockets, US physicist James Van Allen discovered a belt of radiation (later named after him) around the Earth.
1960	The world's first weather satellite, *TIROS 1*, was launched. US geologist Harry Hammond Hess showed that the sea floor spreads out from ocean ridges and descends back into the mantle at deep-sea trenches.
1963	British geophysicists Fred Vine and Drummond Matthews analysed the magnetism of rocks in the Atlantic Ocean floor and found conclusive proof of seafloor spreading.
1985	A British expedition to the Antarctic discovered a hole in the ozone layer above the South Pole.
1991	A borehole in the Kola Peninsula in Arctic Russia, begun in the 1970s, reached a depth of 12,261 m/40,240 ft (where the temperature was found to be 210°C/410°F). It is expected to reach a depth of 15,000 m/49,000 ft by 1995.

the existing one. The old layer then splits, the animal moves free of it, and the new layer expands and hardens.

ECG abbreviation for ◊electrocardiogram.

echinoderm marine invertebrate of the phylum Echinodermata ('spiny-skinned'), characterized by a five-radial symmetry. Echinoderms have a water-vascular system which transports substances around the body. They include starfishes (or sea stars), britlestars, sea-lilies, sea-urchins, and sea-cucumbers. The skeleton is external, made of a series of limy plates, and echinoderms generally move by using tube-feet, small water-filled sacs that can be protruded or pulled back to the body.

echo repetition of a sound wave, or of a ◊radar or ◊sonar signal, by reflection from a surface. By accurately measuring the time taken for an echo to return to the transmitter, and by knowing the speed of a radar signal (the speed of light) or a sonar signal (the speed of sound in water), it is possible to calculate the range of the object causing the echo (echolocation).

echolocation or *biosonar* method used by certain animals, notably bats, whales and dolphins, to detect the positions of objects by using sound. The animal emits a stream of high-pitched sounds, generally at ultrasonic frequencies (beyond the range of human hearing), and listens for the returning echoes reflected off objects to determine their exact location.

echo sounder or *sonar device* device that detects objects under water by means of ◊sonar

– by using reflected sound waves. Most boats are equipped with echo sounders to measure the water depth beneath them. An echo sounder consists of a transmitter, which emits an ultrasonic pulse (see ◊ultrasound), and a receiver, which detects the pulse after reflection from the seabed. The time between transmission and receipt of the reflected signal is a measure of the depth of water. Fishing boats also use echo sounders to detect shoals of fish.

eclipse passage of an astronomical body through the shadow of another. The term is usually employed for solar and lunar eclipses, which may be either partial or total, but also, for example, for eclipses by Jupiter of its satellites. An eclipse of a star by a body in the Solar System is called an occultation.

A *solar eclipse* occurs when the Moon passes in front of the Sun as seen from Earth, and can happen only at new Moon. During a total eclipse the Sun's ◊corona can be seen. A total solar eclipse can last up to 7.5 minutes. When the Moon is at its farthest from Earth it does not completely cover the face of the Sun, leaving a ring of sunlight visible. This is an **annular eclipse** (from the Latin word *annulus* 'ring'). Between two and five solar eclipses occur each year.

A *lunar eclipse* occurs when the Moon passes into the shadow of the Earth, becoming dim until emerging from the shadow. Lunar eclipses may be partial or total, and they can happen only at full Moon. Total lunar eclipses last for up to 100 minutes; the maximum number each year is three.

eclipsing binary binary (double) star in which the two stars periodically pass in front of each other as seen from Earth.

When one star crosses in front of the other the total light received on Earth from the two stars declines. The first eclipsing binary to be noticed was ◊Algol.

ecliptic path, against the background of stars, that the Sun appears to follow each year as the Earth orbits the Sun. It can be thought of as the plane of the Earth's orbit projected on to the ◊celestial sphere (imaginary sphere around the Earth).

ecology study of the relationship among organisms and the environments in which they live, including all living and nonliving components. The term was coined by the biologist Ernst Haeckel 1866.

Ecology may be concerned with individual organisms (for example, behavioural ecology,

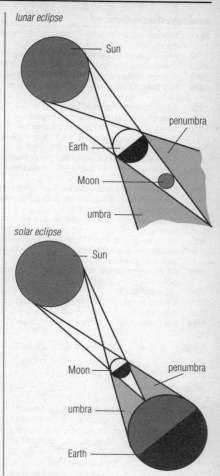

eclipse *The two types of eclipse: lunar and solar. A lunar eclipse occurs when the Moon passes through the shadow of the Earth. A solar eclipse occurs when the Moon passes between the Sun and the Earth, blocking out the Sun's light. During a total solar eclipse, when the Moon completely covers the Sun, the Moon's shadow sweeps across the Earth's surface from west to east at a speed of 3,200 kph/2,000 mph.*

feeding strategies), with populations (for example, population dynamics), or with entire communities (for example, competition between species for access to resources in an ecosystem, or predator–prey relationships).

ecology: chronology

1735	Swedish naturalist Carl Linnaeus developed his system for classifying and naming plants and animals.
1798	English reverend Thomas Malthus produced the earliest theoretical study of population dynamics.
1859	English naturalist Charles Darwin published his *On the Origin of Species*.
1869	German zoologist Ernst Haeckel first defined the term ecology.
1899	US botanist Henry Cowles published his classic paper on succession in sand dunes on Lake Michigan, USA.
1913	British Ecological Society founded.
1915	Ecological Society of America founded.
1916	US ecologist Frederic Clements coined the phrase 'climax communities' for large areas of rather uniform vegetation which he attributed to climactic factors.
1926	Russian botanist N I Vavilov published *Centres of Origin of Cultivated Plants*, concluding that there are relatively few such centres, many of which are located in mountainous areas.
1934	Russian ecologist G F Gause first stated the principles of competitive exclusion, related to a species' niche.
1935	British ecologist Arthur Tansley first coined the term ecosystem.
1938	The coelacanth, a marine fish believed to have become extinct 65 million years ago, was 'rediscovered' in the Indian Ocean.
1940	Population biologist Charles Elton developed the idea of trophic levels in a community of organisms.
1950	The theory that natural selection may favour either individuals with high reproductive rates and rapid development (*r*-selection) or individuals with low reproductive rates and better competitive ability (*k*-selection) is first discussed.
1967	US biologists MacArthur and Wilson proposed their 'Theory of Island Biogeography' which related population and community size to island size. The theory is still widely used in the design of nature reserves today.
1979	English naturalist James Lovelock proposed his Gaia hypothesis, viewing the planet as a single organism.
1993	UN Convention on Biological Diversity came into force.

Applied ecology is concerned with the management and conservation of habitats and the consequences and control of pollution.

ecosystem in ◊ecology, an integrated unit consisting of the ◊community of living organisms and the nonliving, or physical, environment in a particular area. The relationships among species in an ecosystem are usually complex and finely balanced, and removal of any one species may be disastrous. The removal of a major predator, for example, can result in the destruction of the ecosystem through overgrazing by herbivores. Ecosystems can be identified at different scales – for example, the global ecosystem consists of all the organisms living on Earth, the Earth itself (both land and sea), and the atmosphere above; a freshwater pond ecosystem consists of the plants and animals living in the pond, the pondwater and all the substances dissolved or suspended in that water, and the rocks, mud, and decaying matter that make up the pond bottom.

Energy and nutrients pass through organisms in an ecosystem in a particular sequence (see ◊food chain).

ectoparasite ◊parasite that lives on the outer surface of its host.

ectoplasm outer layer of a cell's ◊cytoplasm.

ectotherm 'cold-blooded' animal (see ◊poikilothermy), such as a lizard, that relies on external warmth (ultimately from the Sun) to raise its body temperature so that it can become active. To cool the body, ectotherms seek out a cooler environment.

eddy current electric current induced, in accordance with ◊Faraday's laws, in a conductor located in a changing magnetic field. Eddy currents can cause much wasted energy in the cores of transformers and other electrical machines.

EEG abbreviation for ◊*electroencephalogram*.

EEPROM (acronym for *electrically erasable programmable read-only memory*) computer memory that can record data and retain it indefinitely. The data can be erased with an electrical charge and new data recorded.

efficiency output of a machine (work done by the machine) divided by the input (work put into the machine), usually expressed as a percentage. Because of losses caused by friction, efficiency is always less than 100%, although it can approach this for electrical machines with no moving parts (such as a transformer).

efflorescence loss of water or crystallization of crystals exposed to air, resulting in a dry powdery surface.

effluent liquid discharge of waste from an industrial process, usually into rivers or the sea. Effluent is often toxic but is difficult to control and hard to trace.

EFTPOS (acronym for *electronic funds transfer at point of sale*) a form of electronic funds transfer.

egestion the removal of undigested food or faeces from the gut. In most animals egestion takes place via the anus, although the invertebrate flatworms must use the mouth because their gut has no exit. Egestion is the last part of a complex feeding process that starts with food capture and continues with digestion and assimilation.

egg in animals, the ovum, or female ◊gamete (reproductive cell).

After fertilization by a sperm cell, it begins to divide to form an embryo. Eggs may be deposited by the female (◊ovipary) or they may develop within her body (◊vivipary and ◊ovovivipary). In the oviparous reptiles and birds, the egg is protected by a shell, and well supplied with nutrients in the form of yolk.

eidophor television projection system that produces pictures up to 10 m/33 ft square, at sports events or rock concerts, for example. The system uses three coloured beams of light, one of each primary colour (red, blue, and green), which scan the screen. The intensity of each beam is controlled by the strength of the corresponding colour in the television picture.

einsteinium synthesized, radioactive, metallic element of the actinide series, symbol Es, atomic number 99, relative atomic mass 254.

elastic collision in physics, a collision between two or more bodies in which the total ◊kinetic energy of the bodies is conserved (remains constant); none is converted into any other form of energy. ◊Momentum also is conserved in such collisions. The molecules of a gas may be considered to collide elastically, but large objects may not because some of their kinetic energy will be converted on collision to heat and sound.

elasticity in physics, the ability of a solid to recover its shape once deforming forces (stresses modifying its dimensions or shape) are removed. An elastic material obeys

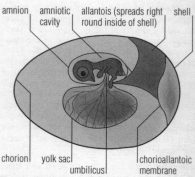

egg *Sections through an unfertilized and a fertilized bird egg. Inside a bird's egg is a complex structure of liquids and membranes designed to meet the needs of the growing embryo. The yolk, which is rich in fat, is gradually absorbed by the embryo. The white of the egg provides protein and water. The chalaza is a twisted band of protein which holds the yolk in place and acts as a shock absorber. The airspace allows gases to be exchanged through the shell. The allantois contains many blood vessels which carry gases between the embryo and the outside.*

◊Hooke's law: that is, its deformation is proportional to the applied stress up to a certain point, called the *elastic limit*, beyond which additional stress will deform it permanently. Elastic materials include metals and rubber; however, all materials have some degree of elasticity.

elastomer any material with rubbery properties, which stretches easily and then quickly returns to its original length when released.

Natural and synthetic rubbers and such materials as polychloroprene and butadiene copolymers are elastomers. The convoluted molecular chains making up these materials are uncoiled by a stretching force, but return to their original position when released because there are relatively few crosslinks between the chains.

E layer (formerly called the Kennelly–Heaviside layer) the lower regions of the ◊ionosphere, which refract radio waves, allowing their reception around the surface of the Earth. The E layer approaches the Earth by day and recedes from it at night.

electrical relay an electromagnetic switch; see ◊relay.

electrical safety measures taken to protect human beings from electric shock or from fires caused by electrical faults. They are of paramount importance in the design of electrical equipment. Safety measures include the fitting of earth wires, and fuses or circuit breakers; the insulation of wires; the double insulation of portable equipment; and the use of residual-current devices (RCDs), which will break a circuit and cut off all currents if there is any imbalance between the currents in the live and neutral wires connected to an appliance (caused, for example, if some current is being conducted through a person).

electric arc a continuous electric discharge of high current between two electrodes, giving out a brilliant light and heat. The phenomenon is exploited in the carbon-arc lamp, once widely used in film projectors. In the electric-arc furnace an arc struck between very large carbon electrodes and the metal charge provides the heating. In arc ◊welding an electric arc provides the heat to fuse the metal. The discharges in low-pressure gases, as in neon and sodium lights, can also be broadly considered as electric arcs.

electric charge property of some bodies that causes them to exert forces on each other. Two bodies both with positive or both with negative charges repel each other, whereas bodies with opposite or 'unlike' charges attract each other, since each is in the ◊electric field of the other. In atoms, ◊electrons possess a negative charge, and ◊protons an equal positive charge. The ◊SI unit of electric charge is the coulomb (symbol C).

Electric charge can be generated by friction induction or chemical change and shows itself as an accumulation of electrons (negative charge) or loss of electrons (positive charge) on an atom or body. Atoms have no charge but can sometimes gain electrons to become negative ions or lose them to become positive ions. So-called ◊static electricity, seen in such phenomena as the charging of nylon shirts when they are pulled on or off, or in brushing hair, is in fact the gain or loss of electrons from the surface atoms. A flow of charge (such as electrons through a copper wire) constitutes an electric current; the flow of current is measured in amperes (symbol A).

electric current the flow of electrically charged particles through a conducting circuit due to the presence of a ◊potential difference. The current at any point in a circuit is the amount of charge flowing per second; its SI unit is the ampere (coulomb per second).

Current carries electrical energy from a power supply, such as a battery of electrical cells, to the components of the circuit, where it is converted into other forms of energy, such as heat, light, or motion. It may be either ◊direct current or ◊alternating current.

heating effect When current flows in a component possessing resistance, electrical energy is converted into heat energy. If the resistance of the component is R ohms and the current through it is I amperes, then the heat energy W (in joules) generated in a time t seconds is given by the formula:

$$W = I^2 R t$$

magnetic effect A ◊magnetic field is created around all conductors that carry a current. When a current-bearing conductor is made into a coil it forms an ◊electromagnet with a magnetic field that is similar to that of a bar magnet, but which disappears as soon as the current is switched off. The strength of the magnetic field is directly proportional to the current in the conductor. The direction of the field created around a conducting wire may be predicted by using ◊Maxwell's screw rule.

motor effect A conductor carrying current in a magnetic field experiences a force, and is impelled to move in a direction perpendicular to both the direction of the current and the direction of the magnetic field. The direction of motion may be predicted by Fleming's left-hand rule (see ◊Fleming's rules). The magnitude of the force experienced depends on the length of the conductor and on the strengths of the current and the magnetic field, and is greatest when the conductor is at right angles to the field. A conductor wound into a coil that can rotate between the poles of a magnet forms the basis of an ◊electric motor.

electric field in physics, a region in which a particle possessing electric charge experiences a force owing to the presence of another electric charge. It is a type of ◊electromagnetic field.

electricity all phenomena caused by ◊electric charge, whether static or in motion. Electric charge is caused by an excess or deficit of electrons in the charged substance, and an electric current by the movement of electrons around a circuit. Substances may be electrical conductors, such as metals, which allow the passage of electricity through them, or insulators, such as rubber, which are extremely poor conductors. Substances with relatively poor conductivities that can be improved by the addition of heat or light are known as ◊semiconductors.

electric motor a machine that converts electrical energy into mechanical energy. There are various types, including direct-current and induction motors, most of which produce rotary motion. A linear induction motor produces linear (sideways) rather than rotary motion.

A simple *direct-current motor* consists of a horseshoe-shaped permanent ◊magnet with a wire-wound coil (◊armature) mounted so that it can rotate between the poles of the magnet. A ◊commutator reverses the current (from a battery) fed to the coil on each half-turn, which rotates because of the mechanical force exerted on a conductor carrying a current in a magnetic field.

An *induction motor* employs ◊alternating current. It comprises a stationary current-carrying coil (stator) surrounding another coil (rotor), which rotates because of the current induced in it by the magnetic field created by the stator; it thus requires no commutator.

electric power the rate at which an electrical machine uses electrical ◊energy or converts it into other forms of energy – for example, light, heat, mechanical energy. Usually measured in watts (equivalent to joules per second), it is equal to the product of the voltage and the current flowing.

electrocardiogram (ECG) graphic recording of the electrical activity of the heart, as detected by electrodes placed on the skin. Electrocardiography is used in the diagnosis of heart disease.

electrochemical series list of chemical elements arranged in descending order of the ease with which they can lose electrons to form cations (positive ions). An element can be displaced (◊displacement reaction) from a compound by any element above it in the series.

electrochemistry the branch of science that studies chemical reactions involving electricity. The use of electricity to produce chemical effects, ◊electrolysis, is employed in many industrial processes, such as the manufacture of chlorine and the extraction of aluminium. The use of chemical reactions to produce electricity is the basis of electrical ◊cells, such as the dry cell and the ◊Leclanché cell.

Since all chemical reactions involve changes to the electronic structure of atoms, all

simple direct-current motor

electric motor *Simple direct current electric motor. Larger DC motors have many coils, each at a slight angle from its neighbour, and the commutator is split into many segments, each connected to one coil. Universal motors, like those found in clothes washing machines, can run on either DC or AC.*

reactions are now recognized as electrochemical in nature. Oxidation, for example, was once defined as a process in which oxygen was combined with a substance, or hydrogen was removed from a compound; it is now defined as a process in which electrons are lost.

Electrochemistry was also used as the basis for methods of destroying toxic organic pollutants 1993. For example, the development of electrochemical cells that operate with supercritical water (see ◊fluid, supercritical) to combust organic waste materials.

electrode any terminal by which an electric current passes in or out of a conducting substance; for example, the anode or cathode in a battery or the carbons in an arc lamp. The terminals that emit and collect the flow of electrons in thermionic ◊valves (electron tubes) are also called electrodes: for example, cathodes, plates, and grids.

electrodynamics the branch of physics dealing with electric currents and associated magnetic forces. ◊Quantum electrodynamics (QED) studies the interaction between charged particles and their emission and absorption of electromagnetic radiation. This field combines quantum theory and relativity theory, making accurate predictions about subatomic processes involving charged particles such as electrons and protons.

electroencephalogram (EEG) graphic record of the electrical discharges of the brain, as detected by electrodes placed on the scalp. The pattern of electrical activity revealed by electroencephalography is helpful in the diagnosis of some brain disorders, in particular epilepsy.

electrolysis in chemistry, the production of chemical changes by passing an electric current through a solution or molten salt (the electrolyte), resulting in the migration of ions to the electrodes: positive ions (cations) to the negative electrode (cathode) and negative ions (anions) to the positive electrode (anode). One application of electrolysis is ◊electroplating.

During electrolysis, the ions react with the electrode, either receiving or giving up electrons. The resultant atoms may be liberated as a gas, or deposited as a solid on the electrode, in amounts that are proportional to the amount of current passed, as discovered by English chemist Michael Faraday (1791–1867). For instance, when acidified water is electrolysed, hydrogen ions (H^+) at the cathode receive electrons to form hydrogen gas; hydroxide ions (OH^-) at the anode give up electrons to form oxygen gas and water.

electrolyte solution or molten substance in which an electric current is made to flow by the movement and discharge of ions in accordance with Faraday's laws of ◊electrolysis.

electromagnet iron bar with coils of wire around it, which acts as a magnet when an electric current flows through the wire. Electromagnets have many uses: in switches, electric bells, solenoids, and metal-lifting cranes.

electromagnetic field in physics, the region in which a particle with an ◊electric charge experiences a force. If it does so only when moving, it is in a pure *magnetic field*; if it does so when stationary, it is in an *electric field*. Both can be present simultaneously.

electromagnetic force one of the four fundamental ◊forces of nature, the other three being gravity, the strong nuclear force, and the weak nuclear force. The ◊elementary particle that is the carrier for the electromagnetic (em) force is the photon.

electromagnetic induction in electronics, the production of an ◊electromotive force (emf) in a circuit by a change of magnetic flux through the circuit or by relative motion of the circuit and the magnetic flux. In a closed circuit an ◊induced current will be produced. All dynamos and generators make use of this effect. When magnetic tape is driven past the playback head (a small coil) of a tape-recorder, the moving magnetic field induces an emf in

electrolysis of acidified water

electrolysis *Passing an electric current through acidified water breaks down the water into its constituent elements – hydrogen and oxygen.*

ELECTROMAGNETIC WAVES

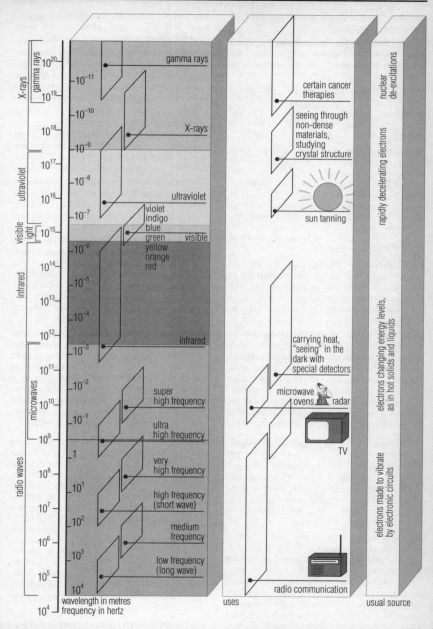

electromagnetic waves *Radio waves have the lowest frequency. Infrared radiation, visible light, ultraviolet radiation, X-rays, and gamma rays have progressively higher frequencies.*

the head, which is then amplified to reproduce the recorded sounds.

electromagnetic spectrum the complete range, over all wavelengths from the lowest to the highest, of ◊electromagnetic waves.

electromagnetic system of units former system of absolute electromagnetic units (emu) based on the ◊c.g.s. system and having, as its primary electrical unit, the unit magnetic pole. It was replaced by ◊SI units.

electromagnetic waves oscillating electric and magnetic fields travelling together through space at a speed of nearly 300,000 km/186,000 mi per second. The (limitless) range of possible wavelengths or ◊frequencies of electromagnetic waves, which can be thought of as making up the *electromagnetic spectrum*, includes radio waves, infrared radiation, visible light, ultraviolet radiation, X-rays, and gamma rays.

electromotive force (emf) loosely, the voltage produced by an electric battery or generator or, more precisely, the energy supplied by a source of electric power in driving a unit charge around an electrical circuit. The unit is the ◊volt.

electron stable, negatively charged ◊elementary particle; it is a constituent of all atoms, and a member of the class of particles known as ◊leptons. The electrons in each atom surround the nucleus in groupings called shells; in a neutral atom the number of electrons is equal to the number of protons in the nucleus. This electron structure is responsible for the chemical properties of the atom (see ◊atomic structure).

Electrons are the basic particles of electricity. Each carries a charge of 1.602192×10^{-19} coulomb, and all electrical charges are multiples of this quantity. A beam of electrons will undergo ◊diffraction (scattering) and produce interference patterns in the same way as ◊electromagnetic waves such as light; hence they may also be regarded as waves.

electronegativity the ease with which an atom can attract electrons to itself. Electronegative elements attract electrons, so forming negative ions.

US chemist Linus Pauling (1901–1994) devised an electronegativity scale to indicate the relative power of attraction of elements for electrons. Fluorine, the most nonmetallic element, has a value of 4.0 on this scale; oxygen, the next most nonmetallic, has a value of 3.5.

In a covalent bond between two atoms of different electronegativities, the bonding electrons will be located close to the more electronegative atom, creating a ◊dipole.

electron gun a part in many electronic devices consisting of a series of ◊electrodes, including a cathode for producing an electron beam. It plays an essential role in ◊cathode-ray tubes (television tubes) and ◊electron microscopes.

electronic flash discharge tube that produces a high-intensity flash of light, used for photography in dim conditions. The tube contains an inert gas such as krypton. The flash lasts only a few thousandths of a second.

electronic mail or *e-mail* ◊telecommunications system that enables the users of a computer network to send messages to other users. Telephone wires are used to send the signals from terminal to terminal.

electronic point of sale (EPOS) system used in retailing in which a bar code on a product is scanned at the cash till and the information relayed to the store computer. The computer will then relay back the price of the item to the cash till. The customer can then be given an itemized receipt while the computer removes the item from stock figures.

electronic publishing the distribution of information using computer-based media such as ◊multimedia and ◊hypertext in the creation of electronic 'books'. Critical technologies in the development of electronic publishing were ◊CD-ROM, with its massive yet compact storage capabilities, and the advent of computer networking with its ability to deliver information instantaneously anywhere in the world.

electronics branch of science that deals with the emission of ◊electrons from conductors and ◊semiconductors, with the subsequent manipulation of these electrons, and with the construction of electronic devices. The first electronic device was the thermionic ◊valve, or vacuum tube, in which electrons moved in a vacuum, and led to such inventions as ◊radio, ◊television, ◊radar, and the digital ◊computer. Replacement of valves with the comparatively tiny and reliable transistor from 1948 revolutionized electronic development. Modern electronic devices are based on minute ◊integrated circuits (silicon chips), wafer-thin crystal slices holding tens of thousands of electronic components.

electron microscope instrument that produces a magnified image by using a beam of ◊electrons instead of light rays, as in an optical

electronics: chronology

1897	The electron was discovered by English physicist John Joseph Thomson.
1904	English physicist Ambrose Fleming invented the diode valve, which allows flow of electricity in one direction only.
1906	The triode electron valve, the first device to control an electric current, was invented by US physicist Lee De Forest.
1947	John Bardeen, William Shockley, and Walter Brattain invented the junction germanium transistor at the Bell Laboratories, New Jersey, USA.
1952	British physicist G W A Dunner proposed the integrated circuit.
1953	Jay Forrester of the Massachusetts Institute of Technology, USA, built a magnetic memory smaller than existing vacuum-tube memories.
1954	The silicon transistor was developed by Gordon Teal of Texas Instruments, USA.
1958	The first integrated circuit, containing five components, was built by US electrical physicist Jack Kilby.
1959	The planar transistor, which is built up in layers, or planes, was designed by Robert Noyce of Fairchild Semiconductor Corporation, USA.
1961	Steven Hofstein designed the field-effect transistor used in integrated circuits.
1971	The first microprocessor, the Intel 4004, was designed by Ted Hoff in the USA; it contained 2,250 components and could add two four-bit numbers in 11-millionths of a second.
1974	The Intel 8080 microprocessor was launched; it contained 4,500 components and could add two eight-bit numbers in 2.5-millionths of a second.
1979	The Motorola 68000 microprocessor was introduced; it contained 70,000 components and could multiply two 16-bit numbers in 3.2-millionths of a second.
1981	The Hewlett-Packard Superchip was introduced; it contained 450,000 components and could multiply two 32-bit numbers in 1.8-millionths of a second.
1985	The Inmos T414 transputer, the first microprocessor designed for use in parallel computers, was launched.
1988	The first optical microprocessor, which uses light instead of electricity, was developed.
1989	Wafer-scale silicon memory chips were introduced: the size of a beer mat, they are able to store 200 million characters.
1990	Memory chips capable of holding 4 million bits of information began to be mass-produced in Japan. The chips can store the equivalent of 520,000 characters, or the contents of a 16-page newspaper. Each chip contains 9 million components packed on a piece of silicon less than 15 mm long by 5 mm wide.
1992	Transistors made from high-temperature superconducting ceramics rather than semiconductors produced in Japan by Sanyo Electric. The new transistors are 10 times faster than semiconductor transistors.
1993	US firm Intel launches the Pentium 64-bit microprocessor, with two separate integer processing units that can run in parallel, promising to be 10 times faster than earlier processors.

◊microscope. An *electron lens* is an arrangement of electromagnetic coils that control and focus the beam. Electrons are not visible to the eye, so instead of an eyepiece there is a fluorescent screen or a photographic plate on which the electrons form an image. The wavelength of the electron beam is much shorter than that of light, so much greater magnification and resolution (ability to distinguish detail) can be achieved. The development of the electron microscope has made possible the observation of very minute organisms, viruses, and even large molecules.

electron probe microanalyser modified ◊electron microscope in which the target emits X-rays when bombarded by electrons. Varying X-ray intensities indicate the presence of different chemical elements. The composition of a specimen can be mapped without the specimen being destroyed.

electrons, delocalized electrons that are not associated with individual atoms or identifiable chemical bonds, but are shared collectively by all the constituent atoms or ions of some chemical substances (such as metals, graphite, and ◊aromatic compounds).

electrons, localized a pair of electrons in a ◊covalent bond that are located in the vicinity of the nuclei of the two contributing atoms. Such electrons cannot move beyond this area.

electron spin resonance in archaeology, a nondestructive dating method applicable to teeth, bone, heat-treated flint, ceramics, sediments, and stalagmitic concretions. It enables electrons, displaced by natural radiation and

electron microscope *The transmission electron microscope. The electromagnetic coils are the 'lenses' of the electron microscope, focusing the electron beam on the specimen. The beam travels through the specimen and is projected on to a screen or imagers. These microscopes can magnify by a million times.*

then trapped in the structure, to be measured; their number indicates the age of the specimen.

electron volt unit (symbol eV) for measuring the energy of a charged particle (◊ion or ◊electron) in terms of the energy of motion an electron would gain from a potential difference of one volt. Because it is so small, more usual units are mega-(million) and giga-(billion) electron volts (MeV and GeV).

electrophoresis the ◊diffusion of charged particles through a fluid under the influence of an electric field. It can be used in the biological sciences to separate ◊molecules of different sizes, which diffuse at different rates. In industry, electrophoresis is used in paint-dipping operations to ensure that paint reaches awkward corners.

electroplating deposition of metals upon metallic surfaces by electrolysis for decorative and/or protective purposes. It is used in the preparation of printers' blocks, 'master' audio discs, and in many other processes.

A current is passed through a bath containing a solution of a salt of the plating metal, the object to be plated being the cathode (negative terminal); the anode (positive terminal) is either an inert substance or the plating metal. Among the metals most commonly used for plating are zinc, nickel, chromium, cadmium, copper, silver, and gold.

electroporation in biotechnology, a technique of introducing foreign ◊DNA into pollen with a strong burst of electricity, used in creating genetically engineered plants.

electropositivity in chemistry, a measure of the ability of elements (mainly metals) to donate electrons to form positive ions. The greater the metallic character, the more electropositive the element.

electrorheological fluid another name for ◊smart fluid, a liquid suspension that gels when an electric field is applied across it.

electroscope apparatus for detecting ◊electric charge. The simple gold-leaf electroscope consists of a vertical conducting (metal) rod ending in a pair of rectangular pieces of gold foil, mounted inside and insulated from an earthed metal case. An electric charge applied to the end of the metal rod makes the gold leaves diverge, because they each receive a similar charge (positive or negative) and so repel each other.

electrostatic precipitator device that removes dust or other particles from air and other gases by electrostatic means. An electric discharge is passed through the gas, giving the impurities a negative electric charge. Positively charged plates are then used to attract the charged particles and remove them from the gas flow. Such devices are attached to the chimneys of coal-burning power stations to remove ash particles.

electrostatics the study of electric charges from stationary sources (not currents). See ◊static electricity.

electrovalent bond another name for an ◊ionic bond, a chemical bond in which the combining atoms lose or gain electrons to form ions.

electrum naturally occurring alloy of gold and silver used by early civilizations to make the first coins, about the 6th century BC.

element substance that cannot be split chemically into simpler substances. The atoms of a particular element all have the same number of protons in their nuclei (their atomic number). Elements are classified in the ◊periodic table of the elements. Of the 109 known elements, 95 are known to occur in nature (those with atomic numbers 1–95). Those from 96 to 109 do not occur in nature and are synthesized only, produced in particle accelerators. Eighty-one of the elements are stable; all the others, which include atomic numbers 43, 61, and from 84 up, are radioactive.

Elements are classified as metals, nonmetals, or metalloids (weakly metallic elements) depending on a combination of their physical and chemical properties; about 75% are metallic. Some elements occur abundantly (oxygen, aluminium); others occur moderately or rarely (chromium, neon); some, in particular the radioactive ones, are found in minute (neptunium, plutonium) or very minute (technetium) amounts.

Symbols are used to denote the elements; the symbol is usually the first letter or letters of the English or Latin name (for example, C for carbon, Ca for calcium, Fe for iron, *ferrum*).

elementary particle in physics, a subatomic particle that is not made up of smaller particles, and so can be considered one of the fundamental units of matter. There are three groups of elementary particles: ◊quarks, ◊leptons, and ◊gauge bosons.

elements, the four earth, air, fire, and water. The Greek philosopher Empedocles (c. 490–430 BC) believed that these four elements made up the fundamental components of all matter and that they were destroyed and renewed through the action of love and discord.

elevation of boiling point raising of the boiling point of a liquid above that of the pure solvent, caused by a substance being dissolved in it. The phenomenon is observed when salt is added to boiling water; the water ceases to boil because its boiling point has been elevated.

How much the boiling point is raised depends on the number of molecules of substance dissolved. For a single solvent, such as pure water, all substances in the same molecular concentration (expressed in ◊moles) produce the same elevation of boiling point. The elevation e produced by the presence of a solute of molar concentration C is given by the equation $e = KC$, where K is a constant (called the ebullioscopic constant) for the solvent concerned.

ellipse curve joining all points (loci) around two fixed points (foci) such that the sum of the distances from those points is always constant.

ellipse *Technical terms used to describe the ellipse; for all points on the ellipse, the sum of the distances from the two foci, F_1 and F_2, is the same.*

The diameter passing through the foci is the major axis, and the diameter bisecting this at right angles is the minor axis. An ellipse is one of a series of curves known as ◊conic sections. A slice across a cone that is not made parallel to, and does not pass through, the base will produce an ellipse.

elongation in astronomy, the angular distance between the Sun and a planet or other solar-system object. This angle is 0° at ◊conjunction, 90° at ◊quadrature, and 180° at ◊opposition.

elution in chemistry, washing of an adsorbed substance from the adsorbing material; it is used, for example, in the separation processes of chromatography and electrophoresis.

e-mail abbreviation for ◊*electronic mail*.

embolism blockage of a blood vessel by an obstruction called an embolus (usually a blood clot, fat particle, or bubble of air).

embryo early developmental stage of an animal or a plant following fertilization of an ovum (egg cell), or activation of an ovum by ◊parthenogenesis. In humans, the term embryo describes the fertilized egg during its first seven weeks of existence; from the eighth week onwards it is referred to as a fetus.

In animals the embryo exists either within an egg (where it is nourished by food contained in the yolk), or in mammals, in the ◊uterus of the mother. In mammals (except marsupials) the embryo is fed through the ◊placenta. The plant embryo is found within the seed in higher plants. It sometimes consists of only a few cells, but usually includes a root, a shoot (or primary bud), and one or two ◊cotyledons, which nourish the growing seedling.

embryology study of the changes undergone by an organism from its conception as a

embryo *The development of a bird and a human embryo. In the human, division of the fertilized egg, or ovum, begins within hours of conception. Within a week, a hollow, fluid-containing ball a blastocyte – with a mass of cells at one end has developed. After the third week, the embryo has changed from a mass of cells into a recognizable shape. At four weeks, the embryo is 3 mm/ 0.1 in long, with a large bulge for the heart and small pits for the ears. At six weeks, the embryo is 1.5 cm/0.6 in with a pulsating heart and ear flaps. By the eighth week, the embryo (now technically a fetus) is 2.5 cm long and recognizably human.*

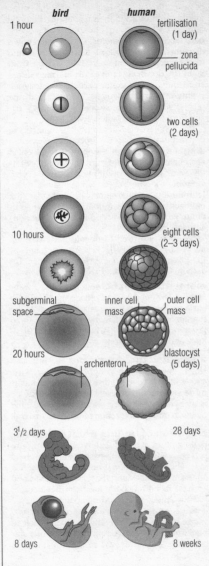

fertilized ovum (egg) to its emergence into the world at hatching or birth. It is mainly concerned with the changes in cell organization in the embryo and the way in which these lead to the structures and organs of the adult (the process of ◊differentiation).

embryo research the study of human embryos at an early stage, in order to detect hereditary disease and genetic defects, and to investigate fertility problems.

embryo sac large cell within the ovule of flowering plants that represents the female ◊gametophyte when fully developed. It typically contains eight nuclei. Fertilization occurs when one of these nuclei, the egg nucleus, fuses with a male ◊gamete.

emerald a clear, green gemstone variety of the mineral ◊beryl. It occurs naturally in Colombia, the Ural Mountains, in Russia, Zimbabwe, and Australia.

emery greyish-black opaque metamorphic rock consisting of ◊corundum and magnetite, together with other minerals such as hematite. It is used as an ◊abrasive.

emf in physics, abbreviation for ◊*electromotive* force.

emission spectroscopy in analytical chemistry, a technique for determining the identity or amount present of a chemical substance by measuring the amount of electromagnetic radiation it emits at specific wavelengths; see ◊spectroscopy.

empirical formula the simplest chemical formula for a compound. Quantitative analysis gives the proportion of each element present, and from this the empirical formula is calculated. It is related to the actual (molecular) formula by the relation:

(empirical formula)$_n$ = molecular formula

where n is a small whole number (1, 2, ...):

emulator in computing, an item of software or firmware that allows one device to imitate the functioning of another. Emulator software is commonly used to allow one make of computer to run programs written for a different make of computer. This allows a user to select from a wider range of ◊applications programs, and perhaps to save money by running programs designed for an expensive computer on a cheaper model.

emulsifier food ◊additive used to keep oils dispersed and in suspension, in products such as mayonnaise and peanut butter. Egg yolk is a naturally occurring emulsifier, but most of the emulsifiers in commercial use today are synthetic chemicals.

emulsion a stable dispersion of a liquid in another liquid – for example, oil and water in some cosmetic lotions.

encephalin a naturally occurring chemical produced by nerve cells in the brain that has the same effect as morphine or other derivatives of opium, acting as a natural painkiller. Unlike morphine, encephalins are quickly degraded by the body, so there is no build-up of tolerance to them, and hence no addiction. Encephalins are a variety of ◊peptides, as are ◊endorphins, which have similar effects.

Encke's comet comet with the shortest known orbital period, 3.3 years. It is named after German mathematician and astronomer Johann Franz Encke (1791–1865), who calculated its orbit in 1819 from earlier sightings.

endangered species plant or animal species whose numbers are so few that it is at risk of becoming extinct. Officially designated endangered species are listed by the International Union for the Conservation of Nature (IUCN).

endocrine gland gland that secretes hormones into the bloodstream to regulate body processes. Endocrine glands are most highly developed in vertebrates, but are also found in other animals, notably insects. In humans the main endocrine glands are the pituitary, thyroid, parathyroid, adrenal, pancreas, ovary, and testis.

endocrinology medical specialty devoted to the diagnosis and treatment of hormone disorders.

endolymph fluid found in the inner ◊ear, filling the central passage of the cochlea as well as the semicircular canals.

endoparasite ◊parasite that lives inside the body of its host.

endoplasm inner, liquid part of a cell's ◊cytoplasm.

endoplasmic reticulum (ER) a membranous system of tubes, channels, and flattened sacs that form compartments within ◊eukaryotic cells. It stores and transports proteins within cells and also carries various enzymes needed for the synthesis of ◊fats. The ◊ribosomes, or the organelles that carry out protein synthesis, are attached to parts of the ER.

endorphin natural substance (a polypeptide) that modifies the action of nerve cells. Endorphins are produced by the pituitary gland and hypothalamus of vertebrates. They lower the perception of pain by reducing the transmission of signals between nerve cells.

endoscopy examination of internal organs or tissues by an instrument allowing direct vision. An endoscope is equipped with an eyepiece, lenses, and its own light source to illuminate the field of vision. The endoscope used to examine the digestive tract is a flexible fibreoptic instrument swallowed by the patient.

endoskeleton the internal supporting structure of vertebrates, made up of cartilage or bone. It provides support, and acts as a system of levers to which muscles are attached to provide movement. Certain parts of the skeleton (the skull and ribs) give protection to vital body organs.

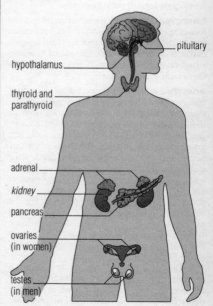

endocrine gland *The main human endocrine glands. These glands produce hormones – chemical messengers – which travel in the bloodstream to stimulate certain cells.*

endosperm nutritive tissue in the seeds of most flowering plants. It surrounds the embryo and is produced by an unusual process that parallels the ◊fertilization of the ovum by a male gamete. A second male gamete from the pollen grain fuses with two female nuclei within the ◊embryo sac. Thus endosperm cells are triploid (having three sets of chromosomes); they contain food reserves such as starch, fat, and protein that are utilized by the developing seedling.

endotherm 'warm-blooded', or homeothermic, animal. Endotherms have internal mechanisms for regulating their body temperatures to levels different from the environmental temperature. See ◊homeothermy.

endothermic reaction chemical reaction that requires an input of energy in the form of heat for it to proceed; the energy is absorbed from the surroundings by the reactants.

end user the user of a computer program; in particular, someone who uses a program to perform a task (such as accounting or playing a computer game), rather than someone who writes programs (a programmer).

Energiya most powerful Soviet space rocket, first launched 15 May 1987. Used to launch the Soviet space shuttle, the Energiya ◊booster is capable, with the use of strap-on boosters, of launching payloads of up to 190 tonnes into Earth orbit.

energy capacity for doing ◊work. Potential energy (PE) is energy deriving from position; thus a stretched spring has elastic PE, and an object raised to a height above the Earth's surface, or the water in an elevated reservoir, has gravitational PE. A lump of coal and a tank of petrol, together with the oxygen needed for their combustion, have chemical energy. Other sorts of energy include electrical and nuclear energy, and light and sound. Moving bodies possess kinetic energy (KE). Energy can be converted from one form to another, but the total quantity stays the same (in accordance with the ◊conservation of energy principle). For example, as an apple falls, it loses gravitational PE but gains KE.

Although energy is never lost, after a number of conversions it tends to finish up as the kinetic energy of random motion of molecules (of the air, for example) at relatively low temperatures. This is 'degraded' energy in that it is difficult to convert it back to other forms.

energy, alternative energy from sources that are renewable and ecologically safe, as

energy level *The arrangement of electrons in an atom of chlorine with three energy levels or shells. As the energy of an electron depends upon its distance from the nucleus, all the electrons shown at the same distance from the nucleus have the same energy*

opposed to sources that are nonrenewable with toxic by-products, such as coal, oil, or gas (fossil fuels), and uranium (for nuclear power). The most important alternative energy source is flowing water, harnessed as ◊hydroelectric power. Other sources include the oceans' tides and waves (see ◊tidal power station and ◊wave power), wind (harnessed by windmills and wind turbines), the Sun (◊solar energy), and the heat trapped in the Earth's crust (◊geothermal energy).

energy level or *shell* or *orbital* the permitted energy that an electron can have in any particular system. Energy levels can be calculated using ◊quantum theory. The permitted energy levels depend mainly on the distance of the electron from the nucleus. See ◊orbital, atomic.

energy of reaction or *enthalpy of reaction* or *heat of reaction* energy released or absorbed during a chemical reaction. In a chemical reaction, the energy stored in the reacting molecules is rarely the same as that stored in the product molecules. Depending on which is the greater, energy is either released (an exothermic reaction) or absorbed (an endothermic reaction) from the surroundings (see ◊conservation of energy). The amount of energy released or absorbed by the quantities of substances represented by the chemical equation is the energy of reaction.

engine device for converting stored energy into useful work or movement. Most engines use a fuel as their energy store. The fuel is burnt to produce heat energy – hence the name 'heat engine' – which is then converted into movement. Heat engines can be classified according to the fuel they use (◊petrol engine or ◊diesel engine), or according to whether the fuel is burnt inside (◊internal combustion engine) or outside (◊steam engine) the engine, or according to whether they produce a reciprocating or rotary motion (◊turbine or ◊Wankel engine).

engineering the application of science to the design, construction, and maintenance of works, machinery, roads, railways, bridges, harbour installations, engines, ships, aircraft and airports, spacecraft and space stations, and the generation, transmission, and use of electrical power. The main divisions of engineering are aerospace, chemical, civil, electrical, electronic, gas, marine, materials, mechanical, mining, production, radio, and structural.

engineering drawing technical drawing that forms the plans for the design and construction of engineering components and structures. Engineering drawings show different projections, or views of objects, with the relevant dimensions, and show how all the separate parts fit together.

enthalpy in chemistry, alternative term for ◊energy of reaction, the heat energy associated with a chemical change.

entomology study of ◊insects.

entropy in ◊thermodynamics, a parameter representing the state of disorder of a system at the atomic, ionic, or molecular level; the greater the disorder, the higher the entropy. Thus the fast-moving disordered molecules of water vapour have higher entropy than those of more ordered liquid water, which in turn have more entropy than the molecules in solid crystalline ice.

E number code number for additives that have been approved for use by the European Commission (EC). The E written before the number stands for European. E numbers do not have to be displayed on lists of ingredients, and the manufacturer may choose to list ◊additives by their name instead. E numbers cover all categories of additives apart from flavourings. Additives, other than flavourings, that are not approved by the EC, but are still used in Britain, are represented by a code number without an E.

environment in ecology, the sum of conditions affecting a particular organism, including physical surroundings, climate, and influences of other living organisms. See also ◊biosphere and ◊habitat.

environment–heredity controversy see ◊*nature–nurture controversy*.

enzyme biological ◊catalyst produced in cells, and capable of speeding up the chemical reactions necessary for life by converting one molecule (substrate) into another. Enzymes are not themselves destroyed by this process. They are large, complex ◊proteins, and are highly specific, each chemical reaction requiring its own particular enzyme. The enzyme fits into a 'slot' (active site) in the substrate molecule, forming an enzyme–substrate complex that lasts until the substrate is altered or split, after which the enzyme can fall away. The substrate may therefore be compared to a lock, and the enzyme to the key required to open it.

Eocene second epoch of the Tertiary period of geological time, 56.5–35.5 million years ago. Originally considered the earliest division of the Tertiary, the name means 'early recent', referring to the early forms of mammals evolv-

enzyme *Lock-and-key model of enzyme activity. Ezymes owe their activity to the three-dimensional shape of their molecules, which allow the molecules of the substances on which they act to fit into a special location on the enzyme molecule. A chemical reaction then takes place and the products are released. This action can be repeated up to 100,000 times each second.*

eotvos unit unit (symbol E) for measuring small changes in the intensity of the Earth's ◊gravity with horizontal distance.

ephemeral plant plant with a very short life cycle, sometimes as little as six or eight weeks. It may complete several generations in one growing season.

epicentre the point on the Earth's surface immediately above the seismic focus of an ◊earthquake. Most damage usually takes place at an earthquake's epicentre. The term sometimes refers to a point directly above or below a nuclear explosion ('at ground zero').

epicyclic gear or *sun-and-planet gear* gear system that consists of one or more gear wheels moving around another. Epicyclic gears are found in bicycle hub gears and in automatic gearboxes.

epidermis outermost layer of ◊cells on an organism's body. In plants and many invertebrates such as insects, it consists of a single layer of cells. In vertebrates, it consists of several layers of cells.

epididymis in male vertebrates, a long coiled tubule in the ◊testis, in which sperm produced in the seminiferous tube are stored. In men, it is a duct about 6 m/20 ft long, convoluted into a small space.

epigeal seed germination in which the ◊cotyledons (seed leaves) are borne above the soil.

epiglottis small flap located behind the root of the tongue in mammals. It closes off the end of the windpipe during swallowing to prevent food from passing into it and causing choking.

epiphyte any plant that grows on another plant or object above the surface of the ground, and has no roots in the soil. An epiphyte does not parasitize the plant it grows on but merely uses it for support. Its nutrients are obtained from rainwater, organic debris such as leaf litter, or from the air.

epithelium in animals, tissue of closely packed cells that forms a surface or lines a cavity or tube. Epithelium may be protective (as in the skin) or secretory (as in the cells lining the wall of the gut).

epoch subdivision of a geological period in the geological time scale. Epochs are sometimes given their own names (such as the Palaeocene, Eocene, Oligocene, Miocene, and Pliocene epochs comprising the Tertiary period), or they are referred to as the late, early, or middle portions of a given period (as the Late Cretaceous or the Middle Triassic epoch).

epoxy resin synthetic ◊resin used as an ◊adhesive and as an ingredient in paints. Household epoxy resin adhesives come in component form as two separate tubes of chemical, one tube containing resin, the other a curing agent (hardener). The two chemicals are mixed just before application, and the mix soon sets hard.

EPROM (acronym for *erasable programmable read-only memory*) computer memory device in the form of an ◊integrated circuit (chip) that can record data and retain it indefinitely. The data can be erased by exposure to ultraviolet light, and new data recorded. Other kinds of computer memory chips are ◊ROM (read-only memory), ◊PROM (programmable read-only memory), and ◊RAM (random-access memory).

Epsilon Aurigae ◊eclipsing binary star in the constellation Auriga. One of the pair is an 'ordinary' star, but the other seems to be an enormous distended object whose exact nature remains unknown. The period (time between eclipses) is 27 years, the longest of its kind. The last eclipse was 1982–84.

equation in mathematics, expression that represents the equality of two expressions involving constants and/or variables, and thus usually includes an equals sign (=). For example, the equation $A = \pi r^2$ equates the area A of a circle of radius r to the product πr^2.

The algebraic equation $y = mx + c$ is the general one in coordinate geometry for a straight line.

equation in chemistry, representation of a chemical reaction by symbols and numbers; see ◊chemical equation.

equations of motion mathematical equations that give the position and velocity of a moving object at any time. Given the mass of an object, the forces acting on it, and its initial position and velocity, the equations of motion are used to calculate its position and velocity at any later time. The equations must be based on ◊Newton's laws of motion or, if speeds near that of light are involved, on the theory of ◊relativity.

equator *terrestrial equator* or the ◊great circle whose plane is perpendicular to the Earth's axis (the line joining the poles). Its length is 40,092 km/24,901.8 mi, divided into

360 degrees of longitude. The equator encircles the broadest part of the Earth, and represents 0° latitude. It divides the Earth into two halves, called the northern and the southern hemispheres.

equilibrium in physics, an unchanging condition in which the forces acting on a particle or system of particles (a body) cancel out, or in which energy is distributed among the particles of a system in the most probable way. In accordance with Newton's first law of motion, a body in equilibrium remains at rest or moves with constant velocity; it does not accelerate.

equinox the points in spring and autumn at which the Sun's path, the ◊ecliptic, crosses the celestial equator, so that the day and night are of approximately equal length. The *vernal equinox* occurs about 21 March and the *autumnal equinox*, 23 Sept.

era any of the major divisions of geological time, each including several periods, but smaller than an eon. The currently recognized eras all fall within the Phanerozoic eon – or the vast span of time, starting about 570 million years ago, when fossils are found to become abundant. The eras in ascending order are the Palaeozoic, Mesozoic, and Cenozoic. We are living in the Recent epoch of the Quaternary period of the Cenozoic era.

erasable optical disc in computing, another name for a ◊floptical disc.

erbium soft, lustrous, greyish, metallic element of the ◊lanthanide series, symbol Er, atomic number 68, relative atomic mass 167.26. It occurs with the element yttrium or as a minute part of various minerals. It was discovered 1843 by Carl Mosander (1797–1858), and named after the town of Ytterby, Sweden, near which the lanthanides (rare-earth elements) were first found.

erg c.g.s. unit of work, replaced in the SI system by the ◊joule. One erg of work is done by a force of one ◊dyne moving through one centimetre.

ergonomics study of the relationship between people and the furniture, tools, and machinery they use at work. The object is to improve work performance by removing sources of muscular stress and general fatigue: for example, by presenting data and control panels in easy-to-view form, making office furniture comfortable, and creating a generally pleasant environment.

Eridanus in astronomy, the sixth largest constellation, which meanders from the celestial equator deep into the southern hemisphere of the sky. Its brightest star is ◊Achernar. Eridanus is represented as a river.

Eros in astronomy, an asteroid, discovered 1898, that can pass 22 million km/14 million mi from the Earth, as observed in 1975. Eros was the first asteroid to be discovered that has an orbit coming within that of Mars. It is elongated, measures about 36 × 12 km/22 × 7 mi, rotates around its shortest axis every 5.3 hours, and orbits the Sun every 1.8 years.

erosion wearing away of the Earth's surface, caused by the breakdown and transportation of particles of rock or soil (by contrast, ◊weathering does not involve transportation). Agents of erosion include the sea, rivers, glaciers, and wind.

Water, consisting of sea waves and currents, rivers, and rain; ice, in the form of glaciers; and wind, hurling sand fragments against exposed rocks and moving dunes along, are the most potent forces of erosion. People also contribute to erosion by bad farming practices and the cutting down of forests, which can lead to the formation of dust bowls.

error in computing, a fault or mistake, either in the software or on the part of the user, that causes a program to stop running (crash) or produce unexpected results. Program errors, or bugs, are largely eliminated in the course of the programmer's initial testing procedure, but some will remain in most programs. All computer operating systems are designed to produce an *error message* (on the display screen, or in an error file or printout) whenever an error is detected, reporting that an error has taken place and, wherever possible, diagnosing its cause.

error detection in computing, the techniques that enable a program to detect incorrect data. A common method is to add a check digit to important codes, such as account numbers and product codes. The digit is chosen so that the code conforms to a rule that the program can verify. Another technique involves calculating the sum (called the ◊hash total) of each instance of a particular item of data, and storing it at the end of the data.

erythrocyte another name for ◊red blood cell.

erythropoietin in biology, a naturally occurring hormone that increases the production of red blood cells, which carry oxygen around the body. It is released in response to a lowered percentage of oxygen in the blood reaching the kidneys, such as in anaemic subjects. Recombinant human erythropoietin is used therapeutically to treat the anaemia associated with chronic kidney failure. It is also used illegally by athletes to enhance their performance.

ESA abbreviation for ◊*European Space Agency*.

escape velocity in physics, minimum velocity with which an object must be projected for it to escape from the gravitational pull of a planetary body. In the case of the Earth, the escape velocity is 11.2 kps/6.9 mps; the Moon 2.4 kps/1.5 mps; Mars 5 kps/3.1 mps; and Jupiter 59.6 kps/37 mps.

ester organic compound formed by the reaction between an alcohol and an acid, with the elimination of water. Unlike ◊salts, esters are covalent compounds.

estradiol alternative spelling of oestradiol, a type of ◊oestrogen (female sex hormone).

esophagus US spelling of ◊oesophagus.

estrogen US spelling of ◊oestrogen.

estrus US spelling of ◊oestrus.

ethanal common name *acetaldehyde* CH_3CHO one of the chief members of the group of organic compounds known as ◊aldehydes. It is a colourless inflammable liquid boiling at 20.8°C/69.6°F. Ethanal is formed by the oxidation of ethanol or ethene and is used to make many other organic chemical compounds.

ethanal trimer common name *paraldehyde* $(CH_3CHO)_3$ colourless liquid formed from ethanal. It is soluble in water.

ethane CH_3CH_3 colourless, odourless gas, the second member of the ◊alkane series of hydrocarbons (paraffins).

ethane-1,2-diol technical name for ◊glycol.

ethanoate common name *acetate* $CH_3CO_2H^{2-}$ negative ion derived from ethanoic (acetic) acid; any salt containing this ion. In textiles, acetate rayon is a synthetic fabric made from modified cellulose (wood pulp) treated with ethanoic acid; in photography, acetate film is a nonflammable film made of cellulose ethanoate.

ethanoic acid common name *acetic acid* CH_3CO_2H one of the simplest fatty acids (a series of organic acids). In the pure state it is a colourless liquid with an unpleasant pungent odour; it solidifies to an icelike mass of crystals at 16.7°C/62.4°F, and hence is often called

ester *Molecular model of the ester ethanoate, $CH_3CH_2COOCH_3$.*

glacial ethanoic acid. Vinegar contains 5% or more ethanoic acid, produced by fermentation.

ethanol common name *ethyl alcohol* C_2H_5OH alcohol found in beer, wine, cider, spirits, and other alcoholic drinks. When pure, it is a colourless liquid with a pleasant odour, miscible with water or ether; it burns in air with a pale blue flame. The vapour forms an explosive mixture with air and may be used in high-compression internal combustion engines. It is produced naturally by the fermentation of carbohydrates by yeast cells.

Industrially, it can be made by absorption of ethene and subsequent reaction with water, or by the reduction of ethanal in the presence of a catalyst, and is widely used as a solvent.

ethanolic solution solution produced when a solute is dissolved in ethanol; for example, ethanolic potassium hydroxide is a solution of potassium hydroxide in ethanol.

ethene common name *ethylene* C_2H_4 colourless, flammable gas, the first member of the ◊alkene series of hydrocarbons. It is the most widely used synthetic organic chemical and is used to produce the plastics polyethene (polyethylene), polychloroethene, and polyvinyl chloride (PVC). It is obtained from natural gas or coal gas, or by the dehydration of ethanol.

Ethene is produced during plant metabolism and is classified as a plant hormone. It is important in the ripening of fruit and in ◊abscission. Small amounts of ethene are often added to the air surrounding fruit to artificially promote ripening.

ether in chemistry, any of a series of organic chemical compounds having an oxygen atom linking the carbon atoms of two hydrocarbon radical groups (general formula R-O-R'); also the common name for ethoxyethane $C_2H_5OC_2H_5$ (also called diethyl ether).

This is used as an anaesthetic and as an external cleansing agent before surgical operations. It is also used as a solvent, and in the extraction of oils, fats, waxes, resins, and alkaloids.

ether or *aether* in the history of science, a hypothetical medium permeating all of space. The concept originated with the Greeks, and has been revived on several occasions to explain the properties and propagation of light. It was supposed that light and other electromagnetic radiation – even in outer space – needed a medium, the ether, in which to travel. The idea was abandoned with the acceptance of ◊relativity.

Ethernet in computing, a protocol for ◊local area networks. Ethernet was developed principally by the Xerox Corporation, but can now be used on many computers. It allows data transfer at rates up to 10 Mbps.

ethnoarchaeology the study of human behaviour and the material culture of living societies, in order to see how materials enter the archaeological record, and that way to provide hypotheses explaining the production, use, and disposal patterns of ancient material culture.

ethology comparative study of animal behaviour in its natural setting. Ethology is concerned with the causal mechanisms (both the stimuli that elicit behaviour and the physiological mechanisms controlling it), as well as the development of behaviour, its function, and its evolutionary history.

ethyl alcohol common name for ◊ethanol.

ethylene common name for ◊ethene.

ethylene glycol alternative name for ◊glycol.

ethyne common name *acetylene* CHCH colourless inflammable gas produced by mixing calcium carbide and water. It is the simplest member of the ◊alkyne series of hydrocarbons. It is used in the manufacture of the synthetic rubber neoprene, and in oxyacetylene welding and cutting.

etiolation in botany, a form of growth seen in plants receiving insufficient light. It is characterized by long, weak stems, small leaves, and a pale yellowish colour (◊chlorosis) owing to a lack of chlorophyll. The rapid increase in height enables a plant that is surrounded by others to quickly reach a source of light, after which a return to normal growth usually occurs.

eugenics study of ways in which the physical and mental characteristics of the human race may be improved. The eugenic principle was abused by the Nazi Party in Germany during the 1930s and early 1940s to justify the attempted extermination of entire social and ethnic groups and the establishment of selective breeding programmes.

Modern eugenics is concerned mainly with the elimination of genetic disease.

eukaryote in biology, one of the two major groupings into which all organisms are divided. Included are all organisms, except bacteria and cyanobacteria (◊blue-green algae), which belong to the ◊prokaryote grouping.

The cells of eukaryotes possess a clearly defined nucleus, bounded by a membrane, within which DNA is formed into distinct chromosomes. Eukaryotic cells also contain mitochondria, chloroplasts, and other structures (organelles) that, together with a defined nucleus, are lacking in the cells of prokaryotes.

Euratom acronym for ◊*European Atomic Energy Commission*, forming part of the European Union organization.

eureka in chemistry, alternative name for the copper-nickel alloy ◊constantan, which is used in electrical equipment.

Europa in astronomy, the fourth-largest moon of the planet Jupiter, diameter 3,140 km/1,950 mi, orbiting 671,000 km/417,000 mi from the planet every 3.55 days. It is covered by ice and criss-crossed by thousands of thin cracks, each some 50,000 km/30,000 mi long.

European Atomic Energy Commission (Euratom) organization established by the second Treaty of Rome 1957, which seeks the cooperation of member states of the European Union in nuclear research and the rapid and large-scale development of nonmilitary nuclear energy.

European Space Agency (ESA) organization of European countries (Austria, Belgium, Denmark, France, Germany, Ireland, Italy, the Netherlands, Norway, Spain, Sweden, Switzerland, and the UK) that engages in space research and technology. It was founded 1975, with headquarters in Paris.

europium soft, greyish, metallic element of the ◊lanthanide series, symbol Eu, atomic number 63, relative atomic mass 151.96. It is used in lasers and as the red phosphor in colour televisions; its compounds are used to make control rods for nuclear reactors. It was named in 1901 by French chemist Eugène Demarçay (1852–1904) after the continent of Europe, where it was first found.

eusociality form of social life found in insects such as honey bees and termites, in which the colony is made up of special castes (for example, workers, drones, and reproductives) whose membership is biologically determined. The worker castes do not usually reproduce. Only one mammal, the naked mole rat, has a social organization of this type. See also ◊social behaviour.

Eustachian tube small air-filled canal connecting the middle ◊ear with the back of the throat. It is found in all land vertebrates and equalizes the pressure on both sides of the eardrum.

eutrophication excessive enrichment of rivers, lakes, and shallow sea areas, primarily by nitrate fertilizers washed from the soil by rain, by phosphates from fertilizers, and from nutrients in municipal sewage, and by sewage itself. These encourage the growth of algae and bacteria which use up the oxygen in the water, thereby making it uninhabitable for fishes and other animal life.

evaporation process in which a liquid turns to a vapour without its temperature reaching boiling point. A liquid left to stand in a saucer eventually evaporates because, at any time, a proportion of its molecules will be fast enough (have enough kinetic energy) to escape through the attractive intermolecular forces at the liquid surface into the atmosphere. The temperature of the liquid tends to fall because the evaporating molecules remove energy from the liquid. The rate of evaporation rises with increased temperature because as the mean kinetic energy of the liquid's molecules rises, so will the number possessing enough energy to escape.

evaporite sedimentary deposit precipitated on evaporation of salt water. With a progressive evaporation of seawater, the most common salts are deposited in a definite sequence: calcite (calcium carbonate), gypsum (hydrous calcium sulphate), halite (sodium chloride), and finally salts of potassium and magnesium.

evergreen in botany, a plant such as pine, spruce, or holly, that bears its leaves all year round. Most ◊conifers are evergreen. Plants that shed their leaves in autumn or during a dry season are described as ◊deciduous.

evolution slow process of change from one form to another, as in the evolution of the universe from its formation in the ◊Big Bang to its present state, or in the evolution of life on Earth. Some Christians and Muslims deny the theory of evolution as conflicting with the belief that God created all things (see ◊creationism).

The idea of continuous evolution can be traced as far back as the 1st century BC, but it did not gain wide acceptance until the 19th century following the work of Scottish geologist Charles Lyell, French naturalist Jean Baptiste Lamarck, English naturalist Charles Darwin, and English biologist Thomas Henry Huxley. Darwin assigned the major role in evolutionary change to ◊natural selection acting on randomly occurring variations (now known to be produced by spontaneous changes or ◊mutations in the genetic material of organisms). Natural selection occurs because those individuals better adapted to their particular environments reproduce more effectively, thus contributing their characteristics (in the form of genes) to future generations. The current theory of evolution, called ◊neo-Darwinism, combines Darwin's theory with Austrian biologist Gregor Mendel's (1822–1884) theories on genetics. Although neither the general concept of evolution nor the importance of natural selection is doubted by biologists, there remains dispute over other possible processes involved in evolutionary change. Besides natural selection and ◊sexual selection, chance may play a large part in deciding which genes become characteristic of a population, a phenomenon called 'genetic drift'. It is now also clear that evolutionary change does not always occur at a constant rate, but that the process can have long periods of relative stability interspersed with periods of rapid change. This has led to new theories, such as the ◊punctuated equilibrium model. See also ◊adaptive radiation.

evolutionary stable strategy (ESS) in ◊sociobiology, an assemblage of behavioural or physical characters (collectively termed a 'strategy') of a population that is resistant to replacement by any forms bearing new traits, because the new traits will not be capable of successful reproduction.

ESS analysis is based on ◊game theory and can be applied both to genetically determined physical characters (such as horn length), and to learned behavioural responses (for example, whether to fight or retreat from an opponent). An ESS may be conditional on the context, as in the rule 'fight if the opponent is smaller, but retreat if the opponent is larger.'

exclusion principle in physics, a principle of atomic structure originated by Austrian–US physicist Wolfgang Pauli (1900–1958). It states that no two electrons in a single atom may have the same set of ◊quantum numbers.

Hence, it is impossible to pack together certain elementary particles, such as electrons, beyond a certain critical density, otherwise they would share the same location and quantum number. A white dwarf star is thus prevented from contracting further by the exclusion principle and never collapses.

excretion in biology, the removal of waste products from the cells of living organisms. In plants and simple animals, waste products are removed by diffusion, but in higher animals they are removed by specialized organs. In mammals the kidneys are the principle organs of excretion. Water and metabolic wastes are also excreted in the faeces and, in humans, through the sweat glands; carbon dioxide and water are removed via the lungs.

executable file in computing, a file – always a program of some kind – that can be run by the computer directly. The file will have been generated from a ◊source program by an ◊assembler or ◊compiler. It will therefore not be coded in ◊ASCII and will not be readable as text. On ◊MS-DOS systems executable files have a .EXE or .COM extension.

exocrine gland gland that discharges secretions, usually through a tube or a duct, onto a surface. Examples include sweat glands which release sweat onto the skin, and digestive glands which release digestive juices onto the walls of the intestine. Some animals also have ◊endocrine glands (ductless glands) that release hormones directly into the bloodstream.

exon in genetics, a sequence of bases in the genetic material ◊DNA that codes for a protein. Exons make up only 2% of the body's total DNA. The remainder is made up of ◊introns. During RNA processing the introns are cut out of the molecule and the exons spliced together.

exoskeleton the hardened external skeleton of insects, spiders, crabs, and other arthropods. It provides attachment for muscles and protection for the internal organs, as well as support. To permit growth it is periodically shed in a process called ◊ecdysis.

exosphere the uppermost layer of the ◊atmosphere. It is an ill-defined zone above the thermosphere, beginning at about 700 km/ 435 mi and fading off into the vacuum of space. The gases are extremely thin, with hydrogen as the main constituent.

exothermic reaction a chemical reaction during which heat is given out (see ◊energy of reaction).

expanded memory in computing, additional memory in an ◊MS-DOS-based computer, usually installed on an expanded-memory board. Expanded memory requires an expanded-memory manager, which gives access to a limited amount of memory at any one time, and is slower to use than ◊extended memory. Software is available under both MS-DOS and ◊Windows to simulate expanded memory for those applications that require it.

expansion in physics, the increase in size of a constant mass of substance caused by, for example, increasing its temperature (◊thermal expansion) or its internal pressure. The *expansivity*, or coefficient of thermal expansion, of a material is its expansion (per unit volume, area, or length) per degree rise in temperature.

expert system computer program for giving advice (such as diagnosing an illness or interpreting the law) that incorporates knowledge derived from human expertise. It is a kind of ◊knowledge-based system containing rules that can be applied to find the solution to a problem. It is a form of ◊artificial intelligence.

Explorer series of US scientific satellites. *Explorer 1*, launched Jan 1958, was the first US satellite in orbit and discovered the Van Allen radiation belts around the Earth.

explosive any material capable of a sudden release of energy and the rapid formation of a large volume of gas, leading when compressed to the development of a high-pressure wave (blast).

exponent or *index* in mathematics, a number that indicates the number of times a term is multiplied by itself; for example $x^2 = x \times x$, $4^3 = 4 \times 4 \times 4$.

exponential in mathematics, descriptive of a ◊function in which the variable quantity is an exponent (a number indicating the power to which another number or expression is raised).

export file in computing, a file stored by the computer in a standard format so that it can be accessed by other programs, possibly running on different makes of computer.

extended memory in computing, memory in an ◊MS-DOS-based system that exceeds the 1 Mbyte that DOS supports. Extended memory is not accessible to the ◊operating system and requires an extended memory manager. ◊Windows and Windows applications require extended memory

extensor a muscle that straightens a limb.

extinction the complete disappearance of a species. In the past, extinctions are believed to have occurred because species were unable to adapt quickly enough to a naturally changing environment. Today, most extinctions are due to human activity. Some species, such as the dodo of Mauritius, the moas of New Zealand, and the passenger pigeon of North America, were exterminated by hunting. Others became extinct when their habitat was destroyed. See also ⟡endangered species.

extracellular matrix strong material naturally occurring in animals and plants, made up of protein and long-chain sugars (polysaccharides) in which cells are embedded. It is often called a 'biological glue', and forms part of ⟡connective tissues such as bone and skin.

extrusion common method of shaping metals, plastics, and other materials. The materials, usually hot, are forced through the hole in a metal die and take its cross-sectional shape. Rods, tubes, and sheets may be made in this way.

eye the organ of vision. In the human eye, the light is focused by the combined action of the curved ***cornea***, the internal fluids, and the ***lens***. The insect eye is compound – made up of many separate facets – known as ***ommatidia***, each of which collects light and directs it separately to a receptor to build up an image. Invertebrates have much simpler eyes, with no lenses. Among molluscs, cephalopods have complex eyes similar to those of vertebrates.

The mantis shrimp's eyes contain ten colour pigments with which to perceive colour; some flies and fishes have five, while the human eye has only three.

eye *The human eye. The retina of the eye contains about 137 million light sensitive cells in an area of about 650 sq mm/1 sq in. There are 130 million rod cells for black and white vision and 7 million cone cells for colour vision. The optic nerve contains about 1 million nerve fibres. The focusing muscles of the eye adjust about 100,000 times a day. To exercise the leg muscles to the same extent would need an 80 km/50 mi walk.*

°F symbol for degrees ◊*Fahrenheit*.

facsimile transmission full name for ◊*fax* or *telefax*.

factor a number that divides into another number exactly. For example, the factors of 64 are 1, 2, 4, 8, 16, 32, and 64. In algebra, certain kinds of polynomials (expressions consisting of several or many terms) can be factorized. For example, the factors of $x^2 + 3x + 2$ are $x + 1$ and $x + 2$, since $x^2 + 3x + 2 = (x + 1)(x + 2)$. This is called factorization. See also ◊prime number.

factorial of a positive number, the product of all the whole numbers (integers) inclusive between 1 and the number itself. A factorial is indicated by the symbol !. Thus $6! = 1 \times 2 \times 3 \times 4 \times 5 \times 6 = 720$. Factorial zero, 0!, is defined as 1.

faeces remains of food and other waste material eliminated from the digestive tract of animals by way of the anus. Faeces consist of quantities of fibrous material, bacteria and other microorganisms, rubbed-off lining of the digestive tract, bile fluids, undigested food, minerals, and water.

Fahrenheit scale temperature scale invented 1714 by Gabriel Fahrenheit which was commonly used in English-speaking countries until the 1970s, after which the ◊Celsius scale was generally adopted, in line with the rest of the world. In the Fahrenheit scale, intervals are measured in degrees (°F); °F = (°C × ⁹⁄₅) + 32.

Fallopian tube or *oviduct* in mammals, one of two tubes that carry eggs from the ovary to the uterus. An egg is fertilized by sperm in the Fallopian tubes, which are lined with cells whose ◊cilia move the egg towards the uterus.

fallout harmful radioactive material released into the atmosphere in the debris of a nuclear explosion and descending to the surface of the Earth. Such material can enter the food chain, cause ◊radiation sickness, and last for hundreds of thousands of years (see ◊half-life).

false-colour imagery graphic technique that displays images in false (not true-to-life) colours so as to enhance certain features. It is widely used in displaying electronic images taken by spacecraft; for example, Earth-survey satellites such as *Landsat*. Any colours can be selected by a computer processing the received data.

falsificationism in philosophy of science, the belief that a scientific theory must be under constant scrutiny and that its merit lies only in how well it stands up to rigorous testing. It was first expounded by philosopher Karl Popper in his *Logic of Scientific Discovery* 1934.

family in biological classification, a group of related genera (see ◊genus). Family names are not printed in italic (unlike genus and species names), and by convention they all have the ending -idae (animals) or -aceae (plants and fungi). For example, the genera of hummingbirds are grouped in the hummingbird family, Trochilidae. Related families are grouped together in an ◊order.

fanjet another name for ◊turbofan, the jet engine used by most airliners.

farad SI unit (symbol F) of electrical capacitance (how much electricity a ◊capacitor can store for a given voltage). One farad is a capacitance of one ◊coulomb per volt. For practical purposes the microfarad (one millionth of a farad) is more commonly used. The farad is named after English scientist Michael Faraday.

faraday unit of electrical charge equal to the charge on one mole of electrons. Its value is 9.648×10^4 coulombs.

Faraday's constant constant (symbol *F*) representing the electric charge carried on one mole of electrons. It is found by multiplying Avogadro's constant by the charge carried on a single electron, and is equal to 9.648×10^4 coulombs per mole.

One *faraday* is this constant used as a unit. The constant is used to calculate the electric charge needed to discharge a particular quantity of ions during ◊electrolysis.

Faraday's laws three laws of electromagnetic induction, and two laws of electrolysis, all proposed originally by English scientist Michael Faraday:

induction (1) a changing magnetic field induces an electromagnetic force in a conductor; (2) the electromagnetic force is proportional to the rate of change of the field; (3) the direction of the induced electromagnetic force depends on the orientation of the field.

electrolysis (1) the amount of chemical change during electrolysis is proportional to the charge passing through the liquid; (2) the amount of chemical change produced in a substance by a given amount of electricity is proportional to the electrochemical equivalent of that substance.

fast breeder or *breeder reactor* alternative names for ◊fast reactor, a type of nuclear reactor.

fast reactor or *fast breeder reactor* ◊nuclear reactor that makes use of fast neutrons to bring about fission. Unlike other reactors used by the nuclear-power industry, it has little or no ◊moderator, to slow down neutrons. The reactor core is surrounded by a 'blanket' of uranium carbide. During operation, some of this uranium is converted into plutonium, which can be extracted and later used as fuel. Fast breeder reactors can extract about 60 times the amount of energy from uranium that thermal reactors do.

fat in the broadest sense, a mixture of ◊lipids – chiefly triglycerides (lipids containing three ◊fatty acid molecules linked to a molecule of glycerol). More specifically, the term refers to a lipid mixture that is solid at room temperature (20°C); lipid mixtures that are liquid at room temperature are called *oils*. The higher the proportion of saturated fatty acids in a mixture, the harder the fat.

Boiling fats in strong alkali forms soaps (saponification). Fats are essential constituents of food for many animals, with a calorific value twice that of carbohydrates; however, eating too much fat, especially fat of animal origin, has been linked with heart disease in humans. In many animals and plants, excess carbohydrates and proteins are converted into fats for storage. Mammals and other vertebrates store fats in specialized connective tissues (◊adipose tissues), which not only act as energy reserves but also insulate the body and cushion its organs.

As a nutrient fat serves five purposes: it is a source of energy (9 kcal/g); makes the diet palatable; provides basic building blocks for cell structure; provides essential fatty acids (linoleic and linolenic); and acts as a carrier for fat-soluble vitamins (A, D, E, and K). Foods rich in fat are butter, lard, margarine, and cooking oils. Products high in monounsaturated or polyunsaturated fats are thought to be less likely to contribute to cardiovascular disease.

fatty acid or *carboxylic acid* organic compound consisting of a hydrocarbon chain, up to 24 carbon atoms long, with a carboxyl group (–COOH) at one end. The covalent bonds between the carbon atoms may be single or double; where a double bond occurs the carbon atoms concerned carry one instead of two hydrogen atoms. Chains with only single

fat *The molecular structure of typical fat. The molecule consists of three fatty acid molecules linked to a molecule of glycerol.*

glucose or *dextrose* or *grape-sugar* $C_6H_{12}O_6$ sugar present in the blood, and found also in honey and fruit juices. It is a source of energy for the body, being produced from other sugars and starches to form the 'energy currency' of many biochemical reactions also involving ◊ATP.

glue type of ◊adhesive.

gluon in physics, a ◊gauge boson that carries the ◊strong nuclear force, responsible for binding quarks together to form the strongly interacting subatomic particles known as ◊hadrons. There are eight kinds of gluon.

gluten protein found in cereal grains, especially wheat and rye. Gluten enables dough to expand during rising. Sensitivity to gluten gives rise to coeliac disease.

glyceride ◊ester formed between one or more acids and glycerol (propan-1,2,3-triol). A glyceride is termed a mono-, di-, or triglyceride, depending on the number of hydroxyl groups from the glycerol that have reacted with the acids.

glycerine another name for ◊glycerol.

glycerol or *glycerine* or *propan-1,2,3-triol* $HOCH_2CH(OH)CH_2OH$ thick, colourless, odourless, sweetish liquid. It is obtained from vegetable and animal oils and fats (by treatment with acid, alkali, superheated steam, or an enzyme), or by fermentation of glucose, and is used in the manufacture of high explosives, in antifreeze solutions, to maintain moist conditions in fruits and tobacco, and in cosmetics.

glycine $CH_2(NH_2)COOH$ the simplest amino acid, and one of the main components of proteins. When purified, it is a sweet, colourless crystalline compound.

glycogen polymer (a polysaccharide) of the sugar ◊glucose made and retained in the liver as a carbohydrate store, for which reason it is sometimes called animal starch. It is a source of energy when needed by muscles, where it is converted back into glucose by the hormone ◊insulin and metabolized.

glycol or *ethylene glycol* or *ethane-1,2-diol* $(CH_2OH)_2$ thick, colourless, odourless, sweetish liquid. It is used in antifreeze solutions, in the preparation of ethers and esters (used for explosives), as a solvent, and as a substitute for glycerol.

glycolysis conversion of glucose to ◊lactic acid. It takes place in the ◊cytoplasm of cells as part of the process of cellular respiration.

gneiss coarse-grained ◊metamorphic rock, formed under conditions of increasing temperature and pressure, and often occurring in association with schists and granites. It has a foliated, laminated structure, consisting of thin bands of micas and/or amphiboles alternating with granular bands of quartz and feldspar. Gneisses are formed during regional metamorphism; *paragneisses* are derived from sedimentary rocks and *orthogneisses* from igneous rocks. Garnets are often found in gneiss.

gold heavy, precious, yellow, metallic element; symbol Au, atomic number 79, relative atomic mass 197.0. It is unaffected by temperature changes and is highly resistant to acids. For manufacture, gold is alloyed with another strengthening metal (such as copper or silver), its purity being measured in ◊carats on a scale of 24.

Golgi apparatus or *Golgi body* stack of flattened membranous sacs found in the cells of ◊eukaryotes. Many molecules travel through the Golgi apparatus on their way to other organelles or to the endoplasmic reticulum. Some are modified and/or assembled inside the sacs. The Golgi apparatus is named after the Italian physician Camillo Golgi.

gonad the part of an animal's body that produces the sperm or egg cells (ova) required for sexual reproduction. The sperm-producing gonad is called a ◊testis, and the egg-producing gonad is called an ◊ovary.

gonadotrophin any hormone that supports and stimulates the function of the gonads (sex glands); some gonadotrophins are used as fertility drugs.

Gondwanaland or *Gondwana* southern landmass formed 200 million years ago by the splitting of the single world continent ◊Pangaea. (The northern landmass was ◊Laurasia.) It later fragmented into the continents of South America, Africa, Australia, and Antarctica, which then drifted slowly to their present positions. The baobab tree found in both Africa and Australia is a relic of this ancient land mass.

Graafian follicle fluid-filled capsule that surrounds and protects the developing egg cell inside the ovary during the ◊menstrual cycle. After the egg cell has been released, the follicle remains and is known as a corpus luteum.

grafting in medicine, the operation by which an organ or other living tissue is removed from one organism and transplanted into the same or a different organism.

In horticulture, grafting is a technique widely used for propagating plants, especially woody species. A bud or shoot on one plant, termed the *scion*, is inserted into another, the *stock*, so that they continue growing together, the tissues combining at the point of union. In this way some of the advantages of both plants are obtained.

grain the smallest unit of mass in the three English systems (avoirdupois, troy, and apothecaries' weights) used in the UK and USA, equal to 0.0648 g. It was reputedly the weight of a grain of wheat. One pound avoirdupois equals 7,000 grains; one pound troy or apothecaries' weight equals 5,760 grains.

gram metric unit of mass; one-thousandth of a kilogram.

grand unified theory (GUT) in physics, a sought-for theory that would combine the theory of the strong nuclear force (called ↓quantum chromodynamics) with the theory of the weak nuclear and electromagnetic forces. The search for the grand unified theory is part of a larger programme seeking a ↓unified field theory, which would combine all the forces of nature (including gravity) within one framework.

granite coarse-grained ↓igneous rock, typically consisting of the minerals quartz, feldspar, and mica. It may be pink or grey, depending on the composition of the feldspar. Granites are chiefly used as building materials.

graphical user interface (GUI) or *WIMP* in computing, a type of ↓user interface in which programs and files appear as icons (small pictures), user options are selected from pull-down menus, and data are displayed in windows (rectangular areas), which the operator can manipulate in various ways. The operator uses a pointing device, typically a ↓mouse, to make selections and initiate actions.

graphic equalizer control used in hi-fi systems that allows the distortions introduced by unequal amplification of different frequencies to be corrected.

graphics used with computers, see ↓computer graphics.

graphics tablet or *bit pad* in computing, an input device in which a stylus or cursor is moved, by hand, over a flat surface. The computer can keep track of the position of the stylus, so enabling the operator to input drawings or diagrams into the computer.

graphite blackish-grey, laminar, crystalline form of ↓carbon. It is used as a lubricant and as the active component of pencil lead.

graphical user interface *The graphical user interface, developed by researchers at Parc Xerox, provides an easy-to-use 'front end' for applications programs on most modern personal computers.*

grass *The parts of a grass flower. The flowers are in a spikelet arranged on a central axis. The floret, on the right, contains three large anthers and feathery stigmas, catching the floating pollen grains like a net.*

The carbon atoms are strongly bonded together in sheets, but the bonds between the sheets are weak so that the sheets are free to slide over one another. Graphite has a very high melting point (3,500°C/6,332°F), and is a good conductor of heat and electricity. In its pure form it is used as a moderator in nuclear reactors.

grass plant of the large family Gramineae of monocotyledons, with about 9,000 species distributed worldwide except in the Arctic regions. The majority are perennial, with long, narrow leaves and jointed, hollow stems; hermaphroditic flowers are borne in spikelets; the fruits are grainlike. Included are bluegrass, wheat, rye, maize, sugarcane, and bamboo.

gravel coarse ◊sediment consisting of pebbles or small fragments of rock, originating in the beds of lakes and streams or on beaches. Gravel is quarried for use in road building, railway ballast, and for an aggregate in concrete. It is obtained from quarries known as gravel pits, where it is often found mixed with sand or clay.

gravimetric analysis in chemistry, a technique for determining, by weighing, the amount of a particular substance present in a sample. It usually involves the conversion of the test substance into a compound of known molecular weight that can be easily isolated and purified.

gravimetry study of the Earth's gravitational field. Small variations in the field (gravimetric anomalies) can be caused by varying densities of rocks and structure beneath the surface. Such variations are measured by a device called a gravimeter, which consists of a weighted spring that is pulled further downwards where the gravity is stronger. Gravimetry is used by geologists to map the subsurface features of the Earth's crust, such as underground masses of heavy rock like granite, or light rock like salt.

gravitational field the region around a body in which other bodies experience a force due to its gravitational attraction. The gravitational field of a massive object such as the Earth is very strong and easily recognized as the force of gravity, whereas that of an object of much smaller mass is very weak and difficult to detect. Gravitational fields produce only attractive forces.

gravitational field strength (symbol g) the strength of the Earth's gravitational field at a particular place. It is defined as the the gravitational force in newtons that acts on a mass of one kilogram. The value of g on the Earth's surface is taken to be 9.806 N kg^{-1}.

gravitational lensing bending of light by a gravitational field, predicted by Einstein's general theory of relativity. The effect was first detected 1917 when the light from stars was found to be bent as it passed the totally eclipsed Sun. More remarkable is the splitting of light from distant quasars into two or more images by intervening galaxies. In 1979 the first double image of a quasar produced by gravitational lensing was discovered and a quadruple image of another quasar was later found.

gravitational potential energy energy stored by an object when it is placed in a position from which, if it were free to do so, it would fall under the influence of gravity. The gravitational potential energy E_p of an object of mass m kg placed at a height h m above the ground is given by the formula:

$$E_p = mgh$$

where ◊g is the gravitational field strength in N kg^{-1} of the Earth at the place.

graviton in physics, the ◊gauge boson that is the postulated carrier of gravity.

gravity force of attraction that arises between objects by virtue of their masses. On

Earth, gravity is the force of attraction between any object in the Earth's gravitational field and the Earth itself. It is regarded as one of the four fundamental forces of nature, the other three being the ◊electromagnetic force, the ◊strong nuclear force, and the ◊weak nuclear force. The gravitational force is the weakest of the four forces, but it acts over great distances. The particle that is postulated as the carrier of the gravitational force is the ◊graviton.

According to Newton's law of gravitation, all objects fall to Earth with the same acceleration, regardless of mass. For an object of mass m_1 at a distance r from the centre of the Earth (mass m_2), the gravitational force of attraction F equals Gm_1m_2/r^2, where G is the gravitational constant. However, according to Newton's second law of motion, F also equals m_1g, where g is the acceleration due to gravity; therefore $g = Gm_2/r^2$ and is independent of the mass of the object; at the Earth's surface it equals 9.806 metres per second per second.

Einstein's general theory of relativity treats gravitation not as a force but as a curvature of space and time around a body.

gray SI unit (symbol Gy) of absorbed radiation dose. It replaces the rad (1 Gy equals 100 rad), and is defined as the dose absorbed when one kilogram of matter absorbs one joule of ionizing radiation. Different types of radiation cause different amounts of damage for the same absorbed dose; the SI unit of *dose equivalent* is the ◊sievert.

Great Bear popular name for the constellation ◊Ursa Major.

Great Red Spot prominent oval feature, 14,000 km/8,500 mi wide and some 30,000 km/20,000 mi long, in the atmosphere of the planet ◊Jupiter, S of the equator. It was first observed in the 19th century. Space probes show it to be an anticlockwise vortex of cold clouds, coloured possibly by phosphorus.

greenhouse effect phenomenon of the Earth's atmosphere by which solar radiation, trapped by the Earth and re-emitted from the surface, is prevented from escaping by various gases in the air. The result is a rise in the Earth's temperature. The main greenhouse gases are carbon dioxide, methane, and ◊chlorofluorocarbons (CFCs). Fossil-fuel consumption and forest fires are the main causes of carbon-dioxide build-up; methane is a by-product of agriculture (rice, cattle, sheep). Water vapour is another greenhouse gas.

grooming in biology, the use by an animal of teeth, tongue, feet, or beak to clean fur or feathers. Grooming also helps to spread essential oils for waterproofing. In many social species, notably monkeys and apes, grooming of other individuals is used to reinforce social relationships.

group in chemistry, a vertical column of elements in the ◊periodic table. Elements in a group have similar physical and chemical properties; for example, the group I elements (the alkali metals: lithium, sodium, potassium, rubidium, caesium, and francium) are all highly reactive metals that form univalent ions. There is a gradation of properties down any group: in group I, melting and boiling points decrease, and density and reactivity increase.

group in mathematics, a finite or infinite set of elements that can be combined by an operation; formally, a group must satisfy certain conditions. For example, the set of all integers (positive or negative whole numbers) forms a group with regard to addition because: (1) addition is associative, that is, the sum of two or more integers is the same regardless of the order in which the integers are added; (2) adding two integers gives another integer; (3) the set includes an identity element 0, which has no effect on any integer to which it is added (for example, $0 + 3 = 3$); and (4) each integer has an inverse (for instance, 7 has the inverse -7), such that the sum of an integer and its inverse is 0. *Group theory* is the study of the properties of groups.

growth in biology, the increase in size and weight during the development of an organism. Growth is an increase in biomass (mass of organic material, excluding water) and is associated with cell division.

growth hormone (GH) or *somatotrophin* hormone from the anterior ◊pituitary gland that promotes growth of long bones and increases protein synthesis. If it is lacking in childhood, dwarfism results; excess GH causes gigantism.

growth ring another name for ◊annual ring.

GRP abbreviation for ◊*glass-reinforced plastic*.

guard cell in plants, a specialized cell on the undersurface of leaves for controlling gas exchange and water loss. Guard cells occur in pairs and are shaped so that a pore, or ◊stomata, exists between them. They can change shape with the result that the pore disappears. During warm weather, when a plant is in

greenhouse effect *The warming effect of the Earth's atmosphere is called the greenhouse effect. Radiation from the Sun enters the atmosphere but is prevented from escaping back into space by gases such as carbon dioxide (produced, for example, by the burning of fossil fuels), nitrogen oxides (from car exhausts), and CFCs (from aerosols and refrigerators). As these gases build-up in the atmosphere, the Earth's average temperature is expected to rise.*

danger of losing excessive water, the guard cells close, cutting down evaporation from the interior of the leaf.

GUI in computing, abbreviation for ◊*graphical user interface*. It was developed in the 1970s and popularized by Apple Macintosh.

gullet another name for the ◊oesophagus.

gum in botany, complex polysaccharides (carbohydrates) formed by many plants and trees, particularly by those from dry regions. They form four main groups: plant exudates (gum arabic); marine plant extracts (agar); seed extracts; and fruit and vegetable extracts. Some are made synthetically.

gum in mammals, the soft tissues surrounding the base of the teeth. Gums are liable to inflammation (gingivitis) or to infection by microbes from food deposits (periodontal disease).

gun metal type of ◊bronze, an alloy high in copper (88%), also containing tin and zinc, so-called because it was once used to cast cannons. It is tough, hard-wearing, and resists corrosion.

gunpowder or *black powder* the oldest known ◊explosive, a mixture of 75% potassium nitrate (saltpetre), 15% charcoal, and 10% sulphur. Sulphur ignites at a low temper-

gut *The human gut, or alimentary canal. When food is swallowed, it is moved down the oesophagus by muscular action to the stomach, where it is mixed with enzymes and strong acid. After several hours the food passes into the small intestine where more enzymes are added to complete digestion. After all nutrients have been absorbed the indigestible waste passes into the large intestine and thence to the rectum.*

ature, charcoal burns readily, and the potassium nitrate provides oxygen for the explosion. Although progressively replaced since the late 19th century by high explosives, gunpowder is still widely used for quarry blasting, fuses, and fireworks.

gut or *alimentary canal* in the ◊digestive system, the part of an animal responsible for processing food and preparing it for entry into the blood.

guttation secretion of water on to the surface of leaves through specialized pores, or ◊hydathodes. The process occurs most frequently during conditions of high humidity when the rate of transpiration is low. Drops of water found on grass in early morning are often the result of guttation, rather than dew. Sometimes the water contains minerals in solution, such as calcium, which leaves a white crust on the leaf surface as it dries.

gymnosperm in botany, any plant whose seeds are exposed, as opposed to the structurally more advanced ◊angiosperms, where they are inside an ovary. The group includes conifers and related plants such as cycads and ginkgos, whose seeds develop in ◊cones. Fossil gymnosperms have been found in rocks about 350 million years old.

gynoecium or *gynaecium* collective term for the female reproductive organs of a flower, consisting of one or more ◊carpels, either free or fused together.

gypsum common ◊mineral, composed of hydrous calcium sulphate, $CaSO_4.2H_2O$. It ranks 2 on the Mohs' scale of hardness. Gypsum is used for making casts and moulds, and for blackboard ◊chalk.

gyroscope mechanical instrument, used as a stabilizing device and consisting, in its simplest form, of a heavy wheel mounted on an axis fixed in a ring that can be rotated about another axis, which is also fixed in a ring capable of rotation about a third axis. Applications of the gyroscope principle include the gyrocompass, the gyropilot for automatic steering, and gyro-directed torpedoes.

ha symbol for ◊*hectare*.

Haber process or *Haber–Bosch process* industrial process by which ammonia is manufactured by direct combination of its elements, nitrogen and hydrogen. The reaction is carried out at 400–500°C/752–932°F and at 200 atmospheres pressure. The two gases, in the proportions of 1:3 by volume, are passed over a ◊catalyst of finely divided iron.

Around 10% of the reactants combine, and the unused gases are recycled. The ammonia is separated either by being dissolved in water or by being cooled to liquid form.

habitat localized ◊environment in which an organism lives, and which provides for all (or almost all) of its needs. The diversity of habitats found within the Earth's ecosystem is enormous, and they are changing all the time. Many can be considered inorganic or physical, for example the Arctic ice cap, a cave, or a cliff face. Others are more complex, for instance a woodland or a forest floor. Some habitats are so precise that they are called *microhabitats*, such as the area under a stone where a particular type of insect lives. Most habitats provide a home for many species.

hacking unauthorized access to a computer, either for fun or for malicious or fraudulent purposes. Hackers generally use microcomputers and telephone lines to obtain access. The term is also used in a wider sense to mean using software for enjoyment or self-education, not necessarily involving unauthorized access. See also computer ◊virus.

hadron in physics, a subatomic particle that experiences the strong nuclear force. Each is made up of two or three indivisible particles called ◊quarks. The hadrons are grouped into the ◊baryons (protons, neutrons, and hyperons) and the ◊mesons (particles with masses between those of electrons and protons).

Hadron–Electron Ring Accelerator (HERA) particle ◊accelerator built under the streets of Hamburg, Germany, occupying a tunnel 6.3 km/3.9 mi in length. It is the world's most powerful collider of protons and electrons, designed to accelerate protons to energies of 820 GeV (820 billion electron volts), and electrons to 30 GeV. HERA began operating 1992.

haemoglobin protein used by all vertebrates and some invertebrates for oxygen transport because the two substances combine reversibly. In vertebrates it occurs in red blood cells (erythrocytes), giving them their colour.

In the lungs or gills where the concentration of oxygen is high, oxygen attaches to haemoglobin to form *oxyhaemoglobin*. This process effectively increases the amount of oxygen that can be carried in the bloodstream. The oxygen is later released in the body tissues where it is at a low concentration, and the deoxygenated blood returned to the lungs or gills. Haemoglobin will combine also with carbon monoxide to form carboxyhaemoglobin, but in this case the reaction is irreversible.

haemolymph circulatory fluid of those molluscs and insects that have an 'open' circulatory system. Haemolymph contains water, amino acids, sugars, salts, and white cells like those of blood. Circulated by a pulsating heart, its main functions are to transport digestive and excretory products around the body. In molluscs, it also transports oxygen and carbon dioxide.

haemolysis destruction of red blood cells. Aged cells are constantly being lysed (broken down), but increased wastage of red cells is seen in some infections and blood disorders. It may result in jaundice (through the release of too much haemoglobin) and in ◊anaemia.

haemophilia any of several inherited diseases in which normal blood clotting is impaired. The sufferer experiences prolonged bleeding from the slightest wound, as well as painful internal bleeding without apparent cause.

hafnium silvery, metallic element, symbol Hf, atomic number 72, relative atomic mass 178.49. It occurs in nature in ores of zirconium, the properties of which it resembles. Hafnium absorbs neutrons better than most metals, so it is used in the control rods of nuclear reactors; it is also used for light-bulb filaments.

hahnium name proposed by US scientists for the element also known as ◊unnilpentium (atomic number 105), in honour of German nuclear physicist Otto Hahn. The symbol is Ha.

hair fine filament growing from mammalian skin. Each hair grows from a pit-shaped follicle embedded in the second layer of the skin, the dermis. It consists of dead cells impregnated with the protein keratin.

half-life during ◊radioactive decay, the time in which the strength of a radioactive source decays to half its original value. In theory, the decay process is never complete and there is always some residual radioactivity. For this reason, the half-life of a radioactive isotope is measured, rather than the total decay time. It may vary from millionths of a second to billions of years.

Radioactive substances decay exponentially; thus the time taken for the first 50% of the isotope to decay will be the same as the time taken by the next 25%, and by the 12.5% after that, and so on. Plutonium-239, one of the most toxic of all radioactive substances, has a half-life of about 24,000 years.

half-lives of some radioactive isotopes

isotope	half-life
(least stable)	
lithium-5	4.4×10^{-22} sec
polonium-213	4.2×10^{-6} sec
lead-211	36 min
lead-209	3.3 hours
uranium-238	4.551×10^9 years
thorium-232	1.39×10^{10} years
tellurium-128	1.5×10^{24} years
(most stable)	

halide any compound produced by the combination of a ◊halogen, such as chlorine or iodine, with a less electronegative element (see ◊electronegativity). Halides may be formed by ◊ionic bonds or by ◊covalent bonds.

halite mineral sodium chloride, NaCl, or common ◊salt. When pure it is colourless and transparent, but it is often pink, red, or yellow. It is soft and has a low density.

Hall effect production of a voltage across a conductor or semiconductor carrying a current at a right angle to a surrounding magnetic field. It was discovered 1897 by the US physicist Edwin Hall (1855–1938). It is used in the *Hall probe* for measuring the strengths of magnetic fields and in magnetic switches.

Halley's comet comet that orbits the Sun about every 76 years, named after Edmond Halley who calculated its orbit. It is the brightest and most conspicuous of the periodic comets. Recorded sightings go back over 2,000 years. It travels around the Sun in the opposite direction to the planets. Its orbit is inclined at almost 20° to the main plane of the Solar System and ranges between the orbits of Venus and Neptune. It will next reappear 2061.

The comet was studied by space probes at its last appearance 1986. The European probe *Giotto* showed that the nucleus of Halley's comet is a tiny and irregularly shaped chunk of ice, measuring some 15 km/10 m long by 8 km/5 m wide, coated by a layer of very dark material, thought to be composed of carbon-rich compounds.

halogen any of a group of five nonmetallic elements with similar chemical bonding properties: fluorine, chlorine, bromine, iodine, and astatine. They form a linked group in the ◊periodic table of the elements, descending from fluorine, the most reactive, to astatine, the least reactive. They combine directly with most metals to form salts, such as common salt (NaCl). Each halogen has seven electrons in its valence shell, which accounts for the chemical similarities displayed by the group.

halon organic chemical compound containing one or two carbon atoms, together with ◊bromine and other ◊halogens. The most commonly used are halon 1211 (bromochlorodifluoromethane) and halon 1301 (bromotrifluoromethane). The halons are gases and are widely used in fire extinguishers. As destroyers of the ◊ozone layer, they are up to ten times more effective than ◊chlorofluorocarbons (CFCs), to which they are chemically related.

Levels in the atmosphere are rising by about 25% each year, mainly through the testing of fire-fighting equipment.

halophyte plant adapted to live where there is a high concentration of salt in the soil, for example, in salt marshes and mud flats.

haploid having a single set of ◊chromosomes in each cell. Most higher organisms are ◊diploid – that is, they have two sets – but their gametes (sex cells) are haploid. Some plants, such as mosses, liverworts, and many seaweeds, are haploid, and male honey bees are haploid because they develop from eggs that have not been fertilized. See also ◊meiosis.

hard disc in computing, a storage device usually consisting of a rigid metal ◊disc coated with a magnetic material. Data are read from and written to the disc by means of a disc drive. The hard disc may be permanently fixed into the drive or in the form of a disc pack that can be removed and exchanged with a different pack. Hard discs vary from large units with capacities of more than 3,000 megabytes, intended for use with mainframe computers, to small units with capacities as low as 20 megabytes, intended for use with microcomputers.

hardening of oils transformation of liquid oils to solid products by ◊hydrogenation.

hardness physical property of materials that governs their use. Methods of heat treatment can increase the hardness of metals. A scale of hardness was devised by German–Austrian mineralogist Friedrich Mohs in the 1800s, based upon the hardness of certain minerals from soft talc (◊Mohs' scale hardness 1) to diamond (10), the hardest of all materials.

hardware the mechanical, electrical, and electronic components of a computer system, as opposed to the various programs, which constitute ◊software.

hard water water that does not lather easily with soap, and produces 'fur' or 'scale' in kettles. It is caused by the presence of certain salts of calcium and magnesium.

Hardy–Weinberg equilibrium in population genetics, the theoretical relative fre- quency of different ◊alleles within a given population of a species, when the stable endpoint of evolution in an undisturbed environment is reached.

Hare's apparatus in physics, a specific kind of ◊hydrometer used to compare the relative densities of two liquids, or to find the density of one if the other is known. It was invented by US chemist Robert Hare (1781–1858).

Harrier the only truly successful vertical takeoff and landing fixed-wing aircraft, often called the *jump jet*. Built in Britain, it made its first flight 1966. It has a single jet engine and a set of swivelling nozzles. These deflect the jet exhaust vertically downwards for takeoff and landing, and to the rear for normal flight. Designed to fly from confined spaces with minimal ground support, it refuels in midair.

hash total in computing, a ◊validation check in which an otherwise meaningless control total is calculated by adding together numbers (such as payroll or account numbers) associated with a set of records. The hash total is checked each time data are input, in order to ensure that no entry errors have been made.

haustorium (plural *haustoria*) specialized organ produced by a parasitic plant or fungus

harmful/irritant

toxic

radioactive

explosive

flammable

corrosive

oxidizing/supports fire

hazard label *International labelling system for hazardous substances.*

that penetrates the cells of its host to absorb nutrients. It may be either an outgrowth of hyphae (see ◊hypha), as in the case of parasitic fungi, or of the stems of flowering parasitic plants, as in dodders (*Cuscuta*). The suckerlike haustoria of a dodder penetrate the vascular tissue of the host plant without killing the cells.

hazard label internationally recognized symbol indicating the potential dangers of handling certain substances.

hazardous substance waste substance, usually generated by industry, which represents a hazard to the environment or to people living or working nearby. Examples include radioactive wastes, acidic resins, arsenic residues, residual hardening salts, lead, mercury, nonferrous sludges, organic solvents, and pesticides.

Their economic disposal or recycling is the subject of research. Greenpeace estimates that

heterostyly in botany, having ◊styles of different lengths.

Certain flowers, such as primroses (*Primula vulgaris*), have different-sized ◊anthers and styles to ensure cross-fertilization (through ◊pollination) by visiting insects.

heterotroph any living organism that obtains its energy from organic substances produced by other organisms. All animals and fungi are heterotrophs, and they include herbivores, carnivores, and saprotrophs (those that feed on dead animal and plant material).

heterozygous in a living organism, having two different ◊alleles for a given trait. In ◊homozygous organisms, by contrast, both chromosomes carry the same allele. In an outbreeding population an individual organism will generally be heterozygous for some genes but homozygous for others.

For example, in humans, alleles for both blue-and brown-pigmented eyes exist, but the 'blue' allele is ◊recessive to the dominant 'brown' allele. Only individuals with blue eyes are predictably homozygous for this trait; brown-eyed people can be either homozygous or heterozygous.

heuristics in computing, a process by which a program attempts to improve its performance by learning from its own experience.

hexachlorophene $(C_6HCl_3OH)_2CH_2$ white, odourless bactericide, used in minute quantities in soaps and surgical disinfectants.

hexadecimal number system number system to the base 16, used in computing. In hex (as it is commonly known) the decimal numbers 0–15 are represented by the characters 0, 1, 2, 3, 4, 5, 6, 7, 8, 9, A, B, C, D, E, F.

Hexadecimal numbers are easy to convert to the computer's internal binary code and are more compact than binary numbers.

HF in physics, the abbreviation for *high ◊frequency*.

hibernation state of dormancy in which certain animals spend the winter. It is associated with a dramatic reduction in all metabolic processes, including body temperature, breathing, and heart rate. It is a fallacy that animals sleep throughout the winter.

Higgs boson or *Higgs particle* postulated ◊elementary particle whose existence would explain why particles have mass. The current theory of elementary particles, called the ◊standard model, cannot explain how mass arises. To overcome this difficulty, Peter Higgs (1929–) of the University of Edinburgh and Thomas Kibble (1932–) of Imperial College, London proposed in 1964 a new particle that binds to other particles and gives them their mass. The Higgs boson has not yet been detected experimentally.

high-definition television (HDTV) ◊television system offering a significantly greater number of scanning lines, and therefore a clearer picture, than that provided by conventional systems. The Japanese HDTV system, or Vision as it is trade-named in Japan, uses 1,125 scanning lines and an aspect ratio of 16:9 instead of the squarish 4:3 that conventional television uses. A European HDTV system, called HD-MAC, using 1,250 lines, is under development. In the USA, a standard incorporating digital techniques is being discussed.

high-level language in computing, a programming language designed to suit the requirements of the programmer; it is independent of the internal machine code of any particular computer. High-level languages are used to solve problems and are often described as *problem-oriented languages* – for example, ◊BASIC was designed to be easily learnt by first-time programmers; ◊COBOL is used to write programs solving business problems; and ◊FORTRAN is used for programs solving scientific and mathematical problems. In contrast, low-level languages, such as ◊assembly languages, closely reflect the machine codes of specific computers, and are therefore described as *machine-oriented languages*.

high-yield variety crop that has been specially bred or selected to produce more than the natural varieties of the same species. During the 1950s and 1960s, new strains of wheat and maize were developed to reduce the food shortages in poor countries (the Green Revolution). Later, IR8, a new variety of rice that increased yields by up to six times, was developed in the Philippines. Strains of crops resistant to drought and disease were also developed. High-yield varieties require large amounts of expensive artificial fertilizers and sometimes pesticides for best results.

hinge joint in vertebrates, a joint where movement occurs in one plane only. Examples are the elbow and knee, which are controlled by pairs of muscles, the ◊flexors and ◊extensors.

Hipparcos (acronym for *high precision parallax collecting satellite*) satellite launched by the European Space Agency Aug 1989. Named after the Greek astronomer Hipparchus, it is the world's first ◊astrometry satellite and is providing precise positions, distances, colours, brightnesses, and apparent motions for over 100,000 stars.

histamine inflammatory substance normally released in damaged tissues, which also accounts for many of the symptoms of ◊allergy. Substances that neutralize its activity are known as antihistamines.

histochemistry study of plant and animal tissue by visual examination, usually with a ◊microscope.

histogram in statistics, a graph showing frequency of data, in which the horizontal axis details discrete units or class boundaries, and the vertical axis represents the frequency. Blocks are drawn such that their areas (rather than their height as in a ◊bar chart) are proportional to the frequencies within a class or across several class boundaries. There are no spaces between blocks.

histology in medicine, the laboratory study of cells and tissues.

HIV (abbreviation for *human immunodeficiency virus*) the infectious agent that is believed to cause ◊AIDS.

HLA system abbreviation for human leucocyte antigen system, a series of gene groups (termed A, B, C and D) that code for proteins on the surface of cells. If two people have the same HLA types, they are histocompatible (having compatible tissue types). HLA types feature in tissue testing before a ◊transplant.

Hoffman's voltameter in chemistry, an apparatus for collecting gases produced by the ◊electrolysis of a liquid.

holdfast organ found at the base of many seaweeds, attaching them to the sea bed. It may be a flattened, suckerlike structure, or dissected and fingerlike, growing into rock crevices and firmly anchoring the plant.

holmium silvery, metallic element of the ◊lanthanide series, symbol Ho, atomic number 67, relative atomic mass 164.93. It occurs in combination with other rare-earth metals and in various minerals such as gadolinite. Its compounds are highly magnetic.

Holocene epoch of geological time that began 10,000 years ago, the second and current epoch of the Quaternary period. The glaciers retreated, the climate became warmer, and humans developed significantly.

holography method of producing three-dimensional (3-D) images by means of ◊laser light. Holography uses a photographic technique (involving the splitting of a laser beam into two beams) to produce a picture, or

holography Recording a transmission hologram. Light from a laser is divided into two beams. One beam goes directly to the photographic plate. The other beam reflects off the object before hitting the photographic plate. The two beams combine to produce a pattern on the plate which contains information about the 3-D shape of the object. If the exposed and developed plate is illuminated by laser light, the pattern can be seen as a 3-D picture of the object.

the systematic naming of simple straight-chain organic molecules

alkane	alcohol	aldehyde	ketone	carboxylic acid	alkene
CH_4 methane	CH_3OH methanol	HCHO methanal	—	HCOOH methanoic acid	—
CH_3CH_3 ethane	CH_3CH_2OH ethanol	CH_3CHO ethanal	—	CH_3COOH ethanoic acid	CH_2CH_2 ethene
$CH_3CH_2CH_3$ propane	$CH_3CH_2CH_2OH$ propanol	CH_3CH_2CHO propanal	CH_3COCH_3 propanone	CH_3CH_2COOH propanoic acid	CH_2CHCH_3 propene
methane	methanol	methanal	propanone	methanoic acid	ethene

homologous series *The systematic naming of simple straight-chain organic molecules depends on two-part names. The first part of a name indicates the number of carbon atoms in the chain: one carbon, meth-; two carbons, eth-; three carbons, prop-; etc The second part of each name indicates the kind of bonding between the carbon atoms, or the atomic group attached to the chain. The name of a molecule containing only single bonds ends in -ane. Molecules with double bonds have names ending with -ene. Molecules containing the OH group have names ending in -anol; those containing -CO- groups have names ending in -anone; those containing the carboxyl group -COOH- have names ending in -anoic acid.*

hologram, that contains 3-D information about the object photographed. Some holograms show meaningless patterns in ordinary light and produce a 3-D image only when laser light is projected through them, but reflection holograms produce images when ordinary light is reflected from them (as found on credit cards).

homeostasis maintenance of a constant internal state in an organism, particularly with regard to pH, salt concentration, temperature, and blood sugar levels. Stable conditions are important for the efficient functioning of the ◊enzyme reactions within the cells, which affect the performance of the entire organism.

homeothermy maintenance of a constant body temperature in endothermic (warm-blooded) animals, by the use of chemical processes to compensate for heat loss or gain when external temperatures change. Such processes include generation of heat by the breakdown of food and the contraction of muscles, and loss of heat by sweating, panting, and other means.

Mammals and birds are homeotherms, whereas invertebrates, fish, amphibians, and reptiles are cold-blooded or poikilotherms. Homeotherms generally have a layer of insulating material to retain heat, such as fur, feathers, or fat (see ◊blubber). Their metabolism functions more efficiently due to homeothermy, enabling them to remain active under most climatic conditions.

homogeneous reaction in chemistry, a reaction where there is no interface between the components. The term applies to all reactions where only gases are involved or where all the components are in solution.

homologous in biology, a term describing an organ or structure possessed by members of different taxonomic groups (for example, species, genera, families, orders) that originally derived from the same structure in a common ancestor. The wing of a bat, the arm of a monkey, and the flipper of a seal are homologous because they all derive from the forelimb of an ancestral mammal.

homologous series any of a number of series of organic chemicals with similar chemical properties in which members differ by a constant relative molecular mass.

homozygous in a living organism, having two identical ◊alleles for a given trait. Individuals homozygous for a trait always breed true; that is, they produce offspring that resemble them in appearance when bred with a genetically similar individual; inbred varieties or species are homozygous for almost all traits. ◊Recessive alleles are only expressed in the homozygous condition. See also ◊heterozygous.

honey sweet syrup produced by honey bees from the nectar of flowers. It is stored in honeycombs and made in excess of their needs as food for the winter. Honey comprises various sugars, mainly laevulose and dextrose, with enzymes, colouring matter, acids, and pollen grains. It has antibacterial properties and was widely used in ancient Egypt, Greece, and Rome as a wound salve. It is still popular for sore throats, in hot drinks or in lozenges.

honey guide in botany, line or spot on the petals of a flower that indicate to pollinating insects the position of the nectaries (see ◊nectar) within the flower. The orange dot on the lower lip of the toadflax flower (*Linaria vulgaris*) is an example. Sometimes the markings reflect only ultraviolet light, which can be seen by many insects although it is not visible to the human eye.

Hooke's law law stating that the deformation of a body is proportional to the magnitude of the deforming force, provided that the body's elastic limit (see ◊elasticity) is not exceeded. If the elastic limit is not reached, the body will return to its original size once the force is removed. The law was discovered by Robert Hooke 1676.

Hope's apparatus in physics, an apparatus used to demonstrate the temperature at which water has its maximum density. It is named after Thomas Charles Hope (1766–1844).

hormone product of the ◊endocrine glands, concerned with control of body functions. The major glands are the thyroid, parathyroid, pituitary, adrenal, pancreas, ovary, and testis. Hormones bring about changes in the functions of various organs according to the body's requirements. The ◊hypothalamus, which adjoins the pituitary gland, at the base of the brain, is a control centre for overall coordination of hormone secretion; the thyroid hormones determine the rate of general body chemistry; the adrenal hormones prepare the organism during stress for 'fight or flight'; and the sexual hormones such as oestrogen govern reproductive functions.

hormone-replacement therapy (HRT) use of ◊oestrogen and progestogen to help limit the unpleasant effects of the menopause in women. The treatment was first used in the 1970s. In 1993 18% of postmenopausal women were treated with HRT.

hornblende green or black rock-forming mineral, one of the ◊amphiboles; it is a hydrous silicate of calcium, iron, magnesium, and aluminium. Hornblende is found in both igneous and metamorphic rocks.

hornfels ◊metamorphic rock formed by rocks heated by contact with a hot igneous body. It is fine-grained and brittle, without foliation.

hornwort nonvascular plant of the class Anthocerotae; it is related to the ◊liverworts and ◊mosses, with which it forms the order Bryophyta. Hornworts are found in warm climates, growing on moist shaded soil. The name is also given to an aquatic flowering plant of the family Ceratophyllaceae. Found in slow-moving water, it has whorls of finely divided leaves and may grow up to 2 m/7 ft long.

horsepower imperial unit (abbreviation hp) of power, now replaced by the ◊watt. It was first used by the engineer James Watt, who employed it to compare the power of steam engines with that of horses.

host in biology, an organism that is parasitized by another. In ◊commensalism, the partner that does not benefit may also be called the host.

hp abbreviation for ◊*horsepower*.

HPGL (abbreviation for *Hewlett Packard Graphics Language*) a file format used in ◊vector graphics. HPGL is often generated by ◊CAD systems.

Hubble's constant in astronomy, a measure of the rate at which the universe is expanding, named after Edwin Hubble. Observations suggest that galaxies are moving apart at a rate of 50–100 kps/30–60 mps for every million ◊parsecs of distance. This means that the universe, which began at one point according to the ◊Big Bang theory, is between 10 billion and 20 billion years old (probably closer to 20).

UK astronomers argued 1993 for a lower value for the constant of between 24 and 54 kps, indicating an older universe.

Hubble's law the law that relates a galaxy's distance from us to its speed of recession as the universe expands, announced in 1929 by Edwin Hubble. He found that galaxies are moving apart at speeds that increase in direct proportion to their distance apart. The rate of expansion is known as Hubble's constant.

Hubble Space Telescope telescope placed into orbit around the Earth, at an altitude of 610 km/380 mi, by the space shuttle *Discovery* in April 1990. The telescope's position in space means that light from distant galaxies is not obscured by the atmosphere, and it therefore performs many times better than any ground-based instrument.

Two months after the launch, the main mirror, 2.4 m/94 in wide, was found to suffer from spherical aberration and so could not be focused properly. An inquiry found both NASA and the contractors to blame. In Jan 1994, after 30 hours of spacewalks by five astronauts and at a cost of $360 million, NASA announced that it had repaired the fault. COSTAR (corrective optics space telescope axial replacement), a device containing ten coin-sized mirrors, now feeds a corrected image from the main mirror to three of the HST's four scientific instruments. The resulting resolution should be at least 10 times better than any ground based instrument.

human body the physical structure of the human being. It develops from the single cell of the fertilized ovum, is born at 40 weeks, and usually reaches sexual maturity between 11 and 18 years of age. The bony framework (skeleton) consists of more than 200 bones, over half of which are in the hands and feet. Bones are held together by joints, some of which allow movement. The circulatory system supplies muscles and organs with blood, which provides oxygen and food and removes carbon dioxide and other waste products. Body functions are controlled by the nervous system and hormones. In the upper part of the trunk is the thorax, which contains the lungs and heart. Below this is the abdomen, containing the digestive system (stomach and intestines); the liver, spleen, and pancreas; the urinary system (kidneys, ureters, and bladder); and, in women, the reproductive organs (ovaries, uterus, and vagina). In men, the prostate gland and seminal vesicles only of the reproductive system are situated in the abdomen, the testes being in the scrotum, which, with the penis, is suspended in front of and below the abdomen. The bladder empties through a small channel (urethra); in the female this opens in the upper end of the vulval cleft, which also contains the opening of the vagina, or birth canal; in the male, the urethra is continued into the penis. In both sexes, the lower bowel terminates in the anus, a ring of strong muscle situated between the buttocks.

Human Genome Project research scheme, begun 1988, to map the complete nucleotide (see ◊nucleic acid) sequence of human ◊DNA. There are approximately 80,000 different ◊genes in the human genome, and one gene may contain more than 2 million nucleotides. The knowledge gained is expected to help prevent or treat many crippling and lethal diseases, but there are potential ethical problems associated with knowledge of an individual's genetic make-up, and fears that it will lead to genetic engineering.

In March 1994 it was estimated that all the genes in the human body should be known by the end of 1995. Work will continue to use that information to understand the organization and expression of genes.

human reproduction an example of ◊sexual reproduction, where the male produces sperm and the female eggs. These gametes contain only half the normal number of chromosomes, 23 instead of 46, so that on fertilization the resulting cell has the correct genetic complement. Fertilization is internal, which increases the chances of conception; unusually for mammals, copulation and pregnancy can occur at any time of the year. Human beings are also remarkable for the length of childhood and for the highly complex systems of parental care found in society. The use of contraception and the development of laboratory methods of insemination and fertilization are issues that make human reproduction more than a merely biological phenomenon.

humerus the upper bone of the forelimb of tetrapods. In humans, the humerus is the bone above the elbow.

humidity the quantity of water vapour in a given volume of the atmosphere (absolute humidity), or the ratio of the amount of water vapour in the atmosphere to the saturation value at the same temperature (relative humidity). At ◊dew point the relative humidity is 100% and the air is said to be saturated. Condensation (the conversion of vapour to liquid) may then occur. Relative humidity is measured by various types of ◊hygrometer.

humus component of ◊soil consisting of decomposed or partly decomposed organic matter, dark in colour and usually richer towards the surface. It has a higher carbon content than the original material and a lower nitrogen content, and is an important source of minerals in soil fertility.

hundredweight imperial unit (abbreviation cwt) of mass, equal to 112 lb (50.8 kg). It is sometimes called the long hundredweight, to distinguish it from the short hundredweight or *cental*, equal to 100 lb (45.4 kg).

Hyades V-shaped cluster of stars that forms the face of the bull in the constellation Taurus.

key
1. brain
2. eye
3. carotid artery
4. jugular vein
5. subclavian artery
6. superior vena cava
7. aorta
8. subclavian vein
9. heart
10. lungs
11. diaphragm
12. liver
13. stomach
14. gall bladder
15. kidney
16. pancreas
17. small intestine
18. large intestine
19. appendix
20. bladder
21. femoral artery
22. femoral vein

key
1. cranium (skull)
2. mandible
3. clavicle
4. scapula
5. sternum
6. rib cage
7. humerus
8. vertebra
9. ulna
10. radius
11. pelvis
12. coccyx
13. metacarpals
14. phalanges
15. femur
16. patella
17. fibula
18. tibia
19. metatarsals
20. phalanges
21. superficial (upper) layer of muscles
22. carpals
23. tarsals

human body *The adult human body has approximately 650 muscles, 100 joints, 100,000 km/60,000 mi of blood vessels and 13,000 nerve cells. There are 206 bones in the adult body, nearly half of them in the hands and feet.*

It is 150 light years away and contains over 200 stars, although only about 12 are visible to the naked eye.

hybrid offspring from a cross between individuals of two different species, or two inbred lines within a species. In most cases, hybrids between species are infertile and unable to reproduce sexually. In plants, however, doubling of the chromosomes (see ♢polyploid) can restore the fertility of such hybrids.

hydathode specialized pore, or less commonly, a hair, through which water is secreted by hydrostatic pressure from the interior of a plant leaf onto the surface. Hydathodes are found on many different plants and are usually situated around the leaf margin at vein endings. Each pore is surrounded by two crescent-shaped cells and resembles an open ◊stoma, but the size of the opening cannot be varied as in a stoma. The process of water secretion through hydathodes is known as ◊guttation.

Hydra in astronomy, the largest constellation, winding across more than a quarter of the sky between Cancer and Libra in the southern hemisphere. Hydra is named after the multiheaded monster slain by Heracles. Despite its size, it is not prominent; its brightest star is second-magnitude Alphard.

hydrate chemical compound that has discrete water molecules combined with it. The water is known as *water of crystallization* and the number of water molecules associated with one molecule of the compound is denoted in both its name and chemical formula: for example, $CuSO_4.5H_2O$ is copper(II) sulphate pentahydrate.

hydration in chemistry, the combination of water and another substance to produce a single product. It is the opposite of ◊dehydration.

hydraulic action in earth science, the erosive force exerted by water (as distinct from the forces exerted by rocky particles carried by water). It can wear away the banks of a river, particularly at the outer curve of a meander (bend in the river), where the current flows most strongly.

hydraulics field of study concerned with utilizing the properties of water and other liquids, in particular the way they flow and transmit pressure, and with the application of these properties in engineering. It applies the principles of ◊hydrostatics and hydrodynamics. The oldest type of hydraulic machine is the *hydraulic press*, invented by Joseph Bramah (1748–1814) in England 1795. The hydraulic principle of pressurized liquid increasing mechanical efficiency is commonly used on vehicle braking systems, the forging press, and the hydraulic systems of aircraft and excavators.

hydride chemical compound containing hydrogen and one other element, and in which the hydrogen is the more electronegative element (see ◊electronegativity).

hydrocarbon any of a class of chemical compounds containing only hydrogen and carbon (for example, the alkanes and alkenes). Hydrocarbons are obtained industrially principally from petroleum and coal tar.

hydrochloric acid HCl solution of hydrogen chloride (a colourless, acidic gas) in water. The concentrated acid is about 35% hydrogen chloride and is corrosive. The acid is a typical strong, monobasic acid forming only one series of salts, the chlorides. It has many industrial uses, including recovery of zinc from galvanized scrap iron and the production of chlorine. It is also produced in the stomachs of animals for the purposes of digestion.

hydrocyanic acid or *prussic acid* solution of hydrogen cyanide gas (HCN) in water. It is a colourless, highly poisonous, volatile liquid, smelling of bitter almonds.

hydrodynamics science of nonviscous fluids (such as water, alcohol, and ether) in motion.

hydroelectric power (HEP) electricity generated by moving water. In a typical HEP scheme, water stored in a reservoir, often created by damming a river, is piped into water ◊turbines, coupled to electricity generators. In ◊pumped storage plants, water flowing through the turbines is recycled. A ◊tidal power station exploits the rise and fall of the tides. About one-fifth of the world's electricity comes from HEP.

hydrofoil wing that develops lift in the water in much the same way that an aeroplane wing develops lift in the air. A hydrofoil boat is one whose hull rises out of the water due to the lift, and the boat skims along on the hydrofoils. The first hydrofoil was fitted to a boat 1906. The first commercial hydrofoil went into operation 1956. One of the most advanced hydrofoil boats is the Boeing ◊jetfoil.

hydrogen colourless, odourless, gaseous, nonmetallic element, symbol H, atomic number 1, relative atomic mass 1.00797. It is the lightest of all the elements and occurs on Earth chiefly in combination with oxygen as water. Hydrogen is the most abundant element in the universe, where it accounts for 93% of the total number of atoms and 76% of the total mass. It is a component of most stars, including the Sun, whose heat and light are produced through the nuclear-fusion process that converts hydrogen into helium. When subjected to a pressure 500,000 times greater than that of the Earth's atmosphere, hydrogen becomes a solid with metallic properties, as in one of the inner zones of Jupiter. Hydrogen's common and industrial uses include the hardening of oils and fats by hydrogenation, the creation of high-temperature flames for welding, and as rocket fuel. It has been proposed as a fuel for road vehicles.

hydrogenation addition of hydrogen to an unsaturated organic molecule (one that contains ◊double bonds or ◊triple bonds). It is widely used in the manufacture of margarine and low-fat spreads by the addition of hydrogen to vegetable oils.

hydrogen bomb bomb that works on the principle of nuclear ◊fusion. Large-scale explosion results from the thermonuclear release of energy when hydrogen nuclei are fused to form helium nuclei. The first hydrogen bomb was exploded at Eniwetok Atoll in the Pacific Ocean by the USA 1952.

hydrogencarbonate or *bicarbonate* compound containing the ion HCO_3^-, an acid salt of carbonic acid (solution of carbon dioxide in water). When heated or treated with dilute acids, it gives off carbon dioxide. The most important compounds are ◊sodium hydrogencarbonate (bicarbonate of soda), and ◊calcium hydrogencarbonate.

hydrogen cyanide HCN poisonous gas formed by the reaction of sodium cyanide with dilute sulphuric acid; it is used for fumigation.

hydrogen sulphate HSO_4^- compound containing the hydrogen sulphate ion. Hydrogen sulphates are ◊acid salts.

hydrogen sulphide H_2S poisonous gas with the smell of rotten eggs. It is found in certain types of crude oil where it is formed by decomposition of sulphur compounds. It is removed from the oil at the refinery and converted to elemental sulphur.

hydrography study and charting of Earth's surface waters in seas, lakes, and rivers.

hydrology study of the location and movement of inland water, both frozen and liquid, above and below ground. It is applied to major civil engineering projects such as irrigation schemes, dams, and hydroelectric power, and in planning water supply.

hydrolysis in chemistry, reaction in which the action of water or its ions breaks down a substance into smaller molecules. Hydrolysis occurs in certain inorganic salts in solution, in nearly all nonmetallic chlorides, in esters, and in other organic substances. It is one of the mechanisms for the breakdown of food by the body, as in the conversion of starch to glucose.

hydrolysis in earth science, a form of chemical ◊weathering caused by the chemical alteration of certain minerals as they react with water. For example, the mineral feldspar in granite reacts with water to from a white clay called ◊china clay.

hydrometer in physics, an instrument used to measure the relative density of liquids (the density compared with that of water). A hydrometer consists of a thin glass tube ending in a sphere that leads into a smaller sphere, the latter being weighted so that the hydrometer floats upright, sinking deeper into less dense liquids than into denser liquids. Hydrometers are used in brewing and to test the strength of acid in car batteries.

hydrophilic in chemistry, a term describing ◊functional groups with a strong affinity for water, such as the carboxyl group (–COOH).

hydrophily type of ◊pollination where the pollen is transported by water. Water-pollinated plants occur in 31 genera in 11 different families. They are found in habitats as diverse as rainforests and seasonal desert pools. Pollen is either dispersed underwater or on the water's surface.

hydrophobic in chemistry, a term describing ◊functional groups that repel water (the opposite of ◊hydrophilic).

hydrophone underwater ◊microphone and ancillary equipment capable of picking up waterborne sounds. It was originally developed to detect enemy submarines but is now also used, for example, for listening to the sounds made by whales.

hydrophyte plant adapted to live in water, or in waterlogged soil.

hydroponics cultivation of plants without soil, using specially prepared solutions of mineral salts. Beginning in the 1930s, large crops were grown by hydroponic methods, at first in California but since then in many other parts of the world.

hydrosphere the water component of the Earth, usually encompassing the oceans, seas, rivers, streams, swamps, lakes, groundwater, and atmospheric water vapour.

hydrostatics in physics, the branch of ◊statics dealing with the mechanical problems of fluids in equilibrium – that is, in a static condition. Practical applications include shipbuilding and dam design.

hydrothermal in geology, pertaining to a fluid whose principal component is hot water, or to a mineral deposit believed to be precipitated from such a fluid.

hydroxide any inorganic chemical compound containing one or more hydroxyl (OH) groups and generally combined with a metal. Hydroxides include sodium hydroxide (caustic soda, NaOH), potassium hydroxide (caustic potash, KOH), and calcium hydroxide (slaked lime, $Ca(OH)_2$).

hydroxyl group an atom of hydrogen and an atom of oxygen bonded together and covalently bonded to an organic molecule. Common compounds containing hydroxyl groups are alcohols and phenols.

In chemical reactions, the hydroxyl group (–OH) frequently behaves as a single entity.

hydroxypropanoic acid technical name for ◊lactic acid.

hygrometer in physics, any instrument for measuring the humidity, or water vapour content, of a gas (usually air). A wet and dry bulb hygrometer consists of two vertical thermometers, with one of the bulbs covered in absorbent cloth dipped into water. As the water evaporates, the bulb cools, producing a temperature difference between the two thermometers. The amount of evaporation, and hence cooling of the wet bulb, depends on the relative humidity of the air.

hygroscopic able to absorb moisture from the air without becoming wet.

hyperbola in geometry, a curve formed by cutting a right circular cone with a plane so that the angle between the plane and the base is greater than the angle between the base and the side of the cone. All hyperbolae are bounded by two asymptotes (straight lines which the hyperbola moves closer and closer to but never reaches).

A hyperbola is a member of the family of curves known as ◊conic sections.

hypercharge in physics, a property of certain ◊elementary particles, analogous to electric charge, that accounts for the absence of some expected behaviour (such as decay) in terms of the short-range strong nuclear force, which holds atomic nuclei together.

◊Protons and ◊neutrons, for example, have a hypercharge of +1, whereas a π meson has a hypercharge of 0.

hyperon in physics, a ◊hadron; any of a group of highly unstable ◊elementary particles that includes all the ◊baryons with a mass greater than the ◊neutron. They are all composed of three quarks. The lambda, xi, sigma, and omega particles are hyperons.

hypertext system for viewing information

hyperbola *The hyperbola is produced when a cone is cut by a plane. It is one of a family of curves called conic sections: the circle, ellipse, and parabola. These curves are produced when the plane cuts the cone at different angles and positions.*

(both text and pictures) on a computer screen in such a way that related items of information can easily be reached. For example, the program might display a map of a country; if the user clicks (with a ◊mouse) on a particular city, the program will display some information about that city.

hypha (plural *hyphae*) delicate, usually branching filament, many of which collectively form the mycelium and fruiting bodies of a ◊fungus. Food molecules and other substances are transported along hyphae by the movement of the cytoplasm, known as 'cytoplasmic streaming'.

hypogeal term used to describe seed germination in which the ◊cotyledons remain below ground. It can refer to fruits that develop underground, such as peanuts *Arachis hypogea*.

hypotenuse the longest side of a right-angled triangle, opposite the right angle. It is of particular application in Pythagoras' theorem (the square of the hypotenuse equals the sum of the squares of the other two sides), and in trigonometry where the ratios ◊sine and ◊cosine are defined as the ratios opposite/hypotenuse and adjacent/hypotenuse respectively.

hypothalamus region of the brain below the ◊cerebrum which regulates rhythmic activity and physiological stability within the body, including water balance and temperature. It regulates the production of the pituitary gland's hormones and controls that part of the ◊nervous system governing the involuntary muscles.

hypothesis in science, an idea concerning an event and its possible explanation. The term is one favoured by the followers of the philosopher Karl Popper, who argue that the merit of a scientific hypothesis lies in its ability to make testable predictions.

hypsometer instrument for testing the accuracy of a thermometer at the boiling point of water. It was originally used for determining altitude by comparing changes in the boiling point with changes in atmospheric pressure.

hysteresis phenomenon seen in the elastic and electromagnetic behaviour of materials, in which a lag occurs between the application or removal of a force or field and its effect.

If the magnetic field applied to a magnetic material is increased and then decreased back to its original value, the magnetic field inside the material does not return to its original value. The internal field 'lags' behind the external field. This behaviour results in a loss of energy, called the *hysteresis loss*, when a sample is repeatedly magnetized and demagnetized. Hence the materials used in transformer cores and electromagnets should have a low hysteresis loss. Similar behaviour is seen in some materials when varying electric fields are applied (*electric hysteresis*). *Elastic hysteresis* occurs when a varying force repeatedly deforms an elastic material. The deformation produced does not completely disappear when the force is removed, and this results in energy loss on repeated deformations.

Hz in physics, the symbol for ◊*hertz*.

produced as an ⟩export file by a different program or make of computer.

imprinting in ⟩ethology, the process whereby a young animal learns to recognize both specific individuals (for example, its mother) and its own species. Imprinting is characteristically an automatic response to specific stimuli at a time when the animal is especially sensitive to those stimuli (the ***sensitive period***).

impulse in mechanics, the product of a force and the time over which it acts. An impulse applied to a body causes its ⟩momentum to change and is equal to that change in momentum. It is measured in newton seconds.

For example, the impulse J given to a football when it is kicked is given by:

$$J = Ft$$

where F is the kick force in newtons and t is the time in seconds for which the boot is in contact with the ball.

in abbreviation for ⟩*inch*, a measure of distance.

inbreeding in ⟩genetics, the mating of closely related individuals. It is considered undesirable because it increases the risk that offspring will inherit copies of rare deleterious ⟩recessive alleles (genes) from both parents and so suffer from disabilities.

incandescence emission of light from a substance in consequence of its high temperature. The colour of the emitted light from liquids or solids depends on their temperature, and for solids generally the higher the temperature the whiter the light. Gases may become incandescent through ⟩ionizing radiation, as in the glowing vacuum ⟩discharge tube.

inch imperial unit of linear measure, a twelfth of a foot, equal to 2.54 centimetres.

incisor sharp tooth at the front of the mammalian mouth. Incisors are used for biting or nibbling, as when a rabbit or a sheep eats grass. Rodents, such as rats and squirrels, have large continually-growing incisors, adapted for gnawing. The elephant tusk is a greatly enlarged incisor. In humans, the incisors are the four teeth at the front centre of each jaw.

inclination angle between the ⟩ecliptic and the plane of the orbit of a planet, asteroid, or comet. In the case of satellites orbiting a planet, it is the angle between the plane of orbit of the satellite and the equator of the planet.

inclusive fitness in ⟩genetics, the success with which a given variant (or allele) of a ⟩gene is passed on to future generations by a particular individual, after additional copies of the allele in the individual's relatives and their offspring have been taken into account.

indeterminacy principle alternative name for ⟩uncertainty principle.

index (plural *indices*) in mathematics, another term for ⟩exponent, the number that indicates the power to which a term should be raised.

indicator in chemistry, a compound that changes its structure and colour in response to its environment. The commonest chemical indicators detect changes in ⟩pH (for example, ⟩litmus), or in the oxidation state of a system (redox indicators).

indicator species plant or animal whose presence or absence in an area indicates certain environmental conditions, such as soil type, high levels of pollution, or, in rivers, low levels of dissolved oxygen. Many plants show a preference for either alkaline or acid soil conditions, while certain trees require aluminium, and are found only in soils where it is present. Some lichens are sensitive to sulphur dioxide in the air, and absence of these species indicates atmospheric pollution.

indium soft, ductile, silver-white, metallic element, symbol In, atomic number 49, relative atomic mass 114.82. It occurs in nature in some zinc ores, is resistant to abrasion, and is

moving a magnet into a coil

induced current *Moving a magnet into a coil of wire starts or induces an electric current in the coil. The faster the movement, the larger the current; when the movement stops, so does the current. Reversing the movement reverses the current.*

used as a coating on metal parts. It was discovered 1863 by German metallurgists Ferdinand Reich (1799–1882) and Hieronymus Richter (1824–1898), who named it after the two indigo lines of its spectrum.

induced current electric current that appears in a closed circuit when there is relative movement of its conductor in a magnetic field. The effect is known as the *dynamo effect*, and is used in all ◊dynamos and generators to produce electricity. See ◊electromagnetic induction.

There is no battery or other source of power in a circuit in which an induced current appears: the energy supply is provided by the relative motion of the conductor and the magnetic field. The magnitude of the induced current depends upon the rate at which the magnetic flux is cut by the conductor, and its direction is given by Fleming's right-hand rule (see ◊Fleming's rules).

inductance in physics, a measure of the capability of an electronic circuit or circuit component to form a magnetic field or store magnetic energy when carrying a current. Its symbol is L, and its unit of measure is the ◊henry.

induction in physics, an alteration in the physical properties of a body that is brought about by the influence of a field. See ◊electromagnetic induction and ◊magnetic induction.

induction coil type of electrical transformer, similar to an ◊ignition coil, that produces an intermittent high-voltage alternating current from a low-voltage direct current supply.

It has a primary coil consisting of a few turns of thick wire wound around an iron core and passing a low voltage (usually from a battery). Wound on top of this is a secondary coil made up of many turns of thin wire. An iron armature and make-and-break mechanism (similar to that in an electric bell) repeatedly interrupts the current to the primary coil, producing a high, rapidly alternating current in the secondary circuit.

inductor device included in an electrical circuit because of its inductance.

inert gas or *noble gas* any of a group of six elements (helium, neon, argon, krypton, xenon, and radon), so named because they were originally thought not to enter into any chemical reactions. This is now known to be incorrect: in 1962, xenon was made to combine with fluorine, and since then, compounds of argon, krypton, and radon with fluorine and/or oxygen have been described.

inertia in physics, the tendency of an object to remain in a state of rest or uniform motion until an external force is applied, as stated by Isaac Newton's first law of motion (see ◊Newton's laws of motion).

inertial navigation navigation system that makes use of gyroscopes and accelerometers to monitor and measure a vehicle's movements. A computer calculates the vehicle's position relative to its starting position using the information supplied by the sensors. Inertial navigation is used in aircraft, submarines, spacecraft, and guided missiles.

infection invasion of the body by disease-causing organisms (pathogens, or germs) that become established, multiply, and produce symptoms. Bacteria and viruses cause most diseases, but diseases are also caused by other microorganisms, protozoans, and other parasites.

inferior planet planet (Mercury or Venus) whose orbit lies within that of the Earth, best observed when at its greatest elongation from the Sun, either at eastern elongation in the evening (setting after the Sun) or at western elongation in the morning (rising before the Sun).

inferno in astrophysics, a unit for describing the temperature inside a star. One inferno is 1 billion K, or approximately 1 billion°C.

infinite series in mathematics, a series of numbers consisting of an infinite sequence of terms. The sequence n, n^2, n^3, ... gives the series $n + n^2 + n^3 + ...$. For example, $1 + 2 + 3 + ...$ is a divergent infinite arithmetic series, and $8 + 4 + 2 + 1 + 1/2 + ...$ is a convergent infinite geometric series that has a sum to infinity of 16.

infinity mathematical quantity that is larger than any fixed assignable quantity; symbol ∞. By convention, the result of dividing any number by zero is regarded as infinity.

inflammation defensive reaction of the body tissues to disease or damage, including

electronic structure of the inert gases

name	symbol	atomic number	electronic arrangement
helium	He	2	2.
neon	Ne	10	2.8.
argon	Ar	18	2.8.8.
krypton	Kr	36	2.8.18.8.
xenon	Xe	54	2.8.18.18.8.
radon	Rn	86	2.8.18.32.18.8.

redness, swelling, and heat. Denoted by the suffix *-itis* (as in appendicitis), it may be acute or chronic, and may be accompanied by the formation of pus. This is an essential part of the healing process.

inflorescence in plants, a branch, or system of branches, bearing two or more individual flowers. Inflorescences can be divided into two main types: cymose (or definite) and racemose (or indefinite). In a *cymose inflorescence*, the tip of the main axis produces a single flower and subsequent flowers arise on lower side branches, as in forget-me-not *Myosotis* and chickweed *Stellaria*; the oldest flowers are, therefore, found at the tip. A *racemose inflorescence* has an active growing region at the tip of its main axis, and bears flowers along its length, as in hyacinth *Hyacinthus*; the oldest flowers are found near the base or, in cases where the inflorescence is flattened, towards the outside.

information superhighway popular collective name for the ◊Internet and other related large scale computer networks. The term was first used 1993 by US vice president Al Gore in a speech outlining plans to build a high-speed national data communications network.

information technology (IT) collective term for the various technologies involved in processing and transmitting information. They include computing, telecommunications, and microelectronics.

infrared absorption spectrometry technique used to determine the mineral or chemical composition of artefacts and organic substances, particularly amber. A sample is bombarded by infrared radiation, which causes the atoms in it to vibrate at frequencies characteristic of the substance present, and absorb energy at those frequencies from the infrared spectrum, thus forming the basis for identification.

infrared astronomy study of infrared radiation produced by relatively cool gas and dust in space, as in the areas around forming stars. In 1983, the Infra-Red Astronomy Satellite (IRAS) surveyed the entire sky at infrared wavelengths. It found five new comets, thousands of galaxies undergoing bursts of star formation, and the possibility of planetary systems forming around several dozen stars.

infrared radiation invisible electromagnetic radiation of wavelength between about 0.75 micrometres and 1 millimetre – that is, between the limit of the red end of the visible spectrum and the shortest microwaves. All bodies above the ◊absolute zero of temperature absorb and radiate infrared radiation. Infrared radiation is used in medical photography and treatment, and in industry, astronomy, and criminology.

ingestion process of taking food into the mouth. The method of food capture varies but may involve biting, sucking, or filtering. Many single-celled organisms have a region of their cell wall that acts as a mouth. In these cases surrounding tiny hairs (cilia) sweep food particles together, ready for ingestion.

inhibition, neural in biology, the process in which activity in one ◊nerve cell suppresses activity in another. Neural inhibition in networks of nerve cells leading from sensory organs, or to muscles, plays an important role in allowing an animal to make fine sensory discriminations and to exercise fine control over movements.

inhibitor or *negative catalyst* ◊catalyst that reduces the rate of a reaction. Inhibitors are widely used in foods, medicines, and toiletries.

ink-jet printer computer printer that creates characters and graphics by spraying very fine jets of quick-drying ink onto paper. Ink-jet printers range in size from small machines designed to work with microcomputers to very large machines designed for high-volume commercial printing.

inoculation injection into the body of dead or weakened disease-carrying organisms or their toxins (◊vaccine) to produce immunity by inducing a mild form of a disease.

inorganic chemistry branch of chemistry dealing with the chemical properties of the elements and their compounds, excluding the more complex covalent compounds of carbon, which are considered in ◊organic chemistry.

The origins of inorganic chemistry lay in observing the characteristics and experimenting with the uses of the substances (compounds and elements) that could be extracted from mineral ores. These could be classified according to their chemical properties: elements could be classified as metals or nonmetals; compounds as acids or bases, oxidizing or reducing agents, ionic compounds (such as salts), or covalent compounds (such as gases). The arrangement of elements into groups possessing similar properties led to Mendeleyev's ◊periodic table of the elements, which prompted chemists to predict the properties of undiscovered elements that might occupy gaps in the table. This, in turn, led to the discovery of new elements, including a number of highly radioactive elements that do not occur naturally.

inorganic compound compound found in organisms that are not typically biological.

Water, sodium chloride, and potassium are inorganic compounds because they are widely found outside living cells. The term is also applied to those compounds that do not contain carbon and that are not manufactured by organisms. However, carbon dioxide is considered inorganic, contains carbon, and is manufactured by organisms during respiration. See ◊organic compound.

input device device for entering information into a computer. Input devices include keyboards, joysticks, mice, light pens, touch-sensitive screens, graphics tablets, speech-recognition devices, and vision systems. Compare ◊output device.

insect any member of the class Insecta among the ◊arthropods or jointed-legged animals. An insect's body is divided into head, thorax, and abdomen. The head bears a pair of feelers or antennae, and attached to the thorax are three pairs of legs and usually two pairs of wings.

The scientific study of insects is termed *entomology*. More than 1 million species are known, and several thousand new ones are discovered every year. Insects vary in size from 0.02 cm/0.007 in to 35 cm/13.5 in in length.

anatomy The skeleton is external and is composed of ◊chitin. It is membranous at the joints, but elsewhere is hard. The head is the feeding and sensory centre. It bears the antennae, eyes, and mouthparts. By means of the antennae, the insect detects odours and experiences the sense of touch. The eyes include compound eyes and simple eyes (ocelli). Compound eyes are formed of a large number of individual facets or lenses; there are about 4,000 lenses to each compound eye in the housefly. The mouthparts include a labrum, or upper lip; a pair of principal jaws, or mandibles; a pair of accessory jaws, or maxillae; and a labium, or lower lip. These mouthparts are modified in the various insect groups, depending on the diet. The thorax is the locomotory centre, and is made up of three segments: the pro-, meso-, and metathorax. Each bears a pair of legs, and, in flying insects, the second and third of these segments also each bear a pair of wings.

Wings are composed of an upper and a lower membrane, and between these two layers they are strengthened by a framework of chitinous tubes known as veins. The abdomen is the metabolic and reproductive centre, where digestion, excretion, and the sexual functions take place. In the female, there is very commonly an egg-laying instrument, or ovipositor, and many insects have a pair of tail feelers, or cerci. Most insects breathe by means of fine airtubes called tracheae, which open to the exterior by a pair of breathing pores, or spiracles. Reproduction is by diverse means. In most insects, mating occurs once only, and death soon follows.

The *classification* of insects is largely based upon characters of the mouthparts, wings, and ◊metamorphosis. Insects are divided into two subclasses (one with two divisions) and 29 orders.

insecticide any chemical pesticide used to kill insects. Among the most effective insecticides are synthetic organic chemicals such as ◊DDT and dieldrin, which are chlorinated hydrocarbons. These chemicals, however, have proved persistent in the environment and are also poisonous to all animal life, including humans, and are consequently banned in many countries. Other synthetic insecticides include organic phosphorus compounds such as malathion. Insecticides prepared from plants, such as derris and pyrethrum, are safer to use but need to be applied frequently and carefully.

insectivore any animal whose diet is made up largely or exclusively of insects. In particular, the name is applied to mammals of the order Insectivora, which includes the shrews, hedgehogs, moles, and tenrecs.

insectivorous plant plant that can capture and digest live prey (normally insects), to obtain nitrogen compounds that are lacking in its usual marshy habitat. Some are passive traps, for example, the pitcher plants *Nepenthes* and *Sarracenia*. One pitcher-plant species has container-traps holding 1.6 l/3.5 pt of the liquid that 'digests' its food, mostly insects but occasionally even rodents. Others, for example, sundews *Drosera*, butterworts *Pinguicula*, and Venus flytraps *Dionaea muscipula*, have an active trapping mechanism. Insectivorous plants have adapted to grow in poor soil conditions where the number of microorganisms recycling nitrogen compounds is very much reduced. In these circumstances other plants cannot gain enough nitrates to grow.

instinct in ◊ethology, behaviour found in all equivalent members of a given species (for example, all the males, or all the females with

instrument landing system (ILS) landing aid for aircraft that uses ◊radio beacons on the ground and instruments on the flight deck. One beacon (localizer) sends out a vertical radio beam along the centre line of the runway. Another beacon (glide slope) transmits a beam in the plane at right angles to the localizer beam at the ideal approach-path angle. The pilot can tell from the instruments how to manoeuvre to attain the correct approach path.

insulator any poor ◊conductor of heat, sound, or electricity. Most substances lacking free (mobile) ◊electrons, such as non-metals, are electrical or thermal insulators. Usually, devices of glass or porcelain, called insulators, are used for insulating and supporting overhead wires.

insulin protein ◊hormone, produced by specialized cells in the islets of Langerhans in the pancreas, that regulates the metabolism (rate of activity) of glucose, fats, and proteins. Insulin was discovered by Canadian physician Frederick Banting (1891–1941), who pioneered its use in treating ◊diabetes.

integer any whole number. Integers may be positive or negative; 0 is an integer, and is often considered positive. Formally, integers are members of the set $Z = \{ \ldots -3, -2, -1, 0, 1, 2, 3, \ldots \}$. Fractions, such as $^1/_2$ and 0.35, are known as non-integral numbers ('not integers').

integral calculus branch of mathematics using the process of ◊integration. It is concerned with finding volumes and areas and summing infinitesimally small quantities.

integrated circuit (IC), popularly called *silicon chip*, a miniaturized electronic circuit produced on a single crystal, or chip, of a semiconducting material – usually silicon. It may contain many thousands of components and yet measure only 5 mm/0.2 in square and 1 mm/0.04 in thick. The IC is encapsulated within a plastic or ceramic case, and linked via gold wires to metal pins with which it is connected to a ◊printed circuit board and the other components that make up such electronic devices as computers and calculators.

Integrated Services Digital Network (ISDN) internationally developed telecommunications system for sending signals in ◊digital format along optical fibres and coaxial cable. It involves converting the 'local loop' the link between the user's telephone (or private automatic branch exchange) and the digital telephone exchange – from an ◊analogue system into a digital system, thereby greatly increasing the amount of information that can be carried. The first large-scale use of ISDN began in Japan 1988.

integration in mathematics, a method in ◊calculus of determining the solutions of definite or indefinite integrals.

An example of a definite integral can be thought of as finding the area under a curve (as represented by an algebraic expression or function) between particular values of the function's variable. In practice, integral calculus provides scientists with a powerful tool for doing calculations that involve a continually varying quantity (such as determining the position at any given instant of a space rocket that is accelerating away from Earth). Its basic principles were discovered in the late 1660s independently by the German philosopher Leibniz and the British scientist Newton.

integument in seed-producing plants, the protective coat surrounding the ovule. In flowering plants there are two, in gymnosperms only one. A small hole at one end, the micropyle, allows a pollen tube to penetrate through to the egg during fertilization.

Intelsat (abbreviation for *International Telecommunications Satellite Organization*) established 1964 to operate a worldwide system of communications satellites. More than 100 countries are members of Intelsat, with headquarters in Washington DC. Intelsat satellites are stationed in geostationary orbit (maintaining their positions relative to the Earth) over the Atlantic, Pacific, and Indian Oceans. The first Intelsat satellite was *Early Bird*, launched 1965.

intensity in physics, the power (or energy per second) per unit area carried by a form of radiation or wave motion. It is an indication of the concentration of energy present and, if measured at varying distances from the source, of the effect of distance on this. For example, the intensity of light is a measure of its brightness, and may be shown to diminish with distance from its source in accordance with the ◊inverse square law (its intensity is inversely proportional to the square of the distance).

interactive describing a computer system that will respond directly to data or commands entered by the user. For example, most popular programs, such as word processors and spreadsheet applications, are interactive. Multimedia programs are usually highly interactive, allowing users to decide what type of information to display (text, graphics, video, or audio) and enabling them (by means of ◊hypertext) to choose an individual route through the information.

interactive computing in computing, a system for processing data in which the operator is in direct communication with the computer, receiving immediate responses to input data. In batch processing, by contrast, the necessary data and instructions are prepared in advance and processed by the computer with little or no intervention from the operator.

interactive video (IV) computer-mediated system that enables the user to interact with and control information (including text, recorded speech, or moving images) stored on video disc. IV is most commonly used for training purposes, using analogue video discs, but has wider applications with digital video systems such as ◊CD-I (Compact Disc Interactive, from Philips and Sony) which are based on the ◊CD-ROM format derived from audio compact discs.

interface in computing, the point of contact between two programs or pieces of equipment. The term is most often used for the physical connection between the computer and a ◊peripheral device, which is used to compensate for differences in such operating characteristics as speed, data coding, voltage, and power consumption. For example, a *printer interface* is the cabling and circuitry used to transfer data from a computer to a printer, and to compensate for differences in speed and coding.

interference in physics, the phenomenon of two or more wave motions interacting and combining to produce a resultant wave of larger or smaller amplitude (depending on whether the combining waves are in or out of ◊phase with each other).

Interference of white light (multiwavelength) results in spectral coloured fringes; for example, the iridescent colours of oil films seen on water or soap bubbles (demonstrated by ◊Newton's rings). Interference of sound waves of similar frequency produces the phenomenon of beats, often used by musicians when tuning an instrument. With monochromatic light (of a single wavelength), interference produces patterns of light and dark bands. This is the basis of ◊holography, for example. Interferometry can also be applied to radio waves, and is a powerful tool in modern astronomy.

interferometer in physics, a device that splits a beam of light into two parts, the parts being recombined after travelling different paths to form an interference pattern of light and dark bands.

Interferometers are used in many branches of science and industry where accurate measurements of distances and angles are needed.

interferon naturally occurring cellular protein that makes up part of the body's defences against viral disease. Three types (alpha, beta, and gamma) are produced by infected cells and enter the bloodstream and uninfected cells, making them immune to virus attack.

Interferons are cytokines, small molecules that carry signals from one cell to another. They can be divided into two main types: *type I* (alpha, beta, tau, and omega) interferons are more effective at bolstering cells' ability to resist infection; *type II* (gamma) interferon is more important to the normal functioning of the immune system.

intermediate technology application of mechanics, electrical engineering, and other technologies, based on inventions and designs developed in scientifically sophisticated cultures, but utilizing materials, assembly, and maintenance methods found in technologically less advanced regions (known as the Third World).

intermediate vector boson alternative name for *weakon*, the elementary particle responsible for carrying the ◊weak nuclear force.

intermolecular force or *van der Waals' force* force of attraction between molecules. Intermolecular forces are relatively weak; hence simple molecular compounds are gases, liquids, or low-melting-point solids.

internal-combustion engine heat engine in which fuel is burned inside the engine, contrasting with an external-combustion engine (such as the steam engine) in which fuel is burned in a separate unit. The ◊diesel engine and ◊petrol engine are both internal-combustion engines. Gas ◊turbines and ◊jet and ◊rocket engines are sometimes also considered to be internal-combustion engines because they burn their fuel inside their combustion chambers.

internal resistance or *source resistance* the resistance inside a power supply, such as a battery of cells, that limits the current that it can supply to a circuit.

International Atomic Energy Agency (IAEA) agency of the United

Nations established 1957 to advise and assist member countries in the development and application of nuclear power, and to guard against its misuse. It has its headquarters in Vienna, and is responsible for research centres in Austria and Monaco, and the International Centre for Theoretical Physics, Trieste, Italy, established 1964.

international biological standards drugs (such as penicillin and insulin) of which the activity for a specific mass (called the international unit, or IU), prepared and stored under specific conditions, serves as a standard for measuring doses. For penicillin, one IU is the activity of 0.0006 mg of the sodium salt of penicillin, so a dose of a million units would be 0.6 g.

International Standards Organization (ISO) international organization founded 1947 to standardize technical terms, specifications, units, and so on. Its headquarters are in Geneva.

Internet global, on-line computer network connecting governments, companies, universities, and many other networks and users. The service offers ◊electronic mail, conferencing and chat services, as well as the ability to access remote computers and send and retrieve files. It began in 1984 and by the end of 1993 was estimated to have over 12 million users in 70 countries.

Janet (joint academic network), a variant of Internet, is used in Britain. SuperJanet, launched 1992, is an extension of this that can carry 1,000 million bits of information per second.

interplanetary matter gas and dust thinly spread through the Solar System. The gas flows outwards from the Sun as the ◊solar wind.

Fine dust lies in the plane of the solar system, scattering sunlight to cause the ◊zodiacal light. Swarms of dust shed by comets enter the Earth's atmosphere to cause ◊meteor showers.

interpreter computer program that translates and executes a program written in a high-level language. Unlike a ◊compiler, which produces a complete machine-code translation of the high-level program in one operation, an interpreter translates the source program, instruction by instruction, each time that program is run.

intersex individual that is intermediate between a normal male and a normal female in its appearance (for example, a genetic male that lacks external genitalia and so resembles a female).

interstellar molecules over 50 different types of molecule existing in gas clouds in our Galaxy. Most have been detected by their radio emissions, but some have been found by the absorption lines they produce in the spectra of starlight. The most complex molecules, many of them based on ◊carbon, are found in the dense clouds where stars are forming. They may be significant for the origin of life elsewhere in space.

intervertebral disc flexible disc of cartilage interposed between two vertebrae. It helps cushion the spine against the impact of movements like running and jumping.

intestine in vertebrates, the digestive tract from the stomach outlet to the anus. The human *small intestine* is 6 m/20 ft long, 4 cm/1.5 in diameter, and consists of the duodenum, jejunum, and ileum; the *large intestine* is 1.5 m/5 ft long, 6 cm/2.5 in diameter, and includes the caecum, colon, and rectum. Both are muscular tubes comprising an inner lining that secretes alkaline digestive juice, a submucous coat containing fine blood vessels and nerves, a muscular coat, and a serous coat covering all, supported by a strong peritoneum, which carries the blood and lymph vessels, and the nerves. The contents are passed along slowly by ◊peristalsis (waves of involuntary muscular action). The term intestine is also applied to the lower digestive tract of invertebrates.

intrauterine device (IUD) or *coil*, a contraceptive device that is inserted into the womb (uterus). It is a tiny plastic object, sometimes containing copper. By causing a mild inflammation of the lining of the uterus it prevents fertilized eggs from becoming implanted.

intron or *junk DNA* in genetics, a sequence of bases in ◊DNA that carries no genetic information. Introns, discovered 1977, make up 98% of DNA (the rest is made up of ◊exons). Their function is unknown.

Invar trademark for an alloy of iron containing 36% nickel, which expands or contracts very little when the temperature changes.

It is used to make precision instruments (such as pendulums and tuning forks) whose dimensions must not alter.

inverse function ◊function that exactly reverses the transformation produced by a function f; it is usually written as f^{-1}. For example $3x + 2$ and $\sqrt{(x-2)/3}$ are mutually inverse functions. Multiplication and division are inverse operations (see ◊reciprocals).

inverse square law in physics, the statement that the magnitude of an effect (usually a force) at a point is inversely proportional to the square of the distance between that point and the point location of its cause.

Light, sound, electrostatic force (Coulomb's law), gravitational force (Newton's law) and magnetic force (see ◊magnetism) all obey the inverse square law.

invertebrate animal without a backbone. The invertebrates comprise over 95% of the million or so existing animal species and include sponges, coelenterates, flatworms, nematodes, annelid worms, arthropods, molluscs, echinoderms, and primitive aquatic chordates, such as sea squirts and lancelets.

in vitro fertilization (IVF) ('fertilization in glass') allowing eggs and sperm to unite in a laboratory to form embryos. The embryos produced are then implanted into the womb of the otherwise infertile mother (an extension of ◊artificial insemination). The first baby to be produced by this method was born 1978 in the UK. In cases where the Fallopian tubes are blocked, fertilization may be carried out by *intra-vaginal culture*, in which egg and sperm are incubated (in a plastic tube) in the mother's vagina, then transferred surgically into the uterus.

in vitro process biological experiment or technique carried out in a laboratory, outside the body of a living organism (literally 'in glass', for example in a test tube). By contrast, an in vivo process takes place within the body of an organism.

in vivo process biological experiment or technique carried out within a living organism; by contrast, an in vitro process takes place outside the organism, in an artificial environment such as a laboratory.

involuntary action behaviour not under conscious control, for example the contractions of the gut during peristalsis or the secretion of adrenaline by the adrenal glands. Breathing and urination reflexes are involuntary, although both can be controlled voluntarily to some extent. These processes are regulated by the ◊autonomic nervous system.

Io in astronomy, the third-largest moon of the planet Jupiter, 3,630 km/2,260 mi in diameter, orbiting in 1.77 days at a distance of 422,000 km/262,000 mi. It is the most volcanically active body in the Solar System, covered by hundreds of vents that erupt not lava but sulphur, giving Io an orange-coloured surface.

iodide compound formed between iodine and another element in which the iodine is the more electronegative element (see ◊electronegativity, ◊halide).

iodide I^- salt of the ◊halide series containing the iodide ion, formed when an iodine atom gains an electron. When a silver nitrate solution is added to any iodide solution containing dilute nitric acid, a yellow precipitate of silver iodide is formed.

$$KI_{(aq)} + AgNO_{3\,(aq)} \rightarrow AgI_{(s)} + KNO_{3\,(aq)}$$

When chlorine is passed into a solution of an iodide salt, the solution turns brown, then forms a black precipitate as iodine is produced by a ◊displacement reaction.

$$Cl_{2\,(g)} + 2I^-_{(aq)} \rightarrow 2Cl^-_{(aq)} + I_{2\,(s)}$$

iodine greyish-black nonmetallic element, symbol I, atomic number 53, relative atomic mass 126.9044. It is a member of the ◊halogen group. Its crystals give off, when heated, a violet vapour with an irritating odour resembling that of chlorine. It only occurs in combination with other elements. Its salts are known as iodides, which are found in sea water. As a mineral nutrient it is vital to the proper functioning of the thyroid gland, where it occurs in trace amounts as part of the hormone thyroxine. Absence of iodine from the diet leads to goitre formation. Iodine is used in photography, in medicine as an antiseptic, and in making dyes.

iodoform (chemical name *triiodomethane*) CHI_3, an antiseptic that crystallizes into yellow hexagonal plates. It is soluble in ether, alcohol, and chloroform, but not in water.

ion atom, or group of atoms, that is either positively charged (◊cation) or negatively charged (◊anion), as a result of the loss or gain of electrons during chemical reactions or exposure to certain forms of radiation.

ion engine rocket engine that uses ◊ions (charged particles) rather than hot gas for propulsion. Ion engines have been successfully tested in space, where they will eventually be used for gradual rather than sudden velocity changes. In an ion engine, atoms of mercury, for example, are ionized (given an electric charge by an electric field) and then accelerated at high speed by a more powerful electric field.

ion exchange process whereby an ion in one compound is replaced by a different ion, of the same charge, from another compound. It is the basis of a type of ◊chromatography in which the components of a mixture of ions in solution are separated according to the ease with which they will replace the ions on the polymer matrix through which they flow. The exchange of positively charged ions is called cation exchange; that of negatively charged ions is called anion exchange.

ionic bond or ***electrovalent bond*** bond produced when atoms of one element donate electrons to atoms of another element, forming positively and negatively charged ◊ions respectively. The electrostatic attraction between the oppositely charged ions constitutes the bond. Sodium chloride (Na^+Cl^-) is a typical ionic compound.

Each ion has the electronic structure of an inert gas (see ◊noble gas structure). The maximum number of electrons that can be gained is usually two.

ionic compound substance composed of oppositely charged ions. All salts, most bases, and some acids are examples of ionic compounds. They possess the following general properties: they are crystalline solids with a high melting point; are soluble in water and insoluble in organic solvents; and always conduct electricity when molten or in aqueous solution. A typical ionic compound is sodium chloride (Na^+Cl^-).

ionization chamber any device for measuring ◊ionizing radiation. The radiation ionizes the gas in the chamber and the ions formed are collected and measured as an electric charge. Ionization chambers are used for determining the intensity of X-rays or the disintegration rate of radioactive materials.

ionization potential measure of the energy required to remove an ◊electron from an ◊atom. Elements with a low ionization potential readily lose electrons to form ◊cations.

ionizing radiation radiation that knocks electrons from atoms during its passage, thereby leaving ions in its path. Alpha and beta particles are far more ionizing in their effect than are neutrons or gamma radiation.

ionosphere ionized layer of Earth's outer ◊atmosphere (60–1,000 km/38–620 mi) that contains sufficient free electrons to modify the way in which radio waves are propagated, for instance by reflecting them back to Earth. The ionosphere is thought to be produced by absorption of the Sun's ultraviolet radiation.

ion plating method of applying corrosion-resistant metal coatings. The article is placed

electronic arrangement, 2.8.1 of a sodium atom,

electronic arrangement, 2.8.7 of a chlorine atom,

becomes a sodium ion, Na^+, with an electron arrangement 2.8

becomes a chlorine ion, Cl^-, with an electron arrangement 2.8.8

ionic bond *The formation of an ionic bond between a sodium atom and a chlorine atom to form a molecule of sodium chloride. The sodium atom transfers an electron from its outer electron shell (becoming the positive ion Na^+) to the chlorine atom (which becomes the negative chloride ion Cl^-). The opposite charges mean that the ions are strongly attracted to each other. The formation of the bond means that each atom becomes more stable, having a full quota of electrons in its outer shell.*

in argon gas, together with some coating metal, which vaporizes on heating and becomes ionized (acquires charged atoms) as it diffuses through the gas to form the coating. It has important applications in the aerospace industry.

ir in physics, abbreviation for ***infrared***.

iridium hard, brittle, silver-white, metallic element, symbol Ir, atomic number 77, relative atomic mass 192.2. It is twice as heavy as lead and is resistant to tarnish and corrosion. It is one of the so-called platinum group of metals; it occurs in platinum ores and as a free metal (◊native metal) with osmium in osmiridium, a natural alloy that includes platinum, ruthenium, and rhodium.

iris in anatomy, the coloured muscular diaphragm that controls the size of the pupil in the vertebrate eye. It contains radial muscle that increases the pupil diameter and circular muscle that constricts the pupil diameter. Both types of muscle respond involuntarily to light intensity.

iron hard, malleable and ductile, silver-grey, metallic element, symbol Fe (from Latin *ferrum*), atomic number 26, relative atomic mass 55.847. It is the fourth most abundant element (the second most abundant metal, after aluminium) in the Earth's crust. Iron occurs in concentrated deposits as the ores hematite (Fe_2O_3), spathic ore ($FeCO_3$), and magnetite (Fe_3O_4). It sometimes occurs as a free metal, occasionally as fragments of iron or iron–nickel meteorites.

iron ore any mineral from which iron is extracted. The chief iron ores are ◊*magnetite*, a black oxide; ◊*hematite*, or kidney ore, a reddish oxide; ◊*limonite*, brown, impure oxyhydroxides of iron; and *siderite*, a brownish carbonate.

iron pyrites or *pyrite* FeS_2 common iron ore. Brassy yellow, and occurring in cubic crystals, it is often called 'fool's gold', since only those who have never seen gold would mistake it.

irradiation subjecting anything to radiation, including cancer tumours. See also ◊food irradiation.

irrational number a number that cannot be expressed as an exact ◊fraction. Irrational numbers include some square roots (for example, $\sqrt{12}$, $\sqrt{13}$, and $\sqrt{15}$ are irrational) and numbers such as π (the ratio of the circumference of a circle to its diameter, which is approximately equal to 3.14159) and e (the base of ◊natural logarithms, approximately 2.71828).

ISDN abbreviation for ◊*Integrated Services Digital Network*, a telecommunications system.

islets of Langerhans groups of cells within the pancreas responsible for the secretion of the hormone insulin. They are sensitive to the blood sugar, producing more hormone when glucose levels rise.

ISO abbreviation for ◊*International Standards Organization*.

isobar line drawn on a map or weather chart linking all places with the same atmospheric pressure (usually measured in millibars).

Where the isobars are drawn close together they indicate strong winds and a ◊depression (area of low pressure).

isochrone on a map, a line that joins places that are equal in terms of the time it takes to reach them.

isoline on a map, a line that joins places of equal value. Examples are contour lines (joining places of equal height), ◊isobars (for equal pressure), isotherms (for equal temperature), and isohyets (for equal rainfall). Isolines are most effective when values change gradually and when there is plenty of data.

isomer chemical compound having the same molecular composition and mass as another, but with different physical or chemical properties owing to the different structural arrangement of its constituent atoms. For example, the organic compounds butane ($CH_3(CH_2)_2CH_3$) and methyl propane ($CH_3CH(CH_3)CH_3$) are isomers, each possessing four carbon atoms and ten hydrogen atoms but differing in the way that these are arranged with respect to each other.

Structural isomers have obviously different constructions, but *geometrical* and *optical isomers* must be drawn or modelled in order to appreciate the difference in their three-dimensional arrangement. Geometrical isomers have a plane of symmetry and arise because of the restricted rotation of atoms around a bond; optical isomers are mirror images of each other.

butane $CH_3(CH_2)_2CH_3$

methyl propane $CH_3CH(CH_3)CH_3$

○ hydrogen atom
● carbon atom
— covalent bond

isomer *The chemicals butane and methyl propane are isomers. Each has the molecular formula $CH_3CH(CH_3)CH_3$, but with different spatial arrangements of atoms in their molecules.*

For instance, 1,1-dichloroethene ($CH_2=CCl_2$) and 1,2-dichloroethene ($CHCl=CHCl$) are structural isomers, but there are two possible geometric isomers of the latter (depending on whether the chlorine atoms are on the same side or on opposite sides of the plane of the carbon-carbon double bond).

isomorphism the existence of substances of different chemical composition but with similar crystalline form.

isoprene $CH_2CHC(CH_3)CH_2$ (technical name *methylbutadiene*) colourless, volatile fluid obtained from petroleum and coal, used to make synthetic rubber.

isotherm line on a map, linking all places having the same temperature at a given time.

isotope one of two or more atoms that have the same atomic number (same number of protons), but which contain a different number of neutrons, thus differing in their atomic masses. They may be stable or radioactive, naturally occurring or synthesized. The term was coined by English chemist Frederick Soddy, pioneer researcher in atomic disintegration.

IT abbreviation for ◊*information technology*.

iteration in computing, a method of solving a problem by performing the same steps repeatedly until a certain condition is satisfied. For example, in one method of ◊sorting, adjacent items are repeatedly exchanged until the data are in the required sequence.

iteration in mathematics, method of solving ◊equations by a series of approximations which approach the exact solution more and more closely. For example, to find the square root of N, start with a guess n_1; calculate $\sqrt{N}/n_1 = x_1$; calculate $\sqrt{(n_1 + x_1/2)} = n_2$; calculate $\sqrt{N}/n_1 = x_2$; calculate $\sqrt{(n_2 + x_2)/2} = n_3$. The sequence n_1, n_2, n_3 approaches the exact square root of N. Iterative methods are particularly suitable for work with computers and programmable calculators.

iteroparity in biology, the repeated production of offspring at intervals throughout the life cycle. It is usually contrasted with ◊semelparity, where each individual reproduces only once during its life.

IUPAC (abbreviation for *International Union of Pure and Applied Chemistry*) organization that recommends the nomenclature to be used for naming substances, the units to be used, and which conventions are to be adopted when describing particular changes.

ivory the hard white substance of which the teeth and tusks of certain mammals are composed. Among the most valuable are elephants' tusks, which are of unusual hardness and density. Ivory is used in carving and other decorative work, and is so valuable that poachers continue to illegally destroy the remaining wild elephant herds in Africa to obtain it.

J in physics, the symbol for *joule*, the SI unit of energy.

jacinth or *hyacinth* red or yellowish-red gem, a variety of ◊zircon.

jack tool or machine for lifting, hoisting, or moving heavy weights, such as motor vehicles. A *screw jack* uses the principle of the screw to magnify an applied effort; in a car jack, for example, turning the handle many times causes the lifting screw to rise slightly, and the effort is magnified to lift heavy weights. A *hydraulic jack* uses a succession of piston strokes to increase pressure in a liquid and force up a lifting ram.

jade semiprecious stone consisting of either jadeite, $NaAlSi_2O_6$ (a pyroxene), or nephrite, $Ca_2(Mg,Fe)_5Si_8O_{22}(OH,F)_2$ (an amphibole), ranging from colourless through shades of green to black according to the iron content. Jade ranks 5.5–6.5 on the Mohs' scale of hardness.

jansky unit of radiation received from outer space, used in radio astronomy. It is equal to 10^{-26} watts per square metre per hertz, and is named after US engineer Karl Jansky.

Jarvik 7 the first successful artificial heart intended for permanent implantation in a human being. Made from polyurethane plastic and aluminium, it is powered by compressed air. Barney Clark became the first person to receive a Jarvik 7, in Salt Lake City, Utah, USA, Dec 1982; it kept him alive for 112 days

jasper hard, compact variety of ◊chalcedony SiO_2, usually coloured red, brown, or yellow. Jasper can be used as a gem.

jaw one of two bony structures that form the framework of the mouth in all vertebrates except lampreys and hagfishes (the agnathous or jawless vertebrates). They consist of the upper jawbone (maxilla), which is fused to the skull, and the lower jawbone (mandible), which is hinged at each side to the bones of the temple by ◊ligaments.

jejunum part of the small intestine connecting the ◊duodenum to the ◊ileum.

jet hard, black variety of lignite, a type of coal. It is cut and polished for use in jewellery and ornaments.

JET (abbreviation for *Joint European Torus*) ◊tokamake machine built in England to conduct experiments on nuclear fusion. It is the focus of the European effort to produce a practical fusion-power reactor. On 9 November 1991, the JET tokamak produced a 1.7 megawatt pulse of power in an experiment that lasted two seconds. This was the first time that a substantial amount of energy has been produced by nuclear power in a controlled experiment

jetfoil advanced type of ◊hydrofoil boat built by Boeing, propelled by water jets. It features horizontal, fully submerged hydrofoils fore and aft and has a sophisticated computerized control system to maintain its stability in all waters.

jet propulsion method of propulsion in which an object is propelled in one direction by a jet, or stream of gases, moving in the other. This follows from Isaac Newton's (1642–1727) third law of motion: 'To every action, there is an equal and opposite reaction.' The most widespread application of the jet principle is in the jet engine, the most common kind of aircraft engine

The *jet engine* is a kind of ◊gas turbine. Air, after passing through a forward-facing intake, is compressed by a compressor, or fan, and fed into a combustion chamber. Fuel (usually kerosene) is sprayed in and ignited. The hot gas produced expands rapidly rearwards, spinning a turbine that drives the compressor before being finally ejected from a rearward-facing tail pipe, or nozzle, at very high speed. Reaction to the jet of gases streaming backwards produces a propulsive thrust forwards, which acts on the aircraft through its engine-mountings, not from any pushing of the hot gas stream against the static air.

Thrust is proportional to the mass of the gas ejected multiplied by the acceleration imparted to it, and is stated in units of kilograms force (kgf) or pounds force (lbf), both now being superseded by the international unit, the Newton (N).

jet stream narrow band of very fast wind (velocities of over 150 kph/95 mph) found at altitudes of 10–16 km/6–10 mi in the upper troposphere or lower stratosphere. Jet streams usually occur about the latitudes of the Westerlies (35°–60°)

joint in any animal with a skeleton, a point of movement or articulation. In vertebrates, it is the point where two bones meet. Some joints allow no motion (the sutures of the skull), others allow a very small motion (the sacroiliac joints in the lower back), but most allow a relatively free motion. Of these, some allow a gliding motion (one vertebra of the spine on another), some have a hinge action (elbow and knee), and others allow motion in all directions (hip and shoulder joints) by means of a ball-and-socket arrangement. The ends of the bones at a moving joint are covered with cartilage for greater elasticity and smoothness, and enclosed in an envelope (capsule) of tough white fibrous tissue lined with a membrane which secretes a lubricating and cushioning ◊synovial fluid. The joint is further strengthened by ligaments. In invertebrates with an ◊exoskeleton, the joints are places where the exoskeleton is replaced by a more flexible outer covering, the arthrodial membrane, which allows the limb (or other body part) to bend at that point.

Joint European Torus experimental nuclear-fusion machine, known as ◊JET.

Josephson junction device used in 'superchips' (large and complex integrated circuits) to speed the passage of signals by a phenomenon called 'electron tunnelling'. Although these superchips respond a thousand times faster than the ◊silicon chip, they have the disadvantage that the components of the Josephson junctions operate only at temperatures close to ◊absolute zero. They are named after English theoretical physicist Brian Josephson.

joule SI unit (symbol J) of work and energy, replacing the ◊calorie (one joule equals 4.2 calories).

It is defined as the work done (energy transferred) by a force of one newton acting over one metre. It can also be expressed as the work done in one second by a current of one ampere at a potential difference of one volt. One watt is equal to one joule per second. It is named after the British physicist James Joule (1818–1889).

Joule–Kelvin effect in physics, the fall in temperature of a gas as it expands adiabatically (without loss or gain of heat to the system) through a narrow jet. It can be felt when, for example, compressed air escapes through the valve of an inflated bicycle tyre. Only hydrogen does not exhibit the effect. It is the basic principle of most refrigerators.

joystick in computing, an input device that signals to a computer the direction and extent of displacement of a hand-held lever. It is similar to the joystick used to control the flight of an aircraft.

joint *The knee joint is the largest and most complex in the human body. It is similar to a hinge joint but with a slight rotation which allows the leg to lock into a rigid position when extended. The ligaments surrounding the joint help prevent overextension. The ligament surrounding the patella, or kneecap, helps to protect the joint.*

jugular vein one of two veins in the necks of vertebrates; they return blood from the head to the superior (or anterior) ◊vena cava and thence to the heart.

junk DNA another name for ◊intron, a region of DNA that contains no genetic information.

Jupiter the fifth planet from the Sun, and the largest in the Solar System (equatorial diameter 142,800 km/88,700 mi), with a mass more than twice that of all the other planets combined, 318 times that of the Earth's. It takes 11.86 years to orbit the Sun, at an average distance of 778 million km/484 million mi, and has at least 16 moons. It is largely composed of hydrogen and helium, liquefied by pressure in its interior, and probably with a rocky core larger than the Earth. Its main feature is the Great Red Spot, a cloud of rising gases, revolving anticlockwise, 14,000 km/8,500 mi wide and some 30,000 km/20,000 mi long.

Jurassic period of geological time 208–146 million years ago; the middle period of the Mesozoic era. Climates worldwide were equable, creating forests of conifers and ferns; dinosaurs were abundant, birds evolved, and limestones and iron ores were deposited.

K

K symbol for *kelvin*, a scale of temperature.

k symbol for *kilo-*, as in kg (kilogram) and km (kilometre).

kaolinite white or greyish ◊clay mineral, hydrated aluminium silicate, $Al_2Si_2O_5(OH)_4$, formed mainly by the decomposition of feldspar in granite. China clay (kaolin) is derived from it. It is mined in France, the UK, Germany, China, and the USA.

karyotype in biology, the set of ◊chromosomes characteristic of a given species. It is described as the number, shape, and size of the chromosomes in a single cell of an organism. In humans for example, the karyotype consists of 46 chromosomes, in mice 40, crayfish 200, and in fruit flies 8.

kayser unit of wave number (number of waves in a unit length), used in spectroscopy. It is expressed as waves per centimetre, and is the reciprocal of the wavelength. A wavelength of 0.1 cm has a wave number of 10 kaysers.

kcal symbol for *kilocalorie* (see ◊calorie).

kelvin scale temperature scale used by scientists. It begins at ◊absolute zero (−273.15°C) and increases by the same degree intervals as the Celsius scale; that is, 0°C is the same as 273K and 100°C is 373K.

Kennelly–Heaviside layer former term for the ◊E layer of the ionosphere.

keratin fibrous protein found in the ◊skin of vertebrates and also in hair, nails, claws, hooves, feathers, and the outer coating of horns.

kermit in computing, a widely used ◊file-transfer protocol, originally developed at Columbia University and made available without charge. Kermit is available as part of most communications packages and available on most operating systems.

kernel the inner, softer part of a ◊nut, or of a seed within a hard shell.

kerosene thin oil obtained from the distillation of petroleum; a highly refined form is used in jet aircraft fuel. Kerosene is a mixture of hydrocarbons of the ◊paraffin series.

ketone member of the group of organic compounds containing the carbonyl group (C=O) bonded to two atoms of carbon (instead of one carbon and one hydrogen as in ◊aldehydes). Ketones are liquids or low-melting-point solids, slightly soluble in water.

keyboard in computing, an input device resembling a typewriter keyboard, used to enter instructions and data. There are many variations on the layout and labelling of keys. Extra numeric keys may be added, as may special-purpose function keys, whose effects can be defined by programs in the computer.

kg symbol for ◊*kilogram*.

kidney in vertebrates, one of a pair of organs responsible for fluid regulation, excretion of waste products, and maintaining the ionic composition of the blood. The kidneys are situated on the rear wall of the abdomen. Each one consists of a number of long tubules; the outer parts filter the aqueous components of blood, and the inner parts selectively reabsorb vital salts, leaving waste products in the remaining fluid (urine), which is passed through the ureter to the bladder.

kiln high-temperature furnace used commercially for drying timber, roasting metal ores, or for making cement, bricks, and pottery. Oil- or gas-fired kilns are used to bake ceramics at up to 1,760°C/3,200°F; electric kilns do not generally reach such high temperatures.

kilo- prefix denoting multiplication by 1,000, as in kilohertz, a unit of frequency equal to 1,000 hertz.

kilobyte (K or KB) in computing, a unit of memory equal to 1,024 ◊bytes. It is sometimes used, less precisely, to mean 1,000 bytes.

kilogram SI unit (symbol kg) of mass equal to 1,000 grams (2.24 lb). It is defined as a mass equal to that of the international prototype, a platinum-iridium cylinder held at the International Bureau of Weights and Measures at Sèvres, France.

kilometre unit of length (symbol km) equal to 1,000 metres, equivalent to 3,280.89 ft or 0.6214 (about $^5/_8$) of a mile.

cross section

kidney *Blood enters the kidney through the renal artery. The blood is filtered through the glomeruli to extract the nitrogenous waste products and excess water that make up urine. The urine flows through the ureter to the bladder; the cleaned blood then leaves the kidney along the renal vein.*

kilowatt unit (symbol kW) of power equal to 1,000 watts or about 1.34 horsepower.

kilowatt-hour commercial unit of electrical energy (symbol kWh), defined as the work done by a power of 1,000 watts in one hour. It is used to calculate the cost of electrical energy taken from the domestic supply.

kimberlite an igneous rock that is ultrabasic (containing very little silica); a type of alkaline ◊peridotite containing mica in addition to olivine and other minerals. Kimberlite represents the world's principal source of diamonds.

kinesis (plural *kineses*) in biology, a nondirectional movement in response to a stimulus; for example, woodlice move faster in drier surroundings. *Taxis* is a similar pattern of behaviour, but there the response is directional.

kinetic energy the energy of a body resulting from motion. It is contrasted with ◊potential energy.

kinetics branch of ◊dynamics dealing with the action of forces producing or changing the motion of a body; *kinematics* deals with motion without reference to force or mass.

kinetics the branch of chemistry that investigates the rates of chemical reactions.

kinetic theory theory describing the physical properties of matter in terms of the behaviour – principally movement – of its component atoms or molecules. The temperature of a substance is dependent on the velocity of movement of its constituent particles, increased temperature being accompanied by increased movement. A gas consists of rapidly moving atoms or molecules and, according to kinetic theory, it is their continual impact on the walls of the containing vessel that accounts for the pressure of the gas. The slowing of molecular motion as temperature falls, according to kinetic theory, accounts for the physical properties of liquids and solids, culminating in the concept of no molecular motion at ◊absolute zero (0K/–273°C).

By making various assumptions about the nature of gas molecules, it is possible to derive from the kinetic theory the various gas laws (such as ◊Avogadro's hypothesis, ◊Boyle's law, and ◊Charles's law).

kingdom the primary division in biological ◊classification. At one time, only two kingdoms were recognized: animals and plants. Today most biologists prefer a five-kingdom system, even though it still involves grouping together organisms that are probably unrelated. One widely accepted scheme is as follows: *Kingdom Animalia* (all multicellular animals); *Kingdom Plantae* (all plants, including seaweeds and other algae, except blue-green); *Kingdom Fungi* (all fungi, including the unicellular yeasts, but not slime moulds); *Kingdom Protista* or *Protoctista* (protozoa, diatoms, dinoflagellates, slime moulds, and various other lower organisms with eukaryotic cells); and *Kingdom Monera* (all prokaryotes – the bacteria and cyanobacteria, or ◊blue-green algae). The first four of these kingdoms make up the eukaryotes.

kin selection in biology, the idea that ◊altruism shown to genetic relatives can be worthwhile, because those relatives share some genes with the individual that is behaving altruistically, and may continue to reproduce. See ▷inclusive fitness.

Kirchhoff's first law law formulated by German physicist Gustav Kirchhoff 1845 that states that the total current entering a junction in a circuit must equal the total current leaving it. It is based on the law of conservation of electric charge: as no charge can be created or destroyed at a junction, the flow of charge (current), into and out of the junction must be the same.

kite quadrilateral with two pairs of adjacent equal sides. The geometry of this figure follows from the fact that it has one axis of symmetry.

km symbol for ▷*kilometre*.

knocking in a spark-ignition petrol engine, a phenomenon that occurs when unburned fuel-air mixture explodes in the combustion chamber before being ignited by the spark. The resulting shock waves produce a metallic knocking sound. Loss of power occurs, which can be prevented by reducing the compression ratio, redesigning the geometry of the combustion chamber, or increasing the octane number of the petrol (usually by the use of lead tetraethyl anti-knock additives).

knot in navigation, unit by which a ship's speed is measured, equivalent to one ▷nautical mile per hour (one knot equals about 1.15 miles per hour). It is also sometimes used in aviation.

knowledge-based system (KBS) computer program that uses an encoding of human knowledge to help solve problems. It was discovered during research into ▷artificial intelligence that adding heuristics (rules of thumb) enabled programs to tackle problems that were otherwise difficult to solve by the usual techniques of computer science.

Königsberg bridge problem long-standing puzzle that was solved by topology (the geometry of those properties of a figure which remain the same under distortion). In the city of Königsberg (now Kaliningrad in Russia), seven bridges connect the banks of the river Pregol'a and the islands in the river. For many years, people were challenged to cross each of the bridges in a single tour and return to their starting point. In 1736 Swiss mathematician Leonhard Euler converted the puzzle into a topological network, in which the islands and river banks were represented as nodes (junctions), and the connecting bridges as lines. By analysing this network he was able to show that it is not traversible – that is, it is impossible to cross each of the bridges once only and return to the point at which one started.

kph or *km/h* symbol for *kilometres per hour*.

Krebs cycle or *citric acid cycle* or *tricarboxylic acid cycle* final part of the chain of biochemical reactions by which organisms break down food using oxygen to release energy (respiration). It takes place within structures called ▷mitochondria in the body's cells, and breaks down food molecules in a series of small steps, producing energy-rich molecules of ▷ATP.

krypton colourless, odourless, gaseous, nonmetallic element, symbol Kr, atomic number 36, relative atomic mass 83.80. It is grouped with the inert gases and was long believed not to enter into reactions, but it is now known to combine with fluorine under certain conditions; it remains inert to all other reagents. It is present in very small quantities in the air (about 114 parts per million). It is used chiefly in fluorescent lamps, lasers, and gas-filled electronic valves.

K-T boundary geologists' shorthand for the boundary between the rocks of the ▷Cretaceous and the ▷Tertiary periods 65 million years ago. It marks the extinction of the dinosaurs and in many places reveals a layer of iridium, possibly deposited by a meteorite that may have caused the extinction by its impact.

Kuiper belt a ring of small, icy bodies orbiting the Sun beyond the outermost planet. The Kuiper belt, named after the American astronomer Gerard Kuiper (1905–1973) who proposed its existence in 1951, is thought to be the source of comets that orbit the Sun with periods of less than 200 years. The first member of the Kuiper belt was seen in 1992.

kurchatovium name proposed by Soviet scientists for the element currently known as ▷unnilquadium (atomic number 104), to honour Soviet nuclear physicist Igor Kurchatov (1903–1960).

kW symbol for ▷*kilowatt*.

kyanite aluminium silicate, Al_2SiO_5, a pale-blue mineral occurring as blade-shaped crystals. It is an indicator of high-pressure conditions in metamorphic rocks formed from clay sediments. Andalusite, kyanite, and sillimanite are all polymorphs.

l symbol for ◊*litre*, a measure of liquid volume.

labelled compound or *tagged compound* chemical compound in which a radioactive isotope is substituted for a stable one. The path taken by such a compound through a system can be followed, for example by measuring the radiation emitted.

labellum lower petal of an orchid flower; it is a different shape from the two lateral petals and gives the orchid its characteristic appearance. The labellum is more elaborate and usually larger than the other petals. It often has distinctive patterning to encourage ◊pollination by insects; sometimes it is extended backwards to form a hollow spur containing nectar.

lacrimal apparatus apparatus that supplies and drains the fluid needed to lubricate and cleanse the surface of the eye. The lacrimal gland in the upper eyelid secretes this fluid, which in quantity is known as tears; it drains through a tiny opening in the inner corner of the eye into the lacrimal canaliculus and from there passes into the nasal cavity by way of the nasolacrimal duct.

lactation secretion of milk in mammals, from the mammary glands. In late pregnancy, the cells lining the lobules inside the mammary glands begin extracting substances from the blood to produce milk. The supply of milk starts shortly after birth with the production of colostrum, a clear fluid consisting largely of water, protein, antibodies, and vitamins. The production of milk continues practically as long as the baby continues to suckle.

lacteal small vessel responsible for absorbing fat in the small intestine. Occurring in the fingerlike villi of the ◊ileum, lacteals have a milky appearance and drain into the lymphatic system.

Before fat can pass into the lacteal, bile from the liver causes its emulsification into droplets small enough for attack by the enzyme lipase. The products of this digestion form into even smaller droplets, which diffuse into the villi. Large droplets reform before entering the lacteal and this causes the milky appearance.

lactic acid or *2-hydroxypropanoic acid* $CH_3CHOHCOOH$ organic acid, a colourless, almost odourless liquid, produced by certain bacteria during fermentation and by active muscle cells when they are exercised hard and are experiencing ◊oxygen debt. An accumulation of lactic acid in the muscles may cause cramp. It occurs in yoghurt, buttermilk, sour cream, poor wine, and certain plant extracts, and is used in food preservation and in the preparation of pharmaceuticals.

lactose white sugar, found in solution in milk; it forms 5% of cow's milk. It is commercially prepared from the whey obtained in cheese-making. Like table sugar (sucrose), it is a disaccharide, consisting of two basic sugar units (monosaccharides), in this case, glucose and galactose. Unlike sucrose, it is tasteless.

Lagrangian points five locations in space where the centrifugal and gravitational forces of two bodies neutralize each other; a third, less massive body located at any one of these points will be held in equilibrium with respect to the other two. Three of the points, L1–L3, lie on a line joining the two large bodies. The other two points, L4 and L5, which are the most stable, lie on either side of this line. Their existence was predicted in 1772 by French mathematician Joseph Louis Lagrange.

Lamarckism theory of evolution, now discredited, advocated during the early 19th century by French naturalist Jean Baptiste Lamarck. Lamarckism is the theory that ◊acquired characteristics were inherited. It differs from the Darwinian theory of evolution.

lambert unit of luminance (the light shining from a surface), equal to one ◊lumen per square centimetre. In scientific work the ◊candela per square metre is preferred.

lamina in flowering plants (◊angiosperms), the blade of the ◊leaf on either side of the midrib. The lamina is generally thin and flattened, and is usually the primary organ of ◊photosynthesis. It has a network of veins through which water and nutrients are conducted. More generally, a lamina is any thin, flat plant structure, such as the ◊thallus of many seaweeds.

lamp, electric device designed to convert electrical energy into light energy.

In a *filament lamp* such as a light bulb an electric current causes heating of a long thin coil of fine high-resistance wire enclosed at low pressure inside a glass bulb. In order to give out a good light the wire must glow white-hot and therefore must be made of a metal, such as tungsten, that has a high melting point. The efficiency of filament lamps is low because most of the electrical energy is converted to heat.

A *fluorescent light* uses an electrical discharge or spark inside a gas-filled tube to produce light. The inner surface of the tube is coated with a fluorescent material that converts the ultraviolet light generated by the discharge into visible light. Although a high voltage is needed to start the discharge, these lamps are far more efficient than filament lamps at producing light.

Landsat series of satellites used for monitoring Earth resources. The first was launched in 1972.

lanolin sticky, purified wax obtained from sheep's wool and used in cosmetics, soap, and leather preparation.

lanthanide any of a series of 15 metallic elements (also known as rare earths) with atomic numbers 57 (lanthanum) to 71 (lutetium). One of its members, promethium, is radioactive. All occur in nature. Lanthanides are grouped because of their chemical similarities (they are all bivalent), their properties differing only slightly with atomic number.

lanthanum soft, silvery, ductile and malleable, metallic element, symbol La, atomic number 57, relative atomic mass 138.91, the first of the lanthanide series. It is used in making alloys. It was named 1839 by Swedish chemist Carl Mosander (1797–1858).

lapis lazuli rock containing the blue mineral lazurite in a matrix of white calcite with small amounts of other minerals. It occurs in silica-poor igneous rocks and metamorphic limestones found in Afghanistan, Siberia, Iran, and Chile. Lapis lazuli was a valuable pigment of the Middle Ages, also used as a gemstone and in inlaying and ornamental work.

laptop computer portable microcomputer, small enough to be used on the operator's lap. It consists of a single unit, incorporating a keyboard, ◊floppy disc or ◊hard disc drives, and a screen. The screen often forms a lid that folds back in use. It uses a liquid-crystal or gas-plasma display, rather than the bulkier and heavier cathode-ray tubes found in most display terminals. A typical laptop computer measures about 210 × 297 mm/8.3 × 11.7 in (A4), is 5 cm/2 in thick, and weighs less than 3 kg/6 lb 9 oz.

Large Electron Positron Collider (LEP) the world's largest particle ◊accelerator, in operation from 1989 at the CERN laboratories near Geneva in Switzerland. It occupies a tunnel 3.8 m/12.5 ft wide and 27 km/16.7 mi long, which is buried 180 m/590 ft underground and forms a ring consisting of eight curved and eight straight sections.

In 1989 the LEP was used to measure the mass and lifetime of the Z particle, carrier of the weak nuclear force.

larva stage between hatching and adulthood in those species in which the young have a different appearance and way of life from the adults. Examples include tadpoles (frogs) and caterpillars (butterflies and moths). Larvae are typical of the invertebrates, some of which (for example, shrimps) have two or more distinct larval stages. Among vertebrates, it is only the amphibians and some fishes that have a larval stage.

larynx in mammals, a cavity at the upper end of the trachea (windpipe) containing the vocal cords. It is stiffened with cartilage and lined with mucous membrane. Amphibians and reptiles have much simpler larynxes, with no vocal cords. Birds have a similar cavity, called the *syrinx*, found lower down the trachea, where it branches to form the bronchii. It is very complex, with well-developed vocal cords.

laser (acronym for *light amplification by stimulated emission of radiation*) a device for producing a narrow beam of light, capable of travelling over vast distances without dispersion, and of being focused to give enormous power densities (10^8 watts per cm^2 for high-energy lasers). The laser operates on a principle similar to that of the ◊maser (a high-frequency microwave amplifier or oscillator). The uses of lasers include communications (a laser beam can carry much more information than can radio waves), cutting, drilling, welding, satellite tracking, medical and biological research, and surgery.

laser mass spectroscopy forensic technique whereby a laser beam is used to scan a minute sample of evidence – such as particles of paint or hair. The beam releases a gas of charged ions, each of which can be identified by mass ◊spectroscopy.

laser printer computer printer in which the image to be printed is formed by the action of a laser on a light-sensitive drum, then transferred to paper by means of an electrostatic charge. Laser printers are page printers, printing a complete page at a time. The printed image, which can take the form of text or pictures, is made up of tiny dots, or ink particles. The quality of the image generated depends on the fineness of these dots – most laser printers can print up to 120 dots per cm/300 dots per in across the page.

latent heat in physics, the heat absorbed or radiated by a substance as it changes state (for example, from solid to liquid) at constant temperature and pressure.

lateral line system system of sense organs in fishes and larval amphibians (tadpoles) that detects water movement. It usually consists of a row of interconnected pores on either side of the body that divide into a system of canals across the head.

laterite red residual soil characteristic of tropical rainforests. It is formed by the weathering of basalts, granites, and shales and contains a high percentage of aluminium and iron hydroxides. It may form an impermeable and infertile layer that hinders plant growth.

latex fluid of some plants (such as the rubber tree and poppy), an emulsion of resins, proteins, and other organic substances. It is used as the basis for making rubber. The name is also applied to a suspension in water of natural or synthetic rubber (or plastic) particles used in rubber goods, paints, and adhesives.

lathe machine tool, used for *turning*. The workpiece to be machined, usually wood or metal, is held and rotated while cutting tools are moved against it. Modern lathes are driven by electric motors, which can drive the spindle carrying the workpiece at various speeds.

latitude and longitude imaginary lines used to locate position on the globe. Lines of latitude are drawn parallel to the equator, with 0° at the equator and 90° at the north and south poles. Lines of longitude are drawn at right angles to these, with 0° (the Prime Meridian) passing through Greenwich, England.

laughing gas popular name for ◊nitrous oxide, an anaesthetic.

Laurasia northern landmass formed 200 million years ago by the splitting of the single world continent ◊Pangaea. (The southern landmass was ◊Gondwanaland.) It consisted of what was to become North America, Greenland, Europe, and Asia, and is believed to have broken up about 100 million years ago with the separation of North America from Europe.

lava molten rock (usually 800–1100°C/1500–2000°F) that erupts from a ◊volcano and cools to form extrusive ◊igneous rock. It differs from magma in that it is molten rock on the surface; magma is molten rock below the surface. Lava that is high in silica is viscous and sticky and does not flow far; it forms a steep-sided conical volcano. Low-silica lava can flow for long distances and forms a broad flat volcano.

law of nature scientific generalization that both explains and predicts physical phenomena; laws of nature are generally assumed to be descriptive of, and applicable to, the world. The three laws of ◊thermodynamics are examples.

lawrencium synthesized, radioactive, metallic element, the last of the actinide series, symbol Lr, atomic number 103, relative atomic mass 262. Its only known isotope, Lr-257, has a half-life of 4.3 seconds and was originally synthesized at the University of California at Berkeley 1961 by bombarding californium with boron nuclei. The original symbol, Lw, was officially changed 1963.

In 1992 the International Unions for Pure and Applied Chemistry and Physics (IUPAC and IUPAP) gave credit to Russian scientists in Dubna for the joint discovery of lawrencium.

lb symbol for ◊*pound* (weight).

LCD abbreviation for ◊*liquid-crystal display*.

L-dopa chemical, normally produced by the body, which is converted by an enzyme to dopamine in the brain. It is essential for integrated movement of individual muscle groups.

LDR abbreviation for ◊*light-dependent resistor*.

leaching process by which substances are washed out of the soil. Fertilizers leached out of the soil drain into rivers, lakes, and ponds and cause water pollution. In tropical areas, leaching of the soil after the destruction of forests removes scarce nutrients and can lead to a dramatic loss of soil fertility. The leaching of soluble minerals in soils can lead to the formation of distinct soil horizons as different minerals are deposited at successively lower levels.

lead heavy, soft, malleable, grey, metallic element, symbol Pb (from Latin *plumbum*), atomic number 82, relative atomic mass 207.19. Usually found as an ore (most often in galena), it occasionally occurs as a free metal (◊native metal), and is the final stable product

LEAD(II) NITRATE

of the decay of uranium. Lead is the softest and weakest of the commonly used metals, with a low melting point; it is a poor conductor of electricity and resists acid corrosion. As a cumulative poison, lead enters the body from lead water pipes, lead-based paints, and leaded petrol. (In humans, exposure to lead shortly after birth is associated with impaired mental health between the ages of two and four.) The metal is an effective shield against radiation and is used in batteries, glass, ceramics, and alloys such as pewter and solder.

lead–acid cell type of ◊accumulator (storage battery).

leaded petrol petrol that contains ◊antiknock, a mixture of the chemicals tetraethyl lead and dibromoethane. The lead from the exhaust fumes enters the atmosphere, mostly as simple lead compounds, which are poisonous to the developing nervous systems of children.

lead(II) nitrate $Pb(NO_3)_2$ one of only two common water-soluble compounds of lead. When heated, it decrepitates (see ◊decrepita

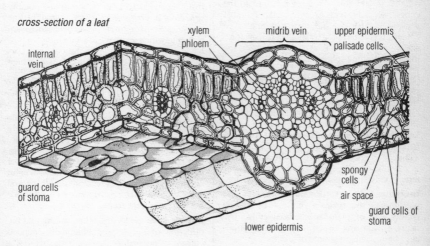

leaf *Leaf shapes and arrangements on the stem are many and varied; in cross-section, a leaf is a complex arrangement of cells surrounded by the epidermis. This is pierced by the stomata through which gases enter and leave.*

tion) and decomposes readily into oxygen, brown nitrogen(IV) oxide gas and the red-yellow solid lead(II) oxide.

$$2Pb(NO_3)_2 \rightarrow 2PbO + 4NO_2 + O_2$$

lead ore any of several minerals from which lead is extracted. The main primary ore is galena or lead sulphite PbS. This is unstable, and on prolonged exposure to the atmosphere it oxidizes into the minerals cerussite $PbCO_3$ and anglesite $PbSO_4$. Lead ores are usually associated with other metals, particularly silver – which can be mined at the same time – and zinc, which can cause problems during smelting.

lead(II) oxide or *lead monoxide* PbO yellow or red solid, an amphoteric oxide (one that reacts with both acids and bases). The other oxides of lead are the brown solid lead(IV) oxide (PbO_2) and red lead (Pb_3O_4).

leaf lateral outgrowth on the stem of a plant, and in most species the primary organ of ◊photosynthesis. The chief leaf types are cotyledons (seed leaves), scale leaves (on underground stems), foliage leaves, and bracts (in the axil of which a flower is produced).

Typically leaves are composed of three parts: the sheath or leaf base, the petiole or stalk, and the lamina or blade. The lamina has a network of veins through which water and nutrients are conducted. Structurally the leaf is made up of ◊mesophyll cells surrounded by the epidermis and usually, in addition, a waxy layer, termed the cuticle, which prevents excessive evaporation of water from the leaf tissues by transpiration. The epidermis is interrupted by small pores, or stomata, through which gas exchange between the plant and the atmosphere occurs.

least action principle in science, the principle that nature 'chooses' the easiest path for moving objects, rays of light, and so on; also known in biology as the *principle of parsimony*.

Le Chatelier's principle or *Le Chatelier-Braun principle* in science, the principle that if a change in conditions is imposed on a system in equilibrium, the system will react to counteract that change and restore the equilibrium.

lecithin lipid (fat), containing nitrogen and phosphorus, that forms a vital part of the cell membranes of plant and animal cells. The name is from the Greek *lekithos* 'egg yolk', eggs being a major source of lecithin.

Leclanché cell disposable, or primary, electrical cell invented 1866 by French engineer Georges Leclanché, which still forms the basis of most dry batteries.

LED abbreviation for ◊*light-emitting diode*.

left-hand rule in physics, a memory aid used to recall the relative directions of motion, magnetic field, and current in an electric motor. It was devised by English physicist John Fleming. See ◊Fleming's rules.

legume plant of the family Leguminosae, which has a pod containing dry seeds. The family includes peas, beans, lentils, clover, and alfalfa (lucerne). Legumes are important in agriculture because of their specialized roots, which have nodules containing bacteria capable of fixing nitrogen from the air and increasing the fertility of the soil. The edible seeds of legumes are called *pulses*.

lek in biology, a closely spaced set of very small ◊territories each occupied by a single male during the mating season. Leks are found in the mating systems of several ground-dwelling birds (such as grouse) and a few antelopes.

lens in optics, a piece of a transparent material, such as glass, with two polished surfaces – one concave or convex, and the other plane, concave, or convex – that modifies rays of light. A convex lens brings rays of light together; a concave lens makes the rays diverge. Lenses are essential to spectacles, microscopes, telescopes, cameras, and almost all optical instruments.

The image formed by a single lens suffers from several defects or ◊aberrations, notably ***spherical aberration*** in which a straight line becomes a curved image, and ***chromatic aberration*** in which an image in white light tends to have coloured edges. Aberrations are corrected by the use of compound lenses, which are built up from two or more lenses of different refractive index.

lensing, gravitational see ◊gravitational lensing.

lenticel small pore on the stems of woody plants or the trunks of trees. Lenticels are means of gas exchange between the stem interior and the atmosphere. They consist of loosely packed cells with many air spaces in between, and are easily seen on smooth-barked trees such as cherries, where they form horizontal lines on the trunk.

Lenz's law in physics, a law stating that the direction of an electromagnetically induced current (generated by moving a magnet near a wire or a wire in a magnetic field) will oppose the motion producing it. It is named after the German physicist Heinrich Friedrich Lenz (1804–1865), who announced it in 1833.

camera) and in switches (such as those that switch on street lights at dusk).

LDRs are made from ◊semiconductors, such as cadmium sulphide.

light-emitting diode (LED) means of displaying symbols in electronic instruments and devices. An LED is made of ◊semiconductor material, such as gallium arsenide phosphide, that glows when electricity is passed through it. The first digital watches and calculators had LED displays, but many later models use ◊liquid-crystal displays.

lightning high-voltage electrical discharge between two charged rainclouds or between a cloud and the Earth, caused by the build-up of electrical charges. Air in the path of lightning ionizes (becomes conducting), and expands; the accompanying noise is heard as thunder. Currents of 20,000 amperes and temperatures of 30,000°C/54,000°F are common.

lightning conductor device that protects a tall building from lightning strike by providing an easier path for current to flow to earth than through the building. It consists of a thick copper strip of very low resistance that connects a metal rod projecting above the building to the ground below. A good connection to the ground is essential and is made by burying a large metal plate deep in the damp earth. In the event of a direct lightning strike, the current in the conductor may be so great as to melt or even vaporize the metal, but the damage to the building will nevertheless be limited.

light pen in computing, a device resembling an ordinary pen, used to indicate locations on a computer screen. With certain computer-aided design (◊CAD) programs, the light pen can be used to instruct the computer to change the shape, size, position, and colours of sections of a screen image.

light second unit of length, equal to the distance travelled by light in one second. It is equal to 2.997925×10^8 m/9.835592×10^8 ft. See ◊light year.

light watt unit of radiant power (brightness of light). One light watt is the power required to produce a perceived brightness equal to that of light at a wavelength of 550 nanometres and 680 lumens.

light year in astronomy, the distance travelled by a beam of light in a vacuum in one year, approximately 9.46 trillion (million million) km/5.88 trillion miles.

lignin naturally occurring substance produced by plants to strengthen their tissues. It is difficult for ◊enzymes to attack lignin, so living organisms cannot digest wood, with the exception of a few specialized fungi and bacteria. Lignin is the essential ingredient of all wood and is, therefore, of great commercial importance.

lignite type of ◊coal that is brown and fibrous, with a relatively low carbon content. In Scandinavia it is burned to generate power.

lime or *quicklime* CaO (technical name *calcium oxide*) white powdery substance used in making mortar and cement. It is made commercially by heating calcium carbonate ($CaCO_3$), obtained from limestone or chalk, in a ◊lime kiln. Quicklime readily absorbs water to become calcium hydroxide (CaOH), known as slaked lime, which is used to reduce soil acidity.

lime kiln oven used to make quicklime (calcium oxide, CaO) by heating limestone (calcium carbonate, $CaCO_3$) in the absence of air. The carbon dioxide is carried away to heat other kilns and to ensure that the reversible reaction proceeds in the right direction.

$$CaCO_3 \rightleftharpoons CaO + CO_2$$

limestone sedimentary rock composed chiefly of calcium carbonate $CaCO_3$, either derived from the shells of marine organisms or precipitated from solution, mostly in the ocean. Various types of limestone are used as building stone.

limewater common name for a dilute solution of slaked lime (calcium hydroxide, $Ca(OH)_2$). In chemistry, it is used to detect the presence of carbon dioxide.

limit in mathematics, in an infinite sequence, the final value towards which the sequence is tending. For example, the limit of the sequence $1/2, 3/4, 7/8, 15/16 \ldots$ is 1, although no member of the sequence will ever exactly equal 1 no matter how many terms are added together.

limiting factor in biology, any factor affecting the rate of a metabolic reaction. Levels of light or of carbon dioxide are limiting factors in ◊photosynthesis because both are necessary for the production of carbohydrates. In experiments, photosynthesis is observed to slow down and eventually stop as the levels of light decrease.

limnology study of lakes and other bodies of open fresh water, in terms of their plant and animal biology, and their physical properties.

limonite iron ore, mostly poorly crystalline iron oxyhydroxide, but usually mixed with

◊hematite and other iron oxides. Also known as brown iron ore, it is often found in bog deposits.

linac contraction of ◊linear accelerator, a type of particle accelerator in which the particles are accelerated along a straight tube.

linear accelerator or *linac* in physics, a type of particle ◊accelerator in which the particles move along a straight tube. Particles pass through a linear accelerator only once – unlike those in a cyclotron (a ring-shaped accelerator), which make many revolutions, gaining energy each time.

linear equation in mathematics, a relationship between two variables that, when plotted on Cartesian axes produces a straight-line graph; the equation has the general form $y = mx + c$, where m is the slope of the line represented by the equation and c is the y-intercept, or the value of y where the line crosses the y-axis in the ◊Cartesian coordinate system. Sets of linear equations can be used to describe the behaviour of buildings, bridges, trusses, and other static structures.

linear motor type of electric motor, an induction motor in which the fixed stator and moving armature are straight and parallel to each other (rather than being circular and one inside the other as in an ordinary induction motor). Linear motors are used, for example, to power sliding doors. There is a magnetic force between the stator and armature; this force has been used to support a vehicle, as in the experimental ◊maglev linear motor train.

linear programming in mathematics and economics, a set of techniques for finding the maxima or minima of certain variables governed by linear equations or inequalities. These maxima and minima are used to represent 'best' solutions in terms of goals such as maximizing profit or minimizing cost.

line of force in physics, an imaginary line representing the direction of force at any point in a magnetic, gravitational, or electrical field.

line printer computer ◊printer that prints a complete line of characters at a time. Line printers can achieve very high printing speeds of up to 2,500 lines a minute, but can print in only one typeface, cannot print graphics, and are very noisy. Until the late 1980s they were the obvious choice for high-volume printing, but high-speed ◊page printers, such as laser printers, are now preferred.

linkage in genetics, the association between two or more genes that tend to be inherited together because they are on the same chromosome.

The closer together they are on the chromosome, the less likely they are to be separated by crossing over (one of the processes of ◊recombination) and they are then described as being 'tightly linked'.

lipase enzyme responsible for breaking down fats into fatty acids and glycerol. It is produced by the ◊pancreas and requires a slightly alkaline environment. The products of fat digestion are absorbed by the intestinal wall.

lipid any of a large number of esters of fatty acids, commonly formed by the reaction of a fatty acid with glycerol (see ◊glyceride). They are soluble in alcohol but not in water. Lipids are the chief constituents of plant and animal waxes, fats, and oils.

lipophilic in chemistry, a term describing ◊functional groups with an affinity for fats and oils.

lipophobic in chemistry, a term describing ◊functional groups that tend to repel fats and oils.

liquefaction the process of converting a gas to a liquid, normally associated with low temperatures and high pressures (see ◊condensation).

liquefied petroleum gas (LPG) liquid form of butane, propane, or pentane, produced by the distillation of petroleum during oil refining. At room temperature these substances are gases, although they can be easily liquefied and stored under pressure in metal containers. They are used for heating and cooking where other fuels are not available: camping stoves and cigarette lighters, for instance, often use liquefied butane as fuel.

liquid state of matter between a ◊solid and a ◊gas. A liquid forms a level surface and assumes the shape of its container. Its atoms do not occupy fixed positions as in a crystalline solid, nor do they have freedom of movement as in a gas. Unlike a gas, a liquid is difficult to compress since pressure applied at one point is equally transmitted throughout (Pascal's principle). ◊Hydraulics makes use of this property.

liquid air air that has been cooled so much that it has liquefied. This happens at temperatures below about −196°C/−321°F. The various constituent gases, including nitrogen, oxygen, argon, and neon, can be separated from liquid air by the technique of ◊fractionation.

liquid-crystal display (LCD) display of numbers (for example, in a calculator) or pictures (such as on a pocket television screen) produced by molecules of a substance in a semiliquid state with some crystalline properties, so that clusters of molecules align in parallel formations. The display is a blank until the application of an electric field, which 'twists' the molecules so that they reflect or transmit light falling on them.

LISP (acronym for *list processing*) high-level computer-programming language designed for manipulating lists of data items. It is used primarily in research into ◊artificial intelligence (AI).

lithium soft, ductile, silver-white, metallic element, symbol Li, atomic number 3, relative atomic mass 6.941. It is one of the ◊alkali metals, has a very low density (far less than most woods), and floats on water (specific gravity 0.57); it is the lightest of all metals. Lithium is used to harden alloys, and in batteries; its compounds are used in medicine to treat manic depression.

lithosphere outermost layer of the Earth's structure, 75 km thick, consisting of the crust and a portion of the upper mantle. It is divided into large rigid areas, or plates, which, according to the ◊plate tectonics model, move relative to each other under the influence of convection currents in the semisolid mantle beneath.

litmus dye obtained from various lichens and used in chemistry as an indicator to test the acidic or alkaline nature of aqueous solutions; it turns red in the presence of acid, and blue in the presence of alkali.

litre metric unit of volume (symbol l), equal to one cubic decimetre (1.76 imperial pints/ 2.11 US pints). It was formerly defined as the volume occupied by one kilogram of pure water at 4°C at standard pressure, but this is slightly larger than one cubic decimetre.

liver large organ of vertebrates, which has many regulatory and storage functions. The human liver is situated in the upper abdomen, and weighs about 2 kg/4.5 lb. It is divided into four lobes. The liver receives the products of digestion, converts glucose to glycogen (a long-chain carbohydrate used for storage), and breaks down fats. It removes excess amino acids from the blood, converting them to urea, which is excreted by the kidneys. The liver also synthesizes vitamins, produces bile and blood-clotting factors, and removes damaged red cells and toxins such as alcohol from the blood.

liquid-crystal display *A liquid-crystal display consists of a liquid crystal sandwiched between polarizing filters similar to polaroid sunglasses. When a segment of the seven-segment display is electrified, the liquid crystal twists the polarized light from the front filter, allowing the light to bounce off the rear reflector and illuminate the segment.*

Trials are likely to begin 1994 on 'living' machines called external liver-assist devices (ELADs), in an attempt to produce a machine that acts as an artificial liver. The simplest ELADs are based on kidney dialysis machines, incorporating liver cells from other animals, such as pigs, because human liver cells are difficult to culture.

liverwort plant of the class Hepaticae, of the bryophyte division of nonvascular plants, related to mosses, found growing in damp places.

local area network (LAN) in computing, a ◊network restricted to a single room or building. Local area networks enable around 500 devices to be connected together.

local bus in computing, an extension of the central processing unit (CPU) ◊bus (electrical pathway), designed to speed up data transfer between the CPU, discs, graphics boards, and other devices. There are two common specifications, VESA and ◊PCI, but PCI is likely to become standard in the 1990s.

Local Group in astronomy, a cluster of about 30 galaxies that includes our own, the Milky Way. Like other groups of galaxies, the Local Group is held together by the gravitational attraction among its members, and does not expand with the expanding universe. Its two largest galaxies are the Milky Way and the Andromeda galaxy; most of the others are small and faint.

local variable in computing, a ◊variable that can be accessed only by the instructions within a particular ◊subroutine.

locomotion the ability to move independently from one place to another, occurring in most animals but not in plants. The development of locomotion as a feature of animal life is closely linked to another vital animal feature, that of nutrition. Animals cannot make their food, as can plants, but must find it first; often the food must be captured and killed, which may require great speed. Locomotion is also important in finding a mate, in avoiding predators, and in migrating to favourable areas.

locus point on a ◊chromosome where a particular ◊gene occurs.

locus in mathematics, traditionally the path traced out by a moving point, but now defined as the set of all points on a curve satisfying given conditions. For example, the locus of a point that moves so that it is always at the same distance from another fixed point is a circle; the locus of a point that is always at the same distance from two fixed points is a straight line that perpendicularly bisects the line joining them.

lodestar or *loadstar* a star used in navigation or astronomy, often ◊Polaris, the Pole Star.

log in mathematics, abbreviation for ◊*logarithm*.

logarithm or *log* the ◊exponent or index of a number to a specified base – usually 10. For example, the logarithm to the base 10 of 1,000 is 3 because $10^3 = 1,000$; the logarithm of 2 is 0.3010 because $2 = 10^{0.3010}$. Before the advent of cheap electronic calculators, multiplication and division could be simplified by being replaced with the addition and subtraction of logarithms.

logic gate or *logic circuit* in electronics, one of the basic components used in building ◊integrated circuits. The five basic types of gate make logical decisions based on the functions NOT, AND, OR, NAND (NOT AND), and NOR (NOT OR). With the exception of the NOT gate, each has two or more inputs.

Information is fed to a gate in the form of

logic gate *Gates are, in essence, switches which turn 'on' (produce an output signal) only if certain combinations of input signals are received. The truth table describes how a gate responds to all possible combinations of input signals.*

binary-coded input signals (logic value 0 stands for 'off' or 'low-voltage pulse', logic 1 for 'on' or 'high-voltage'), and each combination of input signals yields a specific output (logic 0 or 1). An *OR* gate will give a logic 1 output if one or more of its inputs receives a logic 1 signal; however, an *AND* gate will yield a logic 1 output only if it receives a logic 1 signal through both its inputs. The output of a *NOT* or *inverter* gate is the opposite of the signal received through its single input, and a *NOR* or *NAND* gate produces an output signal that is the opposite of the signal that would have been produced by an OR or AND gate respectively. The properties of a logic gate, or of a combination of gates, may be defined and presented in the form of a diagram called a *truth table*, which lists the output that will be triggered by each of the possible combinations of input signals. The process has close parallels in computer programming, where it forms the basis of binary logic.

LOGO high-level computer programming language designed to teach mathematical concepts. Developed about 1970 at the Massachusetts Institute of Technology, it became popular in schools and with home computer users because of its 'turtle graphics' feature. This allows the user to write programs that create line drawings on a computer screen, or drive a small mobile robot (a 'turtle' or 'buggy') around the floor.

lomentum fruit similar to a pod but constricted between the seeds. When ripe, it splits into one-seeded units, as seen, for example, in the fruit of sainfoin *Onobrychis viciifolia* and radish *Raphanus raphanistrum*. It is a type of ◊schizocarp.

lone pair in chemistry, a pair of electrons in the outermost shell of an atom that are not used in bonding. In certain circumstances, they will allow the atom to bond with atoms, ions, or molecules (such as boron trifluoride, BF_3) that are deficient in electrons, forming coordinate covalent (dative) bonds in which they provide both of the bonding electrons.

longitudinal wave ◊wave in which the displacement of the medium's particles is in line with or parallel to the direction of travel of the wave motion.

long-sightedness or *hypermetropia* defect of vision in which a person is able to focus on objects in the distance, but not on close objects. It is caused by the failure of the lens to return to its normal rounded shape, or by the eyeball being too short, with the result that the image is focused on a point behind the retina. Long-sightedness is corrected by wearing spectacles fitted with ◊converging lenses, each of which acts like a magnifying glass.

loudness subjective judgement of the level or power of sound reaching the ear. The human ear cannot give an absolute value to the loudness of a single sound, but can only make comparisons between two different sounds. The precise measure of the power of a sound wave at a particular point is called its ◊intensity.

Accurate comparisons of sound levels may be made using sound-level meters, which are calibrated in units called ◊decibels.

loudspeaker electromechanical device that converts electrical signals into sound waves, which are radiated into the air. The most common type of loudspeaker is the *moving-coil speaker*. Electrical signals from, for example, a radio are fed to a coil of fine wire wound around the top of a cone. The coil is surrounded by a magnet. When signals pass through it, the coil becomes an electromagnet, which by moving causes the cone to vibrate, setting up sound waves.

low-level language in computing, a programming language designed for a particular computer and reflecting its internal ◊machine code; low-level languages are therefore often described as *machine-oriented* languages. They cannot easily be converted to run on a

loudspeaker *A moving-coil loudspeaker. Electrical signals flowing through the wire coil turn it into an electromagnet, which moves as the signals vary. The attached cone vibrates, producing sound waves.*

computer with a different central processing unit, and they are relatively difficult to learn because a detailed knowledge of the internal working of the computer is required. Since they must be translated into machine code by an ◊assembler program, low-level languages are also called ◊assembly languages.

LSI (abbreviation for *large-scale integration*) the technology that enables whole electrical circuits to be etched into a piece of semiconducting material just a few millimetres square.

lubricant substance used between moving surfaces to reduce friction. Carbon-based (organic) lubricants, commonly called grease and oil, are recovered from petroleum distillation.

Extensive research has been carried out on chemical additives to lubricants, which can reduce corrosive wear, prevent the accumulation of 'cold sludge' (often the result of stop-start driving in city traffic jams), keep pace with the higher working temperatures of aviation gas turbines, or provide radiation-resistant greases for nuclear power plants. Silicon-based spray-on lubricants are also used; they tend to attract dust and dirt less than carbon-based ones.

A solid lubricant is graphite, an allotropic

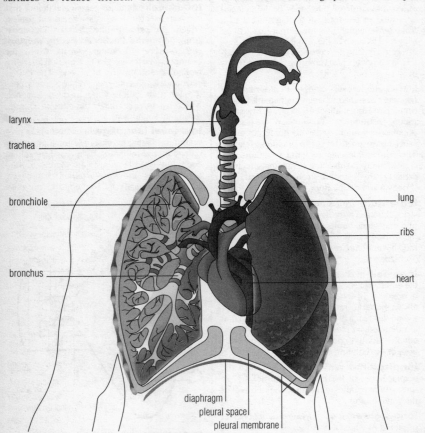

lung *The human lungs contain 300,000 million tiny blood vessels which would stretch for 2,400 km/1,500 mi if laid end to end. A healthy adult at rest breathes 12 times a minute; a baby breathes at twice this rate. Each breath brings 350 millilitres of fresh air into the lungs, and expels 150 millilitres of stale air from the nose and throat.*

form of carbon, either flaked or emulsified (colloidal) in water or oil.

lumen SI unit (symbol lm) of luminous flux (the amount of light passing through an area per second).

luminescence emission of light from a body when its atoms are excited by means other than raising its temperature. Short-lived luminescence is called fluorescence; longer-lived luminescence is called phosphorescence.

luminosity or *brightness* in astronomy, the amount of light emitted by a star, measured in ◊magnitudes. The apparent brightness of an object decreases in proportion to the square of its distance from the observer. The luminosity of a star or other body can be expressed in relation to that of the Sun.

lung large cavity of the body, used for ◊gas exchange. It is essentially a sheet of thin, moist membrane that is folded so as to occupy less space. Most tetrapod (four-limbed) vertebrates have a pair of lungs occupying the thorax. The lung tissue, consisting of multitudes of air sacs and blood vessels, is very light and spongy, and functions by bringing inhaled air into close contact with the blood so that oxygen can pass into the organism and waste carbon dioxide can be passed out. The efficiency of lungs is enhanced by ◊breathing movements, by the thinness and moistness of their surfaces, and by a constant supply of circulating blood.

luteinizing hormone ◊hormone produced by the pituitary gland. In males, it stimulates the testes to produce androgens (male sex hormones). In females, it works with follicle-stimulating hormone to initiate production of egg cells by the ovary. If fertilization occurs, it plays a part in maintaining the pregnancy by controlling the levels of the hormones oestrogen and progesterone.

lutetium silver-white, metallic element, the last of the ◊lanthanide series, symbol Lu, atomic number 71, relative atomic mass 174.97. It is used in the 'cracking', or breakdown, of petroleum and in other chemical processes. It was named by its discoverer, French chemist Georges Urbain (1872–1938), after his native city.

lux SI unit (symbol lx) of illuminance or illumination (the light falling on an object). It is equivalent to one ◊lumen per square metre or to the illuminance of a surface one metre distant from a point source of one ◊candela.

LW abbreviation for *long wave*, a radio wave with a wavelength of over 1,000 m/3,300 ft;

lymph Lymph is the fluid that carries nutrients, and white blood cells to the tissues. Lymph enters the tissue from the capillaries (right) and is drained from the tissues by lymph vessels. The lymph vessels form a network (left) called the lymphatic system. At various points in the lymphatic system, lymph nodes (centre) filter and clean the lymph.

one of the main wavebands into which radio frequency transmissions are divided.

lymph fluid found in the lymphatic system of vertebrates. Lymph is drained from the tissues by lymph capillaries, which empty into larger lymph vessels (lymphatics). These lead to lymph nodes (small, round bodies chiefly situated in the neck, armpit, groin, thorax, and abdomen), which process the ◊lymphocytes produced by the bone marrow, and filter out harmful substances and bacteria. From the lymph nodes, vessels carry the lymph to the thoracic duct and the right lymphatic duct, which drain into the large veins in the neck. Some vertebrates, such as amphibians, have a lymph heart, which pumps lymph through the lymph vessels.

Lymph carries some nutrients, and white blood cells to the tissues, and waste matter away from them. It exudes from capillaries into the tissue spaces between the cells and is similar in composition to blood plasma.

lymph nodes small masses of lymphatic tissue in the body that occur at various points along the major lymphatic vessels. Tonsils and adenoids are large lymph nodes. As the lymph passes through them it is filtered, and bacteria and other microorganisms are engulfed by cells known as macrophages.

lymphocyte type of white blood cell with a large nucleus, produced in the bone marrow. Most occur in the ◊lymph and blood, and around sites of infection. *B lymphocytes* or ◊B cells are responsible for producing ◊antibodies. *T lymphocytes* or ◊T cells have several roles in the mechanism of ◊immunity.

lymphokines chemical messengers produced by lymphocytes that carry messages between the cells of the immune system (see ◊immunity). Examples include interferon, which initiates defensive reactions to viruses, and the interleukins, which activate specific immune cells.

lyophilization technical term for the freeze-drying process used for foods and drugs and in the preservation of organic archaeological remains.

Lyra small but prominent constellation of the northern hemisphere, representing the lyre of Orpheus. Its brightest star is ◊Vega.

lysis in biology, any process that destroys a cell by rupturing its membrane or cell wall (see ◊lysosome).

lysosome membrane-enclosed structure, or organelle, inside a ◊cell, principally found in animal cells. Lysosomes contain enzymes that can break down proteins and other biological substances. They play a part in digestion, and in the white blood cells known as phagocytes the lysosome enzymes attack ingested bacteria.

m symbol for ◊*metre*.

MA abbreviation for ◊*mechanical advantage*.

machine device that allows a small force (the effort) to overcome a larger one (the load). There are three basic machines: the inclined plane (ramp), the lever, and the wheel and axle. All other machines are combinations of these three basic types. Simple machines derived from the inclined plane include the wedge, the gear, and the screw; the spanner is derived from the lever; the pulley from the wheel.

The principal features of a machine are its ◊mechanical advantage, which is the ratio of load to effort, its ◊velocity ratio, and its ◊efficiency, which is the work done by the load divided by the work done by the effort; the latter is expressed as a percentage. In a perfect machine, with no friction, the efficiency would be 100%. All practical machines have efficiencies of less than 100%, otherwise perpetual motion would be possible.

machine code in computing, a set of instructions that a computer's central processing unit (CPU) can understand and obey directly, without any translation. Each type of CPU has its own machine code. Because machine-code programs consist entirely of binary digits (bits), most programmers write their programs in an easy-to-use ◊high-level language. A high-level program must be translated into machine code – by means of a ◊compiler or ◊interpreter program – before it can be executed by a computer.

machine-readable of data, readable directly by a computer without the need for rekeying. The term is usually applied to files on disc or tape, but can also be applied to typed or printed text that can be scanned for ◊optical character recognition.

machine tool automatic or semi-automatic power-driven machine for cutting and shaping metals. Machine tools have powerful electric motors to force cutting tools into the metal: these are made from hardened steel containing heat-resistant metals such as tungsten and chromium. The use of precision machine tools in mass-production assembly methods ensures that all duplicate parts produced are virtually identical.

Mach number ratio of the speed of a body to the speed of sound in the undisturbed medium through which the body travels. Mach 1 is reached when a body (such as an aircraft) has a velocity greater than that of sound ('passes the sound barrier'), namely 331 m/ 1,087 ft per second at sea level. It is named after Austrian physicist Ernst Mach (1838–1916).

Macintosh range of microcomputers produced by Apple Computers. The Apple Macintosh, introduced in 1984, was the first popular microcomputer with a ◊graphical user interface.

macro in computer programming, a new command created by combining a number of existing ones. For example, if a programming language has separate commands for obtaining data from the keyboard and for displaying data on the screen, the programmer might create a macro that performs both these tasks with one command. A *macro key* on the keyboard combines the effects of pressing several individual keys.

macro- a prefix meaning on a very large scale, as opposed to micro.

macromolecule in chemistry, a very large molecule, generally a ◊polymer.

macrophage type of ◊white blood cell, or leucocyte, found in all vertebrate animals. Macrophages specialize in the removal of bacteria and other microorganisms, or of cell debris after injury. Like phagocytes, they engulf foreign matter, but they are larger than phagocytes and have a longer life span. They are found throughout the body, but mainly in the lymph and connective tissues, and especially the lungs, where they ingest dust, fibres, and other inhaled particles.

Magellan NASA space probe to ◊Venus, launched May 1989; it went into orbit around Venus Aug 1990 to make a detailed map of the planet by radar. It revealed volcanoes, meteorite craters, and fold mountains on the planet's surface. In 1993 Magellan ended its life after shifting into a circular orbit round Venus in an attempt to map its gravitational fields.

Magellanic Clouds in astronomy, the two galaxies nearest to our own galaxy. They are irregularly shaped, and appear as detached parts of the ◊Milky Way, in the southern constellations Dorado and Tucana.

The Large Magellanic Cloud is 169,000 light years from Earth, and about a third the diameter of our galaxy; the Small Magellanic Cloud, 180,000 light years away, is about a fifth the diameter of our galaxy. They are named after the navigator Ferdinand Magellan, who first described them.

magic bullet term sometimes used for a drug that is specifically targeted on certain cells or tissues in the body, such as a small collection of cancerous cells (see ◊cancer) or cells that have been invaded by a virus. Such drugs can be made in various ways, but ◊monoclonal antibodies are increasingly being used to direct the drug to a specific target.

magic numbers in atomic physics certain numbers of ◊neutrons or ◊protons (2, 8, 20, 28, 50, 82, 126) in the nuclei of elements of outstanding stability, such as lead and helium.

Such stability is the result of neutrons and protons being arranged in completed 'layers' or 'shells'.

maglev (acronym for *magnetic levitation*) high-speed surface transport using the repellent force of superconductive magnets (see ◊superconductivity) to propel and support, for example, a train above a track.

magma molten rock material beneath the Earth's surface from which ◊igneous rocks are formed. ◊Lava is magma that has reached the surface and solidified, losing some of its components on the way.

magnesia common name for ◊magnesium oxide.

magnesium lightweight, very ductile and malleable, silver-white, metallic element, symbol Mg, atomic number 12, relative atomic mass 24.305. It is one of the ◊alkaline-earth metals, and the lightest of the commonly used metals. Magnesium silicate, carbonate, and chloride are widely distributed in nature. The metal is used in alloys and flash photography. It is a necessary trace element in the human diet, and green plants cannot grow without it since it is an essential constituent of the photosynthetic pigment ◊chlorophyll ($C_{55}H_{72}MgN_4O_5$).

magnesium carbonate $MgCO_3$ white solid that occurs in nature as the mineral magnesite. It is a commercial ◊antacid and the anhydrous form is used as a drying agent in table salt. When rainwater containing dissolved carbon dioxide flows over magnesite rocks, the carbonate dissolves to form magnesium hydrogencarbonate, one of the causes of temporary hardness in water.

$$H_2O + CO_2 + MgCO_3 \rightarrow Mg(HCO_3)_2$$

magnesium oxide or *magnesia* MgO white powder or colourless crystals, formed when magnesium is burned in air or oxygen; a typical basic oxide. It is used to treat acidity of the stomach, and in some industrial processes; for example, as a lining brick in furnaces, because it is very stable when heated (refractory oxide).

magnesium sulphate $MgSO_4$ white solid that occurs in nature in several hydrated forms. The heptahydrate $MgSO_4.7H_2O$ is known as Epsom Salts and is used as a laxative. Industrially it is used in tanning leather and in the manufacture of fertilizers and explosives.

magnet any object that forms a magnetic field (displays ◊magnetism), either permanently or temporarily through induction, causing it to attract materials such as iron, cobalt, nickel, and alloys of these. It always has two ◊magnetic poles, called north and south.

magnetic compass device for determining the direction of the horizontal component of the Earth's magnetic field. It consists of a magnetized needle with its north-seeking pole clearly indicated, pivoted so that it can turn freely in a plane parallel to the surface of the Earth (in a horizontal circle).

The needle will turn so that its north pole points towards the Earth's magnetic north pole. See also ◊compass.

magnetic declination see ◊angle of declination.

magnetic dip see ◊dip, magnetic and ◊angle of dip.

magnetic dipole the pair of north and south magnetic poles, separated by a short distance, that makes up all magnets. Individual magnets are often called 'magnetic dipoles'. Single magnetic poles, or monopoles, have never been observed despite being searched for. See also magnetic ◊domain.

magnetic field region around a permanent magnet, or around a conductor carrying an electric current, in which a force acts on a moving charge or on a magnet placed in the field. The field can be represented by lines of force, which by convention link north and south poles and are parallel to the directions of a small compass needle placed on them. A magnetic field's magnitude and direction are given by the ◊magnetic flux density, expressed in ◊teslas.

magnetic flux measurement of the strength of the magnetic field around electric currents and magnets. Its SI unit is the ◊weber; one weber per square metre is equal to one tesla.

magnetic induction the production of magnetic properties in unmagnetized iron or other ferromagnetic material when it is brought close to a magnet. The material is influenced by the magnet's magnetic field and the two are attracted. The induced magnetism may be temporary, disappearing as soon as the magnet is removed, or permanent depending on the nature of the iron and the strength of the magnet.

magnetic-ink character recognition (MICR) in computing, a technique that enables special characters printed in magnetic ink to be read and input rapidly to a computer. MICR is used extensively in banking because magnetic-ink characters are difficult to forge and are therefore ideal for marking and identifying cheques.

magnetic material one of a number of substances that are strongly attracted by magnets and can be magnetized. These include iron, nickel, and cobalt, and all those ◊alloys that contain a proportion of these metals.

Soft magnetic materials can be magnetized very easily, but the magnetism induced in them (see ◊magnetic induction) is only temporary. They include Stalloy, an alloy of iron with 4% silicon used to make the cores of electromagnets and transformers, and the materials used to make 'iron' nails and paper clips.

Hard magnetic materials can be permanently magnetized by a strong magnetic field. Steel and special alloys such as Alcomax, Alnico, and Ticonal, which contain various amounts of aluminium, nickel, cobalt, and copper, are used to make permanent magnets. The strongest permanent magnets are ceramic, made under high pressure and at high temperature from powders of various metal oxides.

magnetic pole region of a magnet in which its magnetic properties are strongest. Every magnet has two poles, called north and south. The north (or north-seeking) pole is so named because a freely suspended magnet will turn so that this pole points towards the Earth's magnetic north pole. The north pole of one magnet will be attracted to the south pole of another, but will be repelled by its north pole.

Unlike poles may therefore be said to attract, like poles to repel.

magnetic resonance imaging (MRI) diagnostic scanning system based on the principles of nuclear magnetic resonance. MRI yields finely detailed three-dimensional images of structures within the body without exposing the patient to harmful radiation. The technique is invaluable for imaging the soft tissues of the body, in particular the brain and the spinal cord.

magnetic storm in meteorology, a sudden disturbance affecting the Earth's magnetic field, causing anomalies in radio transmissions and magnetic compasses. It is probably caused by ◊sunspot activity.

magnetism phenomena associated with ◊magnetic fields. Magnetic fields are produced by moving charged particles: in electromagnets, electrons flow through a coil of wire connected to a battery; in permanent magnets, spinning electrons within the atoms generate the field.

Substances differ in the extent to which they can be magnetized by an external field (susceptibility). Materials that can be strongly magnetized, such as iron, cobalt, and nickel, are said to be *ferromagnetic*; this is due to the formation of areas called ◊domains in which atoms, weakly magnetic because of their spinning electrons, align to form areas of strong magnetism. Magnetic materials lose their magnetism if heated to the ◊Curie temperature. Most other materials are *paramagnetic*, being only weakly pulled towards a strong magnet. This is because their atoms have a low level of magnetism and do not form domains. *Diamagnetic* materials are weakly repelled by a magnet since electrons within their atoms act as electromagnets and oppose the applied magnetic force. *Antiferromagnetic* materials have a very low susceptibility that increases with temperature; a similar phenomenon in materials such as ferrites is called *ferrimagnetism*.

magnetite black iron ore, iron oxide (Fe_3O_4). Widely distributed, magnetite is found in nearly all igneous and metamorphic rocks. It is strongly magnetic and some deposits, called **lodestone**, are permanently magnetized. Lodestone has been used as a compass since the first millennium BC.

magneto simple electric generator, often used to provide the electricity for the ignition system of motorcycles and used in early cars.

It consists of a rotating magnet that sets up an electric current in a coil, providing the spark.

magnetohydrodynamics (MHD) field of science concerned with the behaviour of ionized gases or liquid in a magnetic field. Systems have been developed that use MHD to generate electrical power.

magnetomotive force or *magnetic potential* the work done in carrying a unit magnetic pole around a magnetic circuit. The concept is analogous to ◊electromotive force in an electric circuit.

magnetosphere volume of space, surrounding a planet, controlled by the planet's magnetic field, and acting as a magnetic 'shell'. The Earth's magnetosphere extends 64,000 km/40,000 mi towards the Sun, but many times this distance on the side away from the Sun.

magnetron thermionic ◊valve (electron tube) for generating very high-frequency oscillations, used in radar and to produce microwaves in a microwave oven. The flow of electrons from the tube's cathode to one or more anodes is controlled by an applied magnetic field.

magnification measure of the enlargement or reduction of an object in an imaging optical system. *Linear magnification* is the ratio of the size (height) of the image to that of the object. *Angular magnification* is the ratio of the angle subtended at the observer's eye by the image to the angle subtended by the object when viewed directly.

magnifying glass the simplest optical instrument, a hand-held converging lens used to produce a magnified, erect and virtual image. The image, being virtual, or an illusion created by the ◊refraction of light rays in the lens, can only be seen by looking through the magnifying glass.

magnitude in astronomy, measure of the brightness of a star or other celestial object. The larger the number denoting the magnitude, the fainter the object. Zero or first magnitude indicates some of the brightest stars. Still brighter are those of negative magnitude, such as Sirius with magnitude −1.46. *Apparent magnitude* is the brightness of an object as seen from Earth; *absolute magnitude* is the brightness at a standard distance of 10 parsecs (32.6 light years).

Magnox early type of nuclear reactor used in the UK, for example in Calder Hall, the world's first commercial nuclear power station. This type of reactor uses uranium fuel encased in tubes of magnesium alloy called Magnox. Carbon dioxide gas is used as a coolant to extract heat from the reactor core. See also ◊nuclear energy.

mail merge in computing, a feature offered by some word-processing packages that enables a list of personal details, such as names and addresses, to be combined with a general document outline to produce individualized documents.

mainframe large computer used for commercial data processing and other large-scale operations. Because of the general increase in computing power, the differences between the mainframe, ◊supercomputer, ◊minicomputer, and ◊microcomputer (personal computer) are becoming less marked.

mains electricity the domestic electricity-supply system. In the UK, electricity is supplied to houses, offices, and most factories as an ◊alternating current at a frequency of 50 hertz and a ◊root-mean-square voltage of 240 volts. An advantage of having an alternating supply is that it may easily be changed, using a ◊transformer, to a lower and safer voltage, such as 9 volts, for operating toys and for recharging batteries.

main sequence in astronomy, the part of the ◊Hertzsprung–Russell diagram that contains most of the stars, including the Sun. It runs diagonally from the top left of the diagram to the lower right. The most massive (and hence brightest) stars are at the top left, with the least massive (coolest) stars at the bottom right.

malachite common ◊copper ore, basic copper carbonate, $Cu_2CO_3(OH)_2$. It is a source of green pigment and is polished for use in jewellery, ornaments, and art objects.

malaria infectious parasitic disease of the tropics transmitted by mosquitoes, marked by periodic fever and an enlarged spleen. When a female mosquito of the *Anopheles* genus bites a human who has malaria, it takes in with the human blood one of four malaria protozoa of the genus *Plasmodium*. This matures within the insect and is then transferred when the mosquito bites a new victim. Malaria affects some 267 million people in 103 countries, claiming more than a million lives a year. Between 1.5 and 2 million children die from malaria and its consequences each year in sub-Saharan Africa alone.

In 1993 research began into using chelating agents, which remove surplus iron from the blood, to treat malaria.

malic acid COOHCH$_2$CH(OH)COOH organic crystalline acid that can be extracted from apples, plums, cherries, grapes, and other fruits, but occurs in all living cells in smaller amounts, being one of the intermediates of the ◊Krebs cycle.

maltase enzyme found in plants and animals that breaks down the disaccharide maltose into glucose.

Maltese-cross tube cathode-ray tube used to demonstrate some of the properties of cathode rays. The cathode rays, or electron streams, emitted by the tube's ◊electron gun are directed towards a fluorescent screen in front of which hangs a metal Maltese cross. Those electrons that hit the screen give up their kinetic energy and cause its phosphor coating to fluoresce. However, the sharply defined cross-shaped shadow cast on the screen shows that electrons are unable to pass through the Maltese cross. Cathode rays are thereby shown to travel in straight lines, and to be unable to pass through metal.

maltose C$_{12}$H$_{22}$O$_{11}$ a ◊disaccharide sugar in which both monosaccharide units are glucose.

mammal animal characterized by having mammary glands in the female; these are used for suckling the young. Other features of mammals are ◊hair (very reduced in some species, such as whales); a middle ear formed of three small bones (ossicles); a lower jaw consisting of two bones only; seven vertebrae in the neck; and no nucleus in the red blood cells.

Mammals are divided into three groups: *placental mammals*, where the young develop inside the ◊uterus, receiving nourishment from the blood of the mother via the ◊placenta;
marsupials, where the young are born at an early stage of development and develop further in a pouch on the mother's body;
monotremes, where the young hatch from an egg outside the mother's body and are then nourished with milk.

The monotremes have been largely displaced by marsupials and placentals, so that there are only a few types surviving (platypus and echidna). Placentals have spread to all parts of the globe and where they have competed with marsupials, the placentals have in general displaced marsupial types. However, marsupials occupy many specialized niches in South America and, especially, Australasia.

The theory that marsupials succeed only where they do not compete with placentals was shaken in 1992, when a tooth, 55 million years old and belonging to a placental mammal, was found in Murgon, Australia, indicating that placental animals appeared in Australia at the same time as the marsupials. The marsupials, however, still prevailed.

mammary gland in female mammals, a milk-producing gland derived from epithelial cells underlying the skin, active only after the production of young. In all but monotremes (egg-laying mammals), the mammary glands terminate in teats which aid infant suckling. The number of glands and their position vary between species. In humans there are 2, in cows 4, and in pigs between 10 and 14.

mandible anatomical name for the lower jawbone.

manganese hard, brittle, grey-white metallic element, symbol Mn, atomic number 25, relative atomic mass 54.9380. It resembles iron (and rusts), but it is not magnetic and is softer. It is used chiefly in making steel alloys, also alloys with aluminium and copper.

It is used in fertilizers, paints, and industrial chemicals. It is a necessary trace element in human nutrition. The name is old, deriving from the French and Italian forms of Latin for *magnesia* (MgO), the white tasteless powder used as an antacid from ancient times.

manganese ore any mineral from which manganese is produced. The main ores are the oxides, such as *pyrolusite*, MnO$_2$; *hausmannite*, Mn$_3$O$_4$; and *manganite*, MnO(OH).

manganese(IV) oxide or *manganese dioxide* MnO$_2$ brown solid that acts as a ◊depolarizer in dry batteries by oxidizing the hydrogen gas produced to water; without this process, the performance of the battery is impaired.

manometer instrument for measuring the pressure of liquids (including human blood pressure) or gases. In its basic form, it is a U-tube partly filled with coloured liquid; pressure of a gas entering at one side is measured by the level to which the liquid rises at the other.

mantissa in mathematics, the decimal part of a ◊logarithm. For example, the logarithm of 347.6 is 2.5411; in this case, the 0.5411 is the

manometer *The manometer indicates gas pressure by the rise of liquid in the tube.*

mantissa, and the integral (whole number) part of the logarithm, the 2, is the ◊**characteristic**.

mantle intermediate zone of the Earth between the crust and the core, accounting for 82% of the Earth's volume. It is thought to consist of silicate minerals such as olivine.

The mantle is subdivided into three shells: the *upper mantle* that extends to 400 km/250 mi beneath the surface; the *transition zone* 400–650 km/250–400 mi down; and the *lower mantle* 650–2,900 km/400–1,800 mi down.

map projection ways of depicting the spherical surface of the Earth on a flat piece of paper. Traditional projections include the *conic*, *azimuthal*, and *cylindrical*. The most famous cylindrical projection is the Mercator projection, which dates from 1569. The weakness of these systems is that countries in different latitudes are disproportionately large, and lines of longitude and latitude appear distorted. In 1973 German historian Arno Peters devised the *Peters projection* in which the countries of the world retain their relative areas.

The US physicist Mitchell Feigenbaum devised the *optimal conformal* projection 1992, using a computer program designed to take data about the boundary of a given area and calculate the projection that produces minimum inaccuracy.

marble metamorphosed ◊limestone that takes and retains a good polish; it is used in building and sculpture. In its pure form it is white and consists almost entirely of calcite $CaCO_3$. Mineral impurities give it various colours and patterns. Carrara, Italy, is known for white marble.

mare (plural *maria*) dark lowland plain on the Moon. The name comes from Latin 'sea', because these areas were once wrongly thought to be water.

Mariner spacecraft series of US space probes that explored the planets Mercury, Venus, and Mars 1962–75.

mark sensing in computing, a technique that enables pencil marks made in predetermined positions on specially prepared forms to be rapidly read and input to a computer. The technique makes use of the fact that pencil marks contain graphite and therefore conduct electricity. A *mark sense reader* scans the form by passing small metal brushes over the paper surface. Whenever a brush touches a pencil mark a circuit is completed and the mark is detected.

marl or *clayey limestone* crumbling sedimentary rock, including various types of calcareous ◊clays and fine-grained ◊limestones. Marls are often laid down in freshwater lakes and are usually soft, earthy, and of a white, grey, or brownish colour. They are used in cement-making and as fertilizer.

Mars fourth planet from the Sun, average distance 227.9 million km/141.6 million mi. It revolves around the Sun in 687 Earth days, and has a rotation period of 24 hr 37 min. It is much smaller than Venus or Earth, with a diameter 6,780 km/4,210 mi, and mass 0.11 that of Earth. Mars is slightly pearshaped, with a low, level northern hemisphere, which is comparatively uncratered and geologically 'young', and a heavily cratered 'ancient' southern hemisphere.

marsh gas gas consisting mostly of ◊methane. It is produced in swamps and marshes by the action of bacteria on dead vegetation.

Mars Observer NASA space probe launched 1992 to orbit Mars and survey the planet, its atmosphere, and the polar caps over two years. The probe was also scheduled to communicate information from the robot vehicles delivered by Russia's Mars 94 mission. The $1 billion project miscarried, however, when the probe unaccountably stopped transmitting Aug 1993, three days before it was due to drop into orbit.

marsupial mammal in which the female has a pouch where she carries her young (born tiny and immature) for a considerable time after birth. Marsupials include omnivorous, herbivorous, and carnivorous species, among them the kangaroo, wombat, opossum, phalanger, bandicoot, dasyure, and wallaby.

The marsupial anteater *Myrmecobius* has no pouch.

maser (acronym for *microwave amplification by stimulated emission of radiation*) in physics, a high-frequency microwave amplifier or oscillator in which the signal to be amplified is used to stimulate unstable atoms into emitting energy at the same frequency. Atoms or molecules are raised to a higher energy level and then allowed to lose this energy by radiation emitted at a precise frequency. The principle has been extended to other parts of the electromagnetic spectrum as, for example, in the ◊laser.

mass in physics, the quantity of matter in a body as measured by its inertia. Mass determines the acceleration produced in a body by a given force acting on it, the acceleration being inversely proportional to the mass of the body. The mass also determines the force exerted on a body by ◊gravity on Earth, although this attraction varies slightly from place to place. In the SI system, the base unit of mass is the kilogram.

mass action, law of in chemistry, a law stating that at a given temperature the rate at which a chemical reaction takes place is proportional to the product of the active masses of the reactants. The active mass is taken to be the molar concentration of the each reactant.

mass–energy equation equation of Albert Einstein (1879–1955), $E = mc^2$, denoting the equivalence of mass and energy, where E is the energy in joules, m is the mass in kilograms, and c is the speed of light, in a vacuum, in metres per second.

mass extinction an event that produced the extinction of many species at about the same time. One notable example is the boundary between the Cretaceous and Tertiary periods (known as the ◊K-T boundary) that saw the extinction of the dinosaurs and other big reptiles, and many of the marine invertebrates as well. Mass extinctions have taken place several times during Earth's history.

mass number or *nucleon number* sum (symbol A) of the numbers of protons and neutrons in the nucleus of an atom. It is used along with the ◊atomic number (the number of protons) in ◊nuclear notation: in symbols that represent nuclear isotopes, such as $^{14}_{6}C$, the lower number is the atomic number, and the upper number is the mass number.

mass spectrometer in physics, an apparatus for analysing chemical composition. Positive ions (charged particles) of a substance are separated by an electromagnetic system, which permits accurate measurement of the relative concentrations of the various ionic masses present, particularly isotopes.

mathematical induction formal method of proof in which the proposition $P(n + 1)$ is proved true on the hypothesis that the proposition $P(n)$ is true. The proposition is then shown to be true for a particular value of n, say k, and therefore by induction the proposition must be true for $n = k + 1, k + 2, k + 3, \ldots$. In many cases $k = 1$, so then the proposition is true for all positive integers.

mathematics science of spatial and numerical relationships. The main divisions of *pure mathematics* include geometry, arithmetic, algebra, calculus, and trigonometry. Mechanics, statistics, numerical analysis, computing, the mathematical theories of astronomy, electricity, optics, thermodynamics, and atomic studies come under the heading of *applied mathematics*.

matrix in biology, usually refers to the ◊extracellular matrix.

matrix in mathematics, a square ($n \times n$) or rectangular ($m \times n$) array of elements (numbers or algebraic variables). They are a means of condensing information about mathematical systems and can be used for, among other things, solving simultaneous linear equations (see ◊simultaneous equations and ◊transformations).

matter in physics, anything that has mass and can be detected and measured. All matter is made up of ◊atoms, which in turn are made up of ◊elementary particles; it exists ordinarily as a solid, liquid, or gas. The history of science and philosophy is largely taken up with accounts of theories of matter, ranging from the hard 'atoms' of Democritus to the 'waves' of modern quantum theory.

maxilla anatomical name for the upper jaw.

maxwell c.g.s. unit (symbol Mx) of magnetic flux (the strength of a ◊magnetic field in an area multiplied by the area). It is now replaced by the SI unit, the ◊weber (one maxwell equals 10^{-8} weber). It is named after the Scottish physicist James Clerk Maxwell.

Maxwell–Boltzmann distribution in physics, a statistical equation describing the distribution of velocities among the molecules of a gas. It is named after James Maxwell and

mathematics: chronology

c. 2500 BC	The people of Mesopotamia (now Iraq) developed a positional numbering (place-value) system, in which the value of a digit depends on its position in a number.
c. 2000 BC	Mesopotamian mathematicians solved quadratic equations (algebraic equations in which the highest power of a variable is 2).
876 BC	A symbol for zero was used for the first time, in India.
c. 550 BC	Greek mathematician Pythagoras formulated a theorem relating the lengths of the sides of a right-angled triangle. The theorem was already known by earlier mathematicians in China, Mesopotamia, and Egypt.
c. 450 BC	Hipparcos of Metapontum discovered that some numbers are irrational (cannot be expressed as the ratio of two integers).
300 BC	Euclid laid out the laws of geometry in his book *Elements*, which was to remain a standard text for 2,000 years.
c. 230 BC	Eratosthenes developed a method for finding all prime numbers.
c. 100 BC	Chinese mathematicians began using negative numbers.
c. 190 BC	Chinese mathematicians used powers of 10 to express magnitudes.
c. AD 210	Diophantus of Alexandria wrote the first book on algebra.
c. 600	A decimal number system was developed in India.
829	Persian mathematician Muhammad ibn-Mūsā al-Khwārizmī published a work on algebra that made use of the decimal number system.
1202	Italian mathematician Leonardo Fibonacci studied the sequence of numbers (1, 1, 2, 3, 5, 8, 13, 21, ...) in which each number is the sum of the two preceding ones.
1550	In Germany, Rheticus published trigonometrical tables that simplified calculations involving triangles.
1614	Scottish mathematician John Napier invented logarithms, which enable lengthy calculations involving multiplication and division to be carried out by addition and subtraction.
1623	Wilhelm Schickard invented the mechanical calculating machine.
1637	French mathematician and philosopher René Descartes introduced coordinate geometry.
1654	In France, Blaise Pascal and Pierre de Fermat developed probability theory.
1666	Isaac Newton developed differential calculus, a method of calculating rates of change.
1675	German mathematician Gottfried Wilhelm Leibniz introduced the modern notation for integral calculus, a method of calculating volumes.
1679	Leibniz introduced binary arithmetic, in which only two symbols are used to represent all numbers.
1684	Leibniz published the first account of differential calculus.
1718	Jakob Bernoulli in Switzerland published his work on the calculus of variations (the study of functions that are close to their minimum or maximum values).
1746	In France, Jean le Rond d'Alembert developed the theory of complex numbers.
1747	D'Alembert used partial differential equations in mathematical physics.
1798	Norwegian mathematician Caspar Wessel introduced the vector representation of complex numbers.
1799	Karl Friedrich Gauss of Germany proved the fundamental theorem of algebra: the number of solutions of an algebraic equation is the same as the exponent of the highest term.
1810	In France, Jean Baptiste Joseph Fourier published his method of representing functions by a series of trigonometric functions.
1812	French mathematician Pierre Simon Laplace published the first complete account of probability theory.
1822	In the UK, Charles Babbage began construction of the first mechanical computer, the difference machine, a device for calculating logarithms and trigonometric functions.
1827	Gauss introduced differential geometry, in which small features of curves are described by analytical methods.
1829	In Russia, Nikolai Ivanonvich Lobachevsky developed hyperbolic geometry, in which a plane is regarded as part of a hyperbolic surface, shaped like a saddle. In France, Evariste Galois introduced the theory of groups (collections whose members obey certain simple rules of addition and multiplication).
1844	French mathematician Joseph Liouville found the first transcendental number, which cannot be expressed as an algebraic equation with rational coefficients. In Germany, Hermann Grassmann studied vectors with more than three dimensions.
1854	George Boole in the UK published his system of symbolic logic, now called Boolean algebra.
1858	English mathematician Arthur Cayley developed calculations using ordered tables called matrices.

1865	August Ferdinand Möbius in Germany described how a strip of paper can have only one side and one edge.
1892	German mathematician Georg Cantor showed that there are different kinds of infinity and studied transfinite numbers.
1895	Jules Henri Poincaré published the first paper on topology, often called 'the geometry of rubber sheets'.
1931	In the USA, Austrian-born mathematician Kurt Gödel proved that any formal system strong enough to include the laws of arithmetic is either incomplete or inconsistent.
1937	English mathematician Alan Turing published the mathematical theory of computing.
1944	John Von Neumann and Oscar Morgenstern developed game theory in the USA.
1945	The first general purpose, fully electronic digital computer, ENIAC (electronic numerator, integrator, analyser, and computer), was built at the University of Pennsylvania, USA.
1961	Meteorologist Edward Lorenz at the Massachusetts Institute of Technology, USA, discovered a mathematical system with chaotic behaviour, leading to a new branch of mathematics – chaos theory.
1962	Benoit Mandelbrot in the USA invented fractal images, using a computer that repeats the same mathematical pattern over and over again.
1975	US mathematician Mitchell Feigenbaum discovered a new fundamental constant (approximately 4.669201609103), which plays an important role in chaos theory.
1980	Mathematicians worldwide completed the classification of all finite and simple groups, a task that took over a hundred mathematicians more than 35 years to complete and whose results took up more than 14,000 pages in mathematical journals.
1989	A team of US computer mathematicians at Amdahl Corporation, California, discovered the highest known prime number (it contains 65,087 digits).
1993	UK mathematician Andrew Wiles published a 1,000-page proof of Fermat's last theorem, one of the most baffling challenges in pure mathematics. It was rejected.

Ludwig Boltzmann, who derived the equation, independently of each other, in the 1860s.

Maxwell's screw rule in physics, a rule formulated by Scottish physicist James Maxwell that predicts the direction of the magnetic field produced around a wire carrying electric current. It states that if a right-handed screw is turned so that it moves forwards in the same direction as the current, its direction of rotation will give the direction of the magnetic field.

mean in mathematics, a measure of the average of a number of terms or quantities. The simple *arithmetic mean* is the average value of the quantities, that is, the sum of the quantities divided by their number. The *weighted mean* takes into account the frequency of the terms that are summed; it is calculated by multiplying each term by the number of times it occurs, summing the results and dividing this total by the total number of occurrences. The *geometric mean* of n quantities is the nth root of their product. In statistics, it is a measure of central tendency of a set of data.

mean deviation in statistics, a measure of the spread of a population from the ◊mean.

mean free path in physics, the average distance travelled by a particle, atom, or molecule between successive collisions. It is of importance in the ◊kinetic theory of gases.

mean life in nuclear physics, the average lifetime of a nucleus of a radioactive isotope. See ◊radioactivity.

mechanical advantage (MA) in physics, the number of times the load moved by a machine is greater than the effort applied to that machine. In equation terms: MA = load/effort.

mechanical equivalent of heat in physics, a constant factor relating the calorie (the c.g.s. unit of heat) to the joule (the unit of mechanical energy), equal to 4.1868 joules per calorie. It is redundant in the SI system of units, which measures heat and all forms of energy in ◊joules (so that the mechanical equivalent of heat is 1).

mechanics branch of physics dealing with the motions of bodies and the forces causing these motions, and also with the forces acting on bodies in ◊equilibrium. It is usually divided into ◊dynamics and ◊statics.

median in mathematics and statistics, the middle number of an ordered group of numbers. If there is no middle number (because there is an even number of terms), the median is the ◊mean (average) of the two middle numbers. For example, the median of the group 2, 3, 7, 11, 12 is 7; that of 3, 4, 7, 9, 11, 13 is 8 (the average of 7 and 9).

In geometry, the term refers to a line from

the vertex of a triangle to the midpoint of the opposite side.

medicine the practice of preventing, diagnosing, and treating disease, both physical and mental; also any substance used in the treatment of disease. The basis of medicine is anatomy (the structure and form of the body) and physiology (the study of the body's functions).

medulla central part of an organ. In the mammalian kidney, the medulla lies beneath the outer cortex and is responsible for the reabsorption of water from the filtrate. In plants, it is a region of packing tissue in the centre of the stem. In the vertebrate brain, the medulla is the posterior region responsible for the coordination of basic activities, such as breathing and temperature control.

medusa the free-swimming phase in the life cycle of a coelenterate, such as a jellyfish or ◊coral. The other phase is the sedentary *polyp*.

meerschaum aggregate of minerals, usually the soft white mineral *sepiolite*, hydrous magnesium silicate. It floats on water and is used for making pipe bowls.

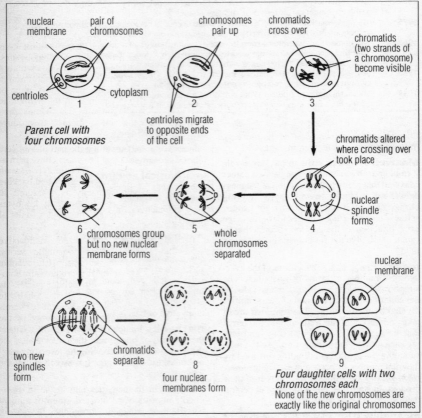

meiosis *Meiosis is a type of cell division that produces gametes (sex cells, sperm and egg). This sequence shows an animal cell but only four chromosomes are present in the parent cell (1). There are two stages in the division process. In the first stage (2–6), the chromosomes come together in pairs and exchange genetic material. This is called crossing over. In the second stage (7–9), the cell divides to produce four gamete cells, each with only one copy of each chromosome from the parent cell.*

mega- prefix denoting multiplication by a million. For example, a megawatt (MW) is equivalent to a million watts.

megabyte (Mb) in computing, a unit of memory equal to 1,024 ◊kilobytes. It is sometimes used, less precisely, to mean 1 million bytes.

megaton one million (10^6) tons. Used with reference to the explosive power of a nuclear weapon, it is equivalent to the explosive force of one million tons of trinitrotoluene (TNT).

meiosis in biology, a process of cell division in which the number of ◊chromosomes in the cell is halved. It only occurs in ◊eukaryotic cells, and is part of a life cycle that involves sexual reproduction because it allows the genes of two parents to be combined without the total number of chromosomes increasing.

In sexually reproducing ◊diploid animals (having two sets of chromosomes per cell), meiosis occurs during formation of the ◊gametes (sex cells, sperm and egg), so that the gametes are ◊haploid (having only one set of chromosomes). When the gametes unite during ◊fertilization, the diploid condition is restored. In plants, meiosis occurs just before spore formation. Thus the spores are haploid and in lower plants such as mosses they develop into a haploid plant called a gametophyte which produces the gametes (see ◊alternation of generations). See also ◊mitosis.

melamine $C_3N_6H_6$ ◊thermosetting ◊polymer based on urea–formaldehyde. It is extremely resistant to heat and is also scratch-resistant. Its uses include synthetic resins.

melanin brown pigment that gives colour to the eyes, skin, hair, feathers, and scales of many vertebrates. In humans, melanin helps protect the skin against ultraviolet radiation from sunlight. Both genetic and environmental factors determine the amount of melanin in the skin.

melanism black coloration of animal bodies caused by large amounts of the pigment melanin. Melanism is of significance in insects, because melanic ones warm more rapidly in sunshine than do pale ones, and can be more active in cool weather. A fall in temperature may stimulate such insects to produce more melanin. In industrial areas, dark insects and pigeons match sooty backgrounds and escape predation, but they are at a disadvantage in rural areas where they do not match their backgrounds. This is known as *industrial melanism*.

meltdown the melting of the core of a nuclear reactor, due to overheating. To prevent such accidents all reactors have equipment intended to flood the core with water in an emergency. The reactor is housed in a strong containment vessel, designed to prevent radiation escaping into the atmosphere. The result of a meltdown is an area radioactively contaminated for 25,000 years or more.

melting point temperature at which a substance melts, or changes from solid to liquid form. A pure substance under standard conditions of pressure (usually one atmosphere) has a definite melting point. If heat is supplied to a solid at its melting point, the temperature does not change until the melting process is complete. The melting point of ice is 0°C or 32°F.

membrane in living things, a continuous layer, made up principally of fat molecules, that encloses a ◊cell or ◊organelles within a cell. Certain small molecules can pass through the cell membrane, but most must enter or leave the cell via channels in the membrane made up of special proteins. The ◊Golgi apparatus within the cell is thought to produce certain membranes.

memory in computing, the part of a system used to store data and programs either permanently or temporarily. There are two main types: immediate access memory and backing storage. Memory capacity is measured in ◊bytes or, more conveniently, in kilobytes (units of 1,024 bytes) or megabytes (units of 1,024 kilobytes).

mendelevium synthesized, radioactive metallic element of the ◊actinide series, symbol Md, atomic number 101, relative atomic mass 258. It was first produced by bombardment of Es-253 with helium nuclei. Its longest-lived isotope, Md-258, has a half-life of about two months. The element is chemically similar to thulium. It was named by the US physicists at the University of California at Berkeley who first synthesized it 1955 after the Russian chemist Mendeleyev, who in 1869 devised the basis for the ◊periodic table of the elements.

Mendelism in genetics, the theory of inheritance originally outlined by Austrian biologist Gregor Mendel. He suggested that, in sexually reproducing species, all characteristics are inherited through indivisible 'factors' (now identified with ◊genes) contributed by each parent to its offspring.

meninges three membranes enclosing the brain and spinal cord. The thickest is the outermost layer, the dura mater; the middle layer is the arachnoid mater and the innermost layer is the pia mater. The ◊cerebrospinal fluid circulates between the two innermost layers.

meniscus in physics, the curved shape of the surface of a liquid in a thin tube, caused by the cohesive effects of ◊surface tension (capillary action). When the walls of the container are made wet by the liquid, the meniscus is concave, but with highly viscous liquids (such as mercury) the meniscus is convex. Meniscus is also the name of a concavo-convex or convexo-concave ◊lens.

menopause in women, the cessation of reproductive ability, characterized by menstruation (see ◊menstrual cycle) becoming irregular and eventually ceasing. The onset is at about the age of 50, but varies greatly. Menopause is usually uneventful, but some women suffer from complications such as flushing, excessive bleeding, and nervous disorders. Since the 1950s, ◊hormone-replacement therapy (HRT), using ◊oestrogen alone or with progestogen, a synthetic form of ◊progesterone, has been developed to counteract such effects.

menstrual cycle cycle that occurs in female mammals of reproductive age, in which the body is prepared for pregnancy. At the beginning of the cycle, a Graafian (egg) follicle develops in the ovary, and the inner wall of the uterus forms a soft spongy lining. The egg is released from the ovary, and the uterus lining (endometrium) becomes vascularized (filled with blood vessels). If fertilization does not occur, the corpus luteum (remains of the Graafian follicle) degenerates, and the uterine lining breaks down, and is shed. This is what causes the loss of blood that marks menstruation. The cycle then begins again. Human menstruation takes place from puberty to menopause, except during pregnancy, occurring about every 28 days.

menu in computing, a list of options, displayed on screen, from which the user may make a choice – for example, the choice of services offered to the customer by a bank cash dispenser: withdrawal, deposit, balance, or statement. Menus are used extensively in ◊graphical user interface (GUI) systems, where the menu options are often selected using a pointing device called a ◊mouse.

Mercalli scale scale used to measure the intensity of an ◊earthquake. It differs from the ◊Richter scale, which measures *magnitude*. It is named after the Italian seismologist Giuseppe Mercalli (1850–1914).

mercury or *quicksilver* heavy, silver-grey, metallic element, symbol Hg (from Latin *hydrargyrum*), atomic number 80, relative atomic mass 200.59. It is a dense, mobile liquid with a low melting point (–38.87°C/ –37.96°F). Its chief source is the mineral

menstrual cycle *From puberty to the menopause, most women produce a regular rhythm of hormones that stimulate the various stages of the menstrual cycle.*

Mercalli scale

intensity value	description
I	Only detected by instrument.
II	Felt by people resting.
III	Felt indoors; hanging objects swing; feels like passing traffic.
IV	Feels like passing heavy traffic; standing cars rock; windows, dishes, and doors rattle; wooden frames creak.
V	Felt outdoors; sleepers are woken; liquids spill; doors swing open.
VI	Felt by everybody; people stagger; windows break; trees and bushes rustle; weak plaster cracks.
VII	Difficult to stand upright; noticed by vehicle drivers; plaster, loose bricks, tiles, and chimneys fall; bells ring.
VIII	Car steering affected; some collapse of masonry; chimney stacks and towers fall; branches break from trees; cracks in wet ground.
IX	General panic; serious damage to buildings; underground pipes break; cracks and subsidence in ground.
X	Most buildings destroyed; landslides; water thrown out of canals.
XI	Rails bent; underground pipes totally destroyed.
XII	Damage nearly total; rocks displaced; objects thrown into the air.

cinnabar, HgS, but it sometimes occurs in nature as a free metal.

Its alloys with other metals are called amalgams (a silver-mercury amalgam is used in dentistry for filling cavities in teeth). Industrial uses include drugs and chemicals, mercury-vapour lamps, arc rectifiers, power-control switches, barometers, and thermometers.

Mercury is a cumulative poison that can contaminate the food chain, and cause intestinal disturbance, kidney and brain damage, and birth defects in humans.

Mercury in astronomy, the closest planet to the Sun, at an average distance of 58 million km/36 million mi. Its diameter is 4,880 km/3,030 mi, its mass 0.056 that of Earth. Mercury orbits the Sun every 88 days, and spins on its axis every 59 days. On its sunward side the surface temperature reaches over 400°C/752°F, but on the 'night' side it falls to −170°C/−274°F. Mercury has an atmosphere with minute traces of argon and helium. In 1974 the US space probe *Mariner 10* discovered that its surface is cratered by meteorite impacts. Mercury has no moons.

mercury fulminate highly explosive compound used in detonators and percussion caps. It is a grey, sandy powder and extremely poisonous.

meridian half a ◊great circle drawn on the Earth's surface passing through both poles and thus through all places with the same longitude. Terrestrial longitudes are usually measured from the Greenwich Meridian.

meristem region of plant tissue containing cells that are actively dividing to produce new tissues (or have the potential to do so). Meristems found in the tip of roots and stems, the apical meristems, are responsible for the growth in length of these organs.

The ◊cambium is a lateral meristem that is responsible for increase in girth in perennial plants. Some plants also have intercalary meristems, as in the stems of grasses, for example. These are responsible for their continued growth after cutting or grazing has removed the apical meristems of the shoots.

mesoglea layer of jellylike noncellular tissue that separates the endoderm and ectoderm in jellyfish and other ◊coelenterates.

meson in physics, an unstable subatomic particle made up of two indivisible elementary particles called ◊quarks. It has a mass intermediate between that of the electron and that of the proton, is found in cosmic radiation, and is emitted by nuclei under bombardment by very high-energy particles. The mesons form a subclass of the hadrons and include the kaons and pions.

mesophyll the tissue between the upper and lower epidermis of a leaf blade (◊lamina), consisting of parenchyma-like cells containing numerous ◊chloroplasts.

In many plants, mesophyll is divided into two distinct layers. The *palisade mesophyll* is usually just below the upper epidermis and is composed of regular layers of elongated cells. Lying below them is the *spongy mesophyll*, composed of loosely arranged cells of irregular shape. This layer contains fewer chloroplasts and has many intercellular spaces for the diffusion of gases (required for ◊respiration and

◊photosynthesis), linked to the outside by means of ◊stomata.

mesosphere layer in the Earth's ◊atmosphere above the stratosphere and below the thermosphere. It lies between about 50 km/31 mi and 80 km/50 mi above the ground.

Mesozoic era of geological time 245–65 million years ago, consisting of the Triassic, Jurassic, and Cretaceous periods. At the beginning of the era, the continents were joined together as Pangaea; dinosaurs and other giant reptiles dominated the sea and air; and ferns, horsetails, and cycads thrived in a warm climate worldwide. By the end of the Mesozoic era, the continents had begun to assume their present positions, flowering plants were dominant, and many of the large reptiles and marine fauna were becoming extinct.

metabolism the chemical processes of living organisms enabling them to grow and to function. It involves a constant alternation of building up (*anabolism*) and breaking down (*catabolism*). For example, green plants build up complex organic substances from water, carbon dioxide, and mineral salts (photosynthesis); by digestion animals partially break down complex organic substances, ingested as food, and subsequently resynthesize them for use in their own bodies.

metal any of a class of chemical elements with certain chemical characteristics (◊metallic character) and physical properties: they are good conductors of heat and electricity; opaque but reflect light well; malleable, which enables them to be cold-worked and rolled into sheets; and ductile, which permits them to be drawn into thin wires.

Metallic elements compose about 75% of the 109 elements shown in the ◊periodic table of the elements. They form alloys with each other, ◊bases with the hydroxyl radical (OH), and replace the hydrogen in an ◊acid to form a salt. The majority are found in nature in the combined form only, as compounds or mineral ores; about 16 of them also occur in the elemental form, as ◊native metals. Their chemical properties are largely determined by the extent to which their atoms can lose one or more electrons and form positive ions (cations).

The following are widely used in commerce: *precious metals*: gold, silver, mercury, and platinum, used principally in jewellery; *heavy metals*: iron, copper, zinc, tin, and lead, the common metals of engineering; *rarer heavy metals*: nickel, cadmium, chromium, tungsten, molybdenum, manganese, cobalt, vanadium, antimony, and bismuth, used principally for alloying with the heavy metals; *light metals*: aluminium and magnesium; *alkali metals*: sodium, potassium, and lithium; and *alkaline-earth metals*: calcium, barium, and strontium, used principally for chemical purposes.

metal detector electronic device for detecting metal, usually below ground, developed from the wartime mine detector. In the head of the metal detector is a coil, which is part of an electronic circuit. The presence of metal causes the frequency of the signal in the circuit to change, setting up an audible note in the headphones worn by the user.

metal fatigue condition in which metals fail or fracture under relatively light loads, when these loads are applied repeatedly. Structures that are subject to flexing, such as the airframes of aircraft, are prone to metal fatigue.

metallic bond the force of attraction operating in a metal that holds the atoms together. In the metal the ◊valency electrons are able to move within the crystal and these electrons are said to be delocalized (see ◊electrons, delocalized). Their movement creates short-lived, positively charged ions. The electrostatic attraction between the delocalized electrons and the ceaselessly forming ions constitutes the metallic bond.

metallic character chemical properties associated with those elements classed as metals. These properties, which arise from the element's ability to lose electrons, are: the displacement of hydrogen from dilute acids; the formation of ◊basic oxides; the formation of ionic chlorides; and their reducing reaction, as in the ◊thermite process (see ◊reduction).

In the periodic table of the elements, metallic character increases down any group and across a period from right to left.

metallic glass substance produced from metallic materials (noncorrosive alloys rather than simple metals) in a liquid state which, by very rapid cooling, are prevented from reverting to their regular metallic structure. Instead they take on the properties of glass, while retaining the metallic properties of malleability and relatively good electrical conductivity.

metalloid or *semimetal* any chemical element having some of but not all the properties of metals; metalloids are thus usually electrically semiconducting. They comprise the elements germanium, arsenic, antimony, and tellurium.

metallurgy the science and technology of producing metals, which includes extraction, alloying, and hardening. *Extractive,* or *process, metallurgy* is concerned with the extraction of metals from their ↓ores and refining and adapting them for use. *Physical metallurgy* is concerned with their properties and application. *Metallography* establishes the microscopic structures that contribute to hardness, ductility, and strength.

metamorphic rock rock altered in structure and composition by pressure, heat, or chemically active fluids after original formation. (If heat is sufficient to melt the original rock, technically it becomes an igneous rock upon cooling.)

metamorphism geological term referring to the changes in rocks of the Earth's crust caused by increasing pressure and temperature. The resulting rocks are metamorphic rocks. All metamorphic changes take place in solid rocks. If the rocks melt and then harden, they become ↓igneous rocks.

metamorphosis period during the life cycle of many invertebrates, most amphibians, and some fish, during which the individual's body changes from one form to another through a major reconstitution of its tissues. For example, adult frogs are produced by metamorphosis from tadpoles, and butterflies are produced from caterpillars following metamorphosis within a pupa.

metazoa another name for animals. It reflects an earlier system of classification, in which there were two main divisions within the animal kingdom, the multicellular animals, or metazoa, and the single-celled 'animals' or protozoa. The ↓protozoa are no longer included in the animal kingdom, so only the metazoa remain.

meteor flash of light in the sky, popularly known as a *shooting* or *falling star*, caused by a particle of dust, a *meteoroid*, entering the atmosphere at speeds up to 70 kps/45 mps and burning up by friction at a height of around 100 km/60 mi. On any clear night, several *sporadic meteors* can be seen each hour.

meteor-burst communications technique for sending messages by bouncing radio waves off the fiery tails of ↓meteors. High-speed computer-controlled equipment is used to sense the presence of a meteor and to broadcast a signal during the short time that the meteor races across the sky.

meteorite piece of rock or metal from space that reaches the surface of the Earth, Moon, or other body. Most meteorites are thought to be fragments from asteroids, although some may be pieces from the heads of comets. Most are stony, although some are made of iron and a few have a mixed rock–iron composition. Meteorites provide evidence for the nature of the Solar System and may be similar to the Earth's core and mantle, neither of which can be observed directly.

meteoroid chunk of rock in interplanetary space. There is no official distinction between meteoroids and asteroids, except that the term asteroid is generally reserved for objects larger than 1.6 km/1 mi in diameter, whereas meteoroids can range anywhere from pebble-size up.

meteorology scientific observation and study of the ↓atmosphere, so that weather can be accurately forecast. Data from meteorological stations and weather satellites are collated by computer at central agencies, and forecast and weather maps based on current readings are issued at regular intervals. Modern analysis can give useful forecasts for up to six days ahead.

meter any instrument used for measurement. The term is often compounded with a prefix to denote a specific type of meter: for example, ammeter, voltmeter, flowmeter, or pedometer.

methanal (common name *formaldehyde*) HCHO gas at ordinary temperatures, condensing to a liquid at $-21°C/-5.8°F$. It has a powerful, penetrating smell. Dissolved in water, it is used as a biological preservative. It is used in the manufacture of plastics, dyes, foam (for example urea–formaldehyde foam, used in insulation), and in medicine.

methane CH_4 the simplest hydrocarbon of the paraffin series. Colourless, odourless, and lighter than air, it burns with a bluish flame and explodes when mixed with air or oxygen. It is the chief constituent of natural gas and also occurs in the explosive firedamp of coal mines. Methane emitted by rotting vegetation forms marsh gas.

Methane causes about 38% of the warming of the globe through the ↓greenhouse effect; the amount of methane in the air is predicted to double over the next 60 years.

methanogenic bacteria one of a group of primitive bacteria (↓archaebacteria). They give off methane gas as a by-product of their metabolism, and are common in sewage treatment plants and hot springs, where the temperature is high and oxygen is absent.

methanoic acid (common name *formic acid*) HCOOH, a colourless, slightly fuming liquid that freezes at 8°C/46.4°F and boils at 101°C/213.8°F. It occurs in stinging ants, nettles, sweat, and pine needles, and is used in dyeing, tanning, and electroplating.

methanol (common name *methyl alcohol*) CH_3OH the simplest of the alcohols. It can be made by the dry distillation of wood (hence it is also known as wood alcohol), but is usually made from coal or natural gas. When pure, it is a colourless, flammable liquid with a pleasant odour, and is highly poisonous.

Methanol is used to produce formaldehyde (from which resins and plastics can be made), methyl-ter-butyl ether (MTB, a replacement for lead as an octane-booster in petrol), vinyl acetate (largely used in paint manufacture), and petrol. In 1993 Japanese engineers built an engine, made largely of ceramics, that runs on methanol. The prototype is lighter and has a cleaner exhaust than comparable metal, petrol-powered engines.

methyl alcohol common name for ◊methanol.

methylated spirit alcohol that has been rendered undrinkable, and is used for industrial purposes, as a fuel for spirit burners or a solvent.

methyl benzene alternative name for ◊toluene.

methyl orange $C_{14}H_{14}N_3NaO_3S$ orange-yellow powder used as an acid–base indicator in chemical tests, and as a stain in the preparation of slides of biological material. Its colour changes with pH; below pH 3.1 it is red, above pH 4.4 it is yellow.

metre SI unit (symbol m) of length, equivalent to 1.093 yards. It is defined by scientists as the length of the path travelled by light in a vacuum during a time interval of 1/299,792,458 of a second.

metric system system of weights and measures developed in France in the 18th century and recognized by other countries in the 19th century.

In 1960 an international conference on weights and measures recommended the universal adoption of a revised International System (Système International d'Unités, or SI), with seven prescribed 'base units': the metre (m) for length, kilogram (kg) for mass, second (s) for time, ampere (A) for electric current, kelvin (K) for thermodynamic temperature, candela (cd) for luminous intensity, and mole (mol) for quantity of matter.

mg symbol for *milligram*.

MHD abbreviation for ◊*magnetohydrodynamics*.

mho SI unit of electrical conductance, now called the ◊siemens; equivalent to a reciprocal ohm.

mi symbol for ◊*mile*.

mica group of silicate minerals that split easily into thin flakes along lines of weakness in their crystal structure (perfect basal cleavage). They are glossy, have a pearly lustre, and are found in many igneous and metamorphic rocks. Their good thermal and electrical insulation qualities make them valuable in industry.

MICR abbreviation for ◊*magnetic-ink character recognition*.

micro- prefix (symbol μ) denoting a one-millionth part (10^{-6}). For example, a micrometre, μm, is one-millionth of a metre.

microbe another name for ◊microorganism.

microbiology the study of microorganisms, mostly viruses and single-celled organisms such as bacteria, protozoa, and yeasts. The practical applications of microbiology are in medicine (since many microorganisms cause disease); in brewing, baking, and other food and beverage processes, where the microorganisms carry out fermentation; and in genetic engineering, which is creating increasing interest in the field of microbiology.

microchip popular name for the silicon chip, or ◊integrated circuit.

microclimate the climate of a small area, such as a woodland, lake, or even a hedgerow. Significant differences can exist between the climates of two neighbouring areas – for example, a town is usually warmer than the surrounding countryside (forming a heat island), and a woodland cooler, darker, and less windy than an area of open land.

microcomputer or *micro* or *personal computer* small desktop or portable computer, typically designed to be used by one person at a time, although individual computers can be linked in a network so that users can share data and programs.

Its central processing unit is a ◊microprocessor, contained on a single integrated circuit.

microfiche sheet of film on which printed text is photographically reduced. See ◊microform.

microform generic name for media on which text or images are photographically reduced. The main examples are *microfilm* (similar to the film in an ordinary camera) and *microfiche* (flat sheets of film, generally 105 mm/4 in × 148 mm/6 in, holding the equivalent of 420 A4 sheets). Microform has the advantage of low reproduction and storage costs, but it requires special devices for reading the text. It is widely used for archiving and for storing large volumes of text, such as library catalogues.

micrometer instrument for measuring minute lengths or angles with great accuracy; different types of micrometer are used in astronomical and engineering work.

micrometre one-millionth of a ◊metre (symbol µm).

microminiaturization reduction in size and weight of electronic components. The first size reduction in electronics was brought about by the introduction of the ◊transistor. Further reductions were achieved with ◊integrated circuits and the ◊silicon chip.

micron obsolete name for the micrometre, one millionth of a metre.

microorganism or *microbe* living organism invisible to the naked eye but visible under a microscope. Microorganisms include viruses and single-celled organisms such as bacteria, protozoa, yeasts, and some algae. The term has no taxonomic significance in biology. The study of microorganisms is known as microbiology.

microphone primary component in a sound-reproducing system, whereby the mechanical energy of sound waves is converted into electrical signals by means of a ◊transducer. One of the simplest is the telephone receiver mouthpiece, invented by Scottish–US inventor Alexander Graham Bell in 1876; other types of microphone are used with broadcasting and sound-film apparatus.

microprocessor complete computer ◊central processing unit contained on a single ◊integrated circuit, or chip. The appearance of the first microprocessor 1971 designed by Intel for a pocket calculator manufacturer heralded the introduction of the microcomputer. The microprocessor has led to a dramatic fall in the size and cost of computers, and ◊dedicated computers can now be found in washing machines, cars, and so on. Examples of microprocessors are the Intel 8086 family and the Motorola 68000 family.

micropropagation the mass production of plants by placing tiny pieces of plant tissue in sterile glass containers along with nutrients. Perfect clones of superplants are produced in sterile cabinets, with filtered air and carefully controlled light, temperature, and humidity. The system is used for the house-plant industry and for forestry – micropropagation gives immediate results, whereas obtaining genetically homogenous tree seed by traditional means would take over 100 years.

micropyle in flowering plants, a small hole towards one end of the ovule. At pollination the pollen tube growing down from the ◊stigma eventually passes through this pore. The male gamete is contained within the tube and is able to travel to the egg in the interior of the ovule. Fertilization can then take place, with subsequent seed formation and dispersal.

microscope instrument for magnification with high resolution for detail. Optical and electron microscopes are the ones chiefly in use; other types include acoustic, ◊scanning tunnelling, and ◊atomic force microscopes. In 1988 a scanning tunnelling microscope was used to photograph a single protein molecule for the first time.

Microsoft US software corporation, now the world's largest supplier. Microsoft's first major product was ◊MS-DOS, written for IBM, but it has increased its hold on the personal computer market with the release of ◊Windows and related applications.

microtubules tiny tubes found in almost all cells with a nucleus. They help to define the shape of a cell by forming scaffolding for cilia and they also form fibres of the mitotic spindle (see ◊mitosis).

microwave ◊electromagnetic wave with a wavelength in the range 0. 3 to 30 cm/0.1 in to 12 in, or 300–300,000 megahertz (between radio waves and ◊infrared radiation). Microwaves are used in radar, in radio broadcasting, and in microwave heating and cooking.

microwave heating heating by means of microwaves. Microwave ovens use this form of heating for the rapid cooking or reheating of foods, where heat is generated throughout the food simultaneously. If food is not heated completely, there is a danger of bacterial growth that may lead to food poisoning. Industrially, microwave heating is used for destroying insects in grain and enzymes in processed food, pasteurizing and sterilizing liquids, and drying timber and paper.

microscope *Terms used to describe an optical microscope. In essence, the optical microscope consists of an eyepiece lens and an objective lens, which are used to produce an enlarged image of a small object by focusing light from a light source. Optical microscopes can achieve magnifications of up to 1,500–2,000. Higher magnifications and resolutions are obtained by electron microscopes.*

midnight sun the constant appearance of the Sun (within the Arctic and Antarctic circles) above the horizon during the summer.

migration the movement, either seasonal or as part of a single life cycle, of certain animals, chiefly birds and fish, to distant breeding or feeding grounds.

Milankovitch hypothesis the combination of factors governing the occurrence of ◊ice ages proposed in 1930 by the Yugoslav geophysicist M Milankovitch (1879–1958). These include the variation in the angle of the Earth's axis, and the geometry of the Earth's orbit around the Sun.

mile imperial unit of linear measure. A statute mile is equal to 1,760 yards (1.60934 km), and an international nautical mile is equal to 2,026 yards (1,852 m).

milk secretion of the ◊mammary glands of female mammals, with which they suckle their young (during ◊lactation). Over 85% is water, the remainder comprising protein, fat, lactose (a sugar), calcium, phosphorus, iron, and vitamins. The milk of cows, goats, and sheep is often consumed by humans, but regular drinking of milk after infancy is principally a Western practice.

Milky Way faint band of light crossing the

night sky, consisting of stars in the plane of our Galaxy. The name Milky Way is often used for the Galaxy itself. It is a spiral ◊galaxy, about 100,000 light years in diameter, containing at least 100 billion stars. The Sun is in one of its spiral arms, about 25,000 light years from the centre.

milli- prefix (symbol m) denoting a one-thousandth part (10^{-3}). For example, a millimetre, mm, is one thousandth of a metre.

millibar unit of pressure, equal to one-thousandth of a ◊bar.

millilitre one-thousandth of a litre (ml), equivalent to one cubic centimetre (cc).

millimetre of mercury (symbol mmHg) unit of pressure, used in medicine for measuring blood pressure defined as the pressure exerted by a column of mercury one millimetre high, under the action of gravity.

mimicry imitation of one species (or group of species) by another. The most common form is *Batesian mimicry* (named after English naturalist H W Bates (1825–1892)), where the mimic resembles a model that is poisonous or unpleasant to eat, and has aposematic, or warning, coloration; the mimic thus benefits from the fact that predators have learned to avoid the model. Hoverflies that resemble bees or wasps are an example. Appearance is usually the basis for mimicry, but calls, songs, scents, and other signals can also be mimicked.

mineral naturally formed inorganic substance with a particular chemical composition and a regularly repeating internal structure. Either in their perfect crystalline form or otherwise, minerals are the constituents of ◊rocks. In more general usage, a mineral is any substance economically valuable for mining (including coal and oil, despite their organic origins).

mineral dressing preparing a mineral ore for processing. Ore is seldom ready to be processed when it is mined; it often contains unwanted rock and dirt. Therefore it is usually crushed into uniform size and then separated from the dirt, or gangue. This may be done magnetically (some iron ores), by washing (gold), by treatment with chemicals (copper ores), or by flotation.

mineral extraction recovery of valuable ores from the Earth's crust. The processes used include open-cast mining, shaft mining, and quarrying, as well as more specialized processes such as those used for oil and sulphur (see, for example, ◊Frasch process).

mineralogy study of minerals. The classification of minerals is based chiefly on their chemical composition and the kind of chemical bonding that holds these atoms together. The mineralogist also studies their crystallographic and physical characters, occurrence, and mode of formation.

mineral oil oil obtained from mineral sources, for example coal or petroleum, as distinct from oil obtained from vegetable or animal sources.

mineral salt in nutrition, a simple inorganic chemical that is required by living organisms. Plants usually obtain their mineral salts from the soil, while animals get theirs from their food. Important mineral salts include iron salts (needed by both plants and animals), magnesium salts (needed mainly by plants, to make chlorophyll), and calcium salts (needed by animals to make bone or shell). A ◊trace element is required only in tiny amounts.

minicomputer multiuser computer with a size and processing power between those of a ◊mainframe and a ◊microcomputer. Nowadays almost all minicomputers are based on ◊microprocessors.

Mini Disc digital audio disc that resembles a computer floppy disc in a 5 cm/2 in square case, with up to an hour's playing time. The system was developed by Sony for release 1993.

minor planet another name for an ◊asteroid.

minute unit of time consisting of 60 seconds; also a unit of angle equal to one sixtieth of a degree.

Miocene fourth epoch of the Tertiary period of geological time, 23.5–5.2 million years ago. At this time grasslands spread over the interior of continents, and hoofed mammals rapidly evolved.

mips (acronym for *million instructions per second*) in computing, a measure of the speed of a processor. It does not equal the computer power in all cases.

Mir Soviet space station, the core of which was launched 20 Feb 1986. *Mir* is intended to be a permanently occupied space station.

Mira or *Omicron Ceti* brightest long-period pulsating ◊variable star, located in the constellation ◊Cetus. Mira was the first star discovered to vary periodically in brightness.

mirage illusion seen in hot climates of water on the horizon, or of distant objects being enlarged. The effect is caused by ◊refraction, or bending, of light.

Light rays from the sky bend as they pass through the hot layers of air near the ground, so that they appear to come from the horizon. Because the light is from a blue sky, the horizon appears blue and watery. If, during the night, cold air collects near the ground, light can be bent in the opposite direction, so that objects below the horizon appear to float above it.

mirror any polished surface that reflects light; often made from 'silvered' glass (in practice, a mercury-alloy coating of glass). A plane (flat) mirror produces a same-size, erect 'virtual' image located behind the mirror at the same distance from it as the object is in front of it. A spherical concave mirror produces a reduced, inverted real image in front or an enlarged, erect virtual image behind it (as in a shaving mirror), depending on how close the object is to the mirror. A spherical convex mirror produces a reduced, erect virtual image behind it (as in a car's rear-view mirror).

Liquid mirrors using, for example mercury, are formed by rotating the liquid so that gravity and centrifugal forces shape it into a perfect parabola. They have a number of advantages over solid mirrors: they do not sag and so can theoretically be made much larger; they are cheaper, and need no polishing. In 1994 a small number of liquid-mirror telescopes had been built for research purposes.

miscarriage spontaneous expulsion of a fetus from the womb before it is capable of independent survival. Often, miscarriages are due to an abnormality in the developing fetus.

missile rocket-propelled weapon, which may be nuclear-armed. Modern missiles are often classified as surface-to-surface missiles (SSM), air-to-air missiles (AAM), surface-to-air missiles (SAM), or air-to-surface missiles (ASM). A *cruise missile* is in effect a pilotless, computer-guided aircraft; it can be sea-launched from submarines or surface ships, or launched from the air or the ground.

Mississippian US term for the Lower or Early ◊Carboniferous period of geological time, 363–323 million years ago. It is named after the state of Mississippi.

mistral cold, dry, northerly wind that occasionally blows during the winter on the Mediterranean coast of France. It has been known to reach a velocity of 145 kph/90 mph.

mitochondria (singular *mitochondrion*) membrane-enclosed organelles within ◊eukaryotic cells, containing enzymes responsible for energy production during ◊aerobic respiration. These rodlike or spherical bodies are thought to be derived from free-living bacteria that, at a very early stage in the history of life, invaded larger cells and took up a symbiotic way of life inside. Each still contains its own small loop of DNA called mitochondrial DNA, and new mitochondria arise by division of existing ones.

mitosis in biology, the process of cell division. The genetic material of ◊eukaryotic cells is carried on a number of ◊chromosomes. To control their movements during cell division so that both new cells get a full complement, a system of protein tubules, known as the spindle, organizes the chromosomes into position in the middle of the cell before they replicate. The spindle then controls the movement of chromosomes as the cell goes through the stages of division: *interphase*, *prophase*, *metaphase*, *anaphase*, and *telophase*. See also ◊meiosis.

mixture in chemistry, a substance containing two or more compounds that still retain their separate physical and chemical properties. There is no chemical bonding between them and they can be separated from each other by physical means (compare ◊compound).

m.k.s. system system of units in which the base units metre, kilogram, and second replace the centimetre, gram, and second of the ◊c.g.s. system. From it developed the SI system (see ◊SI units).

ml symbol for *millilitre*.

mm symbol for *millimetre*.

mmHg symbol for ◊*millimetre of mercury*.

mnemonic in computing, a short sequence of letters used in low-level programming languages (see ◊low-level language) to represent a ◊machine code instruction.

mobile electrons another term for delocalized electrons (see ◊electrons, delocalized). Mobile electrons are found in metals, graphite, and unsaturated molecules.

mobile ion in chemistry, an ion that is free to move; such ions are only found in the aqueous solutions or melts (molten masses) of an ◊electrolyte. The mobility of the ions in an electrolyte is what allows it to conduct electricity.

Möbius strip structure made by giving a half twist to a flat strip of paper and joining the ends together. It has certain remarkable properties, arising from the fact that it has only one

mitosis *The stages of mitosis, the process of cell division that takes place when a plant or animal cell divides for growth or repair. The two daughter cells each receive the same number of chromosomes as were in the original cell.*

edge and one side. If cut down the centre of the strip, instead of two new strips of paper, only one long strip is produced. It was invented by the German mathematician August Möbius.

mode in mathematics, the element that appears most frequently in a given set of data. For example, the mode for the data 0, 0, 9, 9, 9, 12, 87, 87 is 9.

modem (acronym for *modulator/demodulator*) device for transmitting computer data over telephone lines. Such a device is necessary because the ◊digital signals produced by computers cannot, at present, be transmitted directly over the telephone network, which uses ◊analogue signals. The modem converts the digital signals to analogue, and back again. Modems are used for linking remote terminals to central computers and enable computers to communicate with each other anywhere in the world.

moderator in a ◊nuclear reactor, a material such as graphite or heavy water used to reduce the speed of high-energy neutrons. Neutrons produced by nuclear fission are fast-moving and must be slowed to initiate further fission so that nuclear energy continues to be released at a controlled rate.

Slow neutrons are much more likely to cause ◊fission in a uranium-235 nucleus than to be captured in a U-238 (nonfissile uranium) nucleus. By using a moderator, a reactor can thus be made to work with fuel containing only a small proportion of U-235.

modulation in radio transmission, the intermittent change of frequency, or amplitude, of a radio carrier wave, in accordance with the audio characteristics of the speaking voice, music, or other signal being transmitted. See ◊pulse-code modulation, ◊AM (amplitude modulation), and ◊FM (frequency modulation).

modulus in mathematics, a number that divides exactly into the difference between two given numbers. Also, the multiplication factor used to convert a logarithm of one base to a logarithm of another base. Also, another name for ◊absolute value.

Mohorovičić discontinuity also *Moho* or *M-discontinuity* boundary that separates the Earth's crust and mantle, marked by a rapid increase in the speed of earthquake waves. It follows the variations in the thickness of the crust and is found approximately 32 km/20 mi below the continents and about 10 km/6 mi below the oceans. It is named after the Yugoslav geophysicist Andrija Mohorovičić (1857–1936), who suspected its presence after analysing seismic waves from the Kulpa Valley earthquake 1909.

Mohs' scale scale of hardness for minerals (in ascending order): 1 talc; 2 gypsum; 3 calcite; 4 fluorite; 5 apatite; 6 orthoclase; 7 quartz; 8 topaz; 9 corundum; 10 diamond.

molar one of the large teeth found towards the back of the mammalian mouth. The structure of the jaw, and the relation of the muscles, allows a massive force to be applied to molars. In herbivores the molars are flat with sharp ridges of enamel and are used for grinding, an adaptation to a diet of tough plant material. Carnivores have sharp powerful molars called carnassials, which are adapted for cutting meat.

molarity in chemistry, ◊concentration of a solution expressed as the number of ◊moles in grams of solute per cubic decimetre of solution.

molar solution in chemistry, solution that contains one ◊mole of a substance per litre of solvent.

molar volume volume occupied by one ◊mole (the molecular mass in grams) of any gas at standard temperature and pressure, equal to 2.24136×10^{-2} m^3.

mold US spelling of ◊mould.

mole SI unit (symbol mol) of the amount of a substance. It is defined as the amount of a substance that contains as many elementary entities (atoms, molecules, and so on) as there are atoms in 12 g of the ◊isotope carbon-12.

One mole of an element that exists as single atoms weighs as many grams as its ◊atomic number (so one mole of carbon weighs 12 g), and it contains 6.022045×10^{23} atoms, which is ◊Avogadro's number.

molecular biology study of the molecular basis of life, including the biochemistry of molecules such as DNA, RNA, and proteins, and the molecular structure and function of the various parts of living cells.

molecular clock use of rates of ◊mutation in genetic material to calculate the length of time elapsed since two related species diverged from each other during evolution. The method can be based on comparisons of the DNA or of widely occurring proteins, such as haemoglobin.

Since mutations are thought to occur at a constant rate, the length of time that must have elapsed in order to produce the difference between two species can be estimated. This information can be compared with the evidence obtained from palaeontology to reconstruct evolutionary events.

molecular formula in chemistry, formula indicating the actual number of atoms of each element present in a single molecule of a chemical compound. This is determined by two pieces of information: the ◊empirical formula and the ◊relative molecular mass, which is determined experimentally.

molecular mass (also known as ◊relative molecular mass) the mass of a molecule,

Mohs' table

number	defining mineral	other substances compared	
1	talc		
2	gypsum	2½	fingernail
3	calcite	3½	copper coin
4	fluorite		
5	apatite	5½	steel blade
6	orthoclase	5¾	glass
7	quartz	7	steel file
8	topaz		
9	corundum		
10	diamond		

Note: The scale is not regular; diamond, at number 10 the hardest natural substance, is 90 times harder in absolute terms than corundum, number 9.

calculated relative to one-twelfth the mass of an atom of carbon-12. It is found by adding the relative atomic masses of the atoms that make up the molecule.

molecular solid in chemistry, solid composed of molecules that are held together by relatively weak ◊intermolecular forces. Such solids are low-melting and tend to dissolve in organic solvents. Examples of molecular solids are sulphur, ice, sucrose, and solid carbon dioxide.

molecule group of two or more ◊atoms bonded together. A molecule of an element consists of one or more like ◊atoms; a molecule of a compound consists of two or more different atoms bonded together. Molecules vary in size and complexity from the hydrogen molecule (H_2) to the large ◊macromolecules of proteins. They are held together by ionic bonds, in which the atoms gain or lose electrons to form ◊ions, or by covalent bonds, where electrons from each atom are shared in a new molecular orbital.

According to the molecular or ◊kinetic theory of matter, molecules are in a state of constant motion, the extent of which depends on their temperature, and they exert forces on one another.

The shape of a molecule profoundly affects its chemical, physical, and biological properties. Optical ◊isomers (molecules that are mirror images of each other) rotate plane ◊polarized light in opposite directions; isomers of drug molecules may have have different biological effects; and ◊enzyme reactions are crucially dependent on the shape of the enzyme and the substrate on which it acts.

The symbolic representation of a molecule is known as its formula. The presence of more than one atom is denoted by a subscript figure – for example, one molecule of the compound water, having two atoms of hydrogen and one atom of oxygen, is shown as H_2O.

mollusc any invertebrate of the phylum Mollusca with a body divided into three parts, a head, a foot, and a visceral mass. The majority of molluscs are marine animals, but some inhabit fresh water, and a few are terrestrial. They include bivalves, mussels, octopuses, oysters, snails, slugs, and squids. The body is soft, limbless, and cold-blooded. There is no internal skeleton, but many species have a hard shell covering the body.

molybdenite molybdenum sulphide, MoS_2, the chief ore mineral of molybdenum. It possesses a hexagonal crystal structure similar to graphite, has a blue metallic lustre, and is very soft (1–1.5 on Mohs' scale).

molybdenum heavy, hard, lustrous, silver-white, metallic element, symbol Mo, atomic number 42, relative atomic mass 95.94. The chief ore is the mineral molybdenite. The element is highly resistant to heat and conducts electricity easily. It is used in alloys, often to harden steels. It is a necessary trace element in human nutrition. It was named 1781 by Swedish chemist Karl Scheele (1742–1786), after its isolation by P J Hjelm (1746–1813), for its resemblance to lead ore.

moment of a force in physics, measure of the turning effect, or torque, produced by a force acting on a body. It is equal to the product of the force and the perpendicular distance from its line of action to the point, or pivot, about which the body will turn. Its unit is the newton metre.

If the magnitude of the force is F newtons and the perpendicular distance is d metres then the moment is given by:

$$\text{moment} = Fd$$

See also ◊couple.

moment of inertia in physics, the sum of all the point masses of a rotating object multiplied by the squares of their respective distances from the axis of rotation. It is analogous to the ◊mass of a stationary object or one moving in a straight line.

In linear dynamics, Newton's second law of motion states that the force F on a moving object equals the products of its mass m and acceleration a ($F = ma$); the analogous equation in rotational dynamics is $T = I\alpha$, where T is the torque (the turning effect of a force) that causes an angular acceleration α and I is the moment of inertia. For a given object, I depends on its shape and the position of its axis of rotation.

momentum in physics, the product of the mass of a body and its linear velocity. The ***angular momentum*** of an orbiting or rotating body is the product of its moment of inertia and its angular velocity. The momentum of a body does not change unless it is acted on by an external force; angular momentum does not change unless it is acted upon by a turning force, or torque.

The law of conservation of momentum is one of the fundamental concepts of classical physics. It states that the total momentum of all bodies in a closed system is constant and unaffected by processes occurring within the system. Angular momentum is similarly conserved in a closed system.

monazite mineral, $(Ce,La,Th)PO_4$, yellow to red, valued as a source of ◊lanthanides or rare earths, including cerium and europium; generally found in placer deposit (alluvial) sands.

monocarpic or *hapaxanthic* describing plants that flower and produce fruit only once during their life cycle, after which they die. Most ◊annual plants and ◊biennial plants are monocarpic, but there are also a small number of monocarpic ◊perennial plants that flower just once, sometimes after as long as 90 years, dying shortly afterwards, for example, century plant *Agave* and some species of bamboo *Bambusa*. The general biological term related to organisms that reproduce only once during their lifetime is ◊semelparity.

monoclinic sulphur allotropic form of sulphur, formed from sulphur that crystallizes above 96°C. Once formed, it very slowly changes into the rhombic allotrope.

monoclonal antibody (MAB) antibody produced by fusing an antibody-producing lymphocyte with a cancerous myeloma (bone-marrow) cell. The resulting fused cell, called a hybridoma, is immortal and can be used to produce large quantities of a single, specific antibody. By choosing antibodies that are directed against antigens found on cancer cells, and combining them with cytotoxic drugs, it is hoped to make so-called magic bullets that will be able to pick out and kill cancers.

monocotyledon angiosperm (flowering plant) having an embryo with a single cotyledon, or seed leaf (as opposed to ◊dicotyledons, which have two). Monocotyledons usually have narrow leaves with parallel veins and smooth edges, and hollow or soft stems. Their flower parts are arranged in threes. Most are small plants such as orchids, grasses, and lilies, but some are trees such as palms.

monoecious having separate male and female flowers on the same plant. Maize (*Zea mays*), for example, has a tassel of male flowers at the top of the stalk and a group of female flowers (on the ear, or cob) lower down. Monoecism is a way of avoiding self-fertilization. ◊Dioecious plants have male and female flowers on separate plants.

monohybrid inheritance pattern of inheritance seen in simple ◊genetics experiments, where the two animals (or two plants) being crossed are genetically identical except for one gene.

This gene may code for some obvious external features such as seed colour, with one parent having green seeds and the other having yellow seeds. The offspring are monohybrids, that is, hybrids for one gene only, having received one copy of the gene from each parent. Known as the F1 generation, they are all identical, and usually resemble one parent, whose version of the gene (the dominant ◊allele) masks the effect of the other version (the recessive allele). Although the characteristic coded for by the recessive allele (for example, green seeds) completely disappears in this generation, it can reappear in offspring of the next generation if they have two recessive alleles. On average, this will occur in one out of four offspring from a cross between two of the monohybrids. The next generation (called F2) show a 3:1 ratio for the characteristic in question, 75% being like the original parent with the recessive allele. Austrian biologist Gregor Mendel (1822–1884) first carried out experiments of this type (crossing varieties of artificially bred plants, such as peas) and they revealed the principles of genetics. The same basic mechanism underlies all inheritance, but in most plants and animals there are so many genetic differences interacting to produce the external appearance (phenotype) that such simple, clear-cut patterns of inheritance are not evident.

monomer chemical compound composed of simple molecules from which ◊polymers can be made. Under certain conditions the simple molecules (of the monomer) join together (polymerize) to form a very long chain molecule (macromolecule) called a polymer. For example, the polymerization of ethene (ethylene) monomers produces the polymer polyethene (polyethylene).

$$2nCH_2=CH_2 \rightarrow (CH_2-CH_2-CH_2-CH_2)^n$$

monosaccharide or *simple sugar* ◊carbohydrate that cannot be hydrolyzed (split) into smaller carbohydrate units. Examples are glucose and fructose, both of which have the molecular formula $C_6H_{12}O_6$.

monosodium glutamate (MSG) $NaC_5H_8NO_4$ a white, crystalline powder, the sodium salt of glutamic acid (an ◊amino acid found in proteins that plays a role in the metabolism of plants and animals). It is used to enhance the flavour of many packaged and 'fast foods', and in Chinese cooking. Ill effects may arise from its overconsumption, and some people are very sensitive to it, even in small amounts. It is commercially derived from vegetable protein.

monotreme any member of the order Monotremata, the only living egg-laying mammals, found in Australasia. They include the echidnas and the platypus.

moon in astronomy, any natural ◊satellite that orbits a planet. Mercury and Venus are the only planets in the Solar System that do not have moons.

Moon natural satellite of Earth, 3,476 km/ 2,160 mi in diameter, with a mass 0.012 (approximately one-eightieth) that of Earth.

Its surface gravity is only 0.16 (one-sixth) that of Earth. Its average distance from Earth is 384,400 km/238,855 mi, and it orbits in a west-to-east direction every 27.32 days (the *sidereal month*). It spins on its axis with one side permanently turned towards Earth. The Moon has no atmosphere or water.

Moon probe crewless spacecraft used to investigate the Moon. Early probes flew past the Moon or crash-landed on it, but later ones achieved soft landings or went into orbit. Soviet probes included the Luna/Lunik series. US probes (Ranger, Surveyor, Lunar Orbiter) prepared the way for the Apollo crewed flights.

moonstone translucent, pearly variety of potassium sodium ◊feldspar, found in Sri Lanka and Myanmar, and distinguished by a blue, silvery, or red opalescent tint. It is valued as a gem.

morphine narcotic alkaloid $C_{17}H_{19}NO_3$ derived from ◊opium and prescribed only to alleviate severe pain. Its use produces serious side effects, including nausea, constipation, tolerance, and addiction, but it is highly valued for the relief of the terminally ill.

morphing the metamorphosis of one shape or object into another by computer-generated animation. First used in filmmaking 1990, it has transformed cinema special effects. Conventional animation is limited to two dimensions; morphing enables the creation of three-dimensional transformations.

morphogen in medicine, one of a class of substances believed to be present in the growing embryo, controlling its growth pattern.

It is thought that variations in the concentration of morphogens in different parts of the embryo cause them to grow at different rates.

morphology in biology, the study of the physical structure and form of organisms, in particular their soft tissues.

Morse code international code for transmitting messages by wire or radio using signals of short (dots) and long (dashes) duration, originated by US inventor Samuel Morse for use on his invention, the telegraph (see ◊telegraphy).

moss small nonflowering plant of the class Musci (10,000 species), forming with the ◊liverworts and the ◊hornwort the order Bryophyta. The stem of each plant bears ◊rhizoids that anchor it; there are no true roots. Leaves spirally arranged on its lower portion have sexual organs at their tips. Most mosses flourish best in damp conditions where other vegetation is thin. The peat or bog moss *Sphagnum* was formerly used for surgical dressings.

Mössbauer effect the recoil-free emission of gamma rays from atomic nuclei under certain conditions. The effect was discovered 1958 by German physicist Rudolf Mössbauer, and used 1960 to provide the first laboratory test of Einstein's general theory of relativity.

The absorption and subsequent re-emission of a gamma ray by an atomic nucleus usually causes it to recoil, so affecting the wavelength of the emitted ray. Mössbauer found that at low temperatures, crystals will absorb gamma rays of a specific wavelength and resonate so that the crystal as a whole recoils while individual nuclei do not. The wavelength of the re-emitted gamma rays is therefore virtually unaltered by recoil and may be measured to a high degree of accuracy. Changes in the wavelength may therefore be studied as evidence of the effect of, say, neighbouring electrons or gravity. For example, the effect provided the first verification of the general theory of relativity by showing that gamma-ray wavelengths become longer in a gravitational field, as predicted by Einstein.

motherboard ◊printed circuit board that contains the main components of a microcomputer. The power, memory capacity, and capability of the microcomputer may be enhanced by adding expansion boards to the motherboard.

motility the ability to move spontaneously. The term is often restricted to those cells that are capable of independent locomotion, such as spermatozoa. Many single-celled organisms are motile, for example, the amoeba. Research has shown that cells capable of movement, including vertebrate muscle cells, have certain biochemical features in common. Filaments of the proteins actin and myosin are associated with motility, as are the metabolic processes needed for breaking down the energy-rich compound ◊ATP (adenosine triphosphate).

motor anything that produces or imparts motion; a machine that provides mechanical power – for example, an ◊electric motor. Machines that burn fuel (petrol, diesel) are usually called engines, but the internal-combustion engine that propels vehicles has long been called a motor, hence 'motoring' and 'motorcar'. Actually the motor is a part of the car engine.

motor effect tendency of a wire carrying an electric current in a magnetic field to move. The direction of the movement is given by the left-hand rule (see ◊Fleming's rules). This effect is used in the ◊electric motor. It also explains why streams of electrons produced, for instance, in a television tube can be directed by electromagnets.

motor nerve in anatomy, any nerve that transmits impulses from the central nervous system to muscles or organs. Motor nerves cause voluntary and involuntary muscle contractions, and stimulate glands to secrete hormones.

mould mainly saprophytic fungi (see ◊fungus) living on foodstuffs and other organic matter, a few being parasitic on plants, animals, or each other. Many moulds are of medical or industrial importance; for example, penicillin.

moulding use of a pattern, hollow form, or matrix to give a specific shape to something in a plastic or molten state. It is commonly used for shaping plastics, clays, and glass. When metals are used, the process is called ◊casting.

moulting periodic shedding of the hair or fur of mammals, feathers of birds, or skin of reptiles. In mammals and birds, moulting is usually seasonal and is triggered by changes of day length.

mouse in computing, an input device used to control a pointer on a computer screen. It is a feature of ◊graphical user interface (GUI) systems. The mouse is about the size of a pack of playing cards, is connected to the computer by a wire, and incorporates one or more buttons that can be pressed. Moving the mouse across a flat surface causes a corresponding movement of the pointer. In this way, the operator can manipulate objects on the screen and make menu selections.

mouth cavity forming the entrance to the digestive tract. In land vertebrates, air from the nostrils enters the mouth cavity to pass down the trachea. The mouth in mammals is enclosed by the jaws, cheeks, and palate.

moving-coil meter instrument used to detect and measure electrical current. A coil of wire pivoted between the poles of a permanent magnet is turned by the motor effect of an electric current (by which a force acts on a wire carrying a current in a magnetic field). The extent to which the coil turns can then be related to the magnitude of the current.

mp in chemistry, abbreviation for *melting point*.

MPEG (abbreviation for *Moving Picture Experts Group*) group of engineers, formed 1988, representing broadcast organizations from 18 countries. It is responsible for setting standards within broadcasting. The *MPEG-1* is the standard for the digital coding of video pictures for CD recording. *MPEG-2*, set 1993, is a common standard for digital ◊television. In September 1993 work began on *MPEG-4* to devise a standard for mobile television systems and future videophones.

MRI abbreviation for ◊magnetic resonance imaging.

MS-DOS (abbreviation for *Microsoft Disc Operating System*) computer ◊operating system produced by Microsoft Corporation, widely used on ◊microcomputers with Intel X 86 family microprocessors. A version called PC-DOS is sold by IBM specifically for its personal computers. MS-DOS and PC-DOS are usually referred to as DOS. MS-DOS first appeared 1981, and was similar to an earlier system from Digital Research called CP/M.

mucous membrane thin skin lining all animal body cavities and canals that come into contact with the air (for example, eyelids, breathing and digestive passages, genital tract). It secretes mucus, a moistening, lubricating, and protective fluid.

mucus lubricating and protective fluid, secreted by mucous membranes in many different parts of the body. In the gut, mucus smooths the passage of food and keeps potentially damaging digestive enzymes away from the gut lining. In the lungs, it traps airborne particles so that they can be expelled.

mudstone fine-grained sedimentary rock made up of clay-to silt-sized particles (up to 0.0625 mm/0.0025 in).

multimedia computer system that combines audio and video components to create an interactive application that uses text, sound, and graphics (still, animated, and video sequences). For example, a multimedia database of musical instruments may allow a user not only to search and retrieve text about a particular instrument but also to see pictures of it and hear it play a piece of music.

multiple birth in humans, the production of more than two babies from one pregnancy.

Multiple births can be caused by more than two eggs being produced and fertilized (often as the result of hormone therapy to assist pregnancy), or by a single fertilized egg dividing more than once before implantation. See also ◊twin.

multiple proportions, law of in chemistry, the principle that if two elements combine with each other to form more than one compound, then the ratio of the masses of one of them that combine with a particular mass of the other is a small whole number.

multiplexer in telecommunications, a device that allows a transmission medium to carry a number of separate signals at the same time – enabling, for example, several telephone conversations to be carried by one telephone line, and radio signals to be transmitted in stereo.

multistage rocket rocket launch vehicle made up of several rocket stages (often three) joined end to end. The bottom stage fires first, boosting the vehicle to high speed, then it falls away. The next stage fires, thrusting the now lighter vehicle even faster. The remaining stages fire and fall away in turn, boosting the vehicle's payload (cargo) to an orbital speed that can reach 28,000 kph/17,500 mph.

multitasking or *multiprogramming* in computing, a system in which one processor appears to run several different programs (or different parts of the same program) at the same time. All the programs are held in memory together and each is allowed to run for a certain period.

multiuser system or *multiaccess system* in computing, an operating system that enables several users to access the same computer at the same time. Each user has a terminal, which may be local (connected directly to the computer) or remote (connected to the computer via a modem and a telephone line). Multiaccess is usually achieved by *time-sharing*: the computer switches very rapidly between terminals and programs so that each user has sole use of the computer for only a fraction of a second but can work as if she or he had continuous access.

muon an ◊elementary particle similar to the electron except for its mass which is 207 times greater than that of the electron. It has a half-life of 2 millionths of a second, decaying into electrons and ◊neutrinos. The muon was originally thought to be a ◊meson and is thus sometimes called a mu meson, although current opinion is that it is a ◊lepton.

muscle contractile animal tissue that produces locomotion and power, and maintains the movement of body substances. Muscle is

muscle *Muscles make up 35–45% of body weight; there are over 650 skeletal muscles. Muscle cells may be up to 20 cm/0.8 in long; they are arranged in bundles, fibres, myofibrils, and myofilaments.*

made of long cells that can contract to between one-half and one-third of their relaxed length.

Striped (or striated) muscles are activated by ◊motor nerves under voluntary control; their ends are usually attached via tendons to bones. *Involuntary* or *smooth* muscles are controlled by motor nerves of the ◊autonomic nervous system, and are located in the gut, blood vessels, iris, and various ducts. *Cardiac* muscle occurs only in the heart, and is also controlled by the autonomic nervous system.

muscovite white mica, $KAl_2Si_3AlO_{10}(OH,F)_2$, a common silicate mineral. It is colourless to silvery white with shiny surfaces, and like all micas it splits into thin flakes along its one perfect cleavage. Muscovite is a metamorphic mineral occurring mainly in schists; it is also found in some granites, and appears as shiny flakes on bedding planes of some sandstones.

mushroom fruiting body of certain fungi, consisting of an upright stem and a spore-producing cap with radiating gills on the undersurface. There are many edible species belonging to the genus *Agaricus*. See also ◊fungus and ◊toadstool.

mutagen any substance that increases the rate of gene ◊mutation. A mutagen may also act as a ◊carcinogen.

mutation in biology, a change in the genes produced by a change in the ◊DNA that makes up the hereditary material of all living organisms. Mutations, the raw material of evolution, result from mistakes during replication (copying) of DNA molecules. Only a few improve the organism's performance and are therefore favoured by ◊natural selection. Mutation rates are increased by certain chemicals and by radiation.

Common mutations include the omission or insertion of a base (one of the chemical subunits of DNA); these are known as *point mutations*. Larger-scale mutations include removal of a whole segment of DNA or its inversion within the DNA strand. Not all mutations affect the organism, because there is a certain amount of redundancy in the genetic information. If a mutation is 'translated' from DNA into the protein that makes up the organism's structure, it may be in a nonfunctional part of the protein and thus have no detectable effect. This is known as a *neutral mutation*, and is of importance in ◊molecular clock studies because such mutations tend to accumulate gradually as time passes. Some mutations do affect genes that control protein production or functional parts of protein, and most of these are lethal to the organism.

mutual induction in physics, the production of an electromotive force (emf) or voltage in an electric circuit caused by a changing ◊magnetic flux in a neighbouring circuit. The two circuits are often coils of wire, as in a ◊transformer, and the size of the induced emf depends largely on the numbers of turns of wire in each of the coils.

mutualism or ◊*symbiosis* an association between two organisms of different species whereby both profit from the relationship.

mycelium interwoven mass of threadlike filaments or ◊hyphae, forming the main body of most fungi. The reproductive structures, or 'fruiting bodies', grow from the mycelium.

mycorrhiza mutually beneficial (mutualistic) association occurring between plant roots and a soil fungus. Mycorrhizal roots take up nutrients more efficiently than nonmycorrhizal roots, and the fungus benefits by obtaining carbohydrates from the tree.

myelin sheath insulating layer that surrounds nerve cells in vertebrate animals. It serves to speed up the passage of nerve impulses.

Myelin is made up of fats and proteins and is formed from up to a hundred layers, laid down by special cells, the *Schwann cells*.

myoglobin globular protein, closely related to ◊haemoglobin and located in vertebrate muscle. Oxygen binds to myoglobin and is released only when the haemoglobin can no longer supply adequate oxygen to muscle cells.

myopia or *short-sightedness* defect of the eye in which a person can see clearly only those objects that are close up. It is caused either by the eyeball being too long or by the cornea and lens system of the eye being too powerful, both of which cause the images of distant objects to be formed in front of the retina instead of on it. Nearby objects are sharply perceived. Myopia can be corrected by suitable glasses or contact lenses.

myrmecophyte plant that lives in association with a colony of ants and possesses specialized organs in which the ants live. For example, *Myrmecodia*, an epiphytic plant from Malaysia, develops root tubers containing a network of cavities inhabited by ants.

n in mathematics, variable used to denote an indefinite number.

nadir the point on the celestial sphere vertically below the observer and hence diametrically opposite the *zenith*. The term is used metaphorically to mean the low point of a person's fortunes.

nail in biology, a hard, flat, flexible outgrowth of the digits of primates (humans, monkeys, and apes). Nails are composed of ◊keratin.

NAND gate type of ◊logic gate.

nano- prefix used in ◊SI units of measurement, equivalent to a one-billionth part (10^{-9}). For example, a nanosecond is one-billionth of a second.

nanotechnology the building of devices on a molecular scale. Micromachines, such as gears smaller in diameter than a human hair, have been made at the AT&T Bell laboratories in New Jersey, USA. Building large molecules with useful shapes has been accomplished by research groups in the USA. A robot small enough to travel through the bloodstream and into organs of the body, inspecting or removing diseased tissue, was under development in Japan 1990.

naphtha the mixtures of hydrocarbons obtained by destructive distillation of petroleum, coal tar, and shale oil. It is raw material for the petrochemical and plastics industries. The term was originally applied to naturally occurring liquid hydrocarbons.

naphthalene $C_{10}H_8$ solid, white, shiny, aromatic hydrocarbon obtained from coal tar. The smell of moth-balls is due to their napthalene content. It is used in making indigo and certain azo dyes, as a mild disinfectant, and as an insecticide.

narcotic pain-relieving and sleep-inducing drug. The term is usually applied to heroin, morphine, and other opium derivatives, but may also be used for other drugs which depress brain activity, including anaesthetic agents and hypnotics.

NASA (acronym for *National Aeronautics and Space Administration*) US government agency, founded 1958, for spaceflight and aeronautical research. Its headquarters are in Washington DC and its main installation is at the Kennedy Space Center in Florida. NASA's early planetary and lunar programs included Pioneer spacecraft from 1958, which gathered data for the later crewed missions, the most famous of which took the first people to the Moon in *Apollo 11* on 16–24 July 1969.

nastic movement plant movement that is caused by an external stimulus, such as light or temperature, but is directionally independent of its source, unlike ◊tropisms. Nastic movements occur as a result of changes in water pressure within specialized cells or differing rates of growth in parts of the plant.

Examples include the opening and closing of crocus flowers following an increase or decrease in temperature (*thermonasty*), and the opening and closing of evening-primrose *Oenothera* flowers on exposure to dark and light (*photonasty*).

national grid the network of cables, carried overhead on pylons or buried under the ground, that connects consumers of electrical power to power stations, and interconnects the power stations. It ensures that power can be made available to all customers at any time, allowing demand to be shared by several power stations, and particular power stations to be shut down for maintenance work from time to time. Britain has the world's largest grid system, with over 140 power stations able to supply up to 55,000 megawatts.

native metal or *free metal* any of the metallic elements that occur in nature in the chemically uncombined or elemental form (in addition to any combined form). They include bismuth, cobalt, copper, gold, iridium, iron, lead, mercury, nickel, osmium, palladium, platinum, ruthenium, rhodium, tin, and silver. Some are commonly found in the free state, such as gold; others occur almost exclusively in the combined state, but under unusual conditions do occur as native metals, such as mercury. Examples of native nonmetals are carbon and sulphur.

natural frequency the frequency at which a mechanical system will vibrate freely. A pendulum, for example, always oscillates at the same frequency when set in motion. More complicated systems, such as bridges, also vibrate with a fixed natural frequency. If a varying force with a frequency equal to the natural frequency is applied to such an object the vibrations can become violent, a phenomenon known as ◊resonance.

natural gas mixture of flammable gases found in the Earth's crust (often in association with petroleum). It is one of the world's three main fossil fuels (with coal and oil). Natural gas is a mixture of ◊hydrocarbons, chiefly methane, with ethane, butane, and propane. Natural gas is usually transported from its source by pipeline, although it may be liquefied for transport and storage and is, therefore, often used in remote areas where other fuels are scarce and expensive. Prior to transportation, butane and propane are removed and liquefied to form 'bottled gas'.

natural logarithm in mathematics, the ◊exponent of a number expressed to base e, where e represents the ◊irrational number 2.71828... .

Natural ◊logarithms are also called Napierian logarithms, after their inventor, the Scottish mathematician John Napier.

natural radioactivity radioactivity generated by those radioactive elements that exist in the Earth's crust. All the elements from polonium (atomic number 84) to uranium (atomic number 92) are radioactive. ◊Radioisotopes of some lighter elements are also found in nature (for example potassium-40). See ◊background radiation.

natural selection the process whereby gene frequencies in a population change through certain individuals producing more descendants than others because they are better able to survive and reproduce in their environment.

The accumulated effect of natural selection is to produce ◊adaptations such as the insulating coat of a polar bear or the spadelike forelimbs of a mole. The process is slow, relying firstly on random variation in the genes of an organism being produced by ◊mutation and secondly on the genetic ◊recombination of sexual reproduction. It was recognized by Charles Darwin and English naturalist Alfred Russel Wallace as the main process driving ◊evolution.

nature the living world, including plants, animals, fungi, and all microorganisms, and naturally formed features of the landscape, such as mountains and rivers.

nature–nurture controversy or *environment–heredity controversy* long-standing dispute among philosophers and psychologists over the relative importance of environment, that is upbringing, experience and learning ('nurture'), and heredity, that is genetic inheritance ('nature'), in determining the make-up of an organism, as related to human personality and intelligence.

nautical mile unit of distance used in navigation, an internationally agreed-on standard (since 1959) equalling the average length of one minute of arc on a great circle of the Earth, or 1,852 m/6,076.12 ft. The term formerly applied to various units of distance used in navigation. In the UK the nautical mile was formerly defined as 6,082 ft.

navel or *umbilicus* small indentation in the centre of the abdomen of mammals, marking the site of attachment of the ◊umbilical cord, which connects the fetus to the ◊placenta.

navigation the science and technology of finding the position, course, and distance travelled by a ship, plane, or other craft. Traditional methods include the magnetic ◊compass and sextant. Today the gyrocompass is usually used, together with highly sophisticated electronic methods, employing beacons of radio signals, such as ◊Decca and ◊Omega. Satellite navigation uses satellites that broadcast time and position signals.

An Australian prototype for an electronic navigation system ECDIS (Electronic Chart Display Information System) is a single computer-based apparatus that combines information from existing navigational aids, such as charts, radar, sonar and satellites. It is expected to be in use 1995.

navigation, biological the ability of animals or insects to navigate. Although many animals navigate by following established routes or known landmarks, many animals can navigate without such aids; for example, birds can fly several thousand miles back to their nest site, over unknown terrain.

Such feats may be based on compass information derived from the position of the Sun, Moon, or stars, or on the characteristic patterns of Earth's magnetic field.

NBS (abbreviation for *National Bureau of Standards*) the US federal standards organization, on whose technical standards all US weights and measures are based.

nebula cloud of gas and dust in space. Nebulae are the birthplaces of stars, but some nebulae are produced by gas thrown off from dying stars (see ◊planetary nebula; ◊supernova). Nebulae are classified depending on whether they emit, reflect, or absorb light.

neck structure between the head and the trunk in animals. In the back of the neck are the upper seven vertebrae, and there are many powerful muscles that support and move the head. In front, the neck region contains the pharynx and trachea, and behind these the oesophagus. The large arteries (carotid, temporal, maxillary) and veins (jugular) that supply the brain and head are also located in the neck.

nectar sugary liquid secreted by some plants from a nectary, a specialized gland usually situated near the base of the flower. Nectar often accumulates in special pouches or spurs, not always in the same location as the nectary. Nectar attracts insects, birds, bats, and other animals to the flower for ◊pollination and is the raw material used by bees in the production of honey.

Néel temperature the temperature at which the ◊susceptibility of an antiferromagnetic material has a maximum value; see also ◊magnetism.

nematode unsegmented worm of the phylum Nematoda. Nematodes are pointed at both ends, with a tough, smooth outer skin. They include many free-living species found in soil and water, including the sea, but a large number are parasites, such as the roundworms and pinworms that live in humans, or the eelworms that attack plant roots. They differ from ◊flatworms in that they have two openings to the gut (a mouth and an anus).

neo-Darwinism modern theory of ◊evolution, built up since the 1930s by integrating the 19th-century English scientist Charles Darwin's (1809–1822) theory of evolution through natural selection with the theory of genetic inheritance founded on the work of the Austrian biologist Gregor Mendel (1822–1884).

Neo-Darwinism asserts that evolution takes place because the environment is slowly changing, exerting a selection pressure on the individuals within a population. Those with characteristics that happen to adapt to the new environment are more likely to survive and have offspring and hence pass on these favourable characteristics. Over time the genetic make-up of the population changes and ultimately a new species is formed.

neodymium yellowish metallic element of the ◊lanthanide series, symbol Nd, atomic number 60, relative atomic mass 144.24. Its rose-coloured salts are used in colouring glass, and neodymium is used in lasers.

neon colourless, odourless, nonmetallic, gaseous element, symbol Ne, atomic number 10, relative atomic mass 20.183. It is grouped with the ◊inert gases, is nonreactive, and forms no compounds. It occurs in small quantities in the Earth's atmosphere.

neoprene synthetic rubber, developed in the USA 1931 from the polymerization of chloroprene. It is much more resistant to heat, light, oxidation, and petroleum than is ordinary rubber.

neoteny in biology, the retention of some juvenile characteristics in an animal that seems otherwise mature. An example is provided by the axolotl, a salamander that can reproduce sexually although still in its larval form.

neper unit used in telecommunications to express a ratio of powers and currents. It gives the attenuation of amplitudes as the natural logarithm of the ratio.

nephron microscopic unit in vertebrate kidneys that forms *urine*. A human kidney is composed of over a million nephrons. Each nephron consists of a knot of blood capillaries called a glomerulus, contained in the ◊Bowman's capsule, and a long narrow tubule enmeshed with yet more capillaries. Waste materials and water pass from the bloodstream into the tubule, and essential minerals and some water are reabsorbed from the tubule back into the bloodstream. The remaining filtrate (urine) is passed out from the body.

Neptune in astronomy, the eighth planet in average distance from the Sun. Neptune orbits the Sun every 164.8 years at an average distance of 4.497 billion km/2.794 billion mi. It is a giant gas (hydrogen, helium, methane) planet, with a diameter of 48,600 km/30,200 mi and a mass 17.2 times that of Earth. Its rotation period is 16 hours 7 minutes. The methane in its atmosphere absorbs red light and gives the planet a blue colouring. It is believed to have a central rocky core covered by a layer of ice. Neptune has eight known moons.

neptunium silvery, radioactive metallic element of the ◊actinide series, symbol Np, atomic number 93, relative atomic mass 237.048. It occurs in nature in minute amounts in ◊pitchblende and other uranium ores, where it is produced from the decay of neutron-

nephron *The kidney (left) contains more than a million filtering units, or nephrons (right), consisting of the glomerulus, Bowman's capsule, and the loop of Henle. Blood flows through the glomerulus – a tight knot of fine blood vessels from which water and metabolic wastes filter into the tubule. This filtrate flows through the convoluted tubule and loop of Henle where most of the water and useful molecules are reabsorbed into the blood capillaries. The waste materials are passed to the collecting tubule as urine.*

bombarded uranium in these ores. The longest lived isotope, Np-237, has a half-life of 2.2 million years. The element can be produced by bombardment of U-238 with neutrons and is chemically highly reactive.

nerve bundle of nerve cells enclosed in a sheath of connective tissue and transmitting nerve impulses to and from the brain and spinal cord. A single nerve may contain both ◊motor and sensory nerve cells, but they function independently.

nerve cell or *neuron* elongated cell, the basic functional unit of the ◊nervous system that transmits information rapidly between different parts of the body. Each nerve cell has a cell body, containing the nucleus, from which trail processes called dendrites, responsible for receiving incoming signals. The unit of information is the ***nerve impulse***, a travelling wave of chemical and electrical changes involving the membrane of the nerve cell. The cell's longest process, the ◊axon, carries impulses away from the cell body.

The impulse involves the passage of sodium and potassium ions across the nerve-cell membrane. Sequential changes in the permeability of the membrane to positive sodium (Na^+) ions and potassium (K^+) ions produce electrical signals called action potentials. Impulses are received by the cell body and passed, as a pulse of electric charge, along the axon. The axon terminates at the ◊synapse, a specialized area closely linked to the next cell (which may be another nerve cell or a specialized effector cell such as a muscle). On reaching the synapse, the impulse releases a chemical ◊neurotransmitter, which diffuses across to the neighbouring cell and there stimulates another impulse or the action of the effector cell.

nervous system the system of interconnected ◊nerve cells of most invertebrates and all vertebrates. It is composed of the ◊central and ◊autonomic nervous systems. It may be as simple as the nerve net of coelenterates (for example, jellyfishes) or as complex as the mammalian nervous system, with a central nervous system comprising brain and spinal cord, and a peripheral nervous system connecting up with sensory organs, muscles, and glands.

network in computing, a method of connecting computers so that they can share data and ◊peripheral devices, such as printers. The main types are classified by the pattern of the connections – star or ring network, for example – or by the degree of geographical spread allowed; for example, *local area networks* (LANs) for communication within a room or building, and *wide area networks* (WANs) for more remote systems.

neural network artificial network of processors that attempts to mimic the structure of nerve cells (neurons) in the human brain. Neural networks may be electronic, optical, or simulated by computer software.

A basic network has three layers of processors: an input layer, an output layer, and a 'hidden' layer in between. Each processor is connected to every other in the network by a system of 'synapses'; every processor in the top layer connects to every one in the hidden layer, and each of these connects to every processor in the output layer. This means that each nerve cell in the middle and bottom layers receives input from several different sources; only when the amount of input exceeds a critical level does the cell fire an output signal. The chief characteristic of neural networks is their ability to sum up large amounts of imprecise data and decide whether they match a pattern.

neural tube embryological structure from that the brain and spinal cord are derived. It forms when two edges of a groove in a plate of primitive nerve tissue fuse. Incomplete fusion results in neural tube defects, such as spina bifida.

neurohormone chemical secreted by nerve cells and carried by the blood to target cells. The function of the neurohormone is to act as a messenger; for example, the neurohormone ADH (antidiuretic hormone) is secreted by the pituitary gland and carried to the kidney, where it promotes water reabsorption in the kidney tubules.

neuron another name for a ◊nerve cell.

neurotransmitter chemical that diffuses across a ◊synapse, and thus transmits impulses between ◊nerve cells, or between nerve cells and effector organs (for example, muscles). Common neurotransmitters are noradrenaline (which also acts as a hormone) and acetylcholine, the latter being most frequent at junctions between nerve and muscle.

neutralization in chemistry, a process occurring when the excess acid (or excess base) in a substance is reacted with added base (or added acid) so that the resulting substance is neither acidic nor basic.

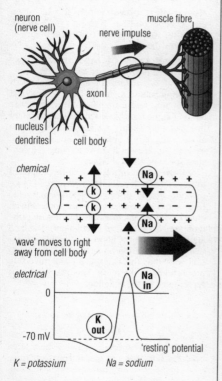

nerve cell or neuron *The anatomy and action of a nerve cell. The nerve cell or neuron consists of a cell body with the nucleus and projections called dendrites which pick up messages. An extension of the cell, the axon, connects one cell to the dendrites of the next. When a nerve cell is stimulated, waves of sodium (Na+) and potassium (K+) ions carry an electrical impulse down the axon.*

In theory neutralization involves adding acid or base as required to achieve ◊pH7. When the colour of an ◊indicator is used to test for neutralization, the final pH may differ from pH7 depending upon the indicator used.

neutral oxide oxide that has neither acidic nor basic properties (see ◊oxide). Neutral oxides are only formed by ◊nonmetals. Examples are carbon monoxide, water, and nitrogen(I) oxide.

neutral solution solution of pH 7, in which the concentrations of $H^+_{(aq)}$ and $OH^-_{(aq)}$ ions are equal.

neutrino in physics, any of three uncharged ◊elementary particles (and their antiparticles) of the ◊lepton class, having a mass too close to zero to be measured. The most familiar type, the antiparticle of the electron neutrino, is emitted in the beta decay of a nucleus. The other two are the muon and tau neutrinos.

neutron one of the three main subatomic particles, the others being the proton and the electron. The neutron is a composite particle, being made up of three ◊quarks, and therefore belongs to the ◊baryon group of the ◊hadrons. Neutrons have about the same mass as protons but no electric charge, and occur in the nuclei of all atoms except hydrogen. They contribute to the mass of atoms but do not affect their chemistry.

neutron activation analysis chemical analysis used to determine the composition of a wide range of materials found in archaeological contexts. A specimen is bombarded with neutrons, which interact with nuclei in the sample to form radioactive isotopes that emit gamma rays as they decay. The energy spectrum of the emitted rays is detected with a counter and constituent elements and concentrations are identified by the characteristic energy spectrum of emitted rays and its intensity.

neutron beam machine nuclear reactor or accelerator producing a stream of neutrons, which can 'see' through metals. It is used in industry to check molecular changes in metal under stress.

neutron bomb small hydrogen bomb for battlefield use that kills by radiation without destroying buildings and other structures.

neutron star very small, 'superdense' star composed mostly of ◊neutrons. They are thought to form when massive stars explode as ◊supernovae, during which the protons and electrons of the star's atoms merge, owing to intense gravitational collapse, to make neutrons. A neutron star may have the mass of up to three Suns, compressed into a globe only 20 km/12 mi in diameter.

new technology collective term applied to technological advances made in such areas as ◊telecommunications, ◊nuclear energy, space ◊satellites and ◊computers.

newton SI unit (symbol N) of ◊force. One newton is the force needed to accelerate an object with mass of one kilogram by one metre per second per second. The weight of a medium size (100 g/3 oz) apple is one newton.

Newtonian physics ◊physics based on the concepts of Isaac Newton (1642–1727), before the formulation of ◊quantum theory or ◊relativity theory.

Newton's laws of motion three laws that form the basis of Newtonian mechanics.

(1) Unless acted upon by an external resultant, or unbalanced, force, an object at rest stays at rest, and a moving object continues moving at the same speed in the same straight line. Put more simply, the law says that, if left alone, stationary objects will not move and moving objects will keep on moving at a constant speed in a straight line.

(2) A resultant or unbalanced force applied to an object produces a rate of change of ◊momentum that is directly proportional to the force and is in the direction of the force. For an object of constant mass m, this law may be rephrased as: a resultant or unbalanced force F applied to an object gives it an acceleration a that is directly proportional to, and in the direction of, the force applied and inversely proportional to the mass. This relationship is represented by the equation:

$$a = F/m$$

which is usually rearranged in the form:

$$F = ma$$

(3) When an object A applies a force to an object B, B applies an equal and opposite force to A; that is, to every action there is an equal and opposite reaction.

Newton's rings in optics, an ◊interference phenomenon seen (using white light) as concentric rings of spectral colours where light passes through a thin film of transparent medium, such as the wedge of air between a large-radius convex lens and a flat glass plate. With monochromatic light (light of a single wavelength), the rings take the form of alternate light and dark bands. They are caused by

niacin one of the 'B group' vitamins see ◊nicotinic acid.

niche in ecology, the 'place' occupied by a species in its habitat, including all chemical, physical, and biological components, such as what it eats, the time of day at which the species feeds, temperature, moisture, the parts of the habitat that it uses (for example, trees or open grassland), the way it reproduces, and how it behaves.

Nichrome trade name for a series of alloys containing mainly nickel and chromium, with small amounts of other substances such as iron, magnesium, silicon, and carbon. Nichrome has a high melting point and is resistant to corrosion. It is therefore used in electrical heating elements and as a substitute for platinum in the ◊flame test.

nickel hard, malleable and ductile, silver-white metallic element, symbol Ni, atomic number 28, relative atomic mass 58.71. It occurs in igneous rocks and as a free metal (◊native metal), occasionally occurring in fragments of iron-nickel meteorites. It is a component of the Earth's core, which is held to consist principally of iron with some nickel. It has a high melting point, low electrical and thermal conductivity, and can be magnetized. It does not tarnish and therefore is much used for alloys, electroplating, and for coinage.

nickel ore any mineral ore from which nickel is obtained. The main minerals are arsenides such as chloanthite ($NiAs_2$), and the sulphides millerite (NiS) and pentlandite ($(Ni,Fe)_9S_8$), the commonest ore. The chief nickel-producing countries are Canada, Russia, Kazakhstan, Cuba, and Australia.

nicotine $C_{10}H_{14}N_2$ ◊alkaloid (nitrogenous compound) obtained from the dried leaves of the tobacco plant *Nicotiana tabacum* and used as an insecticide. A colourless oil, soluble in water, it turns brown on exposure to the air.

nicotinic acid or *niacin* water-soluble ◊vitamin ($C_5H_5N.COOH$) of the B complex, found in meat, fish, and cereals; it can also be formed in small amounts in the body from the essential ◊amino acid tryptophan. Absence of nicotinic acid from the diet leads to the disease ◊pellagra.

nielsbohrium name proposed by Soviet scientists for the element currently known as ◊unnilpentium (atomic number 105), to honour Danish physicist Niels Bohr.

niobium soft, grey-white, somewhat ductile and malleable, metallic element, symbol Nb, atomic number 41, relative atomic mass 92.906. It occurs in nature with tantalum, which it resembles in chemical properties. It is used in making stainless steel and other alloys for jet engines and rockets and for making superconductor magnets.

nitrate salt or ester of nitric acid, containing the NO_3^- ion. Nitrates are used in explosives, in the chemical and pharmaceutical industries, in curing meat (see ◊nitre), and as fertilizers. They are the most water-soluble salts known and play a major part in the nitrogen cycle. Nitrates in the soil, whether naturally occurring or from inorganic or organic fertilizers, can be used by plants to make proteins and nucleic acids. However, runoff from fields can result in nitrate pollution.

nitrate pollution the contamination of water by nitrates. Increased use of artifical fertilizers and land cultivation means that higher levels of nitrates are being washed from the soil into rivers, lakes, and aquifers. There they cause an excessive enrichment of the water (◊eutrophication), leading to a rapid growth of algae, which in turn darkens the water and reduces its oxygen content. The water is expensive to purify and many plants and animals die. High levels are now found in drinking water in arable areas. These may be harmful to newborn babies, and it is possible that they contribute to stomach cancer, although the evidence for this is unproven.

nitre or *saltpetre* potassium nitrate, KNO_3, a mineral found on and just under the ground in desert regions; used in explosives. Nitre occurs in Bihar, India, Iran, and Cape Province, South Africa. The salt was formerly used for the manufacture of gunpowder, but the supply of nitre for explosives is today largely met by making the salt from nitratine (also called Chile saltpetre, $NaNO_3$). Saltpetre is a ◊preservative and is widely used for curing meats.

nitric acid or *aqua fortis* HNO_3 fuming acid obtained by the oxidation of ammonia or the action of sulphuric acid on potassium nitrate. It is a highly corrosive acid, dissolving most metals, and a strong oxidizing agent. It is used in the nitration and esterification of organic substances, and in the making of sulphuric acid, nitrates, explosives, plastics, and dyes.

nitrification process that takes place in soil when bacteria oxidize ammonia, turning it into nitrates. Nitrates can be absorbed by the roots of plants, so this is a vital stage in the ◊nitrogen cycle.

nitrite salt or ester of nitrous acid, containing the nitrite ion (NO_2^-). Nitrites are used as preservatives (for example, to prevent the growth of botulism spores) and as colouring agents in cured meats such as bacon and sausages.

nitrocellulose alternative name for ◊cellulose nitrate.

nitrogen colourless, odourless, tasteless, gaseous, nonmetallic element, symbol N, atomic number 7, relative atomic mass 14.0067. It forms almost 80% of the Earth's atmosphere by volume and is a constituent of all plant and animal tissues (in proteins and nucleic acids). Nitrogen is obtained for industrial use by the liquefaction and fractional distillation of air. Its compounds are used in the manufacture of foods, drugs, fertilizers, dyes, and explosives.

Nitrogen is used in the Haber process to make ammonia, NH_3, and to provide an inert atmosphere for certain chemical reactions.

nitrogen cycle the process of nitrogen passing through the ecosystem. Nitrogen, in the form of inorganic compounds (such as nitrates) in the soil, is absorbed by plants and turned into organic compounds (such as proteins) in plant tissue. A proportion of this nitrogen is eaten by ◊herbivores, with some of this in turn being passed on to the carnivores, which feed on the herbivores. The nitrogen is ultimately returned to the soil as excrement and when organisms die and are converted back to inorganic form by ◊decomposers.

Although about 78% of the atmosphere is nitrogen, this cannot be used directly by most organisms. However, certain bacteria and cyanobacteria (see ◊blue-green algae) are capable of nitrogen fixation. Some nitrogen-fixing bacteria live mutually with leguminous plants (peas and beans) or other plants (for example, alder), where they form characteristic nodules on the roots. The presence of such plants increases the nitrate content, and hence the fertility, of the soil.

nitrogen fixation the process by which nitrogen in the atmosphere is converted into nitrogenous compounds by the action of microorganisms, such as cyanobacteria (see ◊blue-green algae) and bacteria, in conjunction with certain ◊legumes. Several chemical processes duplicate nitrogen fixation to produce fertilizers; see ◊nitrogen cycle.

nitrogen oxide any chemical compound that contains only nitrogen and oxygen. All

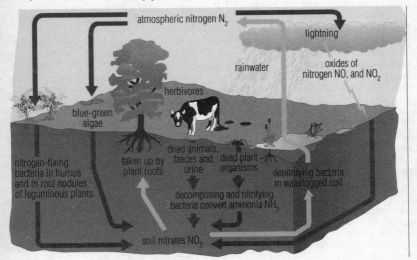

nitrogen cycle *The nitrogen cycle is one of a number of cycles during which the chemicals necessary for life are recycled. The carbon, sulphur, and phosphorus cycles are others. Since there is only a limited amount of these chemicals in the Earth and its atmosphere, the chemicals must be continuously recycled if life is to go on.*

nitrogen oxides are gases. Nitrogen monoxide and nitrogen dioxide contribute to air pollution. See also ◊nitrous oxide.

Nitrogen monoxide (NO), or nitric oxide, is a colourless gas released when metallic copper reacts with concentrated ◊nitric acid. It is also produced when nitrogen and oxygen combine at high temperature. On contact with air it is oxidized to nitrogen dioxide.

Nitrogen dioxide (nitrogen(IV) oxide, NO_2) is a brown, acidic, pungent gas that is harmful if inhaled and contributes to the formation of ◊acid rain, as it dissolves in water to form nitric acid. It is the most common of the nitrogen oxides and is obtained by heating most nitrate salts (for example ◊lead(II) nitrate, $Pb(NO_3)_2$). If liquefied, it gives a colourless solution (N_2O_4). It has been used in rocket fuels.

nitroglycerine $C_3H_5(ONO_2)_3$ flammable, explosive oil produced by the action of nitric and sulphuric acids on glycerol. Although poisonous, it is used in cardiac medicine. It explodes with great violence if heated in a confined space and is used in the preparation of dynamite, cordite, and other high explosives.

nitrous acid HNO_2 weak acid that, in solution with water, decomposes quickly to form nitric acid and nitrogen dioxide.

nitrous oxide or **dinitrogen oxide** N_2O colourless, nonflammable gas that, used in conjunction with oxygen, reduces sensitivity to pain. In higher doses it is an anaesthetic. Well tolerated, it is often combined with other anaesthetic gases to enble them to be used in lower doses. It may be self-administered; for example, in childbirth. It used to be known as 'laughing gas'.

nobelium synthesized, radioactive, metallic element of the ◊actinide series, symbol No, atomic number 102, relative atomic mass 259. It is synthesized by bombarding curium with carbon nuclei.

It was named in 1957 for the Nobel Institute in Stockholm, Sweden, where it was claimed to have been first synthesized. Later evaluations determined that this was in fact not so, as the successful 1958 synthesis at the University of California at Berkeley produced a different set of data. The name was not, however, challenged.

In 1992 the International Unions for Pure and Applied Chemistry and Physics (IUPAC and IUPAP) gave credit to Russian scientists in Dubna for the discovery of nobelium.

Nobel prize annual international prize, first awarded 1901 under the will of Alfred Nobel, Swedish chemist, who invented dynamite. The interest on the Nobel endowment fund is divided annually among the persons who have made the greatest contributions in the fields of physics, chemistry, medicine, literature, and world peace.

noble gas alternative name for ◊inert gas.

noble gas structure the configuration of electrons in noble or ◊inert gases (helium, neon, argon, krypton, xenon, and radon).

node in physics, a position in a ◊standing wave pattern at which there is no vibration. Points at which there is maximum vibration are called ***antinodes***. Stretched strings, for example, can show nodes when they vibrate. Guitarists can produce special effects (harmonics) by touching a sounding string lightly to produce a node.

nodule in geology, a lump of mineral or other matter found within rocks or formed on the seabed surface; mining technology is being developed to exploit them.

nonmetal one of a set of elements (around 20 in total) with certain physical and chemical properties opposite to those of metals. Nonmetals accept electrons (see ◊electronegativity) and are sometimes called electronegative elements.

nonrenewable resource natural resource, such as coal or oil, that takes thousands or millions of years to form naturally and can therefore not be replaced once it is consumed. The main energy sources used by humans are nonrenewable; ◊renewable resources, such as solar, tidal, and geothermal power, have so far been less exploited.

nonvolatile memory in computing, ◊memory that does not lose its contents when the power supply to the computer is disconnected.

noradrenaline in the body, a ◊catecholamine that acts directly on specific receptors to stimulate the sympathetic nervous system. Released by nerve stimulation or by drugs, it slows the heart rate mainly by constricting arterioles (small arteries) and so raising blood pressure. It is used therapeutically to treat shock.

NOR gate in electronics, a type of ◊logic gate.

normal distribution in statistics, a symmetrical bell-shaped distribution whose mean, mode, and median all coincide; its equation may be expressed in terms of the mean (μ) and

⟩standard deviation (σ). Because of this, the area under the normal distribution curve has been tabulated in terms of the number (z) of standard deviations away from the mean. The probability of any score falling between two values may therefore be determined by calculating the area under the curve between those two values. For a normal distribution approximately two-thirds of the data falls within one standard deviation on either side of the mean; approximately 95% of the data falls within two standard deviations of the mean.

Norplant long-lasting hormonal contraceptive available in many countries 1983 (UK 1993), consisting of six silicon-coated rods inserted beneath the skin in a woman's arm. The rods release the progestogen levonorgestrel in controlled doses preventing pregnancy for five years. It has a failure rate of 0.2 pregnancies per 100 women. The main side effects are irregular and breakthrough bleeding.

In some countries Norplant has been used controversially as a means of limiting the fertility of certain groups, notably black teenagers in the USA. In 1991 a Kansas state legislator proposed paying $500 to any woman on social welfare who would use Norplant.

northern lights common name for the ⟩aurora borealis.

nose in humans, the upper entrance of the respiratory tract; the organ of the sense of smell. The external part is divided down the middle by a septum of ⟩cartilage. The nostrils contain plates of cartilage that can be moved by muscles and have a growth of stiff hairs at the margin to prevent foreign objects from entering. The whole nasal cavity is lined with a ⟩mucous membrane that warms and moistens the air as it enters and ejects dirt. In the upper parts of the cavity the membrane contains 50 million olfactory receptor cells (cells sensitive to smell).

nostril in vertebrates, the opening of the nasal cavity, in which cells sensitive to smell are located. (In fish, these cells detect water-borne chemicals, so they are effectively organs of taste.) In vertebrates with lungs, the nostrils also take in air. In humans, and most other mammals, the nostrils are located on the ⟩nose.

notebook computer small ⟩laptop computer. Notebook computers became available in the early 1990s and, even complete with screen and hard-disc drive, are no larger than a standard A4 notebook.

NOT gate or *inverter gate* in electronics, a type of ⟩logic gate.

detail of olfactory epithelium

nose *The structure of the nose. The organs of smell are confined to a small area in the roof of the nasal cavity. The olfactory cells are stimulated when certain molecules reach them. Smell is one of our most subtle senses: tens of thousands of smells can be distinguished. By comparison, taste, although closely related to smell, is a crude sensation. All the subtleties of taste depend upon smell.*

notochord the stiff but flexible rod that lies between the gut and the nerve cord of all embryonic and larval chordates, including the vertebrates. It forms the supporting structure of the adult lancelet, but in vertebrates it is replaced by the vertebral column, or spine.

nova (plural *novae*) faint star that suddenly erupts in brightness by 10,000 times or more. Novae are believed to occur in close ⟩binary star systems, where gas from one star flows to a companion ⟩white dwarf. The gas ignites and is thrown off in an explosion at speeds of 1,500 kps/930 mps or more. Unlike a ⟩supernova, the star is not completely disrupted by the outburst.

nuclear energy: chronology

1896	French physicist Henri Becquerel discovered radioactivity.
1905	In Switzerland, Albert Einstein showed that mass can be converted into energy.
1911	New Zealand physicist Ernest Rutherford proposed the nuclear model of the atom.
1919	Rutherford split the atom, by bombarding a nitrogen nucleus with alpha particles.
1939	Otto Hahn, Fritz Strassman, and Lise Meitner announced the discovery of nuclear fission.
1942	Enrico Fermi built the first nuclear reactor, in a squash court at the University of Chicago, USA.
1946	The first fast reactor, called Clementine, was built at Los Alamos, New Mexico.
1951	The Experimental Breeder Reactor, Idaho, USA, produced the first electricity to be generated by nuclear energy.
1954	The first reactor for generating electricity was built in the USSR, at Obninsk.
1956	The world's first commercial nuclear power station, Calder Hall, came into operation in the UK.
1957	The release of radiation from Windscale (now Sellafield) nuclear power station, Cumbria, England, caused 39 deaths to 1977. In Kyshym, USSR, the venting of plutonium waste caused high but undisclosed casualties (30 small communities were deleted from maps produced in 1958).
1959	Experimental fast reactor built in Dounreay, N Scotland.
1979	Nuclear-reactor accident at Three Mile Island, Pennsylvania, USA.
1986	An explosive leak from a reactor at Chernobyl, the Ukraine, resulted in clouds of radioactive material spreading as far as Sweden; 31 people were killed and thousands of square kilometres were contaminated.
1991	The first controlled and substantial production of nuclear-fusion energy (a two-second pulse of 1.7 MW) was achieved at JET, the Joint European Torus, at Culham, Oxfordshire, England.
1993	The Tokamak Fusion Test Reactor at Princeton University, USA, generated 6.2 MW of power for four seconds – the most energy ever generated by controlled nuclear fusion.

After a few weeks or months it subsides to its previous state; it may erupt many more times.

NTP (abbreviation for *normal temperature and pressure*) the former name for *STP* (◊*standard temperature and pressure*).

NTSC (abbreviation for *National Television System Committee*) the US body responsible for setting up the world's first successful coding system for colour-television broadcasting 1953. See ◊television.

nuclear energy or *atomic energy* energy released from the inner core, or ◊nucleus, of the atom. Energy produced by ◊*nuclear fission* (the splitting of uranium or plutonium nuclei) has been harnessed since the 1950s to generate electricity, and research continues into the possible controlled use of ◊*nuclear fusion* (the fusing, or combining, of atomic nuclei).

In nuclear power stations, fission takes place in a ◊nuclear reactor. The nuclei of uranium or, more rarely, plutonium are induced to split, releasing large amounts of heat energy. The heat is then removed from the core of the reactor by circulating gas or water, and used to produce the steam that drives alternators and turbines to generate electrical power. Unlike fossil fuels, such as coal and oil, which must be burned in large quantities to produce energy, nuclear fuels are used in very small amounts and supplies are therefore unlikely to be exhausted in the foreseeable future. However, the use of nuclear energy has given rise to concern over safety.

nuclear fusion process whereby two atomic nuclei are fused, with the release of a large amount of energy. Very high temperatures and pressures are thought to be required in order for the process to happen. Under these conditions the atoms involved are stripped of all their electrons so that the remaining particles, which together make up a *plasma*, can come close together at very high speeds and overcome the mutual repulsion of the positive charges on the atomic nuclei. At very close range another nuclear force will come into play, fusing the particles together to form a larger nucleus. As fusion is accompanied by the release of large amounts of energy, the process might one day be harnessed to form the basis of commercial energy production. Methods of achieving controlled fusion are therefore the subject of research around the world.

nuclear notation method used for labelling an atom according to the composition of its nucleus. The atoms or isotopes of a particular element are represented by the symbol $^A_Z X$ where A is the mass number of their nuclei, Z is their atomic number, and x is the chemical symbol for that element.

nuclear physics study of the properties of the nucleus of the ◊atom, including the structure of nuclei; nuclear forces; the interactions between particles and nuclei; and the study of radioactive decay. The study of elementary particles is ◊particle physics.

nuclear reaction reaction involving the nuclei of atoms. Atomic nuclei can undergo changes either as a result of radioactive decay, as in the decay of radium to radon (with the emission of an alpha particle) or as a result of particle bombardment in a machine or device, as in the production of cobalt-60 by the bombardment of cobalt-59 with neutrons.

$$^{226}_{88}\text{Ra} \rightarrow ^{222}_{86}\text{Rn} + ^{4}_{2}\text{He}$$
$$^{59}_{27}\text{Co} + ^{1}_{0}\text{n} \rightarrow ^{60}_{27}\text{Co} + \gamma$$

Nuclear ◊fission and nuclear ◊fusion are examples of nuclear reactions. The enormous amounts of energy released arise from the mass–energy relation put forward by Einstein, stating that $E = mc^2$ (where E is energy, m is mass, and c is the velocity of light).

nuclear reactor device for producing ◊nuclear energy in a controlled manner. There are various types of reactor in use, all using nuclear fission. In a ***gas-cooled reactor***, a circulating gas under pressure (such as carbon dioxide) removes heat from the core of the reactor, which usually contains natural uranium. The efficiency of the fission process is increased by slowing neutrons in the core by using a ◊moderator such as carbon. The reaction is controlled with neutron-absorbing rods made of boron. An ***advanced gas-cooled reactor*** (AGR) generally has enriched uranium as its fuel. A ***water-cooled reactor***, such as the steam-generating heavy water (deuterium oxide) reactor, has water circulating through the hot core. The water is converted to steam, which drives turbo-alternators for generating electricity. The most widely used reactor is the ***pressurized-water reactor*** (PWR), which contains a sealed system of pressurized water that is heated to form steam in heat exchangers in an external circuit. The ***fast reactor*** has no moderator and uses fast neutrons to bring about fission. It uses a mixture of plutonium and uranium oxide as fuel. When operating, uranium is converted to plutonium, which can be extracted and used later as fuel. It is also called the fast breeder because it produces more plutonium than it consumes. Heat is removed from the reactor by a coolant of liquid sodium.

nuclear safety measures to avoid accidents in the operation of nuclear reactors and in the production and disposal of nuclear weapons and of ◊nuclear waste. There are no guarantees of the safety of any of the various methods of disposal.

nuclear accidents

Tomsk, Siberia. In April 1993 a tank exploded at a uranium reprocessing plant, sending a cloud of radioactive particles inito the air.

Chernobyl, Ukraine. In April 1986 there was an explosive leak, caused by overheating, from a nonpressurized boiling-water reactor, one of the largest in Europe. The resulting clouds of radioactive material spread as far as the UK; 31 people were killed in the explosion (many more are expected to die or become ill because of the long-term effects of radiation), and thousands of square kilometres of land were contaminated by fallout.

Three Mile Island, Harrisburg, Pennsylvania, USA. In 1979, a combination of mechanical and electrical failure, as well as operator error, caused a pressurized water reactor to leak radioactive matter.

Church Rock, New Mexico, USA. In July 1979, 380 million litres/100 million gallons of radioactive water containing uranium leaked from a pond into the Rio Purco, causing the water to become over 6,500 times as radioactive as safety standards allow for drinking water.

Ticonderoga, 130 km/80 mi off the coast of Japan. In 1965 a US Navy Skyhawk jet bomber fell off the deck of this ship, sinking in 4,900 m/16,000 ft of water. It carried a one-megaton hydrogen bomb. The accident was only revealed in 1989.

Windscale (now Sellafield), Cumbria, England. In 1957, fire destroyed the core of a reactor, releasing large quantities of radioactive fumes into the atmosphere.

nuclear testing the detonation of nuclear devices to verify their reliability, power and destructive abilities. Although carried out secretly in remote regions of the world, such tests are easily detected by the seismic shock waves produced. The first tests were carried out in the atmosphere during the 1950s by the USA, the USSR, and the UK.

nuclear waste the radioactive and toxic by-products of the nuclear-energy and nuclear-weapons industries. Nuclear waste may have an active life of several thousand years. Reactor waste is of three types: ***high-level*** spent fuel, or the residue when nuclear fuel has been removed from a reactor and reprocessed; ***intermediate***, which may be long-or short-lived; and ***low-level***, but bulky, waste from reactors, which has only short-lived radioactivity. Disposal, by burial on land or at sea, has raised problems of safety, environmental pollution, and security. In absolute terms, nuclear waste cannot be safely relocated or disposed of.

The USSR and the Russian Federation between them dumped an estimated 12 trillion becquerels of radioactivity in the sea 1959–93. Russia has no way of treating nuclear waste and in 1993 announced its intention of continuing to dump it in the sea, in violation of international conventions, until 1997.

nucleic acid complex organic acid made up of a long chain of nucleotides, present in the nucleus and sometimes the cytoplasm of the living cell. The two types, known as DNA (deoxyribonucleic acid) and RNA (ribonucleic acid), form the basis of heredity. The nucleotides are made up of a sugar (deoxyribose or ribose), a phosphate group, and one of four purine or pyrimidine bases. The order of the bases along the nucleic acid strand contains the genetic code.

nucleolus in biology, a structure found in the nucleus of eukaryotic cells. It produces the RNA that makes up the ◊ribosomes, from instructions in the DNA.

nucleon in particle physics, either a ◊proton or a ◊neutron, both particles present in the atomic nucleus. *Nucleon number* is an alternative name for the ◊mass number of an atom.

nucleotide organic compound consisting of a purine (adenine or guanine) or a pyrimidine (thymine, uracil, or cytosine) base linked to a sugar (deoxyribose or ribose) and a phosphate group. ◊DNA and ◊RNA are made up of long chains of nucleotides.

nucleus in physics, the positively charged central part of an ◊atom, which constitutes almost all its mass. Except for hydrogen nuclei, which have only protons, nuclei are composed of both protons and neutrons. Surrounding the nuclei are electrons, of equal and opposite charge to that of the protons, thus giving the atom a neutral charge.

nucleus in biology, the central, membrane-enclosed part of a eukaryotic cell, containing the chromosomes.

nuclide in physics, one of two or more atoms having the same atomic number (number of protons) and mass number (number of nucleons); compare ◊isotope.

number symbol used in counting or measuring. In mathematics, there are various kinds of numbers. The everyday number system is the decimal ('proceeding by tens') system, using the base ten. *Real numbers* include all rational numbers (integers, or whole numbers, and fractions) and irrational numbers (those not expressible as fractions). ◊*Complex numbers* include the real and unreal numbers (real-number multiples of the square root of -1). The ◊binary number system, used in computers, has two as its base. The ordinary numerals, 0, 1, 2, 3, 4, 5, 6, 7, 8, and 9, give a counting system that, in the decimal system, continues 10, 11, 12, 13, and so on. These are whole numbers (integers), with fractions represented as, for example, $1/4$, $1/2$, $3/4$, or as decimal fractions (0.25, 0.5, 0.75). They are also *rational numbers*. *Irrational numbers* cannot be represented in this way and require symbols, such as $\sqrt{2}$, π, and e. They can be expressed numerically only as the (inexact) approximations 1.414, 3.142 and 2.718 (to three places of decimals) respectively. The symbols π and e are also examples of *transcendental numbers*, because they (unlike $\sqrt{2}$) cannot be derived by solving a ◊polynomial equation (an equation with one ◊variable quantity) with rational ◊coefficients (multiplying factors). Complex numbers, which include the real numbers as well as unreal numbers, take the general form $a + bi$, where $i = \sqrt{-1}$ (that is, $i^2 = -1$), and a is the real part and bi the unreal part.

number theory in mathematics, the abstract study of the structure of number systems and the properties of positive integers (whole numbers). For example, the theories of factors and prime numbers fall within this area as do the work of mathematicians Giuseppe Peano (1858–1932), Pierre de Fermat, and Karl Gauss.

nut any dry, single-seeded fruit that does not split open to release the seed, such as the chestnut. A nut is formed from more than one carpel, but only one seed becomes fully formed, the remainder aborting. The wall of the fruit, the pericarp, becomes hard and woody, forming the outer shell.

nutation in botany, the spiral movement exhibited by the tips of certain stems during growth; it enables a climbing plant to find a suitable support. Nutation sometimes also occurs in tendrils and flower stalks.

nutation in astronomy, a slight 'nodding' of the Earth in space, caused by the varying gravitational pulls of the Sun and Moon. Nutation changes the angle of the Earth's axial tilt (average 23.5°) by about 9 seconds of arc to either side of its mean position, a complete cycle taking just over 18.5 years.

nutrition the strategy adopted by an organism to obtain the chemicals it needs to live, grow, and reproduce. There are two main types: autotrophic nutrition (plants) and heterotrophic nutrition (animals).

nutrition the science of food, and its effect on human and animal life, health, and disease. Nutrition involves the study of the basic nutrients required to sustain life, their bioavailability in foods and overall diet, and the effects upon them of cooking and storage. It is also concerned with dietary deficiency diseases.

nylon synthetic long-chain polymer similar in chemical structure to protein. Nylon was the first all-synthesized fibre, made from petroleum, natural gas, air, and water by the Du Pont firm in 1938. It is used in the manufacture of moulded articles, textiles, and medical sutures. Nylon fibres are stronger and more elastic than silk and are relatively insensitive to moisture and mildew. Nylon is used for hosiery and woven goods, simulating other materials such as silks and furs; it is also used for carpets.

nymph in entomology, the immature form of insects that do not have a pupal stage; for example, grasshoppers and dragonflies. Nymphs generally resemble the adult (unlike larvae), but do not have fully formed reproductive organs or wings.

object program in computing, the ◊machine code translation of a program written in a ◊source language.

observatory site or facility for observing astronomical or meteorological phenomena. The earliest recorded observatory was in Alexandria, N Africa, built by Ptolemy Soter in about 300 BC. The modern observatory dates from the invention of the telescope. Observatories may be ground-based, carried on aircraft, or sent into orbit as satellites, in space stations, and on the space shuttle.

obsidian black or dark-coloured glassy volcanic rock, chemically similar to ◊granite, but formed by cooling rapidly on the Earth's surface at low pressure.

occluded front weather ◊front formed when a cold front catches up with a warm front. It brings cloud and rain as air is forced to rise upwards along the front, cooling and condensing as it does so.

occultation in astronomy, the temporary obscuring of a star by a body in the solar system. Occultations are used to provide information about changes in an orbit, and the structure of objects in space, such as radio sources.

oceanography study of the oceans. Its subdivisions deal with each ocean's extent and depth, the water's evolution and composition, its physics and chemistry, the bottom topography, currents and wind effects, tidal ranges, the biology, and the various aspects of human use.

OCR abbreviation for ◊*optical character recognition*.

octal number system number system to the base eight, used in computing. The highest digit that can appear in the octal system is seven. Normal decimal, or base-ten, numbers may be considered to be written under column headings based on the number ten. Octal, or base-eight, numbers can be thought of as written under column headings based on the number eight. The octal number 567 is therefore equivalent to the decimal number 375, since $(5 \times 64) + (6 \times 8) + (7 \times 1) = 375$.

octane rating numerical classification of petroleum fuels indicating their combustion characteristics.

Octans constellation in the southern hemisphere containing the southern celestial pole. The closest naked-eye star to the south celestial pole is fifth-magnitude Sigma Octantis.

octet rule in chemistry, rule stating that elements combine in a way that gives them the electronic structure of the nearest ◊inert gas. All the inert gases except helium have eight electrons in their outermost shell, hence the term octet.

oersted c.g.s. unit (symbol Oe) of ◊magnetic field strength, now replaced by the SI unit ampere per metre. The Earth's magnetic field is about 0.5 oersted; the field near the poles of a small bar magnet is several hundred oersteds; and a powerful ◊electromagnet can have a field strength of 30,000 oersteds.

oesophagus muscular tube by which food travels from mouth to stomach. The human oesophagus is about 23 cm/9 in long. Its extends downwards from the ◊pharynx, immediately behind the windpipe. It is lined with a mucous membrane which secretes lubricant fluid to assist the downward movement of food (◊peristalsis).

oestrogen any of a group of hormones produced by the ◊ovaries of vertebrates; the term is also used for various synthetic hormones that mimic their effects. The principal oestrogen in mammals is oestradiol. Oestrogens control female sexual development, promote the growth of female secondary sexual characteristics, stimulate egg production, and, in mammals, prepare the lining of the uterus for pregnancy.

Oestrogens are used therapeutically for some hormone disorders and to inhibit lactation; they also form the basis of oral contraceptives.

oestrus in mammals, the period during a female's reproductive cycle (also known as the oestrus cycle or ◊menstrual cycle) when mating is most likely to occur. It usually coincides with ovulation.

ohm SI unit (symbol Ω) of electrical ◊resistance (the property of a substance that restricts the flow of electrons through it).

ohmic heating method of heating used in the food-processing industry, in which an electric current is passed through foodstuffs to sterilize them before packing. The heating effect is similar to that obtained by microwaves in that electrical energy is transformed into heat throughout the whole volume of the food, not just at the surface.

Ohm's law law that states that the current flowing in a metallic conductor maintained at constant temperature is directly proportional to the potential difference (voltage) between its ends. The law was discovered by German physicist Georg Ohm 1827.

If a current of I amperes flows between two points in a conductor across which the potential difference is V volts, then V/I is a constant called the ◊resistance R ohms between those two points. Hence:

$$V/I = R \text{ or } V = IR$$

oil flammable substance, usually insoluble in water, and composed chiefly of carbon and hydrogen. Oils may be solids (fats and waxes) or liquids. The three main types are: *essential oils*, obtained from plants; *fixed oils*, obtained from animals and plants; and *mineral oils*, obtained chiefly from the refining of ◊petroleum.

oil refinery industrial complex where crude oil is processed into different products. The light volatile parts of the oil form ◊petroleum, while the heavier parts make bitumen and petrochemicals. Oil refineries are often located at deep-water ports or near their industrial markets. They need flat land and water for cooling.

olefin common name for ◊alkene.

Oligocene third epoch of the Tertiary period of geological time, 35.5–3.25 million years ago. The name, from Greek, means 'a little recent', referring to the presence of the remains of some modern types of animals existing at that time.

oligosaccharide ◊carbohydrate comprising a few ◊monosaccharide units linked together. It is a general term used to indicate that a carbohydrate is larger than a simple di-or trisaccharide but not as large as a polysaccharide.

olivenite basic copper arsenate, $Cu_2AsO_4(OH)$, occurring as a mineral in olive-green prisms.

olivine greenish mineral, magnesium iron silicate, $(Mg,Fe)_2SiO_4$. It is a rock-forming mineral, present in, for example, peridotite, gabbro, and basalt. Olivine is called *peridot* when pale green and transparent, and used in jewellery.

Omega navigation system long-range radio-based aid to navigation, giving worldwide coverage. There are eight Omega transmitting stations, located in Norway, Liberia, Hawaii, Réunion, Argentina, Australia, USA, and Japan. The very low-frequency signals from the transmitters are detected by a ship's navigation receiver, and slight differences (phase differences) between the signals indicate the position of the receiver. The system is accurate to within 4 km/2.5 mi during the day and 7 km/ 4 mi at night. See also ◊Decca navigation system.

omnivore animal that feeds on both plant and animal material. Omnivores have digestive adaptations intermediate between those of ◊herbivores and ◊carnivores, with relatively unspecialized digestive systems and gut microorganisms that can digest a variety of foodstuffs.

OMR abbreviation for ◊*optical mark recognition*.

oncogene gene carried by a virus that induces a cell to divide abnormally, giving rise to a cancer. Oncogenes arise from mutations in genes (proto-oncogenes) found in all normal cells. They are usually also found in viruses that are capable of transforming normal cells to tumour cells. Such viruses are able to insert their oncogenes into the host cell's DNA, causing it to divide uncontrollably. More than one oncogene may be necessary to transform a cell in this way.

onco-mouse mouse that has a human ◊oncogene (gene that can cause certain cancers) implanted into its cells by genetic engineering. Such mice are used to test anti-cancer treatments and were patented within the USA by Harvard University in 1988, thereby protecting its exclusive rights to produce the animal and profit from its research.

on-line system in computing, originally a system that allows the computer to work interactively with its users, responding to each instruction as it is given and prompting users for information when necessary. Since almost all the computers people use now work this way, 'on-line system' is now used to refer to large database, electronic mail, and conferencing systems accessed via a dial-up modem. These often have tens or hundreds of users from different places – sometimes from different countries – 'on line' at the same time.

ontogeny process of development of a living organism, including the part of development that takes place after hatching or birth. The idea that 'ontogeny recapitulates phylogeny' (the development of an organism goes through the same stages as its evolutionary history), proposed by the German scientist Ernst Heinrich Haeckel, is now discredited.

onyx semiprecious variety of chalcedonic ◊silica (SiO_2) in which the crystals are too fine to be detected under a microscope, a state known as cryptocrystalline. It has straight parallel bands of different colours: milk-white, black, and red.

oolite limestone made up of tiny spherical carbonate particles called *ooliths*. Ooliths have a concentric structure with a diameter up to 2 mm/0.08 in. They were formed by chemical precipitation and accumulation on ancient sea floors.

Oort cloud spherical cloud of comets beyond Pluto, extending out to about 100,000 astronomical units (1.5 light years) from the Sun. The gravitational effect of passing stars and the rest of our Galaxy disturbs comets from the cloud so that they fall in towards the Sun on highly elongated orbits, becoming visible from Earth. As many as 10 trillion comets may reside in the Oort cloud, named after Dutch astronomer Jan Oort who postulated it 1950.

oosphere another name for the female gamete, or ◊ovum, of certain plants such as algae.

opal form of ◊silica (SiO_2), often occurring as stalactites and found in many types of rock. The common opal is translucent, milk-white, yellow, red, blue, or green, and lustrous. Precious opal is opalescent, the characteristic play of colours being caused by close-packed silica spheres diffracting light rays within the stone.

open-hearth furnace method of steelmaking, now largely superseded by the ◊basic-oxygen process. It was developed in 1864 in England by German-born William and Friedrich Siemens, and improved by Pierre and Emile Martin in France in the same year. In the furnace, which has a wide, saucer-shaped hearth and a low roof, molten pig iron and scrap are packed into the shallow hearth and heated by overhead gas burners using preheated air.

operating system (OS) in computing, a program that controls the basic operation of a computer. A typical OS controls the ◊peripheral devices, organizes the filing system, provides a means of communicating with the operator, and runs other programs.

operational amplifier or *op-amp* electronic circuit that is used to increase the size of an alternating voltage signal without distorting it.

operculum small cap covering the spore-producing body of many mosses. It is pushed aside when the spores are mature and ready to be ejected.

operon group of genes that are found next to each other on a chromosome, and are turned on and off as an integrated unit. They usually produce enzymes that control different steps in the same biochemical pathway. Operons were discovered 1961 (by the French biochemists François Jacob and Jacques Monod) in bacteria.

Ophiuchus large constellation along the celestial equator, known as the serpent bearer because the constellation Serpens is wrapped around it. The Sun passes through Ophiuchus each Dec, but the constellation is not part of the zodiac. Ophiuchus contains ◊Barnard's star.

opiate, endogenous naturally produced chemical in the body that has effects similar to morphine and other opiate drugs; a type of neurotransmitter. Examples include ◊endorphins and ◊encephalins.

opium drug extracted from the unripe seeds of the opium poppy *Papaver somniferum* of SW Asia. An addictive ◊narcotic, it contains several alkaloids, including *morphine*, one of the most powerful natural painkillers and addictive narcotics known, and *codeine*, a milder painkiller.

opposition in astronomy, the moment at which a body in the Solar System lies opposite the Sun in the sky as seen from the Earth and crosses the ◊meridian at about midnight.

optical aberration see ◊aberration, optical.

optical activity in chemistry, the ability of certain crystals, liquids, and solutions to rotate the plane of ◊polarized light as it passes through them. The phenomenon is related to the three-dimensional arrangement of the atoms making up the molecules concerned.

Only substances that lack any form of structural symmetry exhibit optical activity.

optical character recognition (OCR) in computing, a technique for inputting text to a computer by means of a document reader. First, a ◊scanner produces a digital image of the text; then character-recognition software makes use of stored knowledge about the shapes of individual characters to convert the digital image to a set of internal codes that can be stored and processed by computer.

optical disc in computing, a storage medium in which laser technology is used to record and read large volumes of digital data. Types include ◊CD-ROM, ◊WORM, and erasable optical disc.

optical emission spectrometry another term for emission ◊spectroscopy.

optical fibre very fine, optically pure glass fibre through which light can be reflected to transmit images or data from one end to the other. Although expensive to produce and install, optical fibres can carry more data than traditional cables, and are less susceptible to interference.

optical illusion scene or picture that fools the eye. An example of a natural optical illusion is that the Moon appears bigger when it is on the horizon than when it is high in the sky, owing to the ◊refraction of light rays by the Earth's atmosphere.

optical instrument instrument that makes use of one or more lenses or mirrors, or of a combination of these, in order to change the path of light rays and produce an image. Optical instruments such as magnifying glasses, ◊microscopes, and ◊telescopes are used to provide a clear, magnified image of the very small or the very distant. Others, such as cameras, photographic enlargers, and film projectors, may be used to store or reproduce images.

optical mark recognition (OMR) in computing, a technique that enables marks made in predetermined positions on computer-input forms to be detected optically and input to a computer. An *optical mark reader* shines a light beam onto the input document and is able to detect the marks because less light is reflected back from them than from the paler, unmarked paper.

optic nerve large nerve passing from the eye to the brain, carrying visual information. In mammals, it may contain up to a million nerve fibres, connecting the sensory cells of the retina to the optical centres in the brain. Embryologically, the optic nerve develops as an outgrowth of the brain.

optics branch of physics that deals with the study of ◊light and vision – for example, shadows and mirror images, lenses, microscopes, telescopes, and cameras. For all practical purposes light rays travel in straight lines, although Albert Einstein (1879–1955) demonstrated that they may be 'bent' by a gravitational field. On striking a surface they are reflected or refracted with some absorption of energy, and the study of this is known as geometrical optics.

optoelectronics branch of electronics concerned with the development of devices (based on the ◊semiconductor gallium arsenide) that respond not only to the ◊electrons of electronic data transmission, but also to ◊photons.

orbit path of one body in space around another, such as the orbit of Earth around the Sun, or the Moon around Earth. When the two bodies are similar in mass, as in a ◊binary star, both bodies move around their common centre of mass. The movement of objects in orbit follows the laws of Johann Kepler (1571–1630), which apply to artificial satellites as well as to natural bodies.

As stated by the laws, the orbit of one body around another is an ellipse. The ellipse can be highly elongated, as are comet orbits around the Sun, or it may be almost circular, as are those of some planets. The closest point of a planet's orbit to the Sun is called *perihelion*; the most distant point is *aphelion*. (For a body orbiting the Earth, the closest and furthest points of the orbit are called *perigee* and *apogee*.)

orbital, atomic region around the nucleus of an atom (or, in a molecule, around several nuclei) in which an ◊electron is most likely to be found. According to ◊quantum theory, the position of an electron is uncertain; it may be found at any point. However, it is more likely to be found in some places than in others, and it is these that make up the orbital.

order in biological classification, a group of related ◊families. For example, the horse, rhinoceros, and tapir families are grouped in the order Perissodactyla, the odd-toed ungulates, because they all have either one or three toes on each foot. The names of orders are not shown in italic (unlike genus and species names) and by convention they have the ending '-formes' in birds and fish; '-a' in

orbital, atomic *The shapes of atomic orbitals. An atomic orbital is a picture of the 'electron cloud' that surrounds the nucleus of an atom. There are four basic shapes for atomic orbitals: spherical, dumbbell, clover-leaf, and complex (shown at bottom left).*

mammals, amphibians, reptiles, and other animals; and '-ales' in fungi and plants. Related orders are grouped together in a class.

ordered pair in mathematics, any pair of numbers whose order makes a difference to their meaning. Coordinates are an ordered pair because the point (2,3) is not the same as the point (3,2). Vulgar fractions are ordered pairs because the top number gives the quantity of parts while the bottom gives the number of parts into which the unit has been divided.

ordinal number in mathematics, one of the series first, second, third, fourth, ... Ordinal numbers relate to order, whereas ◊cardinal numbers (1, 2, 3, 4, ...) relate to quantity, or count.

ordinate in ◊coordinate geometry, the *y* coordinate of a point; that is, the vertical distance of the point from the horizontal or *x*-axis. For example, a point with the coordinates (3,4) has an ordinate of 4. See ◊abscissa.

Ordovician period of geological time 510–439 million years ago; the second period of the ◊Palaeozoic era. Animal life was confined to the sea: reef-building algae and the first jawless fish are characteristic.

ore body of rock, a vein within it, or a deposit of sediment, worth mining for the economically valuable mineral it contains. The term is usually applied to sources of metals. Occasionally metals are found uncombined (native metals), but more often they occur as compounds such as carbonates, sulphides, or oxides. The ores often contain unwanted impurities that must be removed when the metal is extracted.

organ in biology, part of a living body, such as the liver or brain, that has a distinctive function or set of functions.

organelle discrete and specialized structure in a living cell; organelles include mitochondria, chloroplasts, lysosomes, ribosomes, and the nucleus.

organic chemistry branch of chemistry that deals with carbon compounds. Organic compounds form the chemical basis of life and are more abundant than inorganic compounds. In a typical organic compound, each carbon atom forms bonds covalently with each of its neighbouring carbon atoms in a chain or ring, and additionally with other atoms, commonly hydrogen, oxygen, nitrogen, or sulphur.

The basis of organic chemistry is the ability of carbon to form long chains of atoms, branching chains, rings, and other complex structures. Compounds containing only carbon and hydrogen are known as ***hydrocarbons***.

Organic chemistry is largely the chemistry of a great variety of homologous series – those in which the molecular formulae, when arranged in ascending order, form an arithmetical progression. The physical properties undergo a gradual change from one member to the next.

The linking carbon atoms that form the backbone of an organic molecule may be built up from beginning to end without branching; or may throw off branches at one or more points. Sometimes, however, the ropes of carbon atoms curl round and form rings (***cyclic compounds***), usually of five, six, or seven atoms. Open-chain and cyclic compounds may

formula	name	atomic bonding
CH_3	methyl	H–C(H)(H)–
CH_2CH_3	ethyl	–C(H)(H)–C(H)(H)–H
CC	double bond	C=C
CHO	aldehyde	–C(H)=O
CH_2OH	alcohol	–C(H)(H)–OH
CO	ketone	C=O
COOH	acid	–C(=O)–OH
CH_2NH_2	amine	–C(H)(H)–N(H)(H)
C_6H_6	benzene ring	(hexagonal ring of 6 CH)

organic chemistry *Common organic-molecule groupings. Organic chemistry is the study of carbon compounds, which make up over 90% of all chemical compounds. This diversity arises because carbon atoms can combine in many different ways with other atoms, forming a wide variety of loops and chains.*

be classified as ◊aliphatic or ◊aromatic depending on the nature of the bonds between their atoms. Compounds containing oxygen, sulphur, or nitrogen within a carbon ring are called ***heterocyclic compounds***.

Many organic compounds (such as proteins and carbohydrates) are made only by living organisms and many are derived from petroleum, which represents the chemical remains of millions of microscopic marine organisms.

Hydrocarbons form one of the most prolific of the many organic types; fuel oils are largely made up of hydrocarbons. Typical groups containing only carbon, hydrogen, and oxygen are alcohols, aldehydes, ketones, ethers, esters, and carbohydrates. Among groups containing nitrogen are amides, amines, nitro-compounds, amino acids, proteins, purines, alkaloids, and many others, both natural and artificial. Other organic types contain sulphur, phosphorus, or halogen elements.

organic compound in biochemistry, one of the class of compounds whose behaviour is influenced by the chemistry of carbon. All organic compounds contain carbon, which was once thought to be present only in living things. The original distinction between organic and inorganic compounds was based on the belief that the molecules of living systems were unique, and could not be synthesized in the laboratory. Today it is routine to manufacture thousands of organic chemicals both in research and in the drug industry.

organizer in embryology, a part of the embryo that causes changes to occur in another part, through ◊induction, thus 'organizing' development and ◊differentiation.

OR gate in electronics, a type of ◊logic gate.

orimulsion fuel made by mixing ◊bitumen and water that can be burnt in the same way as heavy oil. It is cheap to make but the smoke produced has a high sulphur content.

Orion in astronomy, a very prominent constellation in the equatorial region of the sky, identified with the hunter of Greek mythology.

It contains the bright stars Betelgeuse and Rigel, as well as a distinctive row of three stars that make up Orion's belt. Beneath the belt, marking the sword of Orion, is the Orion nebula; nearby is one of the most distinctive dark nebulae, the Horsehead.

Orion nebula luminous cloud of gas and dust 1,500 light years away, in the constellation Orion, from which stars are forming. It is about 15 light years in diameter, and contains enough gas to make a cluster of thousands of stars.

ormolu alloy of copper, zinc, and sometimes tin, used for furniture decoration.

ornithology study of birds. It covers scientific aspects relating to their structure and classification, and their habits, song, flight, and value to agriculture as destroyers of insect pests. Worldwide scientific banding (or the fitting of coded rings to captured specimens) has resulted in accurate information on bird movements and distribution. There is an International Council for Bird Preservation with its headquarters at the Natural History Museum, London.

ornithophily ◊pollination of flowers by birds. Ornithophilous flowers are typically brightly coloured, often red or orange. They produce large quantities of thin, watery nectar, and are scentless because most birds do not respond well to smell. They are found mostly in tropical areas, with hummingbirds being important pollinators in North and South America, and the sunbirds in Africa and Asia.

orogeny or *orogenesis* the formation of mountains. It is brought about by the movements of the rigid plates making up the Earth's crust (described by ◊plate tectonics). Where two plates collide at a destructive margin rocks become folded and lifted to form chains of fold mountains (such as the young fold mountains of the Himalayas).

orographic rainfall rainfall that occurs when an airstream is forced to rise over a mountain range. As the air rises, it becomes cooled. The amount of moisture that air can hold decreases with decreasing temperature. So the water vapour in the rising airstream condenses, and rain falls on the winward side of the mountain. The air descending on the leeward side contains less moisture, resulting in a *rainshadow* where there is little or no rain.

OS/2 single-user computer ◊operating system produced jointly by Microsoft Corporation and IBM for use on large microcomputers. Its main features are ◊multitasking and the ability to access large amounts of internal ◊memory.

oscillating universe in astronomy, a theory that states that the gravitational attraction of the mass within the universe will eventually slow down and stop the expansion of the universe. The outward motions of the galaxies will then be reversed, eventually resulting in a 'Big Crunch' where all the matter in the universe would be contracted into a small volume of high density. This could undergo a further ◊Big Bang, thereby creating another expansion phase. The theory suggests that the universe would alternately expand and collapse through alternate Big Bangs and Big Crunches.

oscillation one complete to-and-fro movement of a vibrating object or system. For any particular vibration, the time for one oscillation is called its ◊period and the number of oscillations in one second is called its ◊frequency. The maximum displacement of the vibrating object from its rest position is called the ◊amplitude of the oscillation.

oscillator any device producing a desired oscillation (vibration). There are many types of oscillator for different purposes, involving various arrangements of thermionic ◊valves or components such as ◊transistors, ◊inductors, ◊capacitors, and ◊resistors.

oscillograph instrument for displaying or recording the values of rapidly changing oscillations, electrical or mechanical.

oscilloscope or *cathode-ray oscilloscope* (CRO) instrument used to measure electrical voltages that vary over time and to display the waveforms of electrical oscillations or signals, by means of the deflection of a beam of ◊electrons. Readings are displayed graphically on the screen of a ◊cathode-ray tube.

osmium hard, heavy, bluish-white, metallic element, symbol Os, atomic number 76, relative atomic mass 190.2. It is the densest of the elements, and is resistant to tarnish and corrosion. It occurs in platinum ores and as a free metal (see ◊native metal) with iridium in a natural alloy called osmiridium, containing traces of platinum, ruthenium, and rhodium. Its uses include pen points and light-bulb filaments; like platinum, it is a useful catalyst.

osmoregulation process whereby the water content of living organisms is maintained at a constant level. If the water balance is disrupted, the concentration of salts will be too high or too low, and vital functions, such as nerve conduction, will be adversely affected.

osmosis movement of solvent (liquid) through a semipermeable membrane separating solutions of different concentrations. The solvent passes from a less concentrated solution to a more concentrated solution until the two concentrations are equal. Applying external pressure to the solution on the more concentrated side arrests osmosis, and is a measure of the osmotic pressure of the solution.

Many cell membranes behave as semipermeable membranes, and osmosis is a vital mechanism in the transport of fluids in living organisms – for example, in the transport of water from the roots up the stems of plants.

Fish have protective mechanisms to counteract osmosis, which would otherwise cause fluid transport between the body of the animal and the surrounding water (outwards in saltwater fish, inwards in freshwater ones).

ossification or *osteogenesis* process whereby bone is formed in vertebrate animals by special cells (*osteoblasts*) that secrete layers of ◊extracellular matrix on the surface of the existing ◊cartilage.

Conversion to bone occurs through the deposition of calcium phosphate crystals within the matrix.

osteology part of the science of ◊anatomy, dealing with the structure, function, and development of bones.

Otto cycle alternative name for the ◊four-stroke cycle, introduced by the German engineer Nikolaus Otto (1832–1891) in 1876. It improved on existing piston engines by compressing the fuel mixture in the cylinder before it was ignited.

ounce unit of mass, one-sixteenth of a pound ◊avoirdupois, equal to 437.5 grains (28.35 g); also one-twelfth of a pound troy, equal to 480 grains.

outlier in statistics, a highly atypical data value that distorts the ◊mean within a sample.

output device in computing, any device for displaying, in a form intelligible to the user, the results of processing carried out by a computer.

ovary in female animals, the organ that generates the ◊ovum. In humans, the ovaries are two whitish rounded bodies about 25 mm/1 in by 35 mm/1.5 in, located in the lower abdomen to either side of the uterus. Every month, from puberty to the onset of the menopause, an ovum is released from the ovary. This is called ovulation, and forms part of the ◊menstrual cycle. In botany, an ovary is the expanded basal portion of the ◊carpel of flowering plants, containing one or more ◊ovules. It is hollow with a thick wall to protect the ovules. Following fertilization of the ovum, it develops into the fruit wall or pericarp.

overtone note that has a frequency or pitch that is a multiple of the fundamental frequency,

osmosis *The movement of liquid through a semipermeable membrane separating solutions of different concentrations, is essential for life. Osmosis transports water from the soil into the roots of plants, for example. It is applied in kidney dialysis to remove impurities from the blood.*

the sounding body's ◊natural frequency. Each sound source produces a unique set of overtones, which gives the source its quality or timbre.

ovipary method of animal reproduction in which eggs are laid by the female and develop outside her body, in contrast to ovovivipary and vivipary. It is the most common form of reproduction.

ovovivipary method of animal reproduction in which fertilized eggs develop within the female (unlike ovipary), and the embryo gains no nutritional substances from the female (unlike vivipary). It occurs in some invertebrates, fishes, and reptiles.

ovulation in female animals, the process of releasing egg cells (ova) from the ◊ovary. In mammals it occurs as part of the ◊menstrual cycle.

ovule structure found in seed plants that develops into a seed after fertilization. It consists of an ◊embryo sac containing the female gamete (◊ovum or egg cell), surrounded by nutritive tissue, the nucellus. Outside this there are one or two coverings that provide protection, developing into the testa, or seed coat, following fertilization.

In ◊angiosperms (flowering plants) the ovule is within an ◊ovary, but in ◊gymnosperms (conifers and their allies) the ovules are borne on the surface of an ovuliferous (ovule-bearing) scale, usually within a ◊cone, and are not enclosed by an ovary.

ovum (plural *ova*) female gamete (sex cell) before fertilization. In animals it is called an egg, and is produced in the ovaries. In plants, where it is also known as an egg cell or oosphere, the ovum is produced in an ovule. The ovum is nonmotile. It must be fertilized by a male gamete before it can develop further, except in cases of ◊parthenogenesis.

oxalic acid $(COOH)_2 \cdot 2H_2O$ white, poisonous solid, soluble in water, alcohol, and ether. Oxalic acid is found in rhubarb, and its salts (oxalates) occur in wood sorrel (genus *Oxalis*, family Oxalidaceae) and other plants. It is used in the leather and textile industries, in dyeing ... ing, ink manufacture, metal polishing, removing rust and ink stains.

...hemistry, the loss of ◊electron, or loss of hydrogen by ...olecule during a chemical ...

...e brought about by reaction ...pound (oxidizing agent), ...usly undergoes ◊reduction, ...e anode (positive electrode) ...ell.

...th science, a form of chemical... caused by the chemical ...place between certain iron... ...k and the oxygen in water. ...in the formation of a red... ...deposit. The inside walls of ...bridges often have deposits

...ber Roman numeral often ...ical name, indicating the ...element immediately before ...examples are lead(II) nitrate, ...) oxide, and potassium man-

...ompound of oxygen and another element... ...equently produced by burning the ...t or a compound of it in air or oxygen.

...de film thin film of oxide formed on the ...face of some metals as soon as they are ...xposed to the air. This oxide film makes the metal much more resistant to a chemical attack. The considerable lack of reactivity of aluminium to most reagents arises from this property.

The thickness of the oxide film can be increased by ◊anodizing the aluminium.

oxidizing agent substance that will oxidize another substance (see ◊oxidation).

In a redox reaction, the oxidizing agent is the substance that is itself reduced. Common oxidizing agents include oxygen, chlorine, nitric acid, and potassium manganate(VII).

oxyacetylene torch gas torch that burns ethene (acetylene) in pure oxygen, producing a high-temperature (3,000°C/5,400°F) flame. It is widely used in welding to fuse metals. In the cutting torch, a jet of oxygen burns through metal already melted by the flame.

oxygen colourless, odourless, tasteless, nonmetallic, gaseous element, symbol O, atomic number 8, relative atomic mass 15.9994. It is the most abundant element in the Earth's crust (almost 50% by mass), forms about 21% by volume of the atmosphere, and is present in combined form in water and many other substances. Life on Earth evolved using oxygen, which is a by-product of ◊photosynthesis and the basis for ◊respiration in plants and animals.

Oxygen is very reactive and combines with all other elements except the ◊inert gases and fluorine. It is present in carbon dioxide, silicon dioxide (quartz), iron ore, calcium carbonate (limestone). In nature it exists as a molecule composed of two atoms (O_2); single atoms of oxygen are very short-lived owing to their reactivity. They can be produced in electric sparks and by the Sun's ultraviolet radiation in space, where they rapidly combine with molecular oxygen to form ozone (an allotrope of oxygen).

Oxygen is obtained for industrial use by the fractional distillation of liquid air, by the electrolysis of water, or by heating manganese(IV) oxide with potassium chlorate. It is essential for combustion, and is used with ethyne (acetylene) in high-temperature oxyacetylene welding and cutting torches.

oxygen debt physiological state produced by vigorous exercise, in which the lungs cannot supply all the oxygen that the muscles need.

oxytocin hormone that stimulates the uterus in late pregnancy to initiate and sustain labour. After birth, it stimulates the uterine muscles to contract, reducing bleeding at the site where the placenta was attached.

oz abbreviation for ◊ounce.

ozone O_3 highly reactive pale-blue gas; a type of oxygen. It forms a layer in the upper atmosphere, which protects life on Earth from ultraviolet rays, a cause of skin cancer. At lower atmospheric levels it is an air pollutant and contributes to the ◊greenhouse effect.

ozone depleter any chemical that destroys the ozone in the stratosphere. Most ozone depleters are chemically stable compounds containing chlorine or bromine, which remain unchanged for long enough to drift up to the upper atmosphere. The best known are ◊chlorofluorocarbons (CFCs), but many other ozone depleters are known, including halons, used in some fire extinguishers; methyl chloroform and carbon tetrachloride, both solvents; some CFC substitutes; and the pesticide methyl bromide.

Australian chemists are trialling an alternative to methyl bromide 1993. Carbonyl sulphide is naturally occurring and would have no effect on ozone. In laboratory tests it killed more than 30 pest species.

pacemaker or *sinoatrial node* (SA node) in vertebrates, a group of muscle cells in the wall of the heart that contracts spontaneously and rhythmically, setting the pace for the contractions of the rest of the heart. The pacemaker's intrinsic rate of contraction is increased or decreased, according to the needs of the body, by stimulation from the ◊autonomic nervous system. The term also refers to a medical device implanted under the skin of a patient whose heart beats inefficiently. It delivers minute electric shocks to stimulate the heart muscles at regular intervals and restores normal heartbeat.

packet switching in computing, a method of transmitting data between computers connected in a ◊network. A complete packet consists of the data being transmitted and information about which computer is to receive the data. The packet travels around the network until it reaches the correct destination.

paedomorphosis in biology, an alternative term for ◊neoteny.

paging method of increasing a computer's apparent memory capacity. See ◊virtual memor.

pain sense that gives an awareness of harmful effects on or in the body. It may be triggered by stimuli such as trauma, inflammation, and heat. Pain is transmitted by specialized nerves and also has psychological components controlled by higher centres in the brain. Drugs that control pain are known as painkillers or ◊analgesics.

PAL (abbreviation for *phase alternation by line*) the coding system for colour-television broadcasting adopted by the UK, West Germany, the Netherlands, and Switzerland 1967. NTSC (National Television System Committee) is the system which operates in America. See ◊television.

Palaeocene first epoch of the Tertiary period of geological time, 65–56.5 million years ago. Many types of mammals spread rapidly after the disappearance of the great reptiles of the Mesozoic. Flying mammals replaced the flying reptiles, swimming mammals replaced the swimming reptiles, and all the ecological niches vacated by the reptiles were adopted by mammals.

palaeomagnetism science of the reconstruction of the Earth's ancient magnetic field and the former positions of the continents from the evidence of *remnant magnetization* in ancient rocks; that is, traces left by the Earth's magnetic field in ◊igneous rocks before they cool. Palaeomagnetism shows that the Earth's magnetic field has reversed itself – the magnetic north pole becoming the magnetic south pole, and vice versa – at approximate half-million-year intervals, with shorter reversal periods in between the major spans.

palaeontology the study of ancient life that encompasses the structure of ancient organisms and their environment, evolution, and ecology, as revealed by their ◊fossils. The practical aspects of palaeontology are based on using the presence of different fossils to date particular rock strata and to identify rocks that were laid down under particular conditions, for instance giving rise to the formation of oil.

Palaeozoic era of geological time 570–245 million years ago. It comprises the Cambrian, Ordovician, Silurian, Devonian, Carboniferous, and Permian periods. The Cambrian, Ordovician, and Silurian constitute the Lower or Early Palaeozoic; the Devonian, Carboniferous, and Permian make up the Upper or Late Palaeozoic. The era includes the evolution of hard-shelled multicellular life forms in the sea; the invasion of land by plants and animals; and the evolution of fish, amphibians, and early reptiles. The earliest identifiable fossils date from this era.

palate in mammals, the roof of the mouth. The bony front part is the hard palate, the muscular rear part the soft palate. Incomplete fusion of the two lateral halves of the palate (cleft palate) causes interference with speech.

palisade cell cylindrical cell lying immediately beneath the upper epidermis of a leaf. Palisade cells normally exist as one closely packed row and contain many chloroplasts. During the hours of daylight palisade cells are

palisade cell *Palisade cells are closely packed, columnar cells, lying in the upper surfaces of leaves. They contain many chloroplasts (the structures responsible for photosynthesis) and are well adapted to receive and process the components necessary for photosynthesis – carbon dioxide, water, and sunlight. For instance, their vertical arrangement means that there are fewer cross-walls to interfere with the passage of sunlight.*

photosynthetic, using the energy of the sun to create carbohydrates from water and carbon dioxide.

palladium lightweight, ductile and malleable, silver-white, metallic element, symbol Pd, atomic number 46, relative atomic mass 106.4.

It is one of the so-called platinum group of metals, and is resistant to tarnish and corrosion. It often occurs in nature as a free metal (see ◊native metal) in a natural alloy with platinum. Palladium is used as a catalyst, in alloys of gold (to make white gold) and silver, in electroplating, and in dentistry.

pancreas in vertebrates, an accessory gland of the digestive system located close to the duodenum. When stimulated by the hormone secretin, it releases enzymes into the duodenum that digest starches, proteins, and fats. In humans, it is about 18 cm/7 in long, and lies behind and below the stomach. It contains groups of cells called the *islets of Langerhans*, which secrete the hormones insulin and glucagon that regulate the blood sugar level.

Pangaea or *Pangea* single land mass, made up of all the present continents, believed to have existed between 250 and 200 million years ago; the rest of the Earth was covered by the Panthalassa ocean. Pangaea split into two land masses – ◊Laurasia in the north and ◊Gondwanaland in the south – which subsequently broke up into several continents. These then drifted slowly to their present positions (see ◊continental drift).

Panthalassa ocean that covered the surface of the Earth not occupied by the world continent ◊Pangaea between 250 and 200 million years ago.

pantothenic acid water-soluble ◊vitamin ($C_9H_{17}NO_5$) of the B complex, found in a wide variety of foods. Its absence from the diet can lead to dermatitis, and it is known to be involved in the breakdown of fats and carbohydrates.

paper thin, flexible material made in sheets from vegetable fibres (such as wood pulp) or rags and used for writing, drawing, printing, packaging, and various household needs. The name comes from papyrus, a form of writing material made from water reed, used in ancient Egypt. The invention of true paper, originally made of pulped fishing nets and rags, is credited to Tsai Lun, Chinese minister of agriculture, AD 105.

pappus (plural *pappi*) in botany, a modified ◊calyx comprising a ring of fine, silky hairs, or sometimes scales or small teeth, that persists after fertilization. Pappi are found in members of the daisy family (Compositae) such as the dandelions *Taraxacum*, where they form a parachutelike structure that aids dispersal of the fruit.

parabola in mathematics, a curve formed by cutting a right circular cone with a plane parallel to the sloping side of the cone. A parabola is one of the family of curves known as ◊conic sections. The graph of $y = x^2$ is a parabola.

paraffin common name for ◊alkane, any member of the series of hydrocarbons with the general formula C_nH_{2n+2}. The lower members are gases, such as methane (marsh or natural gas). The middle ones (mainly liquid) form the basis of petrol, kerosene, and lubricating oils, while the higher ones (paraffin waxes) are used in ointment and cosmetic bases.

paraldehyde common name for ◊*ethanal trimer*.

parallax the change in the apparent position of an object against its background when viewed from two different positions. In astronomy, nearby stars show a shift owing to parallax when viewed from different positions on the Earth's orbit around the Sun. A star's parallax is used to deduce its distance.

parallel circuit electrical circuit in which the components are connected side by side. The current flowing in the circuit is shared by the components. The division of the current across each conductor is in the ratio of their resistances.

parallel device in computing, a device that communicates binary data by sending the bits that represent each character simultaneously along a set of separate data lines, unlike a ◊serial device.

parabola *The parabola is a curve produced when a cone is cut by a plane. It is one of a family of curves called conic sections, which also includes the circle, ellipse, and hyperbole. These curves are produced when the plane cuts the cone at different angles and positions.*

parallel lines and parallel planes in mathematics, straight lines or planes that always remain a constant distance from one another no matter how far they are extended. This is a principle of Euclidean geometry. Some non-Euclidean geometries, such as elliptical and hyperbolic geometry, however, reject Euclid's parallel axiom.

parallelogram of forces in physics and applied mathematics, a method of calculating the resultant (combined effect) of two different forces acting together on an object. Because a force has both magnitude and direction it is a ◊vector quantity and can be represented by a straight line. A second force acting at the same point in a different direction can be represented by another line drawn at an angle to the first. By completing the parallelogram (of which the two lines are sides) a diagonal may be drawn from the original angle to the opposite corner to represent the resultant force vector.

parallel processing emerging computer technology that allows more than one computation at the same time. Although in the 1980s this technology enabled only a small number of computer processor units to work in parallel, in theory thousands or millions of processors could be used at the same time.

paralysis loss of voluntary movement due to failure of nerve impulses to reach the muscles involved. It may result from almost any disorder of the nervous system, including brain or spinal cord injury, poliomyelitis, stroke, and progressive conditions such as a tumour or multiple sclerosis. Paralysis may also involve loss of sensation due to sensory nerve disturbance.

paramagnetic material material that is weakly pulled towards a strong magnet. The effect is caused by unpaired electrons in the atoms of the material, which cause the atoms to act as weak magnets. A paramagnetic material has a fairly low magnetic ◊susceptibility that is inversely proportional to temperature. See ◊magnetism.

parameter variable factor or characteristic. For example, length is one parameter of a rectangle; its height is another. In computing, it is frequently useful to describe a program or object with a set of variable parameters rather than fixed values.

paraquat $CH_3(C_5H_4N)_2CH_3 \cdot 2CH_3SO_4$ (technical name *1,1-dimethyl-4,4-dipyridylium*) nonselective herbicide (weedkiller). Although quickly degraded by soil microorganisms, it is deadly to human beings if ingested.

parasite organism that lives on or in another organism (called the host) and depends on it for nutrition, often at the expense of the host's welfare. Parasites that live inside the host, such as liver flukes and tapeworms, are called *endoparasites*; those that live on the exterior, such as fleas and lice, are called *ectoparasites*.

parathyroid one of a pair of small ◊endocrine glands. Most tetrapod vertebrates, including humans, possess two such pairs, located behind the ◊thyroid gland. They secrete parathyroid hormone, which regulates the amount of calcium in the blood.

parenchyma plant tissue composed of loosely packed, more or less spherical cells, with thin cellulose walls. Although parenchyma often has no specialized function, it is usually present in large amounts, forming a packing or ground tissue. It usually has many intercellular spaces.

parental care in biology, the time and energy spent by a parent in order to rear its offspring to maturity. Among animals, it ranges from the simple provision of a food supply for the hatching young at the time the eggs are laid (for example, many wasps) to feeding and protection of the young after hatching or birth, as in birds and mammals. In the more social species, parental care may include the teaching of skills – for example, female cats teach their kittens to hunt.

parental generation in genetic crosses, the set of individuals at the start of a test, providing the first set of gametes from which subsequent generations (known as the F1 and F2) will arise.

parity of a number, the state of being either even or odd. In computing, the term refers to the number of 1s in the binary codes used to represent data. A binary representation has *even parity* if it contains an even number of 1s and *odd parity* if it contains an odd number of 1s.

parotid gland one of two salivary glands located on either side of the mouth just in front of the ear. Their ducts open on the inside of the cheeks.

parsec in astronomy, a unit (symbol pc) used for distances to stars and galaxies. One parsec is equal to 3.2616 ◊light years, 2.063×10^5 ◊astronomical units, and 3.086×10^{13} km.

parthenocarpy in botany, the formation of fruits without seeds. This phenomenon, of no obvious benefit to the plant, occurs naturally in some plants, such as bananas. It can also be induced in some fruit crops, either by breeding or by applying certain plant hormones.

parthenogenesis development of an ovum (egg) without any genetic contribution from a male. Parthenogenesis is the normal means of reproduction in a few plants (for example, dandelions) and animals (for example, certain fish). Some sexually reproducing species, such as aphids, show parthenogenesis at some stage in their life cycle.

particle detector one of a number of instruments designed to detect subatomic particles and track their paths; they include the ◊cloud chamber, ◊bubble chamber, ◊spark chamber, and multiwire chamber.

particle physics study of the particles that make up all atoms, and of their interactions. More than 300 subatomic particles have now been identified by physicists, categorized into several classes according to their mass, electric charge, spin, magnetic moment, and interaction. Subatomic particles include the ◊elementary particles (◊quarks, ◊leptons, and ◊gauge bosons), which are believed to be indivisible and so may be considered the fundamental units of matter; and the ◊hadrons (baryons, such as the proton and neutron, and mesons), which are composite particles, made up of two or three quarks. The proton, electron, and neutrino are the only stable particles (the neutron being stable only when in the atomic nucleus). The unstable particles decay rapidly into other particles, and are known from experiments with particle accelerators and cosmic radiation. See ◊atomic structure.

particle size in chemistry the size of the grains that make up a powder. The grain size has an effect on certain properties of a substance.

Finely divided powders have a greater surface area for contact; they therefore react more quickly, dissolve more readily, and are of increased efficiency as catalysts compared with their larger-sized counterparts.

particle, subatomic in physics, a particle that is smaller than an atom; see ◊particle physics.

PASCAL (French acronym for *program appliqué à la selection et la compilation automatique de la littérature*) a high-level computer-programming language. Designed by Niklaus Wirth (1934–) in the 1960s as an aid to teaching programming, it is still widely used as such in universities, but is also recognized as a good general-purpose programming language. It was named after 17th-century French mathematician Blaise Pascal.

pascal SI unit (symbol Pa) of pressure, equal to one newton per square metre. It replaces ◊bars and millibars (10^5 Pa equals one bar). It is named after the French mathematician Blaise Pascal.

pasteurization treatment of food to reduce the number of microorganisms it contains and so protect consumers from disease. Harmful bacteria are killed and the development of others is delayed. For milk, the method involves heating it to 72°C/161°F for 15 seconds followed by rapid cooling to 10°C/50°F or lower. The process also kills beneficial bacteria and reduces the nutritive property of milk.

patella or *kneecap* a flat bone embedded in the knee tendon of birds and mammals, which protects the joint from injury.

pathogen in medicine, any microorganism that causes disease. Most pathogens are ◊parasites, and the diseases they cause are incidental to their search for food or shelter inside the host. Nonparasitic organisms, such as soil bacteria or those living in the human gut and feeding on waste foodstuffs, can also become pathogenic to a person whose immune system or liver is damaged. The larger parasites that can cause disease, such as nematode worms, are not usually described as pathogens.

pathology medical speciality concerned with the study of disease processes and how these provoke structural and functional changes in the body.

PCI (abbreviation for *p*eripheral *c*omponent *i*nterconnect) form of ◊local bus connection between external devices and the main ◊central processing unit. Developed (but not owned) by ◊Intel, it is currently (1993) 32-bit, but is to be extended to 64-bit.

PCM abbreviation for ◊*pulse-code modulation*.

pd abbreviation for ◊*potential difference*.

pearl shiny, hard, rounded abnormal growth composed of nacre (or mother-of-pearl), a chalky substance. Nacre is secreted by many molluscs, and deposited in thin layers on the inside of the shell around a parasite, a grain of sand, or some other irritant body. After several years of the mantle (the layer of tissue between the shell and the body mass) secreting this nacre, a pearl is formed.

pectoral relating to the upper area of the thorax associated with the muscles and bones used in moving the arms or forelimbs, in vertebrates. In birds, the *pectoralis major* is the very large muscle used to produce a powerful downbeat of the wing during flight.

pedicel the stalk of an individual flower, which attaches it to the main floral axis, often developing in the axil of a bract.

Pegasus in astronomy, a constellation of the northern hemisphere, near Cygnus, and represented as the winged horse of Greek mythology.

pegmatite extremely coarse-grained ◊igneous rock of any composition found in veins; pegmatites are usually associated with large granite masses.

pellagra chronic disease mostly seen in sub-tropical countries in which the staple food is maize. It is caused by deficiency of ◊nicotinic acid (one of the B vitamins), which is contained in protein foods, beans and peas, and yeast. Symptoms include diarrhoea, skin eruptions, and mental disturbances.

Peltier effect in physics, a change in temperature at the junction of two different metals produced when an electric current flows through them. The extent of the change depends on what the conducting metals are, and the nature of change (rise or fall in temperature) depends on the direction of current flow. It is the reverse of the ◊Seebeck effect. It is named after the French physicist Jean Charles Peltier (1785–1845) who discovered it 1834.

pelvis in vertebrates, the lower area of the abdomen featuring the bones and muscles used to move the legs or hindlimbs. The *pelvic girdle* is a set of bones that allows movement of the legs in relation to the rest of the body and provides sites for the attachment of relevant muscles.

pendulum weight (called a 'bob') swinging at the end of a rod or cord. The regularity of a pendulum's swing was used in making the first really accurate clocks in the 17th century. Pendulums can be used for measuring the acceleration due to gravity (an important constant in physics), and in prospecting for oils and minerals.

penicillin any of a group of ◊antibiotic (bacteria killing) compounds obtained from filtrates of moulds of the genus *Penicillium* (especially *P. notatum*) or produced synthetically. Penicillin was the first antibiotic to be discovered (by Alexander Fleming (1881–1955)); it kills a broad spectrum of bacteria, many of which cause disease in humans.

penis male reproductive organ containing the ◊urethra, the channel through which urine and

◊semen are voided. It transfers sperm to the female reproductive tract to fertilize the ovum. In mammals, the penis is made erect by vessels that fill with blood, and in most mammals (but not humans) is stiffened by a bone.

Pennsylvanian US term for the Upper or Late ◊Carboniferous period of geological time, 323–290 million years ago; it is named after the US state.

pentadactyl limb typical limb of the mammals, birds, reptiles, and amphibians. These vertebrates (animals with backbone) are all descended from primitive amphibians whose immediate ancestors were fleshy-finned fish. The limb which evolved in those amphibians had three parts: a 'hand/foot' with five digits (fingers/toes), a lower limb containing two bones, and an upper limb containing one bone.

Pentium the most recent microchip (1993) produced by Intel. The Pentium is designed to take advantage of the ◊PCI ◊local bus architecture, which increases the bandwidth available between devices.

penumbra the region of partial shade between the totally dark part (umbra) of a ◊shadow and the fully illuminated region outside. It occurs when a source of light is only partially obscured by a shadow-casting object. The darkness of a penumbra varies gradually from total darkness at one edge to full brightness at the other. In astronomy, a penumbra is a region of the Earth from which only a partial ◊eclipse of the Sun or Moon can be seen.

pepsin enzyme that breaks down proteins during digestion. It requires a strongly acidic environment and is present in the stomach.

peptide molecule comprising two or more ◊amino acid molecules (not necessarily different) joined by *peptide bonds*, whereby the acid group of one acid is linked to the amino group of the other (–CO.NH). The number of amino acid molecules in the peptide is indicated by referring to it as a di-, tri-, or polypeptide (two, three, or many amino acids).

peptide bond bond that joins two peptides together within a protein. The carboxyl (–COOH) group on one ◊amino acid reacts with the amino (–NH$_2$) group on another amino acid to form a peptide bond (–CO–NH) with the elimination of water.

perborate any salt formed by the action of hydrogen peroxide on borates. Perborates contain the radical BO$_3$.

perennating organ in plants, that part of a ◊biennial plant or herbaceous perennial that allows it to survive the winter; usually a root, tuber, rhizome, bulb, or corm.

perennial plant plant that lives for more than two years. Herbaceous perennials have aerial stems and leaves that die each autumn. They survive the winter by means of an underground storage (perennating) organ, such as a bulb or rhizome. Trees and shrubs or woody perennials have stems that persist above ground throughout the year, and may be either ◊deciduous or ◊evergreen. See also ◊annual plant, ◊biennial plant.

perianth in botany, a collective term for the outer whorls of the flower, which protect the reproductive parts during development. In most ◊dicotyledons the perianth is composed of two distinct whorls, the calyx of ◊sepals and the corolla of ◊petals, whereas in many ◊monocotyledons the sepals and petals are indistinguishable and the segments of the perianth are then known individually as tepals.

periastron in astronomy, the point at which an object travelling in an elliptical orbit around a star is at its closest to the star; the point at which it is furthest is known as the apastron.

pericardium membranous sac surrounding the heart.

pericarp wall of a ◊fruit. It encloses the seeds and is derived from the ◊ovary wall. In fruits such as the acorn, the pericarp becomes dry and hard, forming a shell around the seed. In fleshy fruits the pericarp is typically made up of three distinct layers. The *epicarp*, or *exocarp*, forms the tough outer skin of the fruit, while the *mesocarp* is often fleshy and forms the middle layers. The innermost layer or *endocarp*, which surrounds the seeds, may be membranous or thick and hard, as in the ◊drupe (stone) of cherries, plums, and apricots.

peridot pale-green, transparent gem variety of the mineral ◊olivine.

peridotite rock consisting largely of the mineral olivine; pyroxene and other minerals may also be present. Peridotite is an ultrabasic rock containing less than 45% silica by weight. It is believed to be one of the rock types making up the Earth's upper mantle, and is sometimes brought from the depths to the surface by major movements, or as inclusions in lavas.

perigee the point at which an object, travelling in an elliptical orbit around the Earth, is at its closest to the Earth. The point at which it is furthest from the Earth is the apogee.

perihelion the point at which an object, travelling in an elliptical orbit around the Sun, is at its closest to the Sun. The point at which it is furthest from the Sun is the aphelion.

perimeter or *boundary* line drawn around the edge of an area or shape. For example, the perimeter of a rectangle is the sum of its four sides; the perimeter of a circle is known as its *circumference*.

period in chemistry, a horizontal row of elements in the ◊periodic table. There is a gradation of properties along each period, from metallic (group I, the alkali metals) to nonmetallic (group VII, the halogens).

period in physics, the time taken for one complete cycle of a repeated sequence of events. For example, the time taken for a pendulum to swing from side to side and back again is the period of the pendulum.

periodic table of the elements in chemistry, a table in which the elements are arranged in order of their atomic number. The table summarizes the major properties of the elements and enables predictions to be made about their behaviour.

There are striking similarities in the chemical properties of the elements in each of the vertical columns (called *groups*), which are numbered I–VII and 0 to reflect the number of electrons in the outermost unfilled shell and hence the maximum ◊valency. A gradation of properties may be traced along the horizontal rows (called *periods*). Metallic character increases across a period from right to left, and down a group. A large block of elements, between groups II and III, contains the transition elements, characterized by displaying more than one valency state.

These features are a direct consequence of the electronic (and nuclear) structure of the atoms of the elements. The relationships established between the positions of elements in the periodic table and their major properties has enabled scientists to predict the properties of other elements.

peripheral device in computing, any item of equipment attached to and controlled by a computer. Peripherals are typically for input from and output to the user (for example, a keyboard or printer), storing data (for example, a disc drive), communications (such as a modem), or for performing physical tasks (such as a robot).

periscope optical instrument designed for observation from a concealed position such as from a submerged submarine. In its basic form it consists of a tube with parallel mirrors at each end, inclined at 45° to its axis. The periscope attained prominence in naval and military operations of World War I.

peristalsis wavelike contractions, produced by the contraction of smooth muscle, that pass along tubular organs, such as the intestines. The same term describes the wavelike motion of earthworms and other invertebrates, in which part of the body contracts as another part elongates.

peritoneum membrane lining the abdominal cavity and digestive organs of vertebrates. *Peritonitis*, inflammation within the peritoneum, can occur due to infection or other irritation. It is sometimes seen following a burst appendix and quickly proves fatal if not treated.

permanent hardness hardness of water that cannot be removed by boiling (see ◊hard water).

permeability in physics, the degree to which the presence of a substance alters the magnetic field around it. Most substances have a small constant permeability. When the permeability is less than 1, the material is a ◊diamagnetic material; when it is greater than 1, it is a ◊paramagnetic material. ◊Ferrimagneticism results in very large permeabilities. See also ◊magnetism.

Permian period of geological time 290–245 million years ago, the last period of the Palaeozoic era. Its end was marked by a significant change in marine life, including the extinction of many corals and trilobites. Deserts were widespread, and terrestrial amphibians and mammal-like reptiles flourished. Cone-bearing plants (gymnosperms) came to prominence.

permutation in mathematics, a specified arrangement of a group of objects. It is the arrangement of a distinct objects taken b at a time in all possible orders. It is given by $a!/(a-b)!$, where '!' stands for ◊factorial. For example, the number of permutations of four letters taken from any group of six different letters is $6!/2! = (1 \times 2 \times 3 \times 4 \times 5 \times 6)/(1 \times 2) = 360$. The theoretical number of four-letter 'words' that can be made from an alphabet of 26 letters is $26!/22! = 358,800$.

perpetual motion the idea that a machine can be designed and constructed in such a way that, once started, it will continue in motion indefinitely without requiring any further input of energy (motive power). Such a device contradicts the two laws of thermodynamics that state that (1) energy can neither be created nor destroyed (the law of conservation of energy) and (2) heat cannot by itself flow from a cooler to a hotter object. As a result, all practical (real) machines require a continuous supply of energy, and no heat engine is able to convert all the heat into useful work.

Perseus in astronomy, a constellation of the northern hemisphere, near Cassiopeia, and represented as the mythological hero. The head of the decapitated Gorgon, Medusa, is marked by the variable star Algol. Perseus lies in the Milky Way and contains the Double Cluster, a twin cluster of stars. Every August the *Perseid meteor shower* radiates from its northern part.

Perspex trade name for a clear, lightweight, tough plastic first produced 1930. It is widely used for watch glasses, advertising signs, domestic baths, motorboat windshields, aircraft canopies, and protective shields. Its chemical name is polymethylmethacrylate (PMMA). It is manufactured under other names: Plexiglas (in the USA), Oroglas (in Europe), and Lucite.

perspiration excretion of water and dissolved substances from the ♦sweat glands of the skin of mammals. Perspiration has two main functions: body cooling by the evaporation of water from the skin surface, and excretion of waste products such as salts.

			III	IV	V	VI	VII	0
								2 Helium **He** 4002.60
			5 Boron **B** 10.81	6 Carbon **C** 12.011	7 Nitrogen **N** 14.0067	8 Oxygen **O** 15.9994	9 Fluorine **F** 18.99840	10 Neon **Ne** 20.179
			13 Aluminium **Al** 26.98154	14 Silicon **Si** 28.066	15 Phosphorus **P** 30.9738	16 Sulphur **S** 32.06	17 Chlorine **Cl** 35.453	18 Argon **Ar** 39.948
28 Nickel **Ni** 58.70	29 Copper **Cu** 63.546	30 Zinc **Zn** 65.38	31 Gallium **Ga** 69.72	32 Germanium **Ge** 72.59	33 Arsenic **As** 74.9216	34 Selenium **Se** 78.96	35 Bromine **Br** 79.904	36 Krypton **Kr** 83.80
46 Palladium **Pd** 106.4	47 Silver **Ag** 107.868	48 Cadmium **Cd** 112.40	49 Indium **In** 114.82	50 Tin **Sn** 118.69	51 Antimony **Sb** 121.75	52 Tellurium **Te** 127.75	53 Iodine **I** 126.9045	54 Xenon **Xe** 131.30
78 Platinum **Pt** 195.09	79 Gold **Au** 196.9665	80 Mercury **Hg** 200.59	81 Thallium **Tl** 204.37	82 Lead **Pb** 207.37	83 Bismuth **Bi** 207.2	84 Polonium **Po** 210	85 Astatine **At** 211	86 Radon **Rn** 222.0176

63 Europium **Eu** 151.96	64 Gadolinium **Gd** 157.25	65 Terbium **Tb** 158.9254	66 Dysprosium **Dy** 162.50	67 Holmium **Ho** 164.9304	68 Erbium **Er** 167.26	69 Thulium **Tm** 168.9342	70 Ytterbium **Yb** 173.04	71 Lutetium **Lu** 174.97

95 Americium **Am** 243.0614	96 Curium **Cm** 247.0703	97 Berkelium **Bk** 247	98 Californium **Cf** 251.0786	99 Einsteinium **Es** 252.0828	100 Fermium **Fm** 257.0951	101 Mendelevium **Me** 258.0986	102 Nobelium **No** 259.1009	103 Lawrencium **Lr** 260.1054

pest in biology, any insect, fungus, rodent, or other living organism that has a harmful effect on human beings, other than those that directly cause human diseases. Most pests damage crops or livestock, but the term also covers those that damage buildings, destroy food stores, and spread disease.

pesticide any chemical used in farming, gardening or indoors to combat pests. Pesticides are of three main types: *insecticides* (to kill insects), *fungicides* (to kill fungal diseases), and *herbicides* (to kill plants, mainly those considered weeds). Pesticides cause a number of pollution problems through spray drift onto surrounding areas, direct contamination of users or the public, and as residues on food.

The safest pesticides include those made from plants, such as the insecticides pyrethrum and derris.

petal part of a flower whose function is to attract pollinators such as insects or birds. Petals are frequently large and brightly coloured and may also be scented. Some have a nectary at the base and markings on the petal surface, known as honey guides, to direct pollinators to the source of the nectar. In wind-pollinated plants, however, the petals are usually small and insignificant, and sometimes absent altogether. Petals are derived from modified leaves, and are known collectively as a corolla.

petiole in botany, the stalk attaching the leaf blade, or ◊lamina, to the stem. Typically it is continuous with the midrib of the leaf and attached to the base of the lamina, but occasionally it is attached to the lower surface of the lamina, as in the nasturtium (a peltate leaf). Petioles that are flattened and leaflike are termed phyllodes. Leaves that lack a petiole are said to be sessile.

petrochemical chemical derived from the processing of petroleum (crude oil).

Petrochemical industries are those that obtain their raw materials from the processing of petroleum and natural gas. Polymers, detergents, solvents, and nitrogen fertilizers are all major products of the petrochemical industries. Inorganic chemical products include carbon black, sulphur, ammonia, and hydrogen peroxide.

petrol mixture of hydrocarbons derived from petroleum, mainly used as a fuel for internal combustion engines. It is colourless and highly volatile. *Leaded petrol* contains antiknock (a mixture of tetraethyl lead and dibromoethane), which improves the combustion of petrol and the performance of a car engine. The lead from the exhaust fumes enters the atmosphere, mostly as simple lead compounds. There is strong evidence that it can act as a nerve poison on young children and cause mental impairment. This has prompted a gradual switch to the use of *unleaded petrol* in the UK.

petrol engine the most commonly used source of power for motor vehicles, introduced by the German engineers Gottlieb Daimler and Karl Benz 1885. The petrol engine is a complex piece of machinery made up of about 150 moving parts. It is a reciprocating piston engine, in which a number of pistons move up and down in cylinders. The motion of the pistons rotate a crankshaft, at the end of which is a heavy flywheel. From the flywheel the power is transferred to the car's driving wheels via the transmission system of clutch, gearbox, and final drive.

petroleum or *crude oil* natural mineral oil, a thick greenish-brown flammable liquid found underground in permeable rocks. Petroleum consists of hydrocarbons mixed with oxygen, sulphur, nitrogen, and other elements in varying proportions. It is thought to be derived from ancient organic material that has been converted by, first, bacterial action, then heat and pressure (but its origin may be chemical also). From crude petroleum, various products are made by distillation and other processes; for example, fuel oil, petrol, kerosene, diesel, lubricating oil, paraffin wax,

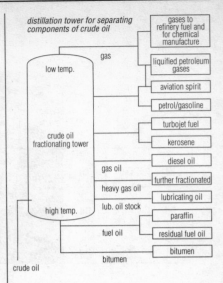

petroleum *Refining petroleum using a distillation column. The crude petroleum is fed in at the bottom of the column where the temperature is high. The gases produced rise up the column, cooling as they travel. At different heights up the column, different gases condense to liquids called fractions, and are drawn off.*

and petroleum jelly. Global consumption of petroleum 1993 was 23 billion barrels.

The organic material in petroleum was laid down millions of years ago (hence it is known as a fossil fuel). Petroleum is found trapped in porous reservoir rocks, such as sandstone or limestone, in anticlines and other traps below impervious rock layers. Oil may flow from wells under natural pressure, causing it to rise up the borehole, but many oil wells require pumping to bring the oil to the surface.

Petroleum products and chemicals are used in large quantities in the manufacture of detergents, artificial fibres, plastics, insecticides, fertilizers, pharmaceuticals, toiletries, and synthetic rubber. Aviation fuel is a volatile form of petrol.

petrology branch of geology that deals with the study of rocks, their mineral compositions, and their origins.

pewter any of various alloys of mostly tin with varying amounts of lead, copper, or antimony. Pewter has been known for centuries and was once widely used for domestic utensils but is now used mainly for ornamental ware.

pH scale from 0 to 14 for measuring acidity or alkalinity. A pH of 7.0 indicates neutrality, below 7 is acid, while above 7 is alkaline. Strong acids, such as those used in car batteries, have a pH of about 2; strong alkalis such as sodium hydroxide are pH 13.

The pH of a solution can be measured by using a broad-range indicator, either in solution or as a paper strip. The colour produced by the indicator is compared with a colour code related to the pH value. An alternative method is to use a pH meter fitted with a glass electrode.

phage another name for a ◊bacteriophage, a virus that attacks bacteria.

phagocyte type of white blood cell, or ◊leucocyte, that can engulf a bacterium or other invading microorganism. Phagocytes are found in blood, lymph, and other body tissues, where they also ingest foreign matter and dead tissue. A ◊macrophage differs in size and life span.

phanerogam obsolete term for a plant that bears flowers or cones and reproduces by means of seeds, that is an ◊angiosperm and ◊gymnosperm, or a ◊seed plant. Plants such as mosses, fungi, and ferns were known as *cryptogams*.

Phanerozoic eon in Earth history, consisting of the most recent 570 million years. It comprises the Palaeozoic, Mesozoic, and Cenozoic eras. The vast majority of fossils come from this eon, owing to the evolution of hard shells and internal skeletons. The name means 'interval of well-displayed life'.

pharmacology study of the properties of drugs and their effects on the human body.

pharmacy the preparation and dispensing of drugs; as an area of study, pharmacy includes an understanding of the origins, nature, and action of drugs. The place where a pharmacist dispenses drugs is also referred to as a pharmacy.

pharynx muscular cavity behind the nose and mouth, extending downwards from the base of the skull. Its walls are made of muscle strengthened with a fibrous layer and lined with mucous membrane. The internal nostrils lead backwards into the pharynx, which continues downwards into the oesophagus and (through the epiglottis) into the windpipe. On each side, a Eustachian tube enters the pharynx from the middle ear cavity.

pH *The pHs of some common substances. The lower the pH the more acidic the substance.*

phase in astronomy, the apparent shape of the Moon or a planet when all or part of its illuminated hemisphere is facing the Earth.

The Moon undergoes a full cycle of phases from new (when between the Earth and the Sun) through first quarter (when at 90° eastern elongation from the Sun), full (when opposite the Sun), and last quarter (when at 90° western elongation from the Sun).

phase in chemistry, a physical state of matter: for example, ice and liquid water are different phases of water; a mixture of the two is termed a two-phase system.

phase in physics, a stage in an oscillatory motion, such as a wave motion: two waves are in phase when their peaks and their troughs coincide. Otherwise, there is a *phase difference*, which has consequences in ◊interference phenomena and ◊alternating current electricity.

phenol member of a group of aromatic chemical compounds with weakly acidic properties, which are characterized by a hydroxyl (OH) group attached directly to an aromatic ring. The simplest of the phenols, derived from benzene, is also known as phenol and has the formula C_6H_5OH. It is sometimes called *carbolic acid* and can be extracted from coal tar.

phenolphthalein acid–base indicator that is clear below pH 8 and red above pH 9.6. It is used in titrating weak acids against strong bases.

phenotype in genetics, visible traits, those actually displayed by an organism. The phenotype is not a direct reflection of the ◊genotype because some alleles are masked by the presence of other, dominant alleles (see ◊dominance). The phenotype is further modified by the effects of the environment (for example, poor nutrition stunts growth).

pheromone chemical signal (such as an odour) that is emitted by one animal and affects the behaviour of others. Pheromones are used by many animal species to attract mates.

phloem tissue found in vascular plants whose main function is to conduct sugars and other food materials from the leaves, where they are produced, to all other parts of the plant.

Phloem is composed of sieve elements and their associated companion cells, together with some ◊sclerenchyma and ◊parenchyma cell types. Sieve elements are long, thin-walled cells joined end to end, forming sieve tubes; large pores in the end walls allow the continuous passage of nutrients. Phloem is usually found in association with ◊xylem, the water-conducting tissue, but unlike the latter it is a living tissue.

phlogiston hypothetical substance formerly believed to have been produced during combustion. The term was invented by the German chemist Georg Stahl. The phlogiston theory was replaced by the theory of oxygen gain and loss, first enunciated by the French chemist Antoine Lavoisier.

Phobos one of the two moons of Mars, discovered 1877 by the US astronomer Asaph Hall (1829–1907). It is an irregularly shaped

phenol *The phenol molecule with its ring of six carbon atoms and a hydroxyl (OH) group attached. Phenol was first extracted from coal tar in 1834. It is used to make phenolic and epoxy resins, explosives, pharmaceuticals, perfumes, and nylon.*

lump of rock, cratered by ◊meteorite impacts. Phobos is 27 × 22 × 19 km/17 × 13 × 12 mi across, and orbits Mars every 0.32 days at a distance of 9,400 km/5,840 mi from the planet's centre. It is thought to be an asteroid captured by Mars' gravity.

phon unit of loudness, equal to the value in decibels of an equally loud tone with frequency 1,000 Hz. The higher the frequency, the louder a noise sounds for the same decibel value; thus an 80-decibel tone with a frequency of 20 Hz sounds as loud as 20 decibels at 1,000 Hz, and the phon value of both tones is 20. An aircraft engine has a loudness of around 140 phons.

phosphate salt or ester of ◊phosphoric acid. Incomplete neutralization of phosphoric acid gives rise to acid phosphates (see ◊acid salts and ◊buffer). Phosphates are used as fertilizers, and are required for the development of healthy root systems. They are involved in many biochemical processes, often as part of complex molecules, such as ◊ATP.

phosphate analysis the regular sampling and chemical analysis of phosphorus levels in the soil around archaeological sites in order to locate concentrations of human bone and excrement, and hence areas of human activity, settlements, and burial grounds.

phospholipid any ◊lipid consisting of a glycerol backbone, a phosphate group, and two long chains. Phospholipids are found everywhere in living systems as the basis for biological membranes.

phosphor any substance that is phosphorescent, that is, gives out visible light when it is illuminated by a beam of electrons or ultraviolet light. The television screen is coated on the inside with phosphors that glow when beams of electrons strike them. Fluorescent lamp tubes are also phosphor-coated. Phosphors are also used in Day-Glo paints, and as optical brighteners in detergents.

phosphorescence in physics, the emission of light by certain substances after they have absorbed energy, whether from visible light, other electromagnetic radiation such as ultraviolet rays or X-rays, or cathode rays (a beam of electrons). When the stimulating energy is removed phosphorescence ceases, although it may persist for a short time after (unlike ◊fluorescence, which stops immediately).

phosphoric acid acid derived from phosphorus and oxygen. Its commonest form (H_3PO_4) is also known as orthophosphoric acid, and is produced by the action of phosphorus pentoxide (P_2O_5) on water. It is used in rust removers and for rust-proofing iron and steel.

phosphorus highly reactive, nonmetallic element, symbol P, atomic number 15, relative atomic mass 30.9738. It occurs in nature as phosphates (commonly in the form of the mineral ◊apatite), and is essential to plant and animal life. Compounds of phosphorus are used in fertilizers, various organic chemicals, for matches and fireworks, and in glass and steel.

PhotoCD picture storage and viewing system developed by Kodak and Philips. The aim of Kodak's PhotoCD is to allow the user to put up to 100 photos onto compact disc: images are transferred from film to a PhotoCD disc and can then be viewed by means of Kodak's own PhotoCD player, which plugs into a television set.

photocell or *photoelectric cell* device for measuring or detecting light or other electromagnetic radiation, since its electrical state is altered by the effect of light. In a *photoemissive* cell, the radiation causes electrons to be emitted and a current to flow (◊photoelectric effect); a *photovoltaic* cell causes an ◊electromotive force to be generated in the presence of light across the boundary of two substances. A *photoconductive* cell, which contains a semiconductor, increases its conductivity when exposed to electromagnetic radiation.

photochemical reaction any chemical reaction in which light is produced or light initiates the reaction. Light can initiate reactions by exciting atoms or molecules and making them more reactive: the light energy becomes converted to chemical energy. Many photochemical reactions set up a ◊chain reaction and produce ◊free radicals.

photodiode semiconductor ◊p–n junction diode used to detect light or measure its intensity. The photodiode is encapsulated in a transparent plastic case that allows light to fall onto the junction. When this occurs, the reverse-bias resistance (high resistance in the opposite direction to normal current-flow) drops and allows a larger reverse-biased current to flow through the device. The increase in current can then be related to the amount of light falling on the junction.

photoelectric cell alternative name for ◊photocell.

photoelectric effect in physics, the emission of ◊electrons from a substance (usually a metallic surface) when it is struck by ◊photons (quanta of electromagnetic radiation), usually those of visible light or ultraviolet radiation.

photoluminescence see ◊luminescence.

photolysis chemical reaction that is driven by light or ultraviolet radiation. For example, the light reaction of ◊photosynthesis (the process by which green plants manufacture carbohydrates from carbon dioxide and water) is a photolytic reaction.

photometer instrument that measures luminous intensity, usually by comparing relative intensities from different sources. Bunsen's grease-spot photometer 1844 compares the intensity of a light source with a known source by each illuminating one half of a translucent area. Modern photometers use ◊photocells, as in a photographer's exposure meter. A photomultiplier can also be used as a photometer.

photomultiplier instrument that detects low levels of electromagnetic radiation (usually visible light or ◊infrared radiation) and amplifies it to produce a detectable signal.

photon in physics, the ◊elementary particle or 'package' (quantum) of energy in which light and other forms of electromagnetic radiation are emitted. The photon has both particle and wave properties; it has no charge, is considered massless but possesses momentum and energy. It is one of the ◊gauge bosons, a particle that cannot be subdivided, and is the carrier of the ◊electromagnetic force, one of the fundamental forces of nature.

According to ◊quantum theory the energy of a photon is given by the formula $E = hf$, where h is Planck's constant and f is the frequency of the radiation emitted.

photoperiodism biological mechanism that determines the timing of certain activities by responding to changes in day length. The flowering of many plants is initiated in this way. Photoperiodism in plants is regulated by a light-sensitive pigment, ***phytochrome***. The breeding seasons of many temperate-zone animals are also triggered by increasing or declining day length, as part of their ◊biorhythms.

photosphere visible surface of the Sun, which emits light and heat. About 300 km/200 mi deep, it consists of incandescent gas at a temperature of 5,800K (5,530°C/9,980°F).

Rising cells of hot gas produce a mottling of the photosphere known as ***granulation***, each granule being about 1,000 km/620 mi in diameter. The photosphere is often marked by large, dark patches called ◊sunspots.

photosynthesis process by which green plants trap light energy and use it to drive a series of chemical reactions, leading to the formation of carbohydrates. All animals ultimately depend on photosynthesis because it is the method by which the basic food (sugar) is created. For photosynthesis to occur, the plant must possess ◊chlorophyll and must have a supply of carbon dioxide and water. Actively photosynthesizing green plants store excess sugar as starch (this can be tested for in the laboratory using iodine).

The chemical reactions of photosynthesis occur in two stages. During the ***light reaction*** sunlight is used to split water (H_2O) into oxygen (O_2), protons (hydrogen ions, H^+), and electrons, and oxygen is given off as a by-product. In the ***dark reaction***, for which sunlight is not required, the protons and electrons are used to convert carbon dioxide (CO_2) into carbohydrates ($C_m(H_2O)_n$). Photosynthesis depends on the ability of chlorophyll to capture the energy of sunlight and to use it to split water molecules.

Other pigments, such as ◊carotenoids, are also involved in capturing light energy and passing it on to chlorophyll. Photosynthesis by cyanobacteria was responsible for the appearance of oxygen in the Earth's atmosphere 2 billion years ago, and photosynthesis by plants maintains the oxygen level today.

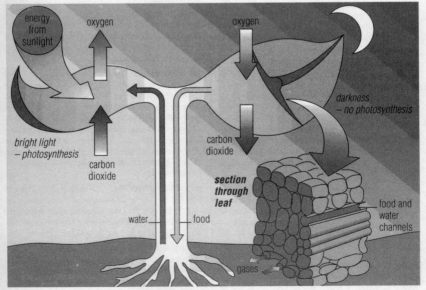

photosynthesis *Process by which green plants and some bacteria manufacture carbohydrates from water and atmospheric carbon dioxide, using the energy of sunlight. Photosynthesis depends on the ability of chlorophyll molecules within plant cells to trap the energy of light to split water molecules, giving off oxygen as a by-product. The hydrogen of the water molecules is then used to reduce carbon dioxide to simple carbohydrates.*

phototropism movement of part of a plant toward or away from a source of light. Leaves are positively phototropic, detecting the source of light and orientating themselves to receive the maximum amount.

phyllite ◊metamorphic rock produced under increasing temperature and pressure, in which mica crystals are aligned so that the rock splits along their plane of orientation, the resulting break being shiny and smooth. It is intermediate between slate and schist.

phyllotaxis the arrangement of leaves on a plant stem. Leaves are nearly always arranged in a regular pattern and in the majority of plants they are inserted singly, either in a *spiral* arrangement up the stem, or on *alternate* sides. Other principal forms are opposite leaves, where two arise from the same node, and whorled, where three or more arise from the same node.

phylogeny historical sequence of changes that occurs in a given species during the course of its evolution. It was once erroneously associated with ontogeny (the process of development of a living organism).

phylum (plural *phyla*) major grouping in biological classification. Mammals, birds, reptiles, amphibians, fishes, and tunicates belong to the phylum Chordata; the phylum Mollusca consists of snails, slugs, mussels, clams, squid, and octopuses; the phylum Porifera contains sponges; and the phylum Echinodermata includes starfish, sea urchins, and sea cucumbers. In classifying plants (where the term 'division' often takes the place of 'phylum'), there are between four and nine phyla depending on the criteria used; all flowering plants belong to a single phylum, Angiospermata, and all conifers to another, Gymnospermata. Related phyla are grouped together in a ◊kingdom; phyla are subdivided into classes.

physical change in chemistry, a type of change that does not produce a new chemical substance, does not involve large energy changes, and that can be easily reversed (the opposite of a ◊chemical change). Boiling and melting are examples of physical change.

physical chemistry branch of chemistry concerned with examining the relationships between the chemical compositions of substances and the physical properties that they display. Most chemical reactions exhibit some physical phenomenon (change of state, temperature, pressure, or volume, or the use or production of electricity), and the measurement and study of such phenomena has led to many chemical theories and laws.

physics branch of science concerned with the laws that govern the structure of the universe, and the forms of matter and energy and their interactions. For convenience, physics is often divided into branches such as nuclear physics, particle physics, solid- and liquid-state physics, electricity, electronics, magnetism, optics, acoustics, heat, and thermodynamics. Before the 20th century, physics was known as *natural philosophy*.

physiology branch of biology that deals with the functioning of living organisms, as opposed to anatomy, which studies their structures.

phytomenadione one form of vitamin K, a fat-soluble chemical found in green vegetables. It is involved in the production of ◊prothrombin, which is essential in blood clotting. It is given to newborns to prevent potentially fatal brain haemorrhages.

pi symbol π, the ratio of the circumference of a circle to its diameter. The value of pi is 3.1415926, correct to seven decimal places. Common approximations to pi are $^{22}/_7$ and 3.14, although the value 3 can be used as a rough estimation.

picric acid $C_6H_2(NO_2)_3OH$ (technical name *2,4,6-trinitrophenol*) strong acid that is used to dye wool and silks yellow, for the treatment of burns, and in the manufacture of explosives. It is a yellow, crystalline solid.

piezoelectric effect property of some crystals (for example, quartz) to develop an electromotive force or voltage across opposite faces when subjected to a mechanical strain, and, conversely, to expand or contract in size when subjected to an electromotive force. Piezoelectric crystal ◊oscillators are used as frequency standards (for example, replacing balance wheels in watches), and for producing ◊ultrasound.

pig iron or *cast iron* the quality of iron produced in a ◊blast furnace. It contains around 4% carbon plus some other impurities.

Pill, the commonly used term for the contraceptive pill, based on female hormones. The combined pill, which contains synthetic hormones similar to oestrogen and progesterone, stops the production of eggs, and makes the mucus produced by the cervix hostile to sperm. It is the most effective form of contraception apart from sterilization, being more than 99% effective.

physics: chronology

c. 400 BC	The first 'atomic' theory was put forward by Democritus.
c. 250	Archimedes' principle of buoyancy was established.
AD 1600	Magnetism was described by William Gilbert.
1608	Hans Lippershey invented the refracting telescope.
c. 1610	The principle of falling bodies descending to earth at the same speed was established by Galileo.
1642	The principles of hydraulics were put forward by Blaise Pascal.
1643	The mercury barometer was invented by Evangelista Torricelli.
1656	The pendulum clock was invented by Christiaan Huygens.
1662	Boyle's law concerning the behaviour of gases was established by Robert Boyle.
c. 1665	Isaac Newton put forward the law of gravity, stating that the Earth exerts a constant force on falling bodies.
1690	The wave theory of light was propounded by Christiaan Huygens.
1704	The corpuscular theory of light was put forward by Isaac Newton.
1714	The mercury thermometer was invented by Daniel Fahrenheit.
1764	Specific and latent heats were described by Joseph Black.
1771	The link between nerve action and electricity was discovered by Luigi Galvani.
c. 1787	Charles's law relating the pressure, volume, and temperature of a gas was established by Jacques Charles.
1795	The metric system was adopted in France.
1798	The link between heat and friction was discovered by Benjamin Rumford.
1800	Alessandro Volta invented the Voltaic cell.
1801	Interference of light was discovered by Thomas Young.
1808	The 'modern' atomic theory was propounded by John Dalton.
1811	Avogadro's hypothesis relating volumes and numbers of molecules of gases was proposed by Amedeo Avogadro.
1814	Fraunhofer lines in the solar spectrum were mapped by Joseph von Fraunhofer.
1815	Refraction of light was explained by Augustin Fresnel.
1819	The discovery of electromagnetism was made by Hans Oersted.
1821	The dynamo principle was described by Michael Faraday; the thermocouple was discovered by Thomas Seebeck.
1822	The laws of electrodynamics were established by André Ampère.
1824	Thermodynamics as a branch of physics was proposed by Sadi Carnot.
1827	Ohm's law of electrical resistance was established by Georg Ohm; Brownian movement resulting from molecular vibrations was observed by Robert Brown.
1829	The law of gaseous diffusion was established by Thomas Graham.
1831	Electromagnetic induction was discovered by Faraday.
1834	Faraday discovered self-induction.
1842	The principle of conservation of energy was observed by Julius von Mayer.
c. 1847	The mechanical equivalent of heat was described by James Joule.
1849	A measurement of speed of light was put forward by French physicist Armand Fizeau (1819–1896).
1851	The rotation of the Earth was demonstrated by Jean Foucault.
1858	The mirror galvanometer, an instrument for measuring small electric currents, was invented by William Thomson (Lord Kelvin).
1859	Spectrographic analysis was made by Robert Bunsen and Gustav Kirchhoff.
1861	Osmosis was discovered.
1873	Light was conceived as electromagnetic radiation by James Maxwell.
1877	A theory of sound as vibrations in an elastic medium was propounded by John Rayleigh.
1880	Piezoelectricity was discovered by Pierre Curie.
1887	The existence of radio waves was predicted by Heinrich Hertz.
1895	X-rays were discovered by Wilhelm Röntgen.
1896	The discovery of radioactivity was made by Antoine Becquerel.
1897	Joseph Thomson discovered the electron.
1899	Ernest Rutherford discovered alpha and beta rays.
1900	Quantum theory was propounded by Max Planck; the discovery of gamma rays was made by French physicist Paul-Ulrich Villard (1860–1934).
1902	Oliver Heaviside discovered the ionosphere.
1904	The theory of radioactivity was put forward by Rutherford and Frederick Soddy.
1905	Albert Einstein propounded his special theory of relativity.

1908	The Geiger counter was invented by Hans Geiger and Rutherford.
1911	The discovery of the atomic nucleus was made by Rutherford.
1913	The orbiting electron atomic theory was propounded by Danish physicist Niels Bohr.
1915	X-ray crystallography was discovered by William and Lawrence Bragg.
1916	Einstein put forward his general theory of relativity; mass spectrography was discovered by William Aston.
1924	Edward Appleton made his study of the Heaviside layer.
1926	Wave mechanics was introduced by Erwin Schrödinger.
1927	The uncertainty principle of atomic physics was established by Werner Heisenberg.
1931	The cyclotron was developed by Ernest Lawrence.
1932	The discovery of the neutron was made by James Chadwick; the electron microscope was developed by Vladimir Zworykin.
1933	The positron, the antiparticle of the electron, was discovered by Carl Anderson.
1934	Artificial radioactivity was developed by Frédéric and Irène Joliot-Curie.
1939	The discovery of nuclear fission was made by Otto Hahn and Fritz Strassmann.
1942	The first controlled nuclear chain reaction was achieved by Enrico Fermi.
1956	The neutrino, an elementary particle, was discovered by Clyde Cowan and Fred Reines.
1960	The Mössbauer effect of atom emissions was discovered by Rudolf Mössbauer; the first maser was developed by US physicist Theodore Maiman (1927–).
1963	Maiman developed the first laser.
1964	Murray Gell-Mann and George Zweig discovered the quark.
1967	Jocelyn Bell (now Burnell) and Antony Hewish discovered pulsars (rapidly rotating neutron stars that emit pulses of energy).
1971	The theory of superconductivity was announced, where electrical resistance in some metals vanishes above absolute zero.
1979	The discovery of the asymmetry of elementary particles was made by US physicists James W Cronin and Val L Fitch.
1982	The discovery of processes involved in the evolution of stars was made by Subrahmanyan Chandrasekhar and William Fowler.
1983	Evidence of the existence of weakons (W and Z particles) was confirmed at CERN, validating the link between the weak nuclear force and the electromagnetic force.
1986	The first high-temperature superconductor was discovered, able to conduct electricity without resistance at a temperature of 238°C/–396°F.
1989	CERN's Large Electron Positron Collider (LEP), a particle accelerator with a circumference of 27 km/16.8 mi, came into operation.
1991	LEP experiments demonstrated the existence of three generations of elementary particles, each with two quarks and two leptons.
1992	Japanese researchers developed a material that becomes superconducting at –103°C/–153°F (about 45°C/80°F warmer than the previous record).
1993	Top quark discovered at Fermilab, the US particle-physics laboratory, near Chicago, USA.

Piltdown man fossil skull fragments 'discovered' by Charles Dawson at Piltdown, E Sussex, England, in 1913, and believed to be the earliest European human remains until proved a hoax in 1953 (the jaw was that of an orang-utan).

pineal body or *pineal gland* a cone-shaped outgrowth of the vertebrate brain. In some lower vertebrates, it develops a rudimentary lens and retina, which show it to be derived from an eye, or pair of eyes, situated on the top of the head in ancestral vertebrates. In fishes that can change colour to match the background, the pineal perceives the light level and controls the colour change. In birds, the pineal detects changes in daylight and stimulates breeding behaviour as spring approaches. Mammals also have a pineal gland, but it is located deeper within the brain. It secretes a hormone, melatonin, thought to influence rhythms of activity. In humans, it is a small piece of tissue attached by a stalk to the rear wall of the third ventricle of the brain.

pinhole camera the simplest type of camera, in which a pinhole rather than a lens is used to form an image. Light passes through the pinhole at one end of a box to form a sharp inverted image on the inside surface of the opposite end. The image is equally sharp for objects placed at different distances from the camera because only one ray from a particular distance or direction can enter through the tiny pinhole, and so only one corresponding point of light will be produced on the image. A photographic film or plate fitted inside the box will, if exposed for a long time, record the image.

pinna in botany, the primary division of a ◊pinnate leaf. In mammals, the pinna is the external part of the ear.

pinnate leaf leaf that is divided up into many small leaflets, arranged in rows along either side of a midrib, as in ash trees (*Fraxinus*). It is a type of compound leaf. Each leaflet is known as a *pinna*, and where the pinnae are themselves divided, the secondary divisions are known as pinnules.

pint imperial dry or liquid measure of capacity equal to 20 fluid ounces, half a quart, one-eighth of a gallon, or 0.568 litre. In the US, a liquid pint is equal to 0.473 litre, while a dry pint is equal to 0.550 litre.

pion or *pi meson* in physics, any of three ◊mesons (positive, negative, neutral) that play a role in binding together the neutrons and protons in the nucleus of an atom. They belong to the ◊hadron class of ◊elementary particles.

Pioneer probe any of a series of US Solar-System space probes 1958–78. The probes *Pioneer 4–9* went into solar orbit to monitor the Sun's activity during the 1960s and early 1970s. *Pioneer 5*, launched 1960, was the first of a series to study the solar wind between the planets. *Pioneer 10*, launched March 1972, was the first probe to reach Jupiter (Dec 1973) and to leave the Solar System 1983. *Pioneer 11*, launched April 1973, passed Jupiter Dec 1974, and was the first probe to reach Saturn (Sept 1979), before also leaving the Solar System.

pioneer species in ecology, those species that are the first to colonize and thrive in new areas. Coal tips, recently cleared woodland, and new roadsides are areas where pioneer species will quickly appear. As the habitat matures other species take over, a process known as *succession*.

pipette device for the accurate measurement of a known volume of liquid, usually for transfer from one container to another, used in chemistry and biology laboratories.

Pisces zodiac constellation, mainly in the northern hemisphere between Aries and Aquarius, near Pegasus. It is represented by two fish tied together by their tails. The Circlet, a delicate ring of stars, marks the head of the western fish in Pisces. The constellation contains the *vernal equinox*, the point at which the Sun's path around the sky (the *ecliptic*) crosses the celestial equator. The Sun reaches this point around 21 March each year as it passes through Pisces from mid-March to late April. In astrology, the dates for Pisces are between about 19 Feb and 20 March.

Piscis Austrinus or *Southern Fish* constellation of the southern hemisphere near Capricornus. Its brightest star is ◊Fomalhaut.

pistil general term for the female part of a flower, either referring to one single ◊carpel or a group of several fused carpels.

piston barrel-shaped device used in reciprocating engines (steam, petrol, diesel oil) to harness power. Pistons are driven up and down in cylinders by expanding steam or hot gases. They pass on their motion via a connecting rod and crank to a crankshaft, which turns the driving wheels. In a pump or compressor, the role of the piston is reversed, being used to move gases and liquids. See also ◊internal-combustion engine.

pitch in chemistry, a black, sticky substance, hard when cold, but liquid when hot, used for waterproofing, roofing, and paving. It is made by the destructive distillation of wood or coal tar, and has been used since antiquity for caulking wooden ships.

pitch in mechanics, the distance between the adjacent threads of a screw or bolt. When a screw is turned through one full turn it moves up or down a distance equal to the pitch of its thread. A screw thread is a simple type of machine, acting like a rolled-up inclined plane, or ramp (as may be illustrated by rolling a long paper triangle around a pencil). A screw has a ◊mechanical advantage greater than one.

pitchblende or *uraninite* brownish-black mineral, the major constituent of uranium ore, consisting mainly of uranium oxide (UO_2). It also contains some lead (the final, stable product of uranium decay) and variable amounts of most of the naturally occurring radioactive elements, which are products of either the decay or the fissioning of uranium isotopes. The uranium yield is 50–80%; it is also a source of radium, polonium, and actinium. Pitchblende was first studied by Pierre and Marie Curie, who found radium and polonium in its residues in 1898.

Pitot tube instrument that measures fluid (gas and liquid) flow. It is used to measure the speed of aircraft, and works by sensing pressure differences in different directions in the airstream.

It was invented in the 1730s by the French scientist Henri Pitot (1695–1771).

pituitary gland major ◊endocrine gland of vertebrates, situated in the centre of the brain. It is attached to the ⟡hypothalamus by a stalk. The pituitary consisits of two lobes. The posterior lobe is an extension of the hypothalamus, and is in effect nervous tissue. It stores two hormones synthesized in the hypothalamus: ⟡ADH and ⟡oxytocin. The anterior lobe secretes six hormones, some of which control the activities of other glands (thyroid, gonads, and adrenal cortex); others are direct-acting hormones affecting milk secretion and controlling growth.

pixel (acronym for *picture element*) single dot on a computer screen. All screen images are made up of a collection of pixels, with each pixel being either off (dark) or on (illuminated, possibly in colour). The number of pixels available determines the screen's resolution. Typical resolutions of microcomputer screens vary from 320 × 200 pixels to 640 × 480 pixels, but screens with 1,024 × 768 pixels are now quite common for high-quality graphic (pictorial) displays.

placenta organ that attaches the developing ⟡embryo or ⟡fetus to the ⟡uterus in placental mammals (mammals other than marsupials, platypuses, and echidnas). Composed of maternal and embryonic tissue, it links the blood supply of the embryo to the blood supply of the mother, allowing the exchange of oxygen, nutrients, and waste products. The two blood systems are not in direct contact, but are separated by thin membranes, with materials diffusing across from one system to the other. The placenta also produces hormones that maintain and regulate pregnancy. It is shed as part of the afterbirth.

placer deposit detrital concentration of an economically important mineral, such as gold, but also other minerals such as cassiterite, chromite, and platinum metals. The mineral grains become concentrated during transport by water or wind because they are more dense than other detrital minerals such as quartz, and (like quartz) they are relatively resistant to chemical breakdown. Examples are the Witwatersrand gold deposits of South Africa, which are gold- and uranium-bearing conglomerates laid down by ancient rivers, and the placer tin deposits of the Malay Peninsula.

place value the value given to a digit because of its position within a number. For example, in the decimal number 2,465 the 2 represents two thousands, the 4 represents four hundreds, the 6 represents six tens, and the 5 represents five units.

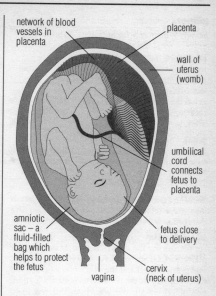

placenta *The placenta is a disc-shaped organ about 25 cm/10 in in diameter and 3 cm/1 in thick. It is connected to the fetus by the umbilical cord.*

Planck's constant in physics, a fundamental constant (symbol h) that is the energy of one quantum of electromagnetic radiation (the smallest possible 'packet' of energy; see ⟡quantum theory) divided by the frequency of its radiation.

Its value is 6.626196×10^{-34} joule seconds.

planet large celestial body in orbit around a star, composed of rock, metal, or gas. There are nine planets in the Solar System: Mercury, Venus, Earth, Mars, Jupiter, Saturn, Neptune, Uranus, and Pluto. The inner four, called the ***terrestrial planets***, are small and rocky, and include the planet Earth. The outer planets, with the exception of Pluto, are called the giant planets, large balls of rock, liquid, and gas; the largest is Jupiter, which contains more than twice as much mass as all the other planets combined. Planets do not produce light, but reflect the light of their parent star.

planetary nebula shell of gas thrown off by a star at the end of its life. Planetary nebulae have nothing to do with planets. They were named by William Herschel, who thought their rounded shape resembled the disc of a planet. After a star such as the Sun has expanded to

planets

planet	main constituents	atmosphere	average distance from Sun in millions of km	time for one orbit in Earth-years	diameter in thousands of km	average density if density of water is 1 unit
Mercury	rocky, ferrous	–	58	0.241	4.88	5.4
Venus	rocky, ferrous	carbon dioxide	108	0.615	12.10	5.2
Earth	rocky, ferrous	nitrogen, oxygen	150	1.00	12.76	5.5
Mars	rocky	carbon dioxide	228	1.88	6.78	3.9
Jupiter	liquid hydrogen, helium	–	778	11.86	142.80	1.3
Saturn	hydrogen, helium	–	1,427	29.46	120.00	0.7
Uranus	icy, hydrogen, helium	hydrogen, helium	2,870	84.00	50.80	1.3
Neptune	icy, hydrogen, helium	hydrogen, helium	4,497	164.80	48.60	1.8
Pluto	icy, rocky	methane	5,900	248.50	2.25	about 2

become a ◊red giant, its outer layers are ejected into space to form a planetary nebula, leaving the core as a ◊white dwarf at the centre.

planimeter simple integrating instrument for measuring the area of a regular or irregular plane surface. It consists of two hinged arms: one is kept fixed and the other is traced around the boundary of the area. This actuates a small graduated wheel; the area is calculated from the wheel's change in position.

plankton small, often microscopic, forms of plant and animal life that live in the upper layers of fresh and salt water, and are an important source of food for larger animals. Marine plankton is concentrated in areas where rising currents bring mineral salts to the surface.

plant organism that carries out ◊photosynthesis, has cellulose cell walls and complex cells, and is immobile. A few parasitic plants have lost the ability to photosynthesize but are still considered to be plants.

Plants are autotrophs, that is, they make carbohydrates from water and carbon dioxide, and are the primary producers in all food chains, so that all animal life is dependent on them. They play a vital part in the carbon cycle, removing carbon dioxide from the atmosphere and generating oxygen. The study of plants is known as botany.

Many of the lower plants (the algae and bryophytes) consist of a simple body, or thallus, on which the organs of reproduction are borne. Simplest of all are the threadlike algae, for example *Spirogyra*, which consist of a chain of cells. The seaweeds (algae) and mosses and liverworts (bryophytes) represent a further development, with simple, multicellular bodies that have specially modified areas in which the reproductive organs are carried. Higher in the morphological scale are the ferns, club mosses, and horsetails (pteridophytes). Ferns produce leaflike fronds bearing sporangia on their undersurface in which the spores are carried. The spores are freed and germinate to produce small independent bodies carrying the sexual organs; thus the fern, like other pteridophytes and some seaweeds, has two quite separate generations in its life cycle (see ◊alternation of generations).

The pteridophytes have special supportive water-conducting tissues, which identify them as vascular plants, a group which includes all seed plants, that is the gymnosperms (conifers, yews, cycads, and ginkgos) and the angiosperms (flowering plants).

The seed plants are the largest group, and structurally the most complex. They are usually divided into three parts: root, stem, and leaves. Stems grow above or below ground. Their cellular structure is designed to carry water and salts from the roots to the leaves in the ◊xylem, and sugars from the leaves to the roots in the ◊phloem. The leaves manufacture the food of the plant by means of photosynthesis, which occurs in the ◊chloroplasts they contain. Flowers and cones are modified leaves arranged in groups, enclosing the reproductive organs from which the fruits and seeds result.

plant classification taxonomy or classification of plants. Originally the plant kingdom included bacteria, diatoms, dinoflagellates, fungi, and slime moulds, but these are not now thought of as plants. The groups that are always classified as plants are the bryophytes (mosses and liverworts), pteridophytes (ferns, horsetails, and club mosses), gymnosperms (conifers, yews, cycads, and ginkgos), and angiosperms (flowering plants). The angiosperms are split into monocotyledons (for example, orchids, grasses, lilies) and

plant *The plant kingdom is divided into two broad sub-kingdoms: the lower plants and the seed plants. The former include algae, lichens, liverworts, horsetails, mosses and ferns; the seed plants include angiosperms (flowering plants) and gymnosperms. Stem and root systems of a plant. The root system can be more extensive than the part of the plant above the ground. In one study, a four-month-old rye plant was found to have 14 billion root hairs, with a total surface area of more than 600 m²/60,000 ft². This was 130 times greater than the surface area of the plant above the ground.*

dicotyledons (for example, oak, buttercup, geranium, and daisy).

plant hormone substance produced by a plant that has a marked effect on its growth, flowering, leaf fall, fruit ripening, or some other process. Examples include ◊auxin, ◊gibberellin, ◊ethylene, and ◊cytokinin.

Unlike animal hormones, these substances are not produced by a particular area of the plant body, and they may be less specific in their effects. It has therefore been suggested that they should not be described as hormones at all.

plant propagation production of plants. Botanists and horticulturalists can use a wide variety of means for propagating plants. There are the natural techniques of ◊vegetative reproduction, together with ◊cuttings, ◊grafting, and ◊micropropagation. The range is wide because most plant tissue, unlike animal tissue, can give rise to the complete range of tissue types within a particular species.

Plaskett's star the most massive ◊binary star known, consisting of two supergiants of about 40 and 50 solar masses, orbiting each other every 14.4 days. Plaskett's star lies in the constellation Monoceros and is named after Canadian astronomer John S Plaskett (1865–1941), who identified it as a binary star and discovered its massive nature in 1922.

plasma in biology, the liquid component of the ◊blood.

plasma in physics, an ionized gas produced at extremely high temperatures, as in the Sun and other stars, which contains positive and negative charges in approximately equal numbers. It is a good electrical conductor. In thermonuclear reactions the plasma produced is confined through the use of magnetic fields.

plasma display type of flat display, which uses an ionized gas between two panels containing grids of wires. When current flows through the wires a ◊pixel is charged causing it to light up.

plasma torch cutting device used mainly in metallurgy. It works by passing a strong electric current through a rarefied gas, ionizing it to produce temperatures on impact of up to 8,000°C/4,500°F.

plasmid small, mobile piece of ◊DNA found in bacteria and used in ◊genetic engineering. Plasmids are separate from the bacterial chromosome but still multiply during cell growth. Their size ranges from 3% to 20% of the size of the chromosome. There is usually only one copy of a single plasmid per cell, but occasionally several are found. Some plasmids carry 'fertility genes' that enable them to move from one bacterium to another and transfer genetic information between strains. Plasmid genes determine a wide variety of bacterial properties including resistance to antibiotics and the ability to produce toxins.

plasmolysis the separation of the plant cell cytoplasm from the cell wall as a result of water loss. As moisture leaves the vacuole the total volume of the cytoplasm decreases while the cell itself, being rigid, hardly changes. Plasmolysis is induced in the laboratory by immersing a plant cell in a strongly saline or sugary solution, so that water is lost by osmosis. Plasmolysis is unlikely to occur in the wild except in severe conditions.

plaster of Paris form of calcium sulphate, obtained from gypsum; it is mixed with water for making casts and moulds.

plastic any of the stable synthetic materials that are fluid at some stage in their manufacture, when they can be shaped, and that later set to rigid or semi-rigid solids. Plastics today are chiefly derived from petroleum. Most are polymers, made up of long chains of identical molecules. See also ◊thermoplastic and ◊thermoset.

plastid general name for a cell ◊organelle of plants that is enclosed by a double membrane and contains a series of internal membranes and vesicles. Plastids contain ◊DNA and are produced by division of existing plastids. They can be classified into two main groups: the *chromoplasts*, which contain pigments such as carotenes and chlorophyll, and the *leucoplasts*, which are colourless; however, the distinction between the two is not always clear-cut.

plate according to ◊plate tectonics, one of a number of slabs of solid rock, about a hundred kilometres thick and often several thousands of kilometres across, making up the Earth's surface. Together, the plates make up the ◊lithosphere.

Plates are made up of two types of crustal material: oceanic crust (sima) and continental crust (sial), both of which are underlain by a solid layer of the ◊mantle. *Oceanic crust* is heavy and consists largely of basalt. It is formed at constructive margins. *Continental crust* is less dense and is rich in granite. It is made up of volcanic islands and folded sediments, and is usually associated with destructive margins.

platelet tiny disc-shaped structure found in the blood, which helps it to clot. Platelets are not true cells, but membrane-bound cell fragments that bud off from large cells in the bone marrow.

plate tectonics theory that explains the phenomena of ◊continental drift and seafloor spreading, and the formation of the major physical features of the Earth's surface. The Earth's outermost layer is regarded as a jigsaw of rigid major and minor plates up to 100 km/62 mi thick, which move relative to each other, probably under the influence of convection currents in the mantle beneath. Major landforms occur at the margins of the plates, where plates are colliding or moving apart – for example, volcanoes, fold mountains, ocean trenches, and ocean ridges.

constructive margin Where two plates are moving apart from each other, molten rock from the mantle wells up in the space left between the plates and hardens to form new crust, usually in the form of an ocean ridge (such as the Mid-Atlantic Ridge). The newly formed crust accumulates on either side of the ocean ridge, causing the seafloor to spread – for example, the floor of the Atlantic Ocean is growing by 5 cm/2 in each year because of the upwelling of new material at the Mid-Atlantic Ridge.

destructive margin Where two plates are moving towards each other, the denser of the two plates may be forced under the other into a region called the subduction zone. The descending plate melts to form a body of magma, which may then rise to the surface through cracks and faults to form volcanoes. If the two plates consist of more buoyant continental crust, subduction does not occur. Instead, the crust crumples gradually to form ranges of young fold mountains, such as the Himalayas in Asia, the Andes in South America, and the Rockies in North America.

conservative margin Sometimes two plates will slide past each other – an example is the San Andreas Fault, California, where the movement of the plates sometimes takes the form of sudden jerks, causing the earthquakes common in the San Francisco–Los Angeles area. Most of the earthquake and volcano zones of the world are, in fact, found in regions where two plates meet or are moving apart.

Pleiades in astronomy, a star cluster about 400 light years away in the constellation Taurus, representing the Seven Sisters of Greek mythology. Its brightest stars (highly luminous, blue-white giants only a few million years old) are visible to the naked eye, but there are many fainter ones.

pleiotropy process whereby a given gene influences several different observed characteristics of an organism. For example, in the fruit fly *Drosophila* the vestigial gene reduces the size of wings, modifies the halteres, changes the number of egg strings in the ovaries, and changes the direction of certain bristles. Many human syndromes are caused by pleiotropic genes, for example Marfan's syndrome where the slender physique, hypermobility of the joints, elongation of the limbs, dislocation of the lens, and susceptibility to heart disease are all caused by one gene.

Pleistocene first epoch of the Quaternary period of geological time, beginning 1.64 million years ago and ending 10,000 years ago. The polar ice caps were extensive and glaciers were abundant during the ice age of this period, and humans evolved into modern *Homo sapiens sapiens* about 100,000 years ago.

pleura covering of the lungs and of the inner surface of the chest wall; the surfaces are lubricated to prevent friction from breathing movements.

Plough, the in astronomy, a popular name for the most prominent part of the constellation ◊Ursa Major.

plumbago alternative name for the mineral ◊graphite.

plumule part of a seed embryo that develops into the shoot, bearing the first true leaves of the plant. In most seeds, for example the sunflower, the plumule is a small conical structure without any leaf structure. Growth of the plumule does not occur until the ◊cotyledons have grown above ground. This is ◊epigeal germination. However, in seeds such as the broad bean, a leaf structure is visible on the plumule in the seed. These seeds develop by the plumule growing up through the soil with the cotyledons remaining below the surface. This is known as ◊hypogeal germination.

Pluto in astronomy, the smallest and, usually, outermost planet of the Solar System. The existence of Pluto was predicted by calculation by Percival Lowell and the planet was located by Clyde Tombaugh in 1930. It orbits the Sun every 248.5 years at an average distance of 5.8 billion km/3.6 billion mi.

Its highly elliptical orbit occasionally takes it within the orbit of Neptune, as in 1979–99. Pluto has a diameter of about 2,300 km/1,400 mi, and a mass about 0.002 of that of Earth. It is of low density, composed of rock and ice, with frozen methane on its surface and a thin atmosphere.

plutonic rock igneous rock derived from magma that has cooled and solidified deep in the crust of the Earth; granites and gabbros are examples of plutonic rocks.

plutonium silvery-white, radioactive, metallic element of the ◊actinide series, symbol Pu, atomic number 94, relative atomic mass 239.13. It occurs in nature in minute quantities in ◊pitchblende and other ores, but is produced in quantity only synthetically. It has six allotropic forms (see ◊allotropy) and is one of three fissile elements (elements capable of splitting into other elements – the others are thorium and uranium). The element has awkward physical properties and is the most toxic substance known.

Because Pu-239 is so easily synthesized from abundant uranium, it has been produced in large quantities by the weapons industry. It has a long half-life (24,000 years) during which time it remains highly toxic. Plutonium is dangerous to handle, difficult to store, and impossible to dispose of.

pneumatic drill drill operated by compressed air, used in mining and tunnelling, for drilling shot holes (for explosives), and in road repairs for breaking up pavements. It contains an air-operated piston that delivers hammer blows to the drill bit many times a second. The French engineer Germain Sommeiller (1815–1871) developed the pneumatic drill 1861 for tunnelling in the Alps.

pneumatophore erect root that rises up above the soil or water and promotes ◊gas exchange. Pneumatophores, or breathing roots, are formed by certain swamp-dwelling trees, such as mangroves, since there is little oxygen available to the roots in waterlogged conditions. They have numerous pores or ◊lenticels over their surface, allowing gas exchange.

pneumothorax the presence of air in the pleural cavity, between a lung and the chest wall. It may be due to a penetrating injury of the lung or to lung disease, or it may occur without apparent cause (spontaneous pneumothorax) in an otherwise healthy person. Prevented from expanding normally, the lung is liable to collapse.

p–n junction diode in electronics, a two-terminal semiconductor device that allows electric current to flow in only one direction, the *forward-bias* direction. A very high resistance prevents current flow in the opposite, or *reverse-bias*, direction. It is used as a ◊rectifier, converting alternating current (AC) to direct current (DC).

pod in botany, a type of fruit that is characteristic of legumes (plants belonging to the Leguminosae family), such as peas and beans. It develops from a single ◊carpel and splits down both sides when ripe to release the seeds.

poikilothermy the condition in which an animal's body temperature is largely dependent on the temperature of the air or water in which it lives. It is characteristic of all animals except birds and mammals, which maintain their body temperatures by ◊homeothermy (they are 'warm-blooded').

poise c.g.s. unit (symbol P) of dynamic ◊viscosity (the property of liquids that determines how readily they flow). It is equal to one dyne-second per square centimetre. For most liquids the centipoise (one hundredth of a poise) is used. Water at 20°C/68°F has a viscosity of 1.002 centipoise.

Poiseuille's formula in physics, a relationship describing the rate of flow of a fluid through a narrow tube. For a capillary (very narrow) tube of length l and radius r with a pressure difference p between its ends, and a liquid of ◊viscosity η, the velocity of flow expressed as the volume per second is $\pi p r^4/8l\eta$. The formula was devised 1843 by French physicist Jean Louis Poiseuille (1799–1869).

poison or *toxin* any chemical substance that, when introduced into or applied to the body, is capable of injuring health or destroying life. The liver removes some poisons from the blood. The majority of poisons may be divided into *corrosives*, such as sulphuric, nitric, and hydrochloric acids; *irritants*, including arsenic and copper sulphate; *narcotics* such as opium, and carbon monoxide; and *narcotico-irritants* from any substances of plant origin including carbolic acid and tobacco.

polar coordinates in mathematics, a way of defining the position of a point in terms of its distance r from a fixed point (the origin) and its angle θ to a fixed line or axis. The coordinates of the point are (r,θ).

Polaris or *Pole Star* or *North Star* the bright star closest to the north celestial pole, and the brightest star in the constellation Ursa Minor. Its position is indicated by the 'pointers' in Ursa Major. Polaris is a yellow ◊supergiant about 500 light years away.

polarized light light in which the electromagnetic vibrations take place in one particular direction. In ordinary (unpolarized) light, the electric and magnetic fields vibrate in all directions perpendicular to the direction of propagation. After reflection from a polished surface or transmission through certain materials (such as Polaroid), the electric and magnetic fields are confined to one direction, and the light is said to be *plane polarized*. In *circularly polarized* and *elliptically polarized* light, the magnetic and electric fields are confined to one direction, but the direction rotates as the light propagates. Polarized light is used to test the strength of sugar solutions, to measure stresses in transparent materials, and to prevent glare.

polarography electrochemical technique for the analysis of oxidizable and reducible compounds in solution. It involves the diffusion of the substance to be analysed onto the surface of a small electrode, usually a bead of mercury, where oxidation or reduction occurs at an electric potential characteristic of that substance.

pole either of the geographic north and south points of the axis about which the Earth rotates. The geographic poles differ from the magnetic poles, which are the points towards which a freely suspended magnetic needle will point.

pole, magnetic see ◊magnetic pole.

Pole Star another name for ◊Polaris, the northern pole star. There is no bright star near the southern celestial pole.

pollen the grains of ◊seed plants that contain the male gametes. In ◊angiosperms (flowering plants) pollen is produced within ◊anthers; in most ◊gymnosperms (cone-bearing plants) it is produced in male cones. A pollen grain is typically yellow and, when mature, has a hard outer wall. Pollen of insect-pollinated plants (see ◊pollination) is often sticky and spiny and larger than the smooth, light grains produced by wind-pollinated species.

pollen tube outgrowth from a pollen grain that grows towards the ◊ovule, following germination of the grain on the ◊stigma. In ◊angiosperms (flowering plants) the pollen tube reaches the ovule by growing down through the ◊style, carrying the male gametes inside. The gametes are discharged into the ovule and one fertilizes the egg cell.

pollination the process by which pollen is transferred from one plant to another. The male ◊gametes are contained in pollen grains, which must be transferred from the anther to the stigma in ◊angiosperms (flowering plants), and from the male cone to the female cone in ◊gymnosperms (cone-bearing plants). Fertilization (not the same as pollination) occurs after the growth of the pollen tube to the ovary. Self-pollination occurs when pollen is transferred to a stigma of the same flower, or to another flower on the same plant; cross-pollination occurs when pollen is transferred to another plant. This involves external pollen-carrying agents, such as wind (see ◊anemophily), water (see ◊hydrophily), insects, birds (see ◊ornithophily), bats, and other small mammals.

pollinium group of pollen grains that is transported as a single unit during pollination. Pollinia are common in orchids.

pollution contamination of the environment caused by human activities. It frequently takes the form of chemicals added to the land, water, or air as a by-product of industry, traffic, or agriculture. Pollutants may enter the ◊food chain and be passed on from one organism to another. They are frequently harmful and may have side effects such as ◊acid rain.

Pollux or *Beta Geminorum* the brightest star in the constellation Gemini (the twins), and the 17th brightest star in the sky. Pollux is a yellowish star with a true luminosity 45 times that of the Sun. It is 35 light years away.

polonium radioactive, metallic element, symbol Po, atomic number 84, relative atomic mass 210. Polonium occurs in nature in small amounts and was isolated from ◊pitchblende. It is the element having the largest number of isotopes (27) and is 5,000 times as radioactive as radium, liberating considerable amounts of heat. It was the first element to have its radioactive properties recognized and investigated.

polychlorinated biphenyl (PCB) any of a group of chlorinated isomers of biphenyl $(C_6H_5)_2$. They are dangerous industrial chemicals, valuable for their fire-resisting qualities. They constitute an environmental hazard because of their persistent toxicity. Since 1973 their use has been limited by international agreement.

polyester synthetic resin formed by the ◊condensation of polyhydric alcohols (alcohols containing more than one hydroxyl group) with dibasic acids (acids containing two replaceable hydrogen atoms). Polyesters are thermosetting ◊plastics, used in making synthetic fibres, such as Dacron and Terylene, and constructional plastics. With glass fibre added as reinforcement, polyesters are used in car bodies and boat hulls.

polyethylene or *polyethene* polymer of the gas ethylene (technically called ethene, C_2H_4). It is a tough, white, translucent, waxy thermoplastic (which means it can be repeatedly softened by heating). It is used for packaging, bottles, toys, wood preservation, electric cable, pipes and tubing.

polygraph technical name for a ◊lie detector.

polymer compound made up of a large longchain or branching matrix composed of many repeated simple units (*monomers*). There are many polymers, both natural (cellulose, chitin, lignin) and synthetic (polyethylene and nylon, types of plastic). Synthetic polymers belong to two groups: thermosoftening and thermosetting (see ◊plastic).

polymerase chain reaction (PCR) technique developed during the 1980s to clone short strands of DNA from the ◊genome of an organism. The aim is to produce enough of the DNA to be able to sequence and identify it.

polymerization chemical union of two or more (usually small) molecules of the same kind to form a new compound. *Addition polymerization* produces simple multiples of the same compound. *Condensation polymerization* joins molecules together with the elimination of water or another small molecule.

polymorphism in genetics, the coexistence of several distinctly different types in a population (groups of animals of one species).

polymerization Addition of ethane molecules to form polythene. A polythene molecule has between 12,000 and 60,000 atoms in its molecules.

Examples include the different blood groups in humans, different colour forms in some butterflies, and snail shell size, length, shape, colour, and stripiness.

polymorphism in mineralogy, the ability of a substance to adopt different internal structures and external forms, in response to different conditions of temperature and/or pressure. For example, diamond and graphite are both forms of the element carbon, but they have very different properties and appearance.

polynomial in mathematics, an algebraic expression that has one or more ◊variables (denoted by letters). A polynomial of degree one, that is, whose highest ◊power of x is 1, as in $2x + 1$, is called a linear polynomial; $3x^2 + 2x + 1$ is quadratic; $4x^3 + 3x^2 + 2x + 1$ is cubic.

polyp in zoology, the sedentary stage in the life cycle of a coelenterate (such as a ◊coral or jellyfish), the other being the free-swimming *medusa*.

polypeptide long-chain ◊peptide.

polyploid in genetics, possessing three or more sets of chromosomes in cases where the normal complement is two sets (◊diploid). Polyploidy arises spontaneously and is common in plants (mainly among flowering plants), but rare in animals. Many crop plants are natural polyploids, including wheat, which has four sets of chromosomes per cell (durum wheat) or six sets (common wheat).

Plant breeders can induce the formation of polyploids by treatment with a chemical, colchicine.

polypropylene plastic made by the polymerization, or linking together, of ◊propene molecules ($CH_2=CH_2$). It is used as a moulding material.

polysaccharide long-chain ◊carbohydrate made up of hundreds or thousands of linked simple sugars (monosaccharides) such as glucose and closely related molecules.

The polysaccharides are natural polymers. They either act as energy-rich food stores in plants (starch) and animals (glycogen), or have structural roles in the plant cell wall (cellulose, pectin) or the tough outer skeleton of insects and similar creatures (chitin). See also ◊carbohydrate.

polystyrene type of ◊plastic used in kitchen utensils or, in an expanded form, in insulation and ceiling tiles. CFCs are used to produce expanded polythene so alternatives are being sought.

polytetrafluoroethene (PTFE) polymer made from the monomer tetrafluoroethene (CF_2CF_2). It is a thermosetting plastic with a high melting point that is used to produce 'non-stick' surfaces on pans and to coat bearings. Its trade name is Teflon.

Polythene trade name for a variety of ◊polyethylene.

polyunsaturate type of ◊fat or oil containing a high proportion of triglyceride molecules whose ◊fatty acid chains contain several double bonds. By contrast, the fatty-acid chains of the triglycerides in saturated fats (such as lard) contain only single bonds. Medical evidence suggests that polyunsaturated fats, used widely in margarines and cooking fats, are less likely to contribute to cardiovascular disease than saturated fats, but there is some evidence that they may also have adverse effects on health.

polyurethane polymer made from the monomer urethane. It is a thermoset ◊plastic, used in liquid form as a paint or varnish, and in foam form for upholstery and in lining materials (where it may be a fire hazard).

polyvinyl chloride (PVC) a type of ◊plastic used for drainpipes, floor tiles, audio discs, shoes, and handbags.

pome type of ◊pseudocarp, or false fruit, typical of certain plants belonging to the Rosaceae

polysaccharide *A typical polysaccharide molecule, glycogen (animal starch), is formed from linked glucose ($C_6H_{12}O_6$) molecules. A glycogen molecule has 100–1,000 linked glucose units.*

family. The outer skin and fleshy tissues are developed from the ◊receptacle (the enlarged end of the flower stalk) after fertilization, and the five ◊carpels (the true fruit) form the pome's core, which surrounds the seeds. Examples of pomes are apples, pears, and quinces.

population cycle in biology, regular fluctuations in the size of a population, as seen in lemmings, for example. Such cycles are often caused by density-dependent mortality: high mortality due to overcrowding causes a sudden decline in the population, which then gradually builds up again. Population cycles may also result from an interaction between a predator and its prey.

population genetics the branch of genetics that studies the way in which the frequencies of different ◊alleles (alternative forms of a gene) in populations of organisms change, as a result of natural selection and other processes.

porphyry any ◊igneous rock containing large crystals in a finer matrix.

port in computing, a socket that enables a computer processor to communicate with an external device. It may be an *input port* (such as a joystick port), or an *output port* (such as a printer port), or both (an *i/o port*).

positron in physics, the antiparticle of the electron; an ◊elementary particle having the same magnitude of mass and charge as an electron but exhibiting a positive charge. The positron was discovered in 1932 by US physicist Carl Anderson at Caltech, USA, its existence having been predicted by the British physicist Paul Dirac 1928.

positron emission tomography (PET) an imaging technique which enables doctors to observe the metabolic activity of the human body by following the progress of a radioactive chemical that has been inhaled or injected. PET scanners pinpoint the location of the chemical by detecting ◊gamma radiation given out when ◊positrons emitted by the chemical are annihilated. The technique has been used to study a wide range of conditions, including schizophrenia, Alzheimer's disease, and Parkinson's disease.

POSIX (acronym for *portable operating system interface* for computers – developed from Uni*x*) ◊American National Standards Institute standard, developed to describe how the programming interfaces and other features of ◊Unix worked, in order to remove control from the developers, AT&T Bell Laboratories. Subsequently many other (proprietary) ◊operating systems were modified in order to become POSIX-compliant, that is, they provide an open systems interface, so that they can communicate with other POSIX-compliant systems, even though the operating systems themselves are internally quite different.

PostScript in computing, a page-description language developed by Adobe that has become a standard. PostScript is primarily a language for printing documents on laser printers, but it can be adapted to produce images on other types of devices.

potash general name for any potassium-containing mineral, most often applied to potassium carbonate (K_2CO_3) or potassium hydroxide (KOH). Potassium carbonate, originally made by roasting plants to ashes in earthenware pots, is commercially produced from the mineral sylvite (potassium chloride, KCl) and is used mainly in making artificial fertilizers, glass, and soap.

potassium soft, waxlike, silver-white, metallic element, symbol K (Latin *kalium*), atomic number 19, relative atomic mass 39.0983. It is one of the ◊alkali metals and has a very low density – it floats on water, and is the second lightest metal (after lithium). It oxidizes rapidly when exposed to air and reacts violently with water. Of great abundance in the Earth's crust, it is widely distributed with other elements and found in salt and mineral deposits in the form of potassium aluminium silicates.

potassium–argon dating isotopic dating method based on the radioactive decay of potassium-40 to the stable isotope argon-40. The method is used primarily to date volcanic layers in stratigraphic sequences with archaeological deposits, and the palaeomagnetic reversal timescale. Ages are based on the known half-life of 40K, and the ratio of 40K/40Ar. The method is routinely applied from about 100,000 to 30 million years ago.

potassium dichromate $K_2Cr_2O_7$ orange, crystalline solid, soluble in water, that is a strong ◊oxidizing agent in the presence of dilute sulphuric acid. As it oxidizes other compounds it is itself reduced to potassium chromate (K_2CrO_4), which is green. Industrially it is used in the manufacture of dyes and glass and in tanning, photography, and ceramics.

potassium manganate(VII) $KMnO_4$ or *potassium permanganate* dark purple, crystalline solid, soluble in water, that is a strong ◊oxidizing agent in the presence of dilute sulphuric acid. In the process of oxidizing other compounds it is itself reduced to manganese(II) salts (containing the Mn^{2+} ion), which are colourless.

potential difference (pd) measure of the electrical potential energy converted to another form for every unit charge moving between two points in an electric circuit (see ◊potential, electric). The unit of potential difference is the volt.

potential divider or *voltage divider* two resistors connected in series in an electrical circuit in order to obtain a fraction of the potential difference, or voltage, across the battery or electrical source. The potential difference is divided across the two resistors in direct proportion to their resistances.

potential, electric in physics, the relative electrical state of an object. A charged conductor, for example, has a higher potential than the Earth, whose potential is taken by convention to be zero. An electric ◊cell (battery) has a potential in relation to emf (◊electromotive force), which can make current flow in an external circuit. The difference in potential between two points – the ◊potential difference – is expressed in ◊volts; that is, a 12 V battery has a potential difference of 12 volts between its negative and positive terminals.

potential energy ◊energy possessed by an object by virtue of its relative position or state (for example, as in a compressed spring). It is contrasted with kinetic energy, the form of energy possessed by moving bodies.

potentiometer in physics, an electrical ◊resistor that can be divided so as to compare, measure, or control voltages. In radio circuits, any rotary variable resistance (such as volume control) is referred to as a potentiometer.

pound imperial unit (abbreviation lb) of mass. The commonly used avoirdupois pound, also called the *imperial standard pound* (7,000 grains/0.45 kg), differs from the *pound troy* (5,760 grains/0.37 kg), which is used for weighing precious metals. It derives from the Roman *libra*, which weighed 0.327 kg.

poundal imperial unit (abbreviation pdl) of force, now replaced in the SI system by the ◊newton. One poundal equals 0.1383 newtons.

powder metallurgy method of shaping heat-resistant metals such as tungsten. Metal is pressed into a mould in powdered form and then sintered (heated to very high temperatures).

power in mathematics, that which is represented by an ◊exponent or index, denoted by a superior small numeral. A number or symbol raised to the power of 2 – that is, multiplied by itself – is said to be squared (for example, 3^2, x^2), and when raised to the power of 3, it is said to be cubed (for example, 2^3, y^3).

power in optics, a measure of the amount by which a lens will deviate light rays. A powerful converging lens will converge parallel rays steeply, bringing them to a focus at a short distance from the lens. The unit of power is the *dioptre*, which is equal to the reciprocal of focal length in metres. By convention, the power of a converging (or convex) lens is positive and that of a diverging (or concave) lens negative.

power in physics, the rate of doing work or consuming energy. It is measured in watts (joules per second) or other units of work per unit time.

power station building where electrical energy is generated from a fuel or from another form of energy. Fuels used include fossil fuels such as coal, gas, and oil, and the nuclear fuel uranium. Renewable sources of energy include ◊gravitational potential energy, used to produce ◊hydroelectric power, and ◊wind power.

power transmission transfer of electrical power from one location, such as a power station, to another. Electricity is conducted along the cables of the ◊national grid at a high voltage (up to 500 kV on the super grid) in order to reduce the current in the wires, and hence minimize the amount of energy wasted from them as heat.

◊Transformers are needed to step down these voltages before power can be supplied to consumers. High voltages require special insulators to prevent current from leaking to the ground and these may clearly be seen on the pylons that carry overhead wires.

praseodymium silver-white, malleable, metallic element of the ◊lanthanide series, symbol Pr, atomic number 59, relative atomic mass 140.907. It occurs in nature in the minerals monzanite and bastnasite, and its green salts are used to colour glass and ceramics. It was named in 1885 by Austrian chemist Carl von Welsbach (1858–1929).

preadaptation in biology, the fortuitous possession of a character that allows an organism to exploit a new situation. In many cases, the character evolves to solve a particular problem that a species encounters in its preferred habitat, but once evolved may allow the organism to exploit an entirely different situation. The ability to extract oxygen directly from the air evolved in some early fishes, probably in response to life in stagnant, deoxygenated pools; this later made it possible for their descendants to spend time on land, so giving rise eventually to the air-breathing amphibians.

Precambrian in geology, the time from the formation of Earth (4.6 billion years ago) up to 570 million years ago. Its boundary with the succeeding Cambrian period marks the time when animals first developed hard outer parts (exoskeletons) and so left abundant fossil remains. It comprises about 85% of geological time and is divided into two periods: the Archaean, in which no life existed, and the Proterozoic, in which there was life in some form.

precession slow wobble of the Earth on its axis, like that of a spinning top. The gravitational pulls of the Sun and Moon on the Earth's equatorial bulge cause the Earth's axis to trace out a circle on the sky every 25,800 years. The position of the celestial poles (see ◊celestial sphere) is constantly changing owing to precession, as are the positions of the equinoxes (the points at which the celestial equator intersects the Sun's path around the sky). The *precession of the equinoxes* means that there is a gradual westward drift in the ecliptic – the path that the Sun appears to follow – and in the coordinates of objects on the celestial sphere.

precipitation in meteorology, water that falls to the Earth from the atmosphere. It includes rain, snow, sleet, hail, dew, and frost.

precipitation in chemistry, the formation of an insoluble solid in a liquid as a result of a reaction within the liquid between two or more soluble substances. If the solid settles, it forms a *precipitate*; if the particles of solid are very small, they will remain in suspension, forming a *colloidal precipitate* (see ◊colloid).

pregnancy in humans, the period during which an embryo grows within the womb. It begins at conception and ends at birth, and the normal length is 40 weeks. Menstruation usually stops on conception. About one in five pregnancies fails, but most of these failures occur very early on, so the woman may notice only that her period is late. After the second month, the breasts become tense and tender, and the areas round the nipples become darker. Enlargement of the uterus can be felt at about the end of the third month, and thereafter the abdomen enlarges progressively. Pregnancy in animals is called ◊gestation.

prehistoric life the diverse organisms that inhabited Earth from the origin of life about 3.5 billion years ago to the time when humans began to keep written records, about 3500 BC. During the course of evolution, new forms of life developed and many other forms, such as the dinosaurs, became extinct. Prehistoric life evolved over this vast timespan from simple bacteria-like cells in the oceans to algae and protozoans and complex multicellar forms such as worms, molluscs, crustaceans, fishes, insects, land plants, amphibians, reptiles, birds, and mammals. On a geological timescale human beings evolved relatively recently, about 4 million years ago, although the exact dating is a matter of some debate. See also ◊geological time.

premolar in mammals, one of the large teeth toward the back of the mouth. In herbivores they are adapted for grinding. In carnivores they may be carnassials. Premolars are present in milk ◊dentition as well as permanent dentition.

presbyopia vision defect, an increasing inability with advancing age to focus on near objects. It is caused by thickening and loss of elasticity in the lens, which is no longer able to relax to the near-spherical shape required for near vision.

preservative substance (◊additive) added to a food in order to inhibit the growth of bacteria, yeasts, moulds, and other microorganisms, and therefore extend its shelf life. The term sometimes refers to ◊antioxidants (substances added to oils and fats to prevent their becoming rancid) as well. All preservatives are potentially damaging to health if eaten in sufficient quantity. Both the amount used, and the foods in which they can be used, are restricted by law.

pressure in physics, the force acting normally (at right angles) to a body per unit surface area. The SI unit of pressure is the pascal (newton per square metre), equal to 0.01 millibars. In a fluid (liquid or gas), pressure increases with depth. At the edge of Earth's atmosphere, pressure is zero, whereas at sea level ◊atmospheric pressure due to the weight of the air above is about 100 kilopascals (1,013 millibars or 1 atmosphere). Pressure is commonly measured by means of a ◊barometer, ◊manometer, or ◊Bourdon gauge.

pressure law law stating that the pressure of a fixed mass of gas at constant volume is directly proportional to its absolute temperature. The law may be expressed as: pressure/temperature = constant or, more usefully, as:

$$P_1/T_1 = P_2/T_2$$

where P_1 and T_1 are the initial pressure and temperature in kelvin of a gas, and P_2 and T_2 are its final pressure and temperature. See also ◊gas laws.

pressurized water reactor (PWR) a ◊nuclear reactor design used in nuclear power stations in many countries, and in nuclear-powered submarines. In the PWR, water under pressure is the coolant and ◊moderator. It circulates through a steam generator, where its heat boils water to provide steam to drive power ◊turbines.

Prestel the ◊viewdata service provided by British Telecom (BT), which provides information on the television screen via the telephone network. BT pioneered the service 1975.

primary sexual characteristic in males, the primary sexual characteristic is the ◊testis; in females it is the ◊ovary. Both are endocrine glands that produce hormones responsible for secondary sexual characteristics, such as facial hair and a deep voice in males and breasts in females.

primate in zoology, any member of the order of mammals that includes monkeys, apes, and humans (together called ***anthropoids***), as well as lemurs, bushbabies, lorises, and tarsiers (together called ***prosimians***). Generally, they have forward-directed eyes, gripping hands and feet, opposable thumbs, and big toes. They tend to have nails rather than claws, with gripping pads on the ends of the digits, all adaptations to the arboreal, climbing mode of life.

pregnancy *The development of a human embryo. Division of the fertilized egg, or ovum, begins within hours of conception. Within a week a ball of cells – a blastocyst – has developed. After the third week, the embryo has changed from a mass of cells into a recognizable shape. At four weeks, the embryo is 3 mm/0.1 in long, with a large bulge for the heart and small pits for the ears. At six weeks, the embryo is 1.5 cm/0.6 in with a pulsating heart and ear flaps. At the eighth week, the embryo is 2.5 cm/1 in long and recognizably human, with eyelids, small fingers, and toes. From the end of the second month, the embryo is almost fully formed and further development is mainly by growth. After this stage, the embryo is termed a fetus.*

prime number number that can be divided only by 1 or itself, that is, having no other factors. There is an infinite number of primes, the first ten of which are 2, 3, 5, 7, 11, 13, 17, 19, 23, and 29 (by definition, the number 1 is excluded from the set of prime numbers). The number 2 is the only even prime because all other even numbers have 2 as a factor.

principal focus in optics, the point at which incident rays parallel to the principal axis of a ◊lens converge, or appear to diverge, after refraction. The distance from the lens to its principal focus is its ◊focal length.

printed circuit board (PCB) electrical circuit created by laying (printing) 'tracks' of a conductor such as copper on one or both sides of an insulating board. The PCB was invented in 1936 by Austrian scientist Paul Eisler, and was first used on a large scale in 1948.

printer in computing, an output device for producing printed copies of text or graphics. Types include the ◊*daisywheel printer*, which produces good-quality text but no graphics; the ◊*dot matrix printer*, which produces text and graphics by printing a pattern of small dots; the ◊*ink-jet printer*, which creates text and graphics by spraying a fine jet of quick-drying ink onto the paper; and the ◊*laser printer*, which uses electrostatic technology very similar to that used by a photocopier to produce high-quality text and graphics.

prion exceptionally small microorganism, a hundred times smaller than a virus. Composed of protein, and without any detectable amount of nucleic acid (genetic material), it is thought to cause diseases such as scrapie in sheep, and certain degenerative diseases of the nervous system in humans. How it can operate without nucleic acid is not yet known.

prism in mathematics, a solid figure whose cross section is constant in planes drawn perpendicular to its axis. A cube, for example, is a rectangular prism with all faces (bases and sides) the same shape and size.

prism in optics, a triangular block of transparent material (plastic, glass, silica) commonly used to 'bend' a ray of light or split a beam into its spectral colours. Prisms are used as mirrors to define the optical path in binoculars, camera viewfinders, and periscopes. The dispersive property of prisms is used in the ◊spectroscope.

probability likelihood, or chance, that an event will occur, often expressed as odds, or in mathematics, numerically as a fraction or decimal.

In general, the probability that n particular events will happen out of a total of m possible events is n/m. A certainty has a probability of 1; an impossibility has a probability of 0. Empirical probability is defined as the number of successful events divided by the total possible number of events.

In tossing a coin, the chance that it will land 'heads' is the same as the chance that it will land 'tails', that is, 1 to 1 or even; mathematically, this probability is expressed as $1/2$ or 0.5. The odds against any chosen number coming up on the roll of a fair die are 5 to 1; the probability is $1/6$ or 0.1666 ... . If two dice are rolled there are $6 \times 6 = 36$ different possible combinations. The probability of a double (two numbers the same) is $6/36$ or $1/6$ since there are six doubles in the 36 events: (1,1), (2,2), (3,3), (4,4), (5,5), and (6,6).

procedure in computing, a small part of a computer program that performs a specific task, such as clearing the screen or sorting a file. A *procedural language*, such as BASIC, is one in which the programmer describes a task in terms of how it is to be done, as opposed to a *declarative language*, such as PROLOG, in which it is described in terms of the required result. See ◊programming.

processing cycle in computing, the sequence of steps performed repeatedly by a computer in the execution of a program. The computer's CPU (central processing unit) continuously works through a loop, involving fetching a program instruction from memory, fetching any data it needs, operating on the data, and storing the result in the memory, before fetching another program instruction.

processor in computing, another name for the ◊central processing unit or ◊microprocessor of a computer.

Procyon or *Alpha Canis Minoris* brightest star in the constellation Canis Minor and the eighth brightest star in the sky. Procyon is a white star 11.4 light years from Earth, with a mass of 1.7 Suns. It has a ◊white dwarf companion that orbits it every 40 years.

productivity, biological in an ecosystem, the amount of material in the food chain produced by the primary producers (plants) that is available for consumption by animals. Plants turn carbon dioxide and water into sugars and other complex carbon compounds by means of photosynthesis. Their net productivity is defined as the quantity of carbon compounds formed, less the quantity used up by the respiration of the plant itself.

progesterone ◊steroid hormone that occurs in vertebrates. In mammals, it regulates the menstrual cycle and pregnancy. Progesterone is secreted by the corpus luteum (the ruptured Graafian follicle of a discharged ovum).

program in computing, a set of instructions that controls the operation of a computer. There are two main kinds: ◊applications programs, which carry out tasks for the benefit of the user – for example, word processing; and ◊systems programs, which control the internal workings of the computer. A ◊utility program is a systems program that carries out specific tasks for the user. Programs can be written in any of a number of ◊programming languages but are always translated into machine code before they can be executed by the computer.

program counter in computing, an alternative name for ◊sequence-control register.

program loop part of a computer program that is repeated several times. The loop may be repeated a fixed number of times (*counter-controlled loop*) or until a certain condition is satisfied (*condition-controlled loop*). For example, a counter-controlled loop might be used to repeat an input routine until exactly ten numbers have been input; a condition-controlled loop might be used to repeat an input routine until the ◊data terminator 'XXX' is entered.

programming writing instructions in a programming language for the control of a computer. *Applications programming* is for end-user programs, such as accounts programs or word-processing packages. *Systems programming* is for operating systems and the like, which are concerned more with the internal workings of the computer.

programming language in computing, a special notation in which instructions for controlling a computer are written. Programming languages are designed to be easy for people to write and read, but must be capable of being mechanically translated (by a ◊compiler or an ◊interpreter) into the ◊machine code that the computer can execute. Programming languages may be classified as ◊high-level languages or ◊low-level languages. See also ◊source language.

progression sequence of numbers each occurring in a specific relationship to its predecessor. An *arithmetic progression* has numbers that increase or decrease by a common sum or difference (for example, 2, 4, 6, 8); a *geometric progression* has numbers each bearing a fixed ratio to its predecessor (for example, 3, 6, 12, 24); and a *harmonic progression* has numbers whose ◊reciprocals are in arithmetical progression, for example 1, $1/2$, $1/3$, $1/4$.

projectile particle that travels with both horizontal and vertical motion in the Earth's gravitational field. The two components of its motion can generally be analysed separately: its vertical motion will be accelerated due to its weight in the gravitational field; its horizontal motion may be assumed to be at constant velocity if the frictional forces of air resistance are ignored. In a uniform gravitational field and in the absence of frictional forces the path of a projectile is a parabola.

prokaryote in biology, an organism whose cells lack organelles (specialized segregated structures such as nuclei, mitochondria, and chloroplasts). Prokaryote DNA is not arranged in chromosomes but forms a coiled structure called a *nucleoid*. The prokaryotes comprise only the *bacteria* and *cyanobacteria* (see ◊blue-green algae); all other organisms are eukaryotes.

PROLOG (acronym for *programming in logic*) high-level computer-programming language based on logic. Invented in 1971 at the University of Marseille, France, it did not achieve widespread use until more than ten years later. It is used mainly for ◊artificial intelligence programming.

PROM (acronym for *programmable read-only memory*) in computing, a memory device in the form of an integrated circuit (chip) that can be programmed after manufacture to hold information permanently. PROM chips are empty of information when manufactured, unlike ROM (read-only memory) chips, which have information built into them. Other memory devices are ◊EPROM (erasable programmable read-only memory) and ◊RAM (random-access memory).

promethium radioactive, metallic element of the ◊lanthanide series, symbol Pm, atomic number 61, relative atomic mass 145.

It occurs in nature only in minute amounts, produced as a fission product/by-product of uranium in ◊pitchblende and other uranium ores; for a long time it was considered not to occur in nature. The longest-lived isotope has a half-life of slightly more than 20 years.

prominence bright cloud of gas projecting from the Sun into space 100,000 km/60,000 mi

programming languages

language	main uses	description
Ada	defence applications	high level
assembler languages	jobs needing detailed control of the hardware, fast execution, and small program sizes	fast and efficient but require considerable effort and skill
BASIC (**b**eginner's **a**ll-purpose **s**ymbolic **i**nstruction **c**ode)	mainly in education, business, and the home, and among non professional programmers, such as engineers systems	easy to learn; early versions lacked the features of other languages
C	systems programming; general programming	fast and efficient; widely used as a general-purpose language; especially popular among professional programmers
COBOL (**co**mmon **b**usiness-**o**riented **l**anguage)	business programming	strongly oriented towards commercial work; easy to learn but very verbose; widely used on mainframes
FORTH	control applications	reverse Polish notation language
FORTRAN (**for**mula **tran**slation)	scientific and computational work	based on mathematical formulae; popular among engineers, scientists, and mathematicians
LISP (**lis**t **p**rocessing)	artificial intelligence	symbolic language with a reputation for being hard to learn; popular in the academic and research communities
Modula-2	systems and real-time programming; general programming	highly structured; intended to replace Pascal for 'real-world' applications
OBERON	general programming	small, compact language incorporating many of the features of PASCAL and Modula-2
PASCAL (**p**rogram **a**ppliqué à la **s**élection et la **c**ompilation **a**utomatique de la **l**ittérature)	general-purpose language	highly structured; widely used for teaching programming in universities
PROLOG (**pro**gramming in **log**ic	artificial intelligence	symbolic-logic programming system, originally intended for theorem solving but now used more generally in artificial intelligence

or more. *Quiescent prominences* last for months, and are held in place by magnetic fields in the Sun's corona. *Surge prominences* shoot gas into space at speeds of 1,000 kps/600 mps.

Loop prominences are gases falling back to the Sun's surface after a solar ◊flare.

proof in mathematics, a set of arguments used to deduce a mathematical theorem from a set of axioms.

proof spirit numerical scale used to indicate the alcohol content of an alcoholic drink. Proof spirit (or 100% proof spirit) acquired its name from a solution of alcohol in water which, when used to moisten gunpowder, contained just enough alcohol to permit it to burn.

propane C_3H_8 gaseous hydrocarbon of the ◊alkane series, found in petroleum and used as fuel.

propanol or *propyl alcohol* third member of the homologous series of ◊alcohols. Propanol is usually a mixture of two isomeric compounds (see ◊isomer): propan-1-ol ($CH_3CH_2CH_2OH$) and propan-2-ol ($CH_3CHOHCH_3$). Both are colourless liquids that can be mixed with water and are used in perfumery.

propanone CH_3COCH_3 (common name *acetone*) colourless flammable liquid used extensively as a solvent, as in nail-varnish remover. It boils at 56.5°C/133.7°F, mixes with water in all proportions, and has a characteristic odour.

propellant substance burned in a rocket for propulsion. Two propellants are used: oxidizer and fuel are stored in separate tanks and pumped independently into the combustion chamber. Liquid oxygen (oxidizer) and liquid hydrogen (fuel) are common propellants, used, for example, in the space-shuttle main engines. The explosive charge that propels a projectile from a gun is also called a propellant.

propeller screwlike device used to propel some ships and aeroplanes. A propeller has a number of curved blades that describe a helical path as they rotate with the hub, and accelerate fluid (liquid or gas) backwards during rotation. Reaction to this backward movement of fluid sets up a propulsive thrust forwards. The marine screw propeller was developed by Francis Pettit Smith in the UK and Swedish-born John Ericson in the USA and was first used 1839.

propene $CH_3CH:CH_2$ (common name *propylene*) second member of the alkene series of hydrocarbons. A colourless, flammable gas, it is widely used by industry to make organic chemicals, including polypropylene plastics.

propenoic acid $H_2C:CHCOOH$ (common name *acrylic acid*) acid obtained from the aldehyde propenal (acrolein) derived from glycerol or fats. Glasslike thermoplastic resins are made by polymerizing ◊esters of propenoic acid or methyl propenoic acid and used for transparent components, lenses, and dentures. Other acrylic compounds are used for adhesives, artificial fibres, and artists' acrylic paint.

proper motion gradual change in the position of a star that results from its motion in orbit around our galaxy, the Milky Way. Proper motions are slight and undetectable to the naked eye, but can be accurately measured on telescopic photographs taken many years apart.

Barnard's Star is the star with the largest proper motion, 10.3 arc seconds per year.

properties in chemistry, the characteristics a substance possesses by virtue of its composition.

Physical properties of a substance can be measured by physical means, for example boiling point, melting point, hardness, elasticity, colour, and physical state. *Chemical properties* are the way it reacts with other substances; whether it is acidic or basic, an oxidizing or a reducing agent, a salt, or stable to heat, for example.

prop root or *stilt root* modified root that grows from the lower part of a stem or trunk down to the ground, providing a plant with extra support. Prop roots are common on some woody plants, such as mangroves, and also occur on a few herbaceous plants, such as maize. *Buttress roots* are a type of prop root found at the base of tree trunks, extended and flattened along the upper edge to form massive triangular buttresses; they are common on tropical trees.

propyl alcohol common name for ◊propanol.

propylene common name for ◊propene.

prosimian or *primitive primate* in zoology, any animal belonging to the suborder Strepsirhin of ◊primates. Prosimians are characterized by a wet nose with slitlike nostrils, the tip of the nose having a prominent vertical groove. Examples are lemurs, pottos, tarsiers, and the aye-aye.

prostaglandin any of a group of complex fatty acids present in the body that act as messenger substances between cells. Effects include stimulating the contraction of smooth muscle (for example, of the womb during birth), regulating the production of stomach acid, and modifying hormonal activity. In excess, prostaglandins may produce inflammatory disorders such as arthritis. Synthetic prostaglandins are used to induce labour in humans and domestic animals.

prostate gland gland surrounding and opening into the ◊urethra at the base of the ◊bladder in male mammals.

The prostate gland produces an alkaline fluid that is released during ejaculation; this fluid activates sperm, and prevents their clumping together. In humans, the prostate often enlarges and obstructs the urethra; this is treated by prostatectomy.

prosthesis artificial device used to substitute for a body part which is defective or missing. Prostheses include artificial limbs, hearing aids, false teeth and eyes, heart ◊pacemakers and plastic heart valves and blood vessels.

protactinium silver–grey, radioactive, metallic element of the ◊actinide series, symbol Pa, atomic number 91, relative atomic mass 231.036. It occurs in nature in very small quantities, in ◊pitchblende and other uranium ores. It has 14 known isotopes; the longest-lived, Pa-231, has a half-life of 32,480 years.

protandry in a flower, the state where the male reproductive organs reach maturity before those of the female. This is a common method of avoiding self-fertilization. See also ◊protogyny.

protease general term for a digestive enzyme capable of splitting proteins. Examples include pepsin, found in the stomach, and trypsin, found in the small intestine.

protein engineering the creation of synthetic proteins designed to carry out specific tasks. For example, an enzyme may be designed to remove grease from soiled clothes and remain stable at the high temperatures in a washing machine.

protein synthesis manufacture, within the cytoplasm of the cell, of the proteins an organism needs. The building blocks of proteins are ◊amino acids, of which there are 20 types. The pattern in which the amino acids are linked decides what kind of protein is produced. In turn it is the genetic code, contained within ◊DNA, that determines the precise order in which the amino acids are linked up during protein manufacture. Interestingly, DNA is found only in the nucleus, yet protein synthesis occurs only in the cytoplasm. The information necessary for making the proteins is carried from the nucleus to the cytoplasm by another nucleic acid, ◊RNA.

Proterozoic eon of geological time, possible 3.5 billion to 570 million years ago, the second division of the Precambrian. It is defined as the time of simple life, since many rocks dating from this eon show traces of biological activity, and some contain the fossils of bacteria and algae.

prothallus short-lived gametophyte of many ferns and other ◊pteridophytes (such as horsetails or clubmosses). It bears either the male or female sex organs, or both. Typically it is a small, green, flattened structure that is anchored in the soil by several ◊rhizoids (slender, hairlike structures, acting as roots) and needs damp conditions to survive. The reproductive organs are borne on the lower surface close to the soil. See also ◊alternation of generations.

prothrombin substance in the plasma from which ◊thrombin is derived during blood clotting.

protist in biology, a single-celled organism which has a eukaryotic cell, but which is not member of the plant, fungal, or animal kingdoms. The main protists are ◊protozoa.

Single-celled photosynthetic organisms, such as diatoms and dinoflagellates, are classified as protists or algae. Recently the term has also been used for members of the kingdom Protista, which features in certain five-kingdom classifications of the living world (see also ◊plant classification). This kingdom may include slime moulds, all algae (seaweeds as well as unicellular forms), and protozoa.

protocol in computing, an agreed set of standards for the transfer of data between different devices. They cover transmission speed, format of data, and the signals required to synchronize the transfer. See also ◊interface.

amino acids, where R is one of many possible side chains

protein *A protein molecule is a long chain of amino acids linked by peptide bonds. The properties are determined by the order, or sequence, of amino acids in its molecule, and by the three-dimensional structure of the molecular chain. The chain folds and twists, often forming a spiral shape.*

protogyny in a flower, the state where the female reproductive organs reach maturity before those of the male. Like protandry, in which the male organs reach maturity first, this is a method of avoiding self-fertilization, but it is much less common.

proton in physics, a positively charged subatomic particle, a constituent of the nucleus of all atoms. It belongs to the ◊baryon subclass of the ◊hadrons. A proton is extremely long-lived, with a lifespan of at least 10^{32} years. It carries a unit positive charge equal to the negative charge of an ◊electron. Its mass is almost 1,836 times that of an electron, or 1.67×10^{-24} g. The number of protons in the atom of an element is equal to the atomic number of that element. Protons are composed of two up ◊quarks and one down quark held together by ◊gluons.

protonema young ◊gametophyte of a moss, which develops from a germinating spore (see ◊alternation of generations). Typically it is a green, branched, threadlike structure that grows over the soil surface bearing several buds that develop into the characteristic adult moss plants.

proton number alternative name for ◊atomic number.

Proton rocket Soviet space rocket introduced 1965, used to launch heavy satellites, space probes, and the Salyut and Mir space stations.

protoplasm contents of a living cell. Strictly speaking it includes all the discrete structures (organelles) in a cell, but it is often used simply to mean the jellylike material in which these float. The contents of a cell outside the nucleus are called ◊cytoplasm.

prototheria subclass of mammals made up of the egg-laying monotremes. It contains only the echidna (spiny anteater) and the platypus.

prototype in technology, any of the first few machines of a new design. Prototypes are tested for performance, reliability, economy, and safety; then the main design can be modified before full-scale production begins.

protozoa group of single-celled organisms without rigid cell walls. Some, such as amoeba, ingest other cells, but most are ◊saprotrophs or parasites. The group is polyphyletic (containing organisms which have different evolutionary origins).

provitamin any precursor substance of a vitamin. Provitamins are injested substances that become converted to active vitamins within the organism. One example is ergosterol (provitamin D_2), which through the action of sunlight is converted to calciferol (vitamin D_2); another example is beta-carotene, which is hydrolysed in the liver to vitamin A.

Proxima Centauri the closest star to the Sun, 4.2 light years away. It is a faint ◊red dwarf, visible only with a telescope, and is a member of the Alpha Centauri triple-star system.

prussic acid former name for ◊hydrocyanic acid.

pseudocarp fruitlike structure that incorporates tissue that is not derived from the ovary wall. The additional tissues may be derived from floral parts such as the ◊receptacle and ◊calyx. For example, the coloured, fleshy part of a strawberry develops from the receptacle and the true fruits are small ◊achenes – the 'pips' embedded in its outer surface. Rose hips are a type of pseudocarp that consists of a hollow, fleshy receptacle containing a number of achenes within. Different types of pseudocarp include pineapples, figs, apples, and pears.

pseudocopulation attempted copulation by a male insect with a flower. It results in ◊pollination of the flower and is common in the orchid family, where the flowers of many species resemble a particular species of female bee. When a male bee attempts to mate with a flower, the pollinia (groups of pollen grains) stick to its body. They are transferred to the stigma of another flower when the insect attempts copulation again.

pseudomorph mineral that has replaced another *in situ* and has retained the external crystal shape of the original mineral.

pt symbol for ◊*pint*.

pteridophyte simple type of ◊vascular plant. The pteridophytes comprise four classes: the Psilosida, including the most primitive vascular plants, found mainly in the tropics; the Lycopsida, including the club mosses; the Sphenopsida, including the horsetails; and the Pteropsida, including the ferns. They do not produce seeds.

They are mainly terrestrial, non-flowering plants characterized by the presence of a vascular system; the possession of true stems, roots, and leaves; and by a marked ◊alternation of generations, with the sporophyte forming the dominant generation in the life cycle. The pteridophytes formed a large and dominant flora during the Carboniferous period, but many are now known only from fossils.

PTFE abbreviation for ◊*polytetrafluoroethene*.

ptomaine any of a group of toxic chemical substances (alkaloids) produced as a result of decomposition by bacterial action on proteins.

puberty stage in human development when the individual becomes sexually mature. It may occur from the age of ten upwards. The sexual organs take on their adult form and pubic hair grows. In girls, menstruation begins, and the breasts develop; in boys, the voice breaks and becomes deeper, and facial hair develops.

pubes lowest part of the front of the human trunk, the region where the external generative organs are situated. The underlying bony structure, the pubic arch, is formed by the union in the midline of the two pubic bones, which are the front portions of the hip bones. In women this is more prominent than in men, to allow more room for the passage of the child's head at birth, and it carries a pad of fat and connective tissue, the *mons veneris* (mount of Venus), for its protection.

pulley simple machine consisting of a fixed, grooved wheel, sometimes in a block, around which a rope or chain can be run. A simple pulley serves only to change the direction of the applied effort (as in a simple hoist for raising loads). The use of more than one pulley results in a mechanical advantage, so that a given effort can raise a heavier load.

pulmonary pertaining to the ◊lungs.

pulsar celestial source that emits pulses of energy at regular intervals, ranging from a few seconds to a few thousandths of a second. Pulsars are thought to be rapidly rotating ◊neutron stars, which flash at radio and other wavelengths as they spin. They were discovered in 1967 by Jocelyn Bell (1943–) (now Burnell) and Antony Hewish (1924–) at the Mullard Radio Astronomy Observatory, Cambridge, England. Over 500 radio pulsars are now known in our Galaxy, although a million or so may exist.

pulse impulse transmitted by the heartbeat throughout the arterial systems of vertebrates. When the heart muscle contracts, it forces blood into the ◊aorta (the chief artery). Because the arteries are elastic, the sudden rise of pressure causes a throb or sudden swelling through them. The actual flow of the blood is about 60 cm/2 ft a second in humans. The average adult pulse rate is generally about 70 per minute. The pulse can be felt where an artery is near the surface, for example in the wrist or the neck.

pulse-code modulation (PCM) in physics, a form of digital ◊modulation in which microwaves or light waves (the carrier waves) are switched on and off in pulses of varying length according to a binary code. It is a relatively simple matter to transmit data that are already in binary code, such as those used

simple pulley (above)
pulley system used for heavy weights (below)

N = newton, a unit of force

pulley *The mechanical advantage of a pulley increases with the number of rope strands. If a pulley system has four ropes supporting the load, the mechanical advantage is four, and a 5 N force will lift a 20 N load.*

by computer, by these means. However, if an analogue audio signal is to be transmitted, it must first be converted to a *pulse-amplitude modulated* signal (PAM) by regular sampling of its amplitude. The value of the amplitude is then converted into a binary code for transmission on the carrier wave.

pumice light volcanic rock produced by the frothing action of expanding gases during the solidification of lava. It has the texture of a hard sponge and is used as an abrasive.

pump any device for moving liquids and gases, or compressing gases.

Some pumps, such as the traditional *lift pump* used to raise water from wells, work by a reciprocating (up-and-down) action. Movement of a piston in a cylinder with a one-way valve creates a partial vacuum in the cylinder, thereby sucking water into it.

pulse-code modulation *Simple pulse-code modulation (PCM) consists of switching the carrier wave off and on to represent the 0s and 1s used in binary code. In pulse-code modulation (PAM), the analogue signal is sampled many times each second. The intensity level of each sample is then translated into binary code before being transmitted. At the receiver, the original signal is reconstituted by translating the binary code and producing an analogue signal whose intensity varies in step with the code.*

pumped storage hydroelectric plant that uses surplus electricity to pump water back into a high-level reservoir. In normal working conditions the water flows from this reservoir through the ◊turbines to generate power for feeding into the grid. At times of low power demand, electricity is taken from the grid to turn the turbines into pumps that then pump the water back again. This ensures that there is always a maximum 'head' of water in the reservoir to give the maximum output when required.

punched card in computing, an early form of data storage and input, now almost obsolete. The 80-column card widely used in the 1960s and 1970s was a thin card, measuring 190 mm × 84 mm/7.5 in × 3.33 in, holding up to 80 characters of data encoded as small rectangular holes.

punctuated equilibrium model evolutionary theory developed by Niles Eldridge and US palaeontologist Stephen Jay Gould 1972 to explain discontinuities in the fossil record. It claims that periods of rapid change alternate with periods of relative stability (stasis), and that the appearance of new lineages is a separate process from the gradual evolution of adaptive changes within a species.

The pattern of stasis and more rapid change is now widely accepted, but the second part of the theory remains unsubstantiated. The *turnover pulse hypothesis* of US biologist Elisabeth Vrba postulates that the periods of rapid evolutionary change are triggered by environmental changes, in particular, climate changes.

punnett square graphic technique used in genetics for determining the likely outcome, in statistical terms, of a genetic cross. It resembles a game of noughts and crosses, in which the genotypes of the parental generation gametes are entered first, so that the subsequent combinations can then be calculated.

pupa nonfeeding, largely immobile stage of some insect life cycles, in which larval tissues are broken down, and adult tissues and structures are formed.

pupil circular aperture in the ◊iris through which light falls onto the lens of the eye.

putrefaction decomposition of organic matter by microorganisms.

PVC abbreviation for *polyvinylchloride*, a type of ◊plastic derived from vinyl chloride (CH_2:CHCl).

PWR abbreviation for ◊*pressurized water reactor*, a type of nuclear reactor.

pyramid of numbers in ecology, a diagram that shows how many plants and animals there are at different levels in a ◊food chain.

There are always far fewer individuals at the top of the chain than at the bottom because only about 10% of the food an animal eats is turned into flesh, so the amount of food flowing through the chain drops at each step. In a pyramid of numbers, the primary producers (usually plants) are represented at the bottom by a broad band, the plant-eaters are shown above by a narrower band, and the animals that prey on them by a narrower band still. At the top of the pyramid are the 'top carnivores' such as lions and sharks, which are present in the smallest number.

pyridine C_5H_5N a heterocyclic compound (see ◊cyclic compounds). It is a liquid with a sickly smell and occurs in coal tar. It is soluble in water, acts as a strong ◊base, and is used as a solvent, mainly in the manufacture of plastics.

pyridoxine or *vitamin B_6* $C_8H_{11}NO_3$ water-soluble ◊vitamin of the B complex. There is no clearly identifiable disease associated with deficiency but its absence from the diet can give rise to malfunction of the central nervous system and general skin disorders. Good sources are liver, meat, milk, and cereal grains. Related compounds may also show vitamin B_6 activity.

pyrite iron sulphide FeS_2; also called *fool's gold* because of its yellow metallic lustre. Pyrite has a hardness of 6–6.5 on the Mohs' scale. It is used in the production of sulphuric acid.

pyroclastic in geology, pertaining to fragments of solidified volcanic magma, ranging in size from fine ash to large boulders, that are extruded during an explosive volcanic eruption; also the rocks that are formed by consolidation of such material. Pyroclastic rocks include tuff (ash deposit) and agglomerate (volcanic breccia).

pyrogallol $C_6H_3OH_3$ (technical name *trihydroxybenzene*) derivative of benzene, prepared from gallic acid. It is used in gas analysis for the measurement of oxygen because its alkaline solution turns black as it rapidly absorbs oxygen. It is also used as a developer in photography.

pyrolysis decomposition of a substance by heating it to a high temperature in the absence of air. The process is used to burn and dispose of old tyres, for example, without contaminating the atmosphere.

pyrometer instrument used for measuring high temperatures.

pyroxene any one of a group of minerals, silicates of calcium, iron, and magnesium with a general formula X,YSi_2O_6, found in igneous and metamorphic rocks. The internal structure is based on single chains of silicon and oxygen. Diopside ($X = Ca$, $Y = Mg$) and augite ($X = Ca$, $Y = Mg,Fe,Al$) are common pyroxenes.

Pythagoras' theorem in geometry, a theorem stating that in a right-angled triangle, the area of the square on the hypotenuse (the longest side) is equal to the sum of the areas of the squares drawn on the other two sides. If the hypotenuse is h units long and the lengths of the other sides are a and b, then $h^2 = a^2 + b^2$.

The theorem provides a way of calculating the length of any side of a right-angled triangle if the lengths of the other two sides are known. It is also used to determine certain trigonometrical relationships such as $\sin^2 \theta + \cos^2 \theta = 1$.

for right-angled triangles

Pythagoras' theorem
$a^2 = b^2 + c^2$

Pythagoras' theorem *Pythagoras' theorem for right-angled triangles is likely to have been known by the ancient Egyptians before Pythagoras' time.*

quadrat in environmental studies, a square structure used to study the distribution of plants in a particular place, for instance a field, rocky shore, or mountainside. The size varies, but is usually 0.5 or 1 metre square, small enough to be carried easily. The quadrat is placed on the ground and the abundance of species estimated. By making such measurements a reliable understanding of species distribution is obtained.

quadratic equation in mathematics, a polynomial equation of second degree (that is, an equation containing as its highest power the square of a variable, such as x^2). The general formula of such equations is $ax^2 + bx + c = 0$, in which a, b, and c are real numbers, and only the coefficient a cannot equal 0.

In ◊coordinate geometry, a quadratic function represents a ◊parabola.

quadrature position of the Moon or an outer planet where a line between it and Earth makes a right angle with a line joining Earth to the Sun.

qualitative analysis in chemistry, a procedure for determining the identity of the component(s) of a single substance or mixture. A series of simple reactions and tests can be carried out on a compound to determine the elements present.

quantitative analysis in chemistry, a procedure for determining the precise amount of a known component present in a single substance or mixture. A known amount of the substance is subjected to particular procedures.

Gravimetric analysis determines the mass of each constituent present; ◊*volumetric analysis* determines the concentration of a solution by ◊titration against a solution of known concentration.

quantum chromodynamics (QCD) in physics, a theory describing the interactions of ◊quarks, the elementary particles that make up all ◊hadrons (subatomic particles such as protons and neutrons). In quantum chromodynamics, quarks are considered to interact by exchanging particles called gluons, which carry the ◊strong nuclear force, and whose role is to 'glue' quarks together. The mathematics involved in the theory is complex, and although a number of successful predictions have been made, as yet the theory does not compare in accuracy with ◊quantum electrodynamics, upon which it is modelled. See ◊elementary particles and ◊forces, fundamental.

quantum electrodynamics (QED) in physics, a theory describing the interaction of charged subatomic particles within electric and magnetic fields. It combines ◊quantum theory and ◊relativity, and considers charged particles to interact by the exchange of photons. QED is remarkable for the accuracy of its predictions; for example, it has been used to calculate the value of some physical quantities to an accuracy of ten decimal places, a feat equivalent to calculating the distance between New York and Los Angeles to within the thickness of a hair. The theory was developed by US physicists Richard Feynman and Julian Schwinger, and by Japanese physicist Sin-Itiro Tomonaga 1948.

quantum number in physics, one of a set of four numbers that uniquely characterize an ◊electron and its state in an ◊atom. The *principal quantum number* n defines the electron's main energy level. The *orbital quantum number* l relates to its angular momentum. The *magnetic quantum number* m describes the energies of electrons in a magnetic field. The *spin quantum number* m_s gives the spin direction of the electron.

The principal quantum number, defining the electron's energy level, corresponds to shells (energy levels) also known by their spectroscopic designations K, L, M, and so on. The orbital quantum number gives rise to a series of subshells designated s, p, d, f, and so on, of slightly different energy levels. The magnetic quantum number allows further subdivision of the subshells (making three subdivisions p_x, p_y, and p_z in the p subshell, for example, of the same energy level). No two electrons in an atom can have the same set of quantum numbers (the ◊exclusion principle).

quantum simulation the use of ◊computer simulation to study the behaviour of matter. Simulations based on just 50 or 60 atoms are able to mimic the properties of real matter. If a small group of atoms is surrounded by other identical groups every atom behaves as if it were part of a larger material. This makes it is possible, for example, to simulate the behaviour of solids and liquids at high temperatures, and to study chemical reactions at surfaces to aid in the design of better catalysts.

quantum theory or *quantum mechanics* in physics, the theory that ◊energy does not have a continuous range of values, but is, instead, absorbed or radiated discontinuously, in multiples of definite, indivisible units called quanta. Just as earlier theory showed how light, generally seen as a wave motion, could also in some ways be seen as composed of discrete particles (◊photons), quantum theory shows how atomic particles such as electrons may also be seen as having wavelike properties. Quantum theory is the basis of particle physics, modern theoretical chemistry, and the solid-state physics that describes the behaviour of the silicon chips used in computers.

quark in physics, the ◊elementary particle that is the fundamental constituent of all ◊hadrons (baryons, such as neutrons and protons, and mesons). There are six types, or 'flavours': up, down, top, bottom, strange, and charm, each of which has three varieties, or 'colours': red, yellow, and blue (visual colour is not meant, although the analogy is useful in many ways). To each quark there is an antiparticle, called an antiquark. See ◊quantum chromodynamics.

The existence of the top quark still remains to be confirmed because of its instability. It is unlikely to last more than a millionth of a billionth of a billionth of a second, but experiments at Fermilab have shown it to be at least as massive as a silver atom, and 20,000 times heavier than an up quark.

quart imperial liquid or dry measure, equal to two pints or 1.136 litres. In the USA, a liquid quart is equal to 0.946 litre, while a dry quart is equal to 1.101 litres.

quartz crystalline form of ◊silica SiO_2, one of the most abundant minerals of the Earth's crust (12% by volume). Quartz occurs in many different kinds of rock, including sandstone and granite. It ranks 7 on the Mohs' scale of hardness and is resistant to chemical or mechanical breakdown. Quartzes vary according to the size and purity of their crystals. Crystals of pure quartz are coarse, colourless, and transparent, and this form is usually called rock crystal. Impure coloured varieties, often used as gemstones, include ◊agate, citrine quartz, and ◊amethyst. Quartz is used in ornamental work and industry, where its reaction to electricity makes it valuable in electronic instruments (see ◊piezoelectric effect). Quartz can also be made synthetically.

quartzite ◊metamorphic rock consisting of pure quartz sandstone that has recrystallized under increasing heat and pressure.

quasar (from *quas*i-stell*ar* object or QSO) one of the most distant extragalactic objects known, discovered 1963. Quasars appear starlike, but each emits more energy than 100 giant galaxies. They are thought to be at the centre of galaxies, their brilliance emanating from the stars and gas falling towards an immense ◊black hole at their nucleus.

quasi-atom particle assemblage resembling an atom, in which particles not normally found in atoms become bound together for a brief period. Quasi-atoms are generally unstable structures, either because they are subject to matter–antimatter annihilation (positronium), or because one or more of their constituents is unstable (muonium).

Quaternary period of geological time that began 1.64 million years ago and is still in process. It is divided into the ◊Pleistocene and ◊Holocene epochs.

quenching ◊heat treatment used to harden metals. The metals are heated to a certain temperature and then quickly plunged into cold water or oil.

quicklime common name for ◊calcium oxide.

quicksilver another name for the element ◊mercury.

quinine antimalarial drug extracted from the bark of the cinchona tree. Peruvian Indians taught French missionaries how to use the bark in 1630, but quinine was not isolated until 1820. It is a bitter alkaloid $C_{20}H_{24}N_2O_2$.

raceme in botany, a type of ◊inflorescence.

rad unit of absorbed radiation dose, now replaced in the SI system by the ◊gray (one rad equals 0.01 gray), but still commonly used. It is defined as the dose when one kilogram of matter absorbs 0.01 joule of radiation energy (formerly, as the dose when one gram absorbs 100 ergs).

radar (acronym for *radio direction and ranging*) device for locating objects in space, direction finding, and navigation by means of transmitted and reflected high-frequency radio waves.

radar astronomy bouncing of radio waves off objects in the Solar System, with reception and analysis of the 'echoes'. Radar contact with the Moon was first made 1945 and with Venus 1961. The travel time for radio reflections allows the distances of objects to be determined accurately. Analysis of the reflected beam reveals the rotation period and allows the object's surface to be mapped. The rotation periods of Venus and Mercury were first determined by radar. Radar maps of Venus were obtained first by Earth-based radar and subsequently by orbiting space probes.

radial circuit circuit used in household electric wiring in which all electrical appliances are connected to cables that radiate out from the main supply point or fuse box. In more modern systems, the appliances are connected in a ring, or ◊ring circuit, with each end of the ring connected to the fuse box.

radian SI unit (symbol rad) of plane angles, an alternative unit to the ◊degree. It is the angle at the centre of a circle when the centre is joined to the two ends of an arc (part of the circumference) equal in length to the radius of the circle. There are 2π (approximately 6.284) radians in a full circle (360°).

radiant heat energy that is radiated by all warm or hot bodies. It belongs to the ◊infrared part of the electromagnetic ◊spectrum and causes heating when absorbed. Radiant heat is invisible and should not be confused with the red glow associated with very hot objects, which belongs to the visible part of the spectrum.

radiation in physics, emission of radiant ◊energy as particles or waves – for example, heat, light, alpha particles, and beta particles (see ◊electromagnetic waves and ◊radioactivity). See also ◊atomic radiation.

radiation biology study of how living things are affected by radioactive (ionizing) emissions (see ◊radioactivity) and by electromagnetic (nonionizing) radiation (◊electromagnetic waves). Both are potentially harmful and can cause mutations as well as leukaemia and other cancers; even low levels of radioactivity are very dangerous. Both are, however, used therapeutically, for example to treat cancer, when the radiation dose is very carefully controlled (◊*radiotherapy* or X-ray therapy).

radiation monitoring system network of monitors to detect raises in background gamma radiation and warn of a major nuclear accident within minutes. Since the accident at Chernobyl 1986 several W European countries have begun to install such systems locally, and in 1994 work began on a pilot system to provide a *gamma curtain*, a dense net of radiation monitors, throughout E and W Europe.

The largest system in operation, the German IMIS monitoring system, uses more than 2,000 sensors. Britain's RIMNET has 92 monitors at sites throughout the country. The RIMNET second phase, scheduled for completion 1993, includes the addition of a software system to predict the path of a detected radioactive plume and plot it on a computer-generated map.

radiation sickness sickness resulting from exposure to radiation, including X-rays, gamma rays, neutrons, and other nuclear radiation, as from weapons and fallout. Such radiation ionizes atoms in the body and causes nausea, vomiting, diarrhoea, and other symptoms. The body cells themselves may be damaged even by very small doses, causing leukaemia; genetic changes may be induced in the embryo, causing infants to be born damaged or mutated.

radiation units units of measurement for radioactivity and radiation doses. Continued use of the units introduced earlier this century (the curie, rad, rem, and roentgen) has been approved while the derived SI units (becquerel, gray, sievert, and coulomb) become familiar. One curie equals 3.7×10^{-10} becquerels (activity); one rad equals 10^{-2} gray (absorbed dose); one rem equals 10^{-2} sievert (dose equivalent); one roentgen equals 2.58×10^{-4} coulomb/kg (exposure to ionizing radiation).

radical in chemistry, a group of atoms forming part of a molecule, which acts as a unit and takes part in chemical reactions without disintegration, yet often cannot exist alone; for example, the methyl radical $-CH_3$, or the carboxyl radical $-COOH$.

radicle part of a plant embryo that develops into the primary root. Usually it emerges from the seed before the embryonic shoot, or ◊plumule, its tip protected by a root cap, or calyptra, as it pushes through the soil. The radicle may form the basis of the entire root system, or it may be replaced by adventitious roots (positioned on the stem).

radio transmission and reception of radio waves. In radio transmission a microphone converts ◊sound waves (pressure variations in the air) into ◊electromagnetic waves that are then picked up by a receiving aerial and fed to a loudspeaker, which converts them back into sound waves.

radioactive decay process of continuous disintegration undergone by the nuclei of radioactive elements, such as radium and various isotopes of uranium and the transuranic elements. This changes the element's atomic number, thus transmuting one element into another, and is accompanied by the emission of radiation. Alpha and beta decay are the most common forms.

radioactive tracer any of various radioactive ◊isotopes used in labelled compounds; see ◊tracer.

radioactive waste any waste that emits radiation in excess of the background level. See ◊nuclear waste.

radioactivity spontaneous alteration of the nuclei of radioactive atoms, accompanied by the emission of radiation. It is the property exhibited by the radioactive ◊isotopes of stable elements and all isotopes of radioactive elements, and can be either natural or induced. See ◊radioactive decay.

Radioactivity establishes an equilibrium in parts of the nuclei of unstable radioactive substances, ultimately to form a stable arrangement of nucleons (protons and neutrons); that is, a nonradioactive (stable) element. This is most frequently accomplished by the emission of ◊alpha particles (helium nuclei); ◊beta particles (electrons and positrons); or ◊gamma radiation (electromagnetic waves of very high frequency). It takes place

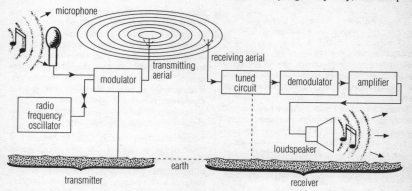

radio Radio transmission and reception. The radio frequency oscillator generates rapidly varying electrical signals, which are sent to the transmitting aerial. In the aerial, the signals produce radio waves (the carrier wave), which spread out at the speed of light. The sound signal is added to the carrier wave by the modulator. When the radio waves fall on the receiving aerial, they induce an electrical current in the aerial. The electrical current is sent to the tuning circuit, which picks out the signal from the particular transmitting station desired. The demodulator separates the sound signal from the carrier wave and sends it, after amplification, to the loudspeaker.

either directly, through a one-step decay, or indirectly, through a number of decays that transmute one element into another. This is called a decay series or chain, and sometimes produces an element more radioactive than its predecessor.

The instability of the particle arrangements in the nucleus of a radioactive atom (the ratio of neutrons to protons and/or the total number of both) determines the lengths of the ◊half-lives of the isotopes of that atom, which can range from fractions of a second to billions of years. All isotopes of relative atomic mass 210 and greater are radioactive. Alpha, beta, and gamma radiation are ionizing in their effect and are therefore dangerous to body tissues, especially if a radioactive substance is ingested or inhaled.

radio astronomy study of radio waves emitted naturally by objects in space, by means of a ◊radio telescope. Radio emission comes from hot gases (***thermal radiation***); electrons spiralling in magnetic fields (***synchrotron radiation***); and specific wavelengths (***lines***) emitted by atoms and molecules in space, such as the 21-cm/8-in line emitted by hydrogen gas.

Radio astronomy has greatly improved our understanding of the evolution of stars, the structure of galaxies, and the origin of the universe. Astronomers have mapped the spiral structure of the Milky Way from the radio waves given out by interstellar gas, and they have detected many individual radio sources within our Galaxy and beyond.

Among radio sources in our Galaxy are the remains of ◊supernova explosions, such as the ◊Crab nebula and ◊pulsars. Short-wavelength radio waves have been detected from complex molecules in dense clouds of gas where stars are forming. Searches have been undertaken for signals from other civilizations in the Galaxy, so far without success.

Strong sources of radio waves beyond our Galaxy include ◊radio galaxies and ◊quasars. Their existence far off in the universe demonstrates how the universe has evolved with time. Radio astronomers have also detected weak ***background radiation*** thought to be from the ◊Big Bang explosion that marked the birth of the universe.

radio beacon radio transmitter in a fixed location, used in marine and aerial ◊navigation. Ships and aircraft pinpoint their positions by reference to continuous signals given out by two or more beacons.

radiocarbon dating or *carbon dating* method of dating organic materials (for example, bone or wood), used in archaeology. Plants take up carbon dioxide gas from the atmosphere and incorporate it into their tissues, and some of that carbon dioxide contains the radioactive isotope of carbon, carbon-14. This decays at a known rate (half of it decays every 5,730 years); the time elapsed since the plant died can therefore be measured in a laboratory. Animals take carbon-14 into their bodies from eating plant tissues and their remains can be similarly dated. After 120,000 years so little carbon-14 is left that no measure is possible (see ◊half-life).

radio, cellular portable telephone system; see ◊cellular phone.

radiochemistry chemical study of radioactive isotopes and their compounds (whether produced from naturally radioactive or irradiated materials) and their use in the study of other chemical processes.

When such isotopes are used in labelled compounds, they enable the biochemical and physiological functioning of parts of the living body to be observed. They can help in the testing of new drugs, showing where the drug goes in the body and how long it stays there. They are also useful in diagnosis for example cancer, fetal abnormalities, and heart disease.

radio galaxy galaxy that is a strong source of electromagnetic waves of radio wavelengths. All galaxies, including our own, emit some radio waves, but radio galaxies are up to a million times more powerful.

radiography branch of science concerned with the use of radiation (particularly ◊X-rays) to produce images on photographic film or fluorescent screens. X-rays penetrate matter according to its nature, density, and thickness. In doing so they can cast shadows on photographic film, producing a radiograph. Radiography is widely used in medicine for examining bones and tissues and in industry for examining solid materials; for example, to check welded seams in pipelines

radioisotope (contraction of *radioactive ◊isotope*) in physics, a naturally occurring or synthetic radioactive form of an element. Most radioisotopes are made by bombarding a stable element with neutrons in the core of a nuclear reactor. The radiations given off by radioisotopes are easy to detect (hence their use as ◊tracers), can in some instances penetrate substantial thicknesses of materials, and have profound effects (such as genetic ◊mutation) on living matter. Although dangerous, radioisotopes are used in the fields of medicine, industry, agriculture, and research.

radioisotope scanning use of radioactive materials (radioisotopes or radionucleides) to pinpoint disease. It reveals the size and shape of the target organ and whether any part of it is failing to take up radioactive material, usually an indication of disease.

radiometric dating method of dating rock by assessing the amount of ◊radioactive decay of naturally occurring ◊isotopes. The dating of rocks may be based on the gradual decay of uranium into lead. The ratio of the amounts of 'parent' to 'daughter' isotopes in a sample gives a measure of the time it has been decaying, that is, of its age. Different elements and isotopes are used depending on the isotopes present and the age of the rocks to be dated. Once-living matter can often be dated by ◊radiocarbon dating, employing the half-life of the isotope carbon-14, which is naturally present in organic tissue.

Radiometric methods have been applied to the decay of long-lived isotopes, such as potassium-40, rubidium-87, thorium-232, and uranium-238 which are found in rocks. These isotopes decay very slowly and this has enabled rocks as old as 3,800 million years to be dated accurately. Carbon dating can be used for material between 100,000 and 1,000 years old. *Potassium* dating is used for material more than 100,000 years old, *rubidium* for rocks more than 10 million years old, and *uranium* and *thorium* dating is suitable for rocks older than 20 million years.

radiosonde balloon carrying a compact package of meteorological instruments and a radio transmitter, used to 'sound', or measure, conditions in the atmosphere. The instruments measure temperature, pressure, and humidity, and the information gathered is transmitted back to observers on the ground. A radar target is often attached, allowing the balloon to be tracked.

radio telescope instrument for detecting radio waves from the universe in ◊radio astronomy. Radio telescopes usually consist of a metal bowl that collects and focuses radio waves the way a concave mirror collects and focuses light waves. Radio telescopes are much larger than optical telescopes, because the wavelengths they are detecting are much longer than the wavelength of light. The largest single dish is 305 m/1,000 ft across, at Arecibo, Puerto Rico.

radiotherapy treatment of disease by ◊radiation from X-ray machines or radioactive sources. Radiation, which reduces the activity of dividing cells, is of special value for its effect on malignant tissues, certain nonmalignant tumours, and some diseases of the skin.

radio wave electromagnetic wave possessing a long wavelength (ranging from about 10^{-3} to 10^4 m) and a low frequency (from about 10^5 to 10^{11} Hz). Included in the radio-wave part of the spectrum are ◊microwaves, used for both communications and for cooking; ultra high- and very high-frequency waves, used for television and FM (◊frequency modulation) radio communications; and short, medium, and long waves, used for AM (◊amplitude modulation) radio communications. Radio waves that are used for communications have all been modulated (see ◊modulation) to carry information. Stars emit radio waves, which may be detected and studied using ◊radio telescopes.

radium white, radioactive, metallic element, symbol Ra, atomic number 88, relative atomic mass 226.02. It is one of the ◊alkaline-earth metals, found in nature in ◊pitchblende and other uranium ores. Of the 16 isotopes, the commonest, Ra-226, has a half-life of 1.622 years. The element was discovered and named in 1898 by Pierre and Marie Curie, who were investigating the residues of pitchblende.

radius in biology, one of the two bones in the lower forearm of tetrapod (four-limbed) vertebrates.

radius a straight line from the centre of a circle to its circumference, or from the centre to the surface of a sphere.

radon colourless, odourless, gaseous, radioactive, nonmetallic element, symbol Rn, atomic number 86, relative atomic mass 222. It is grouped with the ◊inert gases and was formerly considered nonreactive, but is now known to form some compounds with fluorine. Of the 20 known isotopes, only three occur in nature; the longest half-life is 3.82 days.

rainbow arch in the sky displaying the seven colours of the ◊spectrum in bands. It is formed by the refraction, reflection, and dispersion of the Sun's rays through rain or mist. Its cause was discovered by Theodoric of Freiburg in the 14th century.

RAM (acronym for *random-access memory*) in computing, a memory device in the form of a collection of integrated circuits (chips), frequently used in microcomputers. Unlike ◊ROM (read-only memory) chips, RAM chips can be both read from and written to by the computer, but their contents are lost when the power is switched off. Microcomputers of the 1990s may have 16–32 megabytes of RAM.

ramjet simple jet engine (see under ♦jet propulsion) used in some guided missiles. It only comes into operation at high speeds. Air is then 'rammed' into the combustion chamber, into which fuel is sprayed and ignited.

In 1994 a scale model ***scramjet*** (supersonic combustion ramjet) was fired from a light gas gun producing speeds of Mach 8.2 (8.2 times the speed of sound, 9,000 km/h/5,590 mph). The scramjet uses oxygen from the atmosphere to burn its fuel. This makes it lighter than rockets with a closed fuel system that must carry liquid oxygen for fuel combustion. Air-breathing engines usually cannot exceed speeds of Mach 3 as the air hitting the inside of the engine reaches such a high temperature and pressure that combustion is impossible. The scramjet is designed so airflow remains unimpeded and combustion continues. A fullscale prototype scramjet could be tested by the end of the century.

random access in computing, an alternative term for ♦direct access.

random number one of a series of numbers having no detectable pattern. Random numbers are used in ♦computer simulation and ♦computer games. It is impossible for an ordinary computer to generate true random numbers, but various techniques are available for obtaining pseudo-random numbers – close enough to true randomness for most purposes.

rare-earth element alternative name for ♦lanthanide.

rare gas alternative name for ♦inert gas.

raster graphics computer graphics that are stored in the computer memory by using a map to record data (such as colour and intensity) for every ♦pixel that makes up the image. When transformed (enlarged, rotated, stretched, and so on), raster graphics become ragged and suffer loss of picture resolution, unlike ♦vector graphics. Raster graphics are typically used for painting applications, which allow the user to create artwork on a computer screen much as if they were painting on paper or canvas.

ratemeter instrument used to measure the count rate of a radioactive source. It gives a reading of the number of particles emitted from the source and captured by a detector in a unit of time (usually a second).

rate of reaction the speed at which a chemical reaction proceeds. It is usually expressed in terms of the concentration (usually in ♦moles per litre) of a reactant consumed, or product formed, in unit time; so the units would be moles per litre per second (mol l^{-1} s^{-1}). The rate of a reaction may be affected by the concentration of the reactants, the temperature of the reactants, and the presence of a ♦catalyst. If the reaction is entirely in the gas state, the rate is affected by pressure, and, for solids, it is affected by the particle size.

ratio measure of the relative size of two quantities or of two measurements (in similar units), expressed as a proportion. For example, the ratio of vowels to consonants in the alphabet is 5:21; the ratio of 500 m to 2 km is 500:2,000, or 1:4.

rationalized units units for which the defining equations conform to the geometry of the system. Equations involving circular symmetry contain the factor 2π; those involving spherical symmetry 4π. ♦SI units are rationalized, ♦c.g.s. units are not.

rational number in mathematics, any number that can be expressed as an exact fraction (with a denominator not equal to 0), that is, as $a \times b$ where a and b are integers. For example, $2/_1$, $1/_4$, $15/_4$, $-3/_5$ are all rational numbers, whereas π (which represents the constant 3.141592 ...) is not. Numbers such as π are called ♦irrational numbers.

Raunkiaer system method of classification devised by the Danish ecologist Christen Raunkiaer (1860–1938) whereby plants are divided into groups according to the position of their ♦perennating (overwintering) buds in relation to the soil surface. For example, plants in cold areas, such as the tundra, generally have their buds protected below ground, whereas in hot, tropical areas they are above ground and freely exposed. This scheme is useful for comparing vegetation types in different parts of the world.

rayon any of various shiny textile fibres and fabrics made from ♦cellulose. It is produced by pressing whatever cellulose solution is used through very small holes and solidifying the resulting filaments. A common type is ♦viscose, which consists of regenerated filaments of pure cellulose. Acetate and triacetate are kinds of rayon consisting of filaments of cellulose acetate and triacetate.

rays original name for radiation of all types, such as ♦X-rays and gamma rays.

reactance property of an alternating current circuit that together with any ♦resistance makes up the ♦impedance (the total opposition of the circuit to the passage of a current).

The reactance of an inductance L is wL and

(a) rate of reaction decreases with time

(b) concentration of reactant decreases with time

(c) concentration of product increases with time

rate of reaction During a chemical reaction at constant temperature, the rate of reaction and concentration of reactant decreases exponentially with time. This behaviour can be described by equations in which the rate is proportional to some power of the concentration of the reactants. The constant in the rate law is called the rate constant.

that of the capacitance is $1/wC$. Reactance is measured in ◊ohms.

reaction in chemistry, the coming together of two or more atoms, ions, or molecules with the result that a ◊chemical change takes place. The nature of the reaction is portrayed by a chemical equation.

reaction force in physics, the equal and opposite force described by Newton's third law of motion that arises whenever one object applies a force (***action force***) to another. For example, if a magnet attracts a piece of iron, then that piece of iron will also attract the magnet with a force that is equal in magnitude but opposite in direction. When any object rests on the ground the downwards contact force applied to the ground always produces an equal, upwards reaction force.

reactivity series chemical series produced by arranging the metals in order of their ease of reaction with reagents such as oxygen, water, and acids. This arrangement aids the understanding of the properties of metals, helps to explain differences between them, and enables predictions to be made about a metal's behaviour, based on a knowledge of its position or properties.

real number in mathematics, any of the ◊rational numbers (which include the integers) or ◊irrational numbers. Real numbers exclude ◊imaginary numbers, found in ◊complex numbers of the general form $a + bi$ where $i = \sqrt{-1}$, although these do include a real component a.

real-time system in computing, a program that responds to events in the world as they happen. For example, an automatic-pilot program in an aircraft must respond instantly in order to correct deviations from its course. Process control, robotics, games, and many military applications are examples of real-time systems.

receiver, radio component of a radio communication system that receives and processes radio waves. It detects and selects modulated radio waves (see ◊modulation) by means of an aerial and tuned circuit, and then separates the transmitted information from the carrier wave by a process that involves rectification (see ◊rectifier). The receiver device will usually also include the amplifiers that produce the audio signals (signals that are heard).

receptacle the enlarged end of a flower stalk to which the floral parts are attached. Normally the receptacle is rounded, but in some plants it is flattened or cup-shaped. The term is also used for the region on that part of some seaweeds which becomes swollen at certain times of the year and bears the reproductive organs.

recessive gene in genetics, an ◊allele (alternative form of a gene) that will show in the ◊phenotype (observed characteristics of an organism) only if its partner allele on the paired chromosome is similarly recessive. Such an allele will not show if its partner is dominant, that is if the organism is ◊heterozygous for a particular characteristic. Alleles for blue eyes in humans, and for shortness in pea plants are recessive. Most mutant alleles are recessive and therefore are only rarely expressed (see ◊haemophilia and ◊sickle-cell disease).

reciprocal in mathematics, the result of dividing a given quantity into 1. Thus the reciprocal of 2 is $1/2$; of $2/3$ is $3/2$; of x^2 is $1/x^2$ or x^{-2}. Reciprocals are used to replace division by multiplication, since multiplying by the reciprocal of a number is the same as dividing by that number.

recombinant DNA in genetic engineering, ◊DNA formed by splicing together genes from different sources into new combinations.

recombination in genetics, any process that recombines, or 'shuffles', the genetic material, thus increasing genetic variation in the offspring. The two main processes of recombination both occur during meiosis (reduction division of cells). One is ◊*crossing over*, in which chromosome pairs exchange segments; the other is the random reassortment of chromosomes that occurs when each gamete (sperm or egg) receives only one of each chromosome pair.

recording the process of storing information, or the information store itself. Sounds and pictures can be stored on discs or tape. The gramophone record or ◊compact disc stores music or speech as a spiral groove on a plastic disc and the sounds are reproduced by a record or CD player. In ◊tape recording, sounds are stored as a magnetic pattern on plastic tape. The best-quality reproduction is achieved using digital audio tape.

In *digital recording* the signals picked up by the microphone are converted into precise numerical values by computer. These values, which represent the original sound wave form, are recorded on tape or compact disc. When it is played back by ◊laser, the exact values are retrieved.

When the signal is fed via an amplifier to a loudspeaker, sound waves exactly like the original ones are recreated. Pictures can be recorded on magnetic tape in a similar way by using a ◊videotape recorder. The video equivalent of a compact disc is the video disc.

rectifier in electrical engineering, a device used for obtaining one-directional current (DC) from an alternating source of supply (AC). (The process is necessary because almost all electrical power is generated, transmitted, and supplied as alternating current, but many devices, from television sets to electric motors, require direct current.) Types include plate rectifiers, thermionic ◊diodes, and ◊semiconductor diodes.

rectum lowest part of the large intestine of animals, which stores faeces prior to elimination (defecation).

recursion in computing and mathematics, a technique whereby a ◊function or ◊procedure calls itself into use in order to enable a complex problem to be broken down into simpler steps. For example, a function that finds the factorial of a number n (calculates the product of all the whole numbers between 1 and n) would obtain its result by multiplying n by the factorial of $n - 1$.

red blood cell or *erythrocyte* the most common type of blood cell, responsible for transporting oxygen around the body. It contains haemoglobin, which combines with oxygen from the lungs to form oxyhaemoglobin. When transported to the tissues, these cells are able to release the oxygen because the oxyhaemoglobin splits into its original constituents.

Mammalian erythrocytes are disc-shaped with a depression in the centre and no nucleus; they are manufactured in the bone marrow and, in humans, last for only four months before being destroyed in the liver and spleen. Those of other vertebrates are oval and nucleated.

red dwarf any star that is cool, faint, and small (about one-tenth the mass and diameter of the Sun). Red dwarfs burn slowly, and have estimated lifetimes of 100 billion years. They may be the most abundant type of star, but are difficult to see because they are so faint. Two of the closest stars to the Sun, ◊Proxima Centauri and ◊Barnard's Star, are red dwarfs.

red giant any large bright star with a cool surface. It is thought to represent a late stage in the evolution of a star like the Sun, as it runs out of hydrogen fuel at its centre. Red giants have diameters between 10 and 100 times that of the Sun. They are very bright because they are so large, although their surface temperature is lower than that of the Sun, about 2,000–3,000K (1,700–2,700°C/3,000°–5,000°F).

redox reaction chemical change where one reactant is reduced and the other reactant oxidized. The reaction can only occur if both reactants are present and each changes simultaneously. For example, hydrogen reduces copper(II) oxide to copper while it is itself oxidized to water. The corrosion of iron and the reactions taking place in electric and electrolytic cells are just a few instances of redox reactions.

red shift *The red shift causes lines in the spectra of galaxies to be shifted towards the red end of the spectrum. More distant galaxies have greater red shifts than closer galaxies. The red shift indicates that distant galaxies are moving apart rapidly, as the universe expands.*

red shift in astronomy, the lengthening of the wavelengths of light from an object as a result of the object's motion away from us. It is an example of the ◊Doppler effect. The red shift in light from galaxies is evidence that the universe is expanding.

reducing agent substance that reduces another substance (see ◊reduction). In a redox reaction, the reducing agent is the substance that is itself oxidized. Strong reducing agents include hydrogen, carbon monoxide, carbon, and metals.

reduction in chemistry, the gain of electrons, loss of oxygen, or gain of hydrogen by an atom, ion, or molecule during a chemical reaction.

reed switch switch containing two or three thin iron strips, or reeds, inside a sealed glass tube. When a magnet is brought near the switch, it induces magnetism in the reeds, which then attract or repel each other according to the positions of their induced magnetic poles. Electric current is switched on or off as the reeds make or break contact.

refining any process that purifies or converts something into a more useful form. Metals usually need refining after they have been extracted from their ores by such processes as ◊smelting. Petroleum, or crude oil, needs refining before it can be used; the process involves fractional ◊distillation, the separation of the substance into separate components or 'fractions'.

reflex in animals, a very rapid involuntary response to a particular stimulus. It is controlled by the ◊nervous system. A reflex involves only a few nerve cells, unlike the slower but more complex responses produced by the many processing nerve cells of the brain.

A *simple reflex* is entirely automatic and involves no learning. Examples of such reflexes include the sudden withdrawal of a hand in response to a painful stimulus, or the jerking of a leg when the kneecap is tapped. Sensory cells (receptors) in the knee send signals to the spinal cord along a sensory nerve cell. Within the spine a *reflex arc* switches the

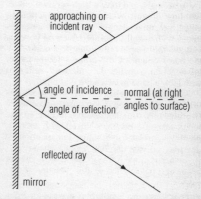

reflection *The law of reflection: the angle of incidence of a light beam equals the angle of reflection of the beam.*

REFRACTION

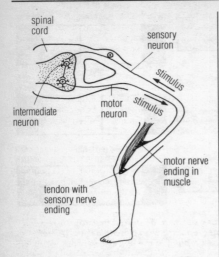

reflex *The knee-jerk reflex arc. When the knee-cap is tapped, the sensory nerve in the tendon produces a stimulus signal which travels to the spinal cord. Here, the signal is switched back to the motor nerve in the leg muscle, and the leg jerks.*

signals straight back to the muscles of the leg (effectors) via an intermediate nerve cell and then a motor nerve cell; contraction of the leg occurs, and the leg kicks upwards. Only three nerve cells are involved, and the brain is only aware of the response after it has taken place. Such reflex arcs are particularly common in lower animals, and have a high survival value, enabling organisms to take rapid action to avoid potential danger. In higher animals (those with a well-developed ◊central nervous system) the simple reflex can be modified by the involvement of the brain – for instance, humans can override the automatic reflex to withdraw a hand from a source of pain.

A *conditioned reflex* involves the modification of a reflex action in response to experience (learning). A stimulus that produces a simple reflex response becomes linked with another, possibly unrelated, stimulus. For example, a dog may salivate (a reflex action) when it sees its owner remove a tin-opener from a drawer because it has learned to associate that stimulus with the stimulus of being fed.

refraction the bending of a wave of light, heat, or sound when it passes from one medium to another. It occurs because waves travel at different velocities in different media.

refractive index measure of the refraction of a ray of light as it passes from one transparent medium to another. If the angle of incidence is i and the angle of refraction is r, the refractive index $n = \sin i/\sin r$. It is also equal to the speed of light in the first medium divided by the speed of light in the second, and it varies with the wavelength of the light.

refractory (of a material) able to resist high temperature, for example ceramics made from clay, minerals, or other earthy materials. Furnaces are lined with refractory materials such as silica and dolomite.

refrigeration use of technology to transfer heat from cold to warm, against the normal temperature gradient, so that a body can remain substantially colder than its surroundings. Refrigeration equipment is in ◊food technology, and in air conditioners and industrial processes.

regelation phenomenon in which water refreezes to ice after it has been melted by pressure at a temperature below the freezing point of water. Pressure makes an ice skate, for example, form a film of water that freezes once again after the skater has passed.

regeneration in biology, regrowth of a new organ or tissue after the loss or removal of the original. It is common in plants, where a new individual can often be produced from a 'cutting' of the original. In animals, regeneration of major structures is limited to lower organisms; certain lizards can regrow their tails if

refraction *Refraction is the bending of a light beam when it passes from one transparent medium to another. This is why a spoon appears bent when standing in a glass of water and pools of water appear shallower than they really are. The quantity sin i/ sin r has a constant value, for each material, called the refractive index.*

these are lost, and new flatworms can grow from a tiny fragment of an old one. In mammals, regeneration is limited to the repair of tissue in wound healing and the regrowth of peripheral nerves following damage.

register in computing, a memory location that can be accessed rapidly; it is often built into the computer's central processing unit.

Some registers are reserved for special tasks – for example, an *instruction register* is used to hold the machine-code command that the computer is currently executing, while a *sequence-control register* keeps track of the next command to be executed. Other registers are used for holding frequently used data and for storing intermediate results.

Regulus or *Alpha Leonis* the brightest star in the constellation Leo, and the 21st brightest star in the sky. Regulus has a true luminosity 100 times that of the Sun, and is 69 light years from Earth.

relational database ◊database in which data are viewed as a collection of linked tables. It is the most popular of the three basic database models, the others being *network* and *hierarchical*.

relative in computing (of a value), variable and calculated from a base value. For example, a *relative address* is a memory location that is found by adding a variable to a base (fixed) address, and a *relative cell reference* locates a cell in a spreadsheet by its position relative to a base cell – perhaps directly to the left of the base cell or three columns to the right of the base cell. The opposite of relative is absolute.

relative atomic mass the mass of an atom relative to one-twelfth the mass of an atom of carbon-12. It depends on the number of protons and neutrons in the atom, the electrons having negligible mass. If more than one ◊isotope of the element is present, the relative atomic mass is calculated by taking an average that takes account of the relative proportions of each isotope, resulting in values that are not whole numbers. The term *atomic weight*, although commonly used, is strictly speaking incorrect.

relative biological effectiveness (RBE) the relative damage caused to living tissue by different types of radiation. Some radiations do much more damage than others; alpha particles, for example, cause 20 times as much destruction as electrons (beta particles).

relative density or *specific gravity* the density (at 20°C/68°F) of a solid or liquid relative to (divided by) the maximum density of water (at 4°C/39.2°F). The relative density of a gas is its density divided by the density of hydrogen (or sometimes dry air) at the same temperature and pressure.

relative humidity the concentration of water vapour in the air. It is expressed as the percentage that its moisture content represents of the maximum amount that the air could contain at the same temperature and pressure. The higher the temperature, the more water vapour the air can hold.

relative molecular mass the mass of a molecule, calculated relative to one-twelfth the mass of an atom of carbon-12. It is found by adding the relative atomic masses of the atoms that make up the molecule.

The term *molecular weight* is often used, but strictly this is incorrect.

relativity in physics, the theory of the relative rather than absolute character of motion and mass, and the interdependence of matter, time, and space, as developed by German physicist Albert Einstein (1879–1955) in two phases:

special theory (1905) Starting with the premises that (1) the laws of nature are the same for all observers in unaccelerated motion, and (2) the speed of light is independent of the motion of its source, Einstein postulated that the time interval between two events was longer for an observer in whose frame of reference the events occur in different places than for the observer for whom they occur at the same place.

general theory of relativity (1915) The geometrical properties of space-time were to be conceived as modified locally by the presence of a body with mass. A planet's orbit around the Sun (as observed in three-dimensional space) arises from its natural trajectory in modified space-time; there is no need to invoke, as Isaac Newton did, a force of ◊gravity coming from the Sun and acting on the planet.

Einstein's theory predicted slight differences in the orbits of the planets from Newton's theory, which were observable in the case of Mercury. The new theory also said light rays should bend when they pass by a massive object, owing to the object's effect on local space-time. The predicted bending of starlight was observed during the eclipse of the Sun 1919, when light from distant stars passing close to the Sun was not masked by sunlight.

relay in electrical engineering, an electromagnetic switch. A small current passing through a coil of wire wound around an iron core attracts an ◊armature whose movement closes a pair of sprung contacts to complete a secondary circuit, which may carry a large current or activate other devices. The solid-state equivalent is a ◊thyristor switching device.

relay neuron nerve cell in the spinal cord, connecting motor neurons to sensory neurons. Relay neurons allow information to pass straight through the spinal cord, bypassing the brain. In humans such reflex actions, which are extremely rapid, cause the sudden removal of a limb from a painful stimulus.

rem acronym of *roentgen equivalent man* SI unit of radiation dose equivalent.

remote sensing gathering and recording information from a distance. Space probes have sent back photographs and data about planets as distant as Neptune. In archaeology, surface survey techniques provide information without disturbing subsurface deposits.

remote terminal in computing, a terminal that communicates with a computer via a modem (or acoustic coupler) and a telephone line.

REM sleep (acronym for *rapid-eye-movement* sleep) phase of sleep that recurs several times nightly in humans and is associated with dreaming. The eyes flicker quickly beneath closed lids.

renal pertaining to the ◊kidneys.

renewable energy power from any source that replenishes itself. Most renewable systems rely on ◊solar energy directly or through the weather cycle as ◊wave power, ◊hydroelectric power, or wind power via ◊wind turbines, or solar energy collected by plants (alcohol fuels, for example). In addition, the gravitational force of the Moon can be harnessed through ◊tidal power stations, and the heat trapped in the centre of the Earth is used via ◊geothermal energy systems.

renewable resource natural resource that is replaced by natural processes in a reasonable amount of time. Soil, water, forests, plants, and animals are all renewable resources as long as they are properly conserved. Solar, wind, wave, and geothermal energies are based on renewable resources.

rennin or *chymase* enzyme found in the gastric juice of young mammals, used in the digestion of milk.

repellent anything whose smell, taste, or other properties discourages nearby creatures.

Insect repellent is usually a chemical substance that keeps, for example, mosquitoes at bay; natural substances include citronella, lavender oil, and eucalyptus oils. A device that emits ultrasound waves is also claimed to repel insects and small mammals.

replication in biology, production of copies of the genetic material DNA; it occurs during cell division (◊mitosis and ◊meiosis). Most mutations are caused by mistakes during replication.

reproduction in biology, process by which a living organism produces other organisms similar to itself. There are two kinds: ◊asexual reproduction and ◊sexual reproduction.

reproduction rate or *fecundity* in ecology, the rate at which a population or species reproduces itself.

reptile any member of a class (Reptilia) of vertebrates. Unlike amphibians, reptiles have hard-shelled, yolk-filled eggs that are laid on land and from which fully formed young are born. Some snakes and lizards retain their eggs and give birth to live young. Reptiles are cold-blooded and produced from eggs, and the skin is usually covered with scales. The metabolism is slow, and in some cases (certain large snakes) intervals between meals may be months. Reptiles date back over 300 million years.

residual current device or *earth leakage circuit breaker* device that protects users of electrical equipment from electric shock by interrupting the electricity supply if a short circuit or current leakage occurs.

residue in chemistry, a substance or mixture of substances remaining in the original container after the removal of one or more components by a separation process.

resin substance exuded from pines, firs, and other trees in gummy drops that harden in air. Varnishes are common products of the hard resins, and ointments come from the soft resins.

resistance in physics, that property of a substance that restricts the flow of electricity through it, associated with the conversion of electrical energy to heat; also the magnitude of this property. Resistance depends on many factors, such as the nature of the material, its temperature, dimensions, and thermal properties; degree of impurity; the nature and state of illumination of the surface; and the frequency and magnitude of the current. The SI unit of resistance is the ohm.

resistivity in physics, a measure of the ability of a material to resist the flow of an electric current. It is numerically equal to the ◊resistance of a sample of unit length and unit cross-sectional area, and its unit is the ohm metre. A good conductor has a low resistivity (1.7×10^{-8} ohm metres for copper); an insulator has a very high resistivity (10^{15} ohm metres for polyethane).

resistor in physics, any component in an electrical circuit used to introduce ◊resistance to a current. Resistors are often made from wire-wound coils or pieces of carbon. ◊Rheostats and ◊potentiometers are variable resistors.

resolution in computing, the number of dots per unit length in which an image can be reproduced on a screen or printer. A typical screen resolution for colour monitors is 75 dpi (dots per inch). A ◊laser printer will typically have a printing resolution of 300 dpi, and ◊dot matrix printers typically have resolutions from 60 dpi to 180 dpi. Photographs in books and magazines have a resolution of 1,200 dpi or 2,400 dpi.

resolution of forces in mechanics, the division of a single force into two parts that act at right angles to each other. The two parts of a resolved force, called its *components*, have exactly the same effect on an object as the single force which they replace.

For example, the weight W of an object on a slope, tilted at an angle θ, can be resolved into two components: one acting at a right angle to the slope, equal to $W \cos \theta$, and one acting parallel to and down the slope, equal to $W \sin \theta$. The component acting down the slope (minus any friction force that may be acting in the opposite direction) is responsible for the acceleration of the object.

resonance rapid and uncontrolled increase in the size of a vibration when the vibrating object is subject to a force varying at its ◊natural frequency. In a trombone, for example, the length of the air column in the instrument is adjusted until it resonates with the note being sounded. Resonance effects are also produced by many electrical circuits. Tuning a radio, for example, is done by adjusting the natural frequency of the receiver circuit until it coincides with the frequency of the radio waves falling on the aerial.

respiration biochemical process whereby food molecules are progressively broken down (oxidized) to release energy in the form of ◊ATP. In most organisms this requires oxygen, but in some bacteria the oxidant is the nitrate or sulphate ion instead. In all higher organisms, respiration occurs in the ◊mitochondria. Respiration is also used to mean breathing, in which oxygen is exchanged for carbon dioxide in the lung alveoli, though this is more accurately described as a form of ◊gas exchange.

respiratory surface area used by an organism for the exchange of gases, for example the lungs, gills or, in plants, the leaf interior. The gases oxygen and carbon dioxide are both usually involved in respiration and photosynthesis. Although organisms have evolved different types of respiratory surface according to need, there are certain features in common. These include thinness and moistness, so that the gas can dissolve in a membrane and then diffuse into the body of the organism. In many animals the gas is then transported away from the surface and towards interior cells by the blood system.

response any change in an organism occurring as a result of a stimulus. There are many different types of response, some involving the entire organism, others only groups of cells or tissues. Examples include the muscular contractions in an animal, the movement of leaves towards the light, and the onset of hibernation by small mammals at the start of winter.

response time in computing, the delay between entering a command and seeing its effect.

rest mass in physics, the mass of a body when its velocity is zero. For subatomic particles, it is their mass at rest or at velocities considerably below that of light. According to the theory of ◊relativity, at very high velocities, there is a relativistic effect that increases the mass of the particle.

restriction enzyme bacterial ◊enzyme that breaks a chain of ◊DNA into two pieces at a specific point; used in ◊genetic engineering. The point along the DNA chain at which the enzyme can work is restricted to places where a specific sequence of base pairs occurs. Different restriction enzymes will break a DNA chain at different points. The overlap between the fragments is used in determining the sequence of base pairs in the DNA chain.

resultant force in mechanics, a single force acting on a particle or body whose effect is equivalent to the combined effects of two or more component forces. The resultant of two forces acting at one point on an object can be found using the ◊parallelogram of forces method.

retina light-sensitive area at the back of the ◊eye connected to the brain by the optic nerve. It has several layers and in humans contains over a million rods and cones, sensory cells capable of converting light into nervous messages that pass down the optic nerve to the brain.

The *rod cells*, about 120 million in each eye, are distributed throughout the retina. They are sensitive to low levels of light, but do not provide detailed or sharp images, nor can they detect colour. The *cone cells*, about 6 million in number, are mostly concentrated in a central region of the retina called the *fovea*, and provide both detailed and colour vision. The cones of the human eye contain three visual pigments, each of which responds to a different primary colour (red, green, or blue). The brain can interpret the varying signal levels from the three types of cone as any of the different colours of the visible spectrum.

retinol or *vitamin A* fat-soluble chemical derived from β-carotene and found in milk, butter, cheese, egg yolk, and liver. Lack of retinol in the diet leads to the eye disease ◊xerophthalmia.

retrovirus any of a family of ◊viruses (Retroviridae) containing the genetic material ◊RNA rather than the more usual ◊DNA.

Retroviruses are used as vectors in ◊gene therapy, but they cannot be used to target specific sites on the chromosome. Instead they incorporate their genes at random sites.

reverberation in acoustics, the multiple reflections, or echoes, of sounds inside a building that merge and persist a short while (up to a few seconds) before fading away. At each reflection some of the sound energy is absorbed, causing the amplitude of the sound wave and the intensity of the sound to reduce a little.

reverse osmosis the movement of solvent (liquid) through a semipermeable membrane from a more concentrated solution to a more dilute solution. The solvent's direction of movement is opposite to that which it would experience during ◊osmosis, and is achieved by applying an external pressure to the solution on the more concentrated side of the membrane. The technique is used in desalination plants, when water (the solvent) passes from brine (a salt solution) into fresh water via a semipermeable filtering device.

reversible reaction chemical reaction that proceeds in both directions at the same time, as the product decomposes back into reactants as it is being produced. Such reactions do not run to completion, provided that no substance leaves the system. Examples include the manufacture of ammonia from hydrogen and nitrogen, and the oxidation of sulphur dioxide to sulphur trioxide.

Reynolds number number used in ◊fluid mechanics to determine whether fluid flow in a particular situation (through a pipe or around an aircraft body) will be turbulent or smooth. The Reynolds number is calculated using the flow velocity, viscosity of the fluid, and the dimensions of the flow channel. It is named after British engineer Osborne Reynolds.

RGB (abbreviation of *red–green–blue*) method of connecting a colour screen to a computer, involving three separate signals: red, green, and blue. All the colours displayed by the screen can be made up from these three component colours.

rhe unit of fluidity equal to the reciprocal of the ◊poise.

rhenium heavy, silver-white, metallic element, symbol Re, atomic number 75, relative atomic mass 186.2. It has chemical properties similar to those of manganese and a very high melting point (3,180°C/5,756°F), which makes it valuable as an ingredient in alloys.

rheostat in physics, a variable ◊resistor, usually consisting of a high-resistance wire-wound coil with a sliding contact. It is used to vary electrical resistance without interrupting the current (for example, when dimming lights). The circular type in electronics (which can be used, for example, as the volume control of an amplifier) is also known as a ◊potentiometer.

rhesus factor group of ◊antigens on the surface of red blood cells of humans which characterize the rhesus blood group system. Most individuals possess the main rhesus factor (Rh+), but those without this factor (Rh−) produce ◊antibodies if they come into contact with it. The name comes from rhesus monkeys, in whose blood rhesus factors were first found.

If an Rh− mother carries an Rh+ fetus, she may produce antibodies if fetal blood crosses the ◊placenta. This is not normally a problem with the first infant because antibodies are only produced slowly. However, the antibodies continue to build up after birth, and a second Rh+ child may be attacked by antibodies passing from mother to fetus, causing the child to contract anaemia, heart failure, or brain damage. In such cases, the blood of the infant has to be changed for Rh− blood; a badly

affected fetus may be treated in the womb. The problem can be circumvented by giving the mother anti-Rh globulin just after the first pregnancy, preventing the formation of antibodies.

rhizoid hairlike outgrowth found on the ◊gametophyte generation of ferns, mosses and liverworts. Rhizoids anchor the plant to the substrate and can absorb water and nutrients.

They may be composed of many cells, as in mosses, where they are usually brownish, or may be unicellular, as in liverworts, where they are usually colourless. Rhizoids fulfil the same functions as the ◊roots of higher plants but are simpler in construction.

rhizome or ◊*rootstock* horizontal underground plant stem. It is a ◊perennating organ in some species, where it is generally thick and fleshy, while in other species it is mainly a means of ◊vegetative reproduction, and is therefore long and slender, with buds all along it that send up new plants. The potato is a rhizome that has two distinct parts, the tuber being the swollen end of a long, cordlike rhizome.

rhm (abbreviation of *roentgen–hour–metre*) the unit of effective strength of a radioactive source that produces gamma rays. It is used for substances for which it is difficult to establish radioactive disintegration rates.

rhodium hard, silver-white, metallic element, symbol Rh, atomic number 45, relative atomic mass 102.905. It is one of the so-called platinum group of metals and is resistant to tarnish, corrosion, and acid. It occurs as a free metal in the natural alloy osmiridium and is used in jewellery, electroplating, and thermocouples.

rhombic sulphur allotropic form of sulphur. At room temperature, it is the stable ◊allotrope, unlike ◊monoclinic sulphur.

rhyolite ◊igneous rock, the fine-grained volcanic (extrusive) equivalent of granite.

rib long, usually curved bone that extends laterally from the ◊spine in vertebrates. Most fishes and many reptiles have ribs along most of the spine, but in mammals they are found only in the chest area. In humans, there are 12 pairs of ribs. The ribs protect the lungs and heart, and allow the chest to expand and contract easily.

riboflavin or *vitamin B_2* ◊vitamin of the B complex important in cell respiration. It is obtained from eggs, liver, and milk. A deficiency in the diet causes stunted growth.

Richter scale

magnitude	relative amount of energy released	examples
1		
2		
3		
4	1	Carlisle, England, 1979
5	30	San Francisco and New England, USA, 1979
		Wrexham, Wales, 1990
6	100	San Fernando, California, 1971
7	30,000	Chimbote, Peru, 1970
		San Francisco, 1989
8	1,000,000	Tangshan, China, 1976
		San Francisco, 1906
		Lisbon, Portugal, 1755
		Alaska, 1964
		Gansu, China, 1920

ribonucleic acid full name of ◊RNA.

ribosome in biology, the protein-making machinery of the cell. Ribosomes are located on the endoplasmic reticulum (ER) of eukaryotic cells, and are made of proteins and a special type of ◊RNA, ribosomal RNA. They receive messenger RNA (copied from the ◊DNA) and ◊amino acids, and 'translate' the messenger RNA by using its chemically coded instructions to link amino acids in a specific order, to make a strand of a particular protein.

Richter scale scale based on measurement of seismic waves, used to determine the magnitude of an ◊earthquake at its epicentre. The magnitude of an earthquake differs from its intensity, measured by the ◊Mercalli scale, which is subjective and varies from place to place for the same earthquake. The scale is named after US seismologist Charles Richter.

rickets defective growth of bone in children due to an insufficiency of calcium deposits. The bones, which do not harden adequately, are bent out of shape. It is usually caused by a lack of vitamin D and insufficient exposure to sunlight. Renal rickets, also a condition of malformed bone, is associated with kidney disease.

rickettsia any of a group of small bacteria that are spread to humans by invertebrates such as ticks, fleas, and lice. The diseases they cause include epidemic typhus, and the milder endemic typhus and scrub typhus, Rocky Mountain spotted fever, rickettsial pox, trench fever and Q fever.

ridge of high pressure elongated area of high atmospheric pressure extending from an anticyclone. On a synoptic weather chart it is shown as a pattern of lengthened isobars. The weather under a ridge of high pressure is the same as that under an anticyclone.

Rigel or *Beta Orionis* brightest star in the constellation Orion. It is a blue-white supergiant, with an estimated diameter 50 times that of the Sun. It is 900 light years from Earth, and is about 100,000 times more luminous than our Sun. It is the seventh-brightest star in the sky.

right ascension in astronomy, the coordinate on the ◊celestial sphere that corresponds to longitude on the surface of the Earth. It is measured in hours, minutes, and seconds eastwards from the point where the Sun's path, the ecliptic, once a year intersects the celestial equator; this point is called the *vernal equinox*.

right-hand rule in physics, a memory aid used to recall the relative directions of motion, magnetic field, and current in an electric generator. It was devised by English physicist John Fleming. See ◊Fleming's rules.

ring circuit household electrical circuit in which appliances are connected in series to form a ring with each end of the ring connected to the power supply. It superseded the ◊radial circuit.

ripple tank in physics, shallow water-filled tray used to demonstrate various properties of waves, such as reflection, refraction, diffraction, and interference, by programming and manipulating their movement.

RISC (acronym for *reduced instruction-set computer*) in computing, a microprocessor (processor on a single chip) that carries out fewer instructions than other (◊CISC) microprocessors in common use in the 1990s. Because of the low number of ◊machine code instructions, the processor carries out those instructions very quickly.

ritualization in ethology, a stereotype that occurs in certain behaviour patterns when these are incorporated into displays. For example, the exaggerated and stylized head toss of the goldeneye drake during courtship is a ritualization of the bathing movement used to wet the feathers; its duration and form have become fixed. Ritualization may make displays clearly recognizable, so ensuring that individuals mate only with members of their own species.

RNA (abbreviation for *ribonucleic acid*) nucleic acid involved in the process of translating the genetic material ◊DNA into proteins. It is usually single-stranded, unlike the double-stranded DNA, and consists of a large number of nucleotides strung together, each of which comprises the sugar ribose, a phosphate group, and one of four bases (uracil, cytosine, adenine, or guanine). RNA is copied from DNA by the formation of ◊base pairs, with uracil taking the place of thymine.

RNA occurs in three major forms, each with a different function in the synthesis of protein molecules. *Messenger RNA* (mRNA) acts as the template for protein synthesis. Each ◊codon (a set of three bases) on the RNA molecule is matched up with the corresponding amino acid, in accordance with the ◊genetic code. This process (translation) takes place in the ribosomes, which are made up of proteins and *ribosomal RNA* (rRNA). *Transfer RNA* (tRNA) is responsible for combining with specific amino acids, and then matching up a special 'anticodon' sequence of its own with a codon on the mRNA. This is how the genetic code is translated. RNA makes up the hereditary material itself in some viruses, such as ◊retroviruses.

robot any computer-controlled machine that can be programmed to move or carry out work. Robots are often used in industry to transport materials or to perform repetitive tasks. For instance, robotic arms, may be used to paint machine parts or assemble electronic circuits. Other robots are designed to work in situations that would be dangerous to humans – for example, in defusing bombs or in space and deep-sea exploration.

rock hard solid mass composed of mineral particles that have become cemented together. Rocks may be classified into three main groups: ◊igneous, ◊sedimentary, and ◊metamorphic rocks.

rocket projectile driven by the reaction of gases produced by a fast-burning fuel. Unlike jet engines, which are also reaction engines, modern rockets carry their own oxygen supply to burn their fuel and do not require any surrounding atmosphere. For warfare, rocket heads carry an explosive device.

The only form of propulsion available that can function in a vacuum, rockets are essential to exploration in outer space. ◊Multistage rockets have to be used, consisting of a number of rockets joined together. Most rockets have liquid propellants. Liquid hydrogen and kerosene are common fuels, while liquid oxygen is the most common oxygen provider, or oxidizer. One of the biggest rockets ever built, the Saturn V moon rocket, was a rocket *The three-stage Saturn V rocket used in the Apollo moonshots of the 1960s and 1970s. It stood 111 m/365 ft high, as tall as a 30-storey skyscraper, weighed 2,700 tonnes/3,000 tons when loaded with fuel, and developed a power equivalent to 50 Boeing 747 jumbo jets.*

three-stage design, standing 111 m/365 ft high, weighed more than 2,700 tonnes/3,000 tons on the launch pad, developed a takeoff thrust of some 3.4 million kg/7.5 million lb, and could place almost 140 tonnes/150 tons into low Earth orbit. In the early 1990s, the most powerful rocket system was the Soviet Energiya, capable of placing 100 tonnes/110 tons into low Earth orbit.

rod a type of light-sensitive cell in the ◊retina of most vertebrates. Rods are highly sensitive and provide only black and white vision. They are used when lighting conditions are poor and are the only type of visual cell found in animals active at night.

rodent any mammal of the worldwide order Rodentia, making up nearly half of all mammal species. Besides ordinary 'cheek teeth', they have a single front pair of incisor teeth in both upper and lower jaw, which continue to grow as they are worn down.

roentgen or *röntgen* unit (symbol R) of radiation exposure, used for X-rays and gamma rays. It is defined in terms of the number of ions produced in one cubic centimetre of air by the radiation. Exposure to 1,000 roentgens gives rise to an absorbed dose of about 870 rads (8.7 grays), which is a dose equivalent of 870 rems (8.7 sieverts).

ROM (acronym for *read-only memory*) in computing, a memory device in the form of a collection of integrated circuits (chips), frequently used in microcomputers. ROM chips are loaded with data and programs during manufacture and, unlike ◊RAM (random-access memory) chips, can subsequently only be read, not written to, by computer. However, the contents of the chips are not lost when the power is switched off, as happens in RAM.

röntgen alternative spelling for ◊roentgen.

root the part of a plant that is usually underground, and whose primary functions are anchorage and the absorption of water and dissolved mineral salts. Roots usually grow downwards and towards water (that is, they are positively geotropic and hydrotropic; see ◊tropism). Plants such as epiphytic orchids, which grow above ground, produce aerial roots that absorb moisture from the atmosphere. Others, such as ivy, have climbing roots arising from the stems, which serve to attach the plant to trees and walls.

The absorptive area of roots is greatly increased by the numerous, slender root hairs formed near the tips. A calyptra, or root cap, protects the tip of the root from abrasion as it grows through the soil.

Symbiotic associations occur between the roots of certain plants, such as clover, and various bacteria that fix nitrogen from the air (see ◊nitrogen fixation). Other modifications of roots include ◊contractile roots, ◊pneumatophores, ◊taproots, and ◊prop roots.

root of an equation, a value that satisfies the equality. For example, $x = 0$ and $x = 5$ are roots of the equation $x^2 - 5x = 0$.

root directory in computing, the top directory in a ◊tree-and-branch filing system. It contains all the other directories.

root hair tiny hairlike outgrowth on the surface cells of plant roots that greatly increases the area available for the absorption of water and other materials. It is a delicate structure, which survives for a few days only and does not develop into a root.

root mean square (rms) in mathematics, value obtained by taking the ◊square root of the mean of the squares of a set of values; for example the rms value of four quantities a, b, c, and d is $\sqrt{[(a^2 + b^2 + c^2 + d^2)/4]}$.

root nodule clearly visible swelling that develops in the roots of members of the bean family, the Leguminosae. The cells inside this tumorous growth have been invaded by the bacteria Rhizobium, a soil microbe capable of converting gaseous nitrogen into nitrate. The nodule is therefore an association between a plant and a bacterium, with both partners benefiting. The plant obtains nitrogen compounds while the bacterium obtains nutrition and shelter.

ROSAT joint US/German/UK satellite launched 1990 to study cosmic sources of X-rays and extremely short ultraviolet wavelengths, named after Wilhelm Röntgen, the discoverer of X-rays.

rotifer any of the tiny invertebrates, also called 'wheel animalcules', of the phylum Rotifera. Mainly freshwater, some marine, rotifers have a ring of ◊cilia that carries food to the mouth and also provides propulsion. They are the smallest of multicellular animals – few reach 0.05 cm/0.02 in.

roughage alternative term for dietary ◊fibre, material of plant origin that cannot be digested by enzymes normally present in the human ◊gut.

roundworm parasitic ◊nematode worm, genus *Toxocara*, found in dogs, cats, and other animals.

router device that enables dissimilar networks (such as ◊Ethernet and Token Ring) to interconnect, provided there is a common protocol. Routers can be programmed to determine the fastest or most cost-effective route for data to travel. See also ◊bridge and ◊brouter.

RS-232 interface standard type of computer ◊interface used to connect computers to serial devices. It is used for modems, mice, screens, and serial printers.

rubber coagulated latex of a variety of plants, mainly from the New World. Most important is Para rubber, which derives from the tree *Hevea brasiliensis* of the spurge family. It was introduced from Brazil to SE Asia, where most of the world supply is now produced, the chief exporters being Peninsular Malaysia, Indonesia, Sri Lanka, Cambodia, Thailand, Sarawak, and Brunei. At about seven years the tree, which may grow to 20 m/60 ft, is ready for 'tapping'. Small incisions are made in the trunk and the latex drips into collecting cups. In pure form, rubber is white and has the formula $(C_5H_8)_n$.

rubidium soft, silver-white, metallic element, symbol Rb, atomic number 37, relative atomic mass 85.47. It is one of the ◊alkali metals, ignites spontaneously in air, and reacts violently with water. It is used in photocells and vacuum-tube filaments.

ruby the red transparent gem variety of the mineral ◊corundum Al_2O_3, aluminium oxide. Small amounts of chromium oxide, Cr_2O_3, substituting for aluminium oxide, give ruby its colour. Natural rubies are found mainly in Myanmar (Burma), but rubies can also be produced artificially and such synthetic stones are used in ◊lasers.

ruminant any even-toed hoofed mammal with a rumen, the 'first stomach' of its complex digestive system. Plant food is stored and fermented before being brought back to the mouth for chewing (chewing the cud) and then is swallowed to the next stomach. Ruminants include cattle, antelopes, goats, deer, and giraffes, all with a four-chambered stomach. Camels are also ruminants, but they have a three-chambered stomach.

runner or *stolon* in botany, aerial stem that produces new plants.

rust reddish-brown oxide of iron formed by the action of moisture and oxygen on the metal. It consists mainly of hydrated iron(III) oxide ($Fe_2O_3.H_2O$) and iron(III) hydroxide ($Fe(OH)_3$).

ruthenium hard, brittle, silver-white, metallic element, symbol Ru, atomic number 44, relative atomic mass 101.07. It is one of the so-called platinum group of metals; it occurs in platinum ores as a free metal and in the natural alloy osmiridium. It is used as a hardener in alloys and as a catalyst; its compounds are used as colouring agents in glass and ceramics.

rutherfordium name proposed by US scientists for the element currently known as ⟡unnilquadium (atomic number 104), to honour New Zealand physicist Ernest Rutherford.

rutile TiO_2, titanium oxide, a naturally occurring ore of titanium. It is usually reddish brown to black, with a very bright (adamantine) surface lustre. It crystallizes in the tetragonal system. Rutile is common in a wide range of igneous and metamorphic rocks and also occurs concentrated in sands; the coastal sands of E and W Australia are a major source. It is also used as a pigment that gives a brilliant white to paint, paper, and plastics.

Rydberg constant in physics, a constant that relates atomic spectra to the ⟡spectrum of hydrogen. Its value is 1.0977×10^7 per metre.

sabin unit of sound absorption, used in acoustical engineering. One sabin is the absorption of one square foot (0.093 square metre) of a perfectly absorbing surface (such as an open window).

saccharide another name for a ◊sugar molecule.

saccharin or *ortho-sulpho benzimide* $C_7H_5NO_3S$ sweet, white, crystalline solid derived from coal tar and substituted for sugar. Since 1977 it has been regarded as potentially carcinogenic. Its use is not universally permitted and it has been largely replaced by other sweetening agents.

safety lamp portable lamp designed for use in places where flammable gases such as methane may be encountered; for example, in coal mines. The electric head lamp used as a miner's working light has the bulb and contacts in protected enclosures. The flame safety lamp, now used primarily for gas detection, has the wick enclosed within a strong glass cylinder surmounted by wire gauzes. English chemist Humphrey Davy (1778–1829) 1815 and English engineer George Stephenson (1781–1848) each invented flame safety lamps.

Sagittarius zodiac constellation in the southern hemisphere, represented as a centaur aiming a bow and arrow at neighbouring Scorpius. The Sun passes through Sagittarius from mid-Dec to mid-Jan, including the winter solstice, when it is farthest south of the equator. The constellation contains many nebulae and ◊globular clusters, and open ◊star clusters. Kaus Australis and Nunki are its brightest stars. The centre of our Galaxy, the Milky Way, is marked by the radio source Sagittarius A. In astrology, the dates for Sagittarius are between about 22 Nov and 21 Dec.

St Elmo's fire bluish, flamelike electrical discharge that sometimes occurs above ships' masts and other pointed objects or about aircraft in stormy weather. Although high voltage, it is low current and therefore harmless. St Elmo (or St Erasmus) is the patron saint of sailors.

sal ammoniac former name for ◊ammonium chloride.

salicylic acid HOC_6H_4COOH the active chemical constituent of aspirin, an analgesic drug. The acid and its salts (salicylates) occur naturally in many plants; concentrated sources include willow bark and oil of wintergreen.

saliva in vertebrates, an alkaline secretion from the salivary glands that aids the swallowing and digestion of food in the mouth. In mammals, it contains the enzyme amylase, which converts starch to sugar. The salivary glands of mosquitoes and other blood-sucking insects produce ◊anticoagulants.

Salmonella very varied group of bacteria that colonize the intestines of humans and some animals. Some strains cause typhoid and paratyphoid fevers, while others cause salmonella ◊food poisoning, which is characterized by stomach pains, vomiting, diarrhoea, and headache. It can be fatal in elderly people, but others usually recover in a few days without antibiotics. Most cases are caused by contaminated animal products, especially poultry meat.

In 1993 a law was passed in Germany that all chicks must be immunized against salmonella. This is effected by adding a live weakened strain of *Salmonella enteriditis* to their drinking water.

salt in chemistry, any compound formed from an acid and a base through the replacement of all or part of the hydrogen in the acid by a metal or electropositive radical. ***Common salt*** is sodium chloride (see ◊salt, common).

A salt may be produced by chemical reaction between an acid and a base, or by the displacement of hydrogen from an acid by a metal (see ◊displacement activity). As a solid, the ions normally adopt a regular arrangement to form crystals. Some salts only form stable crystals as hydrates (when combined with water). Most inorganic salts readily dissolve in water to give an electrolyte (a solution that conducts electricity).

saltation in biology, the idea that an abrupt genetic change can occur in an individual,

common ions that form salts

positive ions	negative ions
silver Ag+	bromide Br−
aluminium Al³⁺	chloride Cl−
barium Ba²⁺	carbonate CO_3^{2-}
calcium Ca²⁺	fluoride F−
copper Cu²⁺	hydrogencarbonate HCO_3^-
iron(II) Fe²⁺	hydrogensulphate HSO_4^-
iron(III) Fe³⁺	iodide I−
hydrogen H+	nitrate NO_3^-
potassium K+	oxide O²−
lithium Li+	hydroxide OH−
magnesium Mg²⁺	sulphide S²−
sodium Na+	sulphite SO_3^{2-}
ammonium NH_4^+	sulphate SO_4^{2-}
lead Pb²⁺	
zinc Zn²⁺	

which then gives rise to a new species. The idea has now been largely discredited, although the appearance of ◊polyploid individuals can be considered an example.

salt, common or *sodium chloride* NaCl white crystalline solid, found dissolved in sea water and as rock salt (halite) in large deposits and salt domes. Common salt is used extensively in the food industry as a preservative and for flavouring, and in the chemical industry in the making of chlorine and sodium.

saltpetre former name for potassium nitrate (KNO₃), the compound used in making gunpowder (from about 1500). It occurs naturally, being deposited during dry periods in places with warm climates, such as India.

sal volatile another name for ◊smelling salts.

Salyut series of seven space stations launched by the USSR 1971–82. Salyut was cylindrical in shape, 15 m/50 ft long, and weighed 19 tonnes/21 tons. It housed two or three cosmonauts at a time, for missions lasting up to eight months.

Crews observed Earth and the sky, and carried out processing of materials in weightlessness. The last in the series, *Salyut 7*, crashed onto Earth in Feb 1991, scattering debris in Argentina.

samara in botany, a winged fruit, a type of ◊achene.

samarium hard, brittle, grey-white, metallic element of the ◊lanthanide series, symbol Sm, atomic number 62, relative atomic mass 150.4. It is widely distributed in nature and is obtained commercially from the minerals monzanite and bastnasite. It is used only occasionally in industry, mainly as a catalyst in organic reactions. Samarium was discovered by spectroscopic analysis of the mineral samarskite and named in 1879 by French chemist Paul Lecoq de Boisbaudran (1838–1912) after its source.

sand loose grains of rock, sized 0.0625–2.00 mm/0.0025–0.08 in in diameter, consisting chiefly of ◊quartz, but owing their varying colour to mixtures of other minerals. Sand is used in cement-making, as an abrasive, in glass-making, and for other purposes.

sandstone ◊sedimentary rocks formed from the consolidation of sand, with sand-sized grains (0.0625–2 mm/0.0025–0.08 in) in a matrix or cement. Their principal component is quartz. Sandstones are commonly permeable and porous, and may form freshwater ◊aquifers. They are mainly used as building materials.

sap the fluids that circulate through ◊vascular plants, especially woody ones. Sap carries water and food to plant tissues. Sap contains alkaloids, protein, and starch; it can be milky (as in rubber trees), resinous (as in pines), or syrupy (as in maples).

saponification in chemistry, the ◊hydrolysis (splitting) of an ◊ester by treatment with a strong alkali, resulting in the liberation of the alcohol from which the ester had been derived and a salt of the constituent fatty acid. The process is used in the manufacture of soap.

sapphire deep-blue, transparent gem variety of the mineral ◊corundum Al₂O₃, aluminium oxide. Small amounts of iron and titanium give it its colour. A corundum gem of any colour except red (which is a ruby) can be called a sapphire; for example, yellow sapphire.

saprophyte in botany, an obsolete term for a ◊saprotroph, an organism that lives in dead or decaying matter.

saprotroph (formerly *saprophyte*) organism that feeds on the excrement or the dead bodies or tissues of others. They include most fungi (the rest being parasites); many bacteria and protozoa; animals such as dung beetles and vultures; and a few unusual plants, including several orchids. Saprotrophs cannot make food for themselves, so they are a type of ◊heterotroph. They are useful scavengers, and in sewage farms and refuse dumps break down organic matter into nutrients easily assimilable by green plants.

sarcoma malignant ◊tumour arising from the fat, muscles, bones, cartilage, or blood and lymph vessels and connective tissues. Sarcomas are much less common than ◊carcinomas.

sard or *sardonyx* yellow or red-brown variety of ◊onyx.

satellite any small body that orbits a larger one, either natural or artificial. Natural satellites that orbit planets are called moons. The first *artificial satellite*, *Sputnik 1*, was launched into orbit around the Earth by the USSR 1957. Artificial satellites are used for scientific purposes, communications, weather forecasting, and military applications. The largest artificial satellites can be seen by the naked eye.

satellite applications the uses to which artificial satellites are put. These include:

scientific experiments and observation Many astronomical observations are best taken above the disturbing effect of the atmosphere. Satellite observations have been carried out by *IRAS* (*Infrared Astronomical Satellite*, 1983) which made a complete infrared survey of the skies, and *Solar Max* 1980, which observed solar flares. The *Hipparcos* satellite, launched 1989, measured the positions of many stars. In 1992, the COBE (Cosmic Background Explorer) satellite detected details of the Big Bang that mark the first stage in the formation of galaxies. Medical experiments have been carried out aboard crewed satellites, such as the Soviet *Mir* and the US *Skylab*.

reconnaissance, land resource, and mapping applications Apart from military use and routine mapmaking, the US *Landsat*, the French *SPOT*, and equivalent USSR satellites have provided much useful information about water sources and drainage, vegetation, land use, geological structures, oil and mineral locations, and snow and ice.

weather monitoring The US NOAA series of satellites, and others launched by the European space agency, Japan, and India, provide continuous worldwide observation of the atmosphere.

navigation The US Global Positioning System, when complete in 1993, will feature 24 Navstar satellites that will enable users (including walkers and motorists) to find their position to within 100 m/328 ft. The US military will be able to make full use of the system, obtaining accuracy to within 1.5 m/4 ft 6 in. The Transit system, launched in the 1960s, with 12 satellites in orbit, locates users to within 100 m/328 ft.

communications A complete worldwide communications network is now provided by satellites such as the US-run ◊Intelsat system.

satellite television transmission of broadcast signals through artificial communications satellites. Mainly positioned in ◊geostationary orbit, satellites have been used since the 1960s to relay television pictures around the world.

Higher-power satellites have more recently been developed to broadcast signals to cable systems or directly to people's homes.

saturated compound organic compound, such as propane, that contains only single covalent bonds. Saturated organic compounds can only undergo further reaction by ◊substitution reactions, as in the production of chloropropane from propane.

saturated fatty acid ◊fatty acid in which there are no double bonds in the hydrocarbon chain.

saturated solution in physics, a solution obtained when a solvent (liquid) can dissolve no more of a solute (usually a solid) at a particular temperature. Normally, a slight fall in temperature causes some of the solute to crystallize out of solution. If this does not happen the phenomenon is called supercooling, and the solution is said to be *supersaturated*.

Saturn in astronomy, the second-largest planet in the Solar System, sixth from the Sun, and encircled by bright and easily visible equatorial rings. Viewed through a telescope it is

largest natural planetary satellites

planet	satellite	diameter in km	mean distance from centre of primary in km/mi	orbital period in days	reciprocal mass (planet = 1)
Jupiter	Ganymede	5,262/3,300	1,070,000	7.16	12,800
Saturn	Titan	5,150/3,200	1,221,800	15.95	4,200
Jupiter	Callisto	4,800/3,000	1,883,000	16.69	17,700
Jupiter	Io	3,630/2,240	421,600	1.77	21,400
Earth	Moon	3,476/2,160	384,400	27.32	81.3
Jupiter	Europa	3,138/1,900	670,900	3.55	39,700
Neptune	Triton	2,700/1,690	354,300	5.88	770

ochre. Saturn orbits the Sun every 29.46 years at an average distance of 1,427,000,000 km/886,700,000 mi. Its equatorial diameter is 120,000 km/75,000 mi, but its polar diameter is 12,000 km/7,450 mi smaller, a result of its fast rotation and low density, the lowest of any planet.

Saturn rocket family of large US rockets, developed by Wernher von Braun for the ◊Apollo project. The two-stage *Saturn IB* was used for launching Apollo spacecraft into orbit around the Earth. The three-stage *Saturn V* sent Apollo spacecraft to the Moon, and launched the ◊*Skylab* space station. The liftoff thrust of a *Saturn V* was 3,500 tonnes. After Apollo and *Skylab*, the Saturn rockets were retired in favour of the ◊space shuttle.

scalar quantity in mathematics and science, a quantity that has magnitude but no direction, as distinct from a ◊vector quantity, which has a direction as well as a magnitude. Temperature, mass, and volume are scalar quantities.

scaler instrument that counts the number of radioactive particles passing through a radiation detector such as a Geiger–Muller tube (the scaler and tube together form a ◊Geiger counter). It gives the total, or cumulative, number of particles counted whereas a ◊ratemeter gives the number of particles detected in a unit of time.

scandium silver-white, metallic element of the ◊lanthanide series, symbol Sc, atomic number 21, relative atomic mass 44.956. Its compounds are found widely distributed in nature, but only in minute amounts. The metal has little industrial importance.

scanner device, usually electronic, used to sense and reproduce an image.

scanner in computing, a device that can produce a digital image of a document for input and storage in a computer. It uses technology similar to that of a photocopier. Small scanners can be passed over the document surface by hand; larger versions have a flat bed, like that of a photocopier, on which the input document is placed and scanned.

scanning electron microscope (SEM) electron microscope that produces three-dimensional images, magnified 10–200,000 times. A fine beam of electrons, focused by electromagnets, is moved, or scanned, across the specimen. Electrons reflected from the specimen are collected by a detector, giving rise to an electrical signal, which is then used to generate a point of brightness on a television-like screen. As the point moves rapidly over the screen, in phase with the scanning electron beam, an image of the specimen is built up.

scanning transmission electron microscope (STEM) electron microscope that combines features of the ◊scanning electron microscope (SEM) and the ◊transmission electron microscope (TEM). First built in the USA in 1966, the microscope has both the SEM's contrast characteristics and lack of aberrations and the high resolution of the TEM. Magnifications of over 90 million times can be achieved, enough to image single atoms.

scanning tunnelling microscope (STM) microscope that produces a magnified image by moving a tiny tungsten probe across the surface of the specimen. The tip of the probe is so fine that it may consist of a single atom, and it moves so close to the specimen surface that electrons jump (or tunnel) across the gap between the tip and the surface.

scapolite group of white or greyish minerals, silicates of sodium, aluminium, and calcium, common in metamorphosed limestones and forming at high temperatures and pressures.

scapula or *shoulder blade* large bone forming part of the pectoral girdle, assisting in the articulation of the arm with the chest region. Its flattened shape allows a large region for the attachment of muscles.

scattering in physics, the random deviation or reflection of a stream of particles or of a beam of radiation such as light.

scent gland gland that opens onto the outer surface of animals, producing odorous compounds that are used for communicating between members of the same species (◊pheromones), or for discouraging predators.

schist ◊metamorphic rock containing ◊mica or another platy or elongate mineral, whose crystals are aligned to give a foliation (planar texture) known as schistosity. Schist may contain additional minerals such as ◊garnet.

schizocarp dry ◊fruit that develops from two or more carpels and splits, when mature, to form separate one-seeded units known as mericarps.

science any systematic field of study or body of knowledge that aims, through experiment, observation, and deduction, to produce reliable explanation of phenomena, with reference to the material and physical world.

Science is divided into separate areas of study, such as astronomy, biology, geology, chemistry, physics, and mathematics, although more recently attempts have been made to combine traditionally separate disciplines under such headings as ◊life sciences and ◊earth sciences. These areas are usually jointly referred to as the *natural sciences*. The *physical sciences* comprise mathematics, physics, and chemistry. The application of science for practical purposes is called *technology*.

scientific law in science, principles that are taken to be universally applicable.

scientific method in science, the belief that experimentation and observation, properly understood and applied, can avoid the influence of cultural and social values and so build up a picture of a reality independent of the observer.

Improved techniques and mechanical devices, which improve the reliability of measurements, may seem to support this theory; but the realization that observations of subatomic particles influence their behaviour has undermined the view that objectivity is possible in science (see ◊uncertainty principle).

scintillation counter instrument for measuring very low levels of radiation. The radiation strikes a scintillator (a device that emits a unit of light when a charged elementary particle collides with it), whose light output is 'amplified' by a ◊photomultiplier; the current pulses of its output are in turn counted or added by a scaler to give a numerical reading.

sclerenchyma plant tissue whose function is to strengthen and support, composed of thick-walled cells that are heavily lignified (toughened). On maturity the cell inside dies, and only the cell walls remain.

Sclerenchyma may be made up of one or two types of cells: *sclereids*, occurring singly or in small clusters, are often found in the hard shells of fruits and in seed coats, bark, and the stem cortex; *fibres*, frequently grouped in bundles, are elongated cells, often with pointed ends, associated with the vascular tissue (◊xylem and ◊phloem) of the plant.

Scorpius or *Scorpio* zodiacal constellation in the southern hemisphere between Libra and Sagittarius, represented as a scorpion. The Sun passes briefly through Scorpius in the last week of Nov. The heart of the scorpion is marked by the red supergiant star Antares. Scorpius contains rich Milky Way star fields, plus the strongest ◊X-ray source in the sky, Scorpius X-1. In astrology, the dates for Scorpius are between about 24 Oct and 21 Nov.

scrambling circuit in radiotelephony, a transmitting circuit that renders signals unintelligible unless received by the corresponding unscrambling circuit.

scramjet (acronym for *supersonic combustion ramjet*) supersonic jet engine under development in the early 1990s that uses oxygen from the atmosphere rather than liquid oxygen to burn its fuel. See ◊jet propulsion, ◊ramjet.

screen or *monitor* output device on which a computer displays information for the benefit of the operator. The commonest type is the ◊cathode-ray tube (CRT), which is similar to a television screen. Portable computers often use ◊liquid crystal display (LCD) screens. These are harder to read than CRTs, but require less power, making them suitable for battery operation.

screen dump in computing, the process of making a printed copy of the current VDU screen display. The screen dump is sometimes stored as a data file instead of being printed immediately.

scrolling in computing, the action by which data displayed on a VDU screen are automatically moved upwards and out of sight as new lines of data are added at the bottom.

scrotum paired sac enclosing the testes and the epididymes (coiled tubes in which the sperm mature).

scuba (acronym for *self-contained underwater breathing apparatus*) another name for ◊aqualung.

scurvy disease caused by deficiency of vitamin C (ascorbic acid), which is contained in fresh vegetables and fruit. The signs are weakness and aching joints and muscles, progressing to bleeding of the gums and other spontaneous haemorrhage, and drying-up of the skin and hair. It is reversed by giving the vitamin.

seafloor spreading growth of the ocean ◊crust outwards (sideways) from ocean ridges. The concept of seafloor spreading has been combined with that of continental drift and incorporated into ◊plate tectonics.

searching in computing, extracting a specific item from a large body of data, such as a file or table. The method used depends on how the data are organized. For example, a binary search, which requires the data to be in sequence, involves first deciding which half of the data contains the required item, then which quarter, then which eighth, and so on until the item is found.

sebaceous gland gland in the skin producing ◊sebum. It opens into a hair follicle. Distribution of sebaceous glands varies in different parts of the body.

sebum oily secretion from the sebaceous glands that acts as a skin lubricant. Acne is caused by inflammation of the sebaceous glands and over-secretion of sebum.

sec or **s** abbreviation for *second*, a unit of time.

SECAM abbreviation for *sequential and memory*, the coding system for colour-television broadcasting used in France and E Europe; see ◊television.

secant in trigonometry, the function of a given angle in a right-angled triangle, obtained by dividing the length of the hypotenuse (the longest side) by the length of the side adjacent to the angle. It is the ◊reciprocal of the ◊cosine (sec = 1/cos).

second basic SI unit (symbol sec or s) of time, one-sixtieth of a minute. It is defined as the duration of 9,192,631,770 periods of the radiation corresponding to the transition between two hyperfine levels of the ground state of the caesium-133 isotope. In mathematics, the second is a unit (symbol ″) of angular measurement, equalling one-sixtieth of a minute, which in turn is one-sixtieth of a degree.

secondary emission in physics, an emission of electrons from the surface of certain substances when they are struck by high-speed electrons or other particles from an external source. It can be detected with a ◊photomultiplier.

secondary growth or *secondary thickening* increase in diameter of the roots and stems of certain plants (notably shrubs and trees) that results from the production of new cells by the ◊cambium. It provides the plant with additional mechanical support and new conducting cells, the secondary ◊xylem and ◊phloem. Secondary growth is generally confined to ◊gymnosperms and, among the ◊angiosperms, to the dicotyledons. With just a few exceptions, the monocotyledons (grasses, lilies) exhibit only primary growth, resulting from cell division at the apical ◊meristems.

secondary sexual characteristic in biology, an external feature of an organism that is indicative of its gender (male or female), but

secant *The secant of an angle is a function used in the mathematical study of the triangle. If the secant of angle B is known, then the hypotenuse can be found given the length of the adjacent side, or the adjacent side can be found from the hypotenuse.*

not the reproductive organs themselves. They include facial hair in men and breasts in women, combs in cockerels, brightly coloured plumage in many male birds, and manes in male lions. In many cases, they are involved in displays and contests for mates and have evolved by ◊sexual selection. Their development is stimulated by sex hormones.

secretin ◊hormone produced by the small intestine of vertebrates that stimulates the production of digestive secretions by the pancreas and liver.

secretion in biology, any substance (normally a fluid) produced by a cell or specialized gland, for example, sweat, saliva, enzymes, and hormones. The process whereby the substance is discharged from the cell is also known as secretion.

sector in geometry, part of a circle enclosed by two radii and the arc that joins them.

secular variable in astronomy, a star that, according to comparisons with ancient observations, has either increased or decreased substantially (and permanently) in brightness over the intervening centuries.

sedative any drug that has a calming effect, reducing anxiety and tension. Sedatives will induce sleep in larger doses. Examples are barbiturates, ◊narcotics, and benzodiazepines.

sediment any loose material that has 'settled' – deposited from suspension in water, ice, or air, generally as the water current or wind speed decreases. Typical sediments are, in order of increasing coarseness, clay, mud, silt, sand, gravel, pebbles, cobbles, and boulders.

sedimentary rock rock formed by the accumulation and cementation of deposits that have been laid down by water, wind, ice, or gravity. Sedimentary rocks cover more than two-thirds of the Earth's surface and comprise three major categories: clastic, chemically precipitated, and organic (or biogenic). Clastic sediments are the largest group and are composed of fragments of pre-existing rocks; they include clays, sands, and gravels.

Chemical precipitates include some limestones and evaporated deposits such as gypsum and halite (rock salt). Coal, oil shale, and limestone made of fossil material are examples of organic sedimentary rocks.

Seebeck effect in physics, the generation of a voltage in a circuit containing two different metals, or semiconductors, by keeping the junctions between them at different temperatures. Discovered by the German physicist Thomas Seebeck (1770–1831), it is also called the thermoelectric effect, and is the basis of the ◊thermocouple. It is the opposite of the ◊Peltier effect (in which current flow causes a temperature difference between the junctions of different metals).

seed the reproductive structure of higher plants (◊angiosperms and ◊gymnosperms). It develops from a fertilized ovule and consists of an embryo and a food store, surrounded and protected by an outer seed coat, called the testa. The food store is contained either in a specialized nutritive tissue, the ◊endosperm, or in the ◊cotyledons of the embryo itself. In angiosperms the seed is enclosed within a ◊fruit, whereas in gymnosperms it is usually naked and unprotected, once shed from the female cone. Following ◊germination the seed develops into a new plant.

Seeds may be dispersed from the parent plant in a number of different ways. Agents of dispersal include animals, as with ◊burs and fleshy edible fruits, and wind, where the seed or fruit may be winged or plumed. Water can disperse seeds or fruits that float, and various mechanical devices may eject seeds from the fruit, as in the pods of some leguminous plants (see ◊legume).

There may be a delay in the germination of some seeds to ensure that growth occurs under favourable conditions (see ◊after-ripening, ◊dormancy). Most seeds remain viable for at least 15 years if dried to about 5% water and kept at −20°C/−4°F, although 20% of them will not survive this process.

seed plant any seed-bearing plant; also known as a *spermatophyte*. The seed plants are subdivided into two classes: the ◊angiosperms, or flowering plants, and the ◊gymnosperms, principally the cycads and conifers. Together, they comprise the major types of vegetation found on land.

segment in geometry, part of a circle cut off by a straight line or ◊chord, running from one point on the circumference to another. All angles in the same segment are equal.

seismic gap theory theory that aims to predict the location of ◊earthquakes. When records of past earthquakes are studied and plotted onto a map, it becomes possible to identify areas along a fault or plate margin where an earthquake should be due – such areas are called *seismic gaps*. According to the theory, an area that has not had an earthquake for some time will have a great deal of stress building up, which must eventually be released in the form of an earthquake.

seismic wave energy wave generated by an ◊earthquake; it is responsible for the shaking of the ground and the conversion of soft deposits, such as clay, to a jellylike state (◊liquefaction).

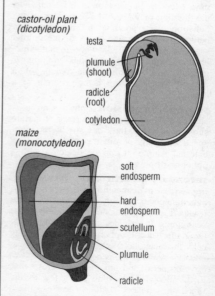

seed *The structure of seeds. The castor is a dicotyledon, a plant in which the developing plant has two leaves, developed from the cotyledon. In maize, a monocotyledon, there is a single leaf developed from the scutellum.*

seismogram or *seismic record* trace, or graph, of an earthquake's activity over time, recorded by a seismograph. It is used to determine the magnitude and duration of an earthquake.

seismograph instrument used to record the activity of an ◊earthquake. A heavy inert weight is suspended by a spring and attached to this is a pen that is in contact with paper on a rotating drum. During an earthquake the instrument frame and drum move, causing the pen to record a zigzag line on the paper; the pen does not move.

seismology study of earthquakes and how their shock waves travel through the Earth. By examining the global pattern of waves produced by an earthquake, seismologists can deduce the nature of the materials through which they have passed. This leads to an understanding of the Earth's internal structure.

selenium grey, nonmetallic element, symbol Se, atomic number 34, relative atomic mass 78.96. It belongs to the sulphur group and occurs in several allotropic forms that differ in their physical and chemical properties. It is an essential trace element in human nutrition. Obtained from many sulphide ores and selenides, it is used as a red colouring for glass and enamel.

self-induction or *self-inductance* in physics, the creation of a counter emf (◊electromotive force) in a coil because of variations in the current flowing through it.

semelparity in biology, the occurrence of a single act of reproduction during an organism's lifetime. Most semelparous species produce very large numbers of offspring when they do reproduce, and normally die soon afterwards. Examples include the Pacific salmon and the pine looper moth. Many plants are semelparous, or ◊monocarpic. Repeated reproduction is called ◊iteroparity.

semen fluid containing ◊sperm from the testes and secretions from various sex glands (such as the prostate gland) that is ejaculated by male animals during copulation. The secretions serve to nourish and activate the sperm cells, and prevent them clumping together.

semicircular canal one of three looped tubes that form part of the labyrinth in the inner ◊ear. They are filled with fluid and detect changes in the position of the head, contributing to the sense of balance.

semiconductor crystalline material with an electrical conductivity between that of metals (good) and insulators (poor). The conductivity of semiconductors can usually be improved by minute additions of different substances or by other factors. Silicon, for example, has poor conductivity at low temperatures, but this is improved by the application of light, heat, or voltage; hence silicon is used in ◊transistors, rectifiers, and ◊integrated circuits (silicon chips).

seminiferous tubule in male vertebrates, one of a number of tightly packed, highly coiled tubes in the testis. The tubes are lined with germinal epithelium cells, from which ◊sperm are produced by cell division (◊meiosis) and in the midst of which the sperm mature, nourished and protected in the folds of large Sertoli cells. After maturation, the sperm are stored in a long tube called the epididymis.

Semtex plastic explosive, manufactured in the Czech Republic. It is safe to handle (it can only be ignited by a detonator) and difficult to trace, since it has no smell. It has been used by extremist groups in the Middle East and by the IRA in Northern Ireland.

senescence in biology, the deterioration in physical and (sometimes) mental capacities that occurs with ◊ageing.

sense organ any organ that an animal uses to gain information about its surroundings. All sense organs have specialized receptors (such as light receptors in the eye) and some means of translating their response into a nerve impulse that travels to the brain. The main human sense organs are the eye, which detects light and colour (different wavelengths of light); the ear, which detects sound (vibrations of the air) and gravity; the nose, which detects some of the chemical molecules in the air; and the tongue, which detects some of the chemicals in food, giving a sense of taste. There are also many small sense organs in the skin, including pain, temperature, and pressure sensors, contributing to our sense of touch.

sensitivity the ability of an organism, or part of an organism, to detect changes in the environment. Although all living things are capable of some sensitivity, evolution has led to the formation of highly complex mechanisms for detecting light, sound, chemicals, and other stimuli. It is essential to an animal's survival that it can process this type of information and make an appropriate response.

sensor in computing, a device designed to detect a physical state or measure a physical quantity, and produce an input signal for a computer. For example, a sensor may detect the fact that a printer has run out of paper or may measure the temperature in a kiln.

sepal part of a flower, usually green, that surrounds and protects the flower in bud. The sepals are derived from modified leaves, and are collectively known as the ♢calyx.

sequence-control register or *program counter* in computing, a special memory location used to hold the address of the next instruction to be fetched from the immediate access memory for execution by the computer (see ♢fetch-execute cycle). It is located in the control unit of the ♢central processing unit.

sequencing in biochemistry, determining the sequence of chemical subunits within a large molecule. Techniques for sequencing amino acids in proteins were established in the 1950s, insulin being the first for which the sequence was completed. Efforts are now being made to determine the sequence of base pairs within ♢DNA.

sere plant ♢succession developing in a particular habitat. A *lithosere* is a succession starting on the surface of bare rock. A *hydrosere* is a succession in shallow freshwater, beginning with planktonic vegetation and the growth of pondweeds and other aquatic plants, and ending with the development of swamp. A *plagiosere* is the sequence of communities that follows the clearing of the existing vegetation.

serial device in computing, a device that communicates binary data by sending the bits that represent each character one by one along a single data line, unlike a ♢parallel device.

series circuit electrical circuit in which the components are connected end to end, so that the current flows through them all one after the other.

serotonin or *5-hydroxytryptamine (5-HT)* chemical widely distributed in the body tissues – in the blood platelets, the wall of the intestine and the ♢central nervous system. It is believed to be implicated in the inflammatory process and, in the nervous system, it acts as a ♢neurotransmitter, controlling sleep.

Serpens constellation of the equatorial region of the sky, represented as a serpent coiled around the body of Ophiuchus. It is the only constellation divided into two halves: *Serpens Caput*, the head (on one side of Ophiuchus), and *Serpens Cauda*, the tail (on the other side). Its main feature is the Eagle nebula.

serpentine group of minerals, hydrous magnesium silicate, $Mg_3Si_2O_5(OH)_4$, occurring in soft ♢metamorphic rocks and usually dark green. The fibrous form **chrysotile** is a source of ♢asbestos; other forms are *antigorite* and *lizardite* Serpentine minerals are formed by hydration of ultrabasic rocks during metamorphism. Rare snake-patterned forms are used in ornamental carving.

serum clear fluid that separates out from clotted blood. It is blood plasma with the anticoagulant proteins removed, and contains ♢antibodies and other proteins, as well as the fats and sugars of the blood. It can be produced synthetically, and is used to protect against disease.

sessile in botany, a leaf, flower, or fruit that lacks a stalk and sits directly on the stem, as with the sessile acorns of certain oaks. In

Venn diagram of two intersecting sets

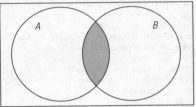

Venn diagram showing the whole numbers from 1 to 20 and the subsets of the prime and odd

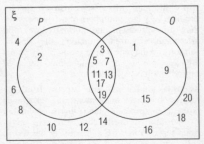

ξ = set of whole numbers from 1 to 20
O = set odd numbers
P = set of prime numbers
the intersection of P and O ($P \cap O$) contains all the prime numbers that are also odd

set *Venn diagrams are used to represent sets.*

zoology, it is an animal that normally stays in the same place, such as a barnacle or mussel. The term is also applied to the eyes of ◊crustaceans when these lack stalks and sit directly on the head.

set or *class* in mathematics, any collection of defined things (elements), provided the elements are distinct and that there is a rule to decide whether an element is a member of a set. It is usually denoted by a capital letter and indicated by curly brackets { }.

For example, *L* may represent the set that consists of all the letters of the alphabet. The symbol ∈ stands for 'is a member of'; thus *p* ∈ *L* means that *p* belongs to the set consisting of all letters, and 4 ∈ *L* means that 4 does not belong to the set consisting of all letters.

There are various types of sets. A *finite set* has a limited number of members, such as the letters of the alphabet; an *infinite set* has an unlimited number of members, such as all whole numbers; an *empty* or *null set* has no members, such as the number of people who have swum across the Atlantic Ocean, written as { } or ∅; a *single-element set* has only one member, such as days of the week beginning with M, written as {Monday}. *Equal sets* have the same members; for example, if *W* = {days of the week} and *S* = {Sunday, Monday, Tuesday, Wednesday, Thursday, Friday, Saturday}, it can be said that *W* = *S*. Sets with the same number of members are *equivalent sets*. Sets with some members in common are *intersecting sets*; for example, if *R* = {red playing cards} and *F* = {face cards}, then *R* and *F* share the members that are red face cards. Sets with no members in common are *disjoint sets*. Sets contained within others are *subsets*; for example, *V* = {vowels} is a subset of *L* = {letters of the alphabet}. Sets and their interrelationships are often illustrated by a ◊Venn diagram.

sex chromosome chromosome that differs between the sexes and that serves to determine the sex of the individual. In humans, females have two X chromosomes and males have an X and a Y chromosome.

sex determination process by which the sex of an organism is determined. In many species, the sex of an individual is dictated by the two sex chromosomes (X and Y) it receives from its parents. In mammals, some plants, and a few insects, males are XY, and females XX; in birds, reptiles, some amphibians, and butterflies the reverse is the case. In bees and wasps, males are produced from unfertilized eggs, females from fertilized eggs.

Environmental factors can affect some fish and reptiles, such as turtles, where sex is influenced by the temperature at which the eggs develop. A six-year study completed 1994 showed global warming to be affecting sex determination in that crocodile and turtle populations are producing many more female hatchlings as a result of warmer summers.

In 1991 it was shown that maleness is caused by a single gene, 14 base pairs long, on the Y chromosome.

female reproductive system

male reproductive system

sexual reproduction *The human reproductive organs. In the female, gametes called ova are produced regularly in the ovaries after puberty. The Fallopian tubes carry the ova to the uterus or womb, in which the baby will develop. In the male, sperm is produced inside the testes after puberty; about 10 million sperm cells are produced each day, enough to populate the world in six months. The sperm duct or vas deferens, a continuation of the epididymis, carries sperm to the urethra during ejaculation.*

reproductive organs in flowering plants

sexual reproduction *Reproductive organs in flowering plants. The stamens are the male parts of the plant. Each consists of a stalklike filament topped by an anther. The anther contains four pollen sacs which burst to releases tiny grains of pollen, the male sex cells. The carpels are the female reproductive parts. Each carpel has a stigma which catches the pollen grain. The style connects the stigma to the ovary. The ovary contains one or more ovules, the female sex cells. Buttercups have many ovaries; the lupin has only one.*

sex hormone steroid hormone produced and secreted by the gonads (testes and ovaries). Sex hormones control development and reproductive functions and influence sexual and other behaviour.

sex linkage in genetics, the tendency for certain characteristics to occur exclusively, or predominantly, in one sex only. Human examples include red-green colour blindness and haemophilia, both found predominantly in males. In both cases, these characteristics are ◊recessive and are determined by genes on the ◊X chromosome.

Since females possess two X chromosomes, any such recessive ◊allele on one of them is likely to be masked by the corresponding allele on the other. In males (who have only one X chromosome paired with a largely inert ◊Y chromosome) any gene on the X chromosome will automatically be expressed. Colour blindness and haemophilia can appear in females, but only if they are ◊homozygous for these traits, due to inbreeding, for example.

sexual reproduction reproductive process in organisms that requires the union, or ◊fertilization, of gametes (such as eggs and sperm). These are usually produced by two different individuals, although self-fertilization occurs in a few ◊hermaphrodites such as tapeworms. Most organisms other than bacteria and cyanobacteria (◊blue-green algae) show some sort of sexual process. Except in some lower organisms, the gametes are of two distinct types called eggs and sperm. The organisms producing the eggs are called females, and those producing the sperm, males. The fusion of a male and female gamete produces a *zygote*, from which a new individual develops.

sexual selection process similar to ◊natural selection but relating exclusively to success in finding a mate for the purpose of sexual reproduction and producing offspring. Sexual selection occurs when one sex (usually but not always the female) invests more effort in producing young than the other. Members of the other sex compete for access to this limited resource (usually males competing for the chance to mate with females).

Seyfert galaxy galaxy whose small, bright centre is caused by hot gas moving at high speed around a massive central object, possibly a ◊black hole. Almost all Seyferts are spiral galaxies. They seem to be closely related to ◊quasars, but are about 100 times fainter. They are named after their discoverer Carl Seyfert (1911–1960).

SGML abbreviation for ◊*Standard Generalized Markup Language*.

shadow area of darkness behind an opaque object that cannot be reached by some or all of the light coming from a light source in front. Its presence may be explained in terms of light rays travelling in straight lines and being unable to bend round obstacles. A point source of light produces an ◊umbra, a completely black shadow with sharp edges. An extended source of light produces both a central umbra and a ◊penumbra, a region of semidarkness with blurred edges where darkness gives way to light.

shale fine-grained and finely layered ◊sedimentary rock composed of silt and clay. It is a weak rock, splitting easily along bedding planes to form thin, even slabs (by contrast, mudstone splits into irregular flakes). Oil shale contains kerogen, a solid bituminous material that yields ◊petroleum when heated.

shark any member of various orders of cartilaginous fishes (class Chondrichthyes), found throughout the oceans of the world. There are about 400 known species of shark. They have tough, usually grey, skin covered in denticles (small toothlike scales). A shark's streamlined body has side pectoral fins, a high dorsal fin, and a forked tail with a large upper lobe. Five open gill slits are visible on each side of the generally pointed head. Most sharks are fish-eaters, and a few will attack humans. They range from several feet in length to the great *white shark Carcharodon carcharias*, 9 m/30 ft long, and the harmless plankton-feeding *whale shark Rhincodon typus*, over 15 m/50 ft in length.

point source of light

extended source of light

shadow *A point source of light produces an evenly dark shadow with a sharp edge. An extended source of light produces a shadow with a darker inner region, called the umbra, and a lighter outer area, called the penumbra.*

shell the hard outer covering of a wide variety of invertebrates. The covering is usually mineralized, normally with large amounts of calcium. The shell of birds' eggs is also largely made of calcium.

SHF in physics, the abbreviation for *super-high ◊frequency*.

shock absorber in technology, any device for absorbing the shock of sudden jarring actions or movements. Shock absorbers are used in conjunction with coil springs in most motor-vehicle suspension systems and are usually of the telescopic type, consisting of a piston in an oil-filled cylinder. The resistance to movement of the piston through the oil creates the absorbing effect.

shock wave narrow band or region of high pressure and temperature in which the flow of a fluid changes form subsonic to supersonic.

Shock waves are produced when an object moves through a fluid at a supersonic speed. See ◊sonic boom.

shoot in botany, the parts of a ◊vascular plant growing above ground, comprising a stem bearing leaves, buds, and flowers. The shoot develops from the ◊plumule of the embryo.

shooting star another name for a ◊meteor.

short circuit direct connection between two points in an electrical circuit. Its relatively low resistance means that a large current flows through it, bypassing the rest of the circuit, and this may cause the circuit to overheat dangerously.

short-sightedness nontechnical term for ◊myopia.

shrub perennial woody plant that typically produces several separate stems, at or near ground level, rather than the single trunk of most trees. A shrub is usually smaller than a tree, but there is no clear distinction between large shrubs and small trees.

shunt in electrical engineering, a conductor of very low resistance that is connected in parallel to an ◊ammeter in order to enable it to measure larger electric currents. Its low resistance enables it to act like a bypass, diverting the extra current through itself and away from the ammeter, and thereby reducing the instrument's sensitivity.

SI abbreviation for *Système International [d'Unités]* (French 'International System [of Metric Units]'); see ◊SI units.

sial in geochemistry and geophysics, the substance of the Earth's continental ◊crust, as distinct from the ◊sima of the ocean crust. The name, now used rarely, is derived from *si*lica and *al*umina, its two main chemical constituents. Sial is often rich in granite.

sick building syndrome malaise diagnosed in the early 1980s among office workers and thought to be caused by such pollutants as formaldehyde (from furniture and insulating materials), benzene (from paint), and the solvent trichloroethene, concentrated in air-conditioned buildings. Symptoms include headache, sore throat, tiredness, colds, and flu. Studies have found that it can cause a 40% drop in productivity and a 30% rise in absenteeism.

sickle-cell disease hereditary chronic blood disorder common among people of black African descent; also found in the E Mediterranean, parts of the Persian Gulf, and in NE India. It is characterized by distortion and fragility of the red blood cells, which are lost too rapidly from the circulation. This often results in ◊anaemia.

People with this disease have abnormal red blood cells (sickle cells), containing a defective ◊haemoglobin. The presence of sickle cells in the blood, with or without accompanying anaemia, is called *sicklemia*. It confers a degree of protection against ◊malaria because fewer normal red blood cells are available to the parasites for infection.

sidereal period the orbital period of a planet around the Sun, or a moon around a planet, with reference to a background star. The sidereal period of a planet is in effect a 'year'. A ◊synodic period is a full circle as seen from Earth.

siemens SI unit (symbol S) of electrical conductance, the reciprocal of the ◊impedance of an electrical circuit. One siemens equals one ampere per volt. It was formerly called the mho or reciprocal ohm.

sievert SI unit (symbol Sv) of radiation dose equivalent. It replaces the rem (1 Sv equals 100 rem). Some types of radiation do more damage than others for the same absorbed dose – for example, an absorbed dose of alpha radiation causes 20 times as much biological damage as the same dose of beta radiation. The equivalent dose in sieverts is equal to the absorbed dose of radiation in rays multiplied by the relative biological effectiveness. Humans can absorb up to 0.25 Sv without immediate ill effects; 1 Sv may produce radiation sickness; and more than 8 Sv causes death.

sight the detection of light by an ◊eye, which can form images of the outside world.

Sigma Octantis the star closest to the south celestial pole (see ◊celestial sphere), in effect the southern equivalent of ◊Polaris, although far less conspicuous. Situated just less than 1° from the south celestial pole in the constellation Octans, Sigma Octantis is 120 light years away.

signal-to-noise ratio ratio of the power of an electrical signal to that of the unwanted noise accompanying the signal. It is expressed in ◊decibels.

In general, the higher the signal-to-noise ratio, the better. For a telephone, an acceptable ratio is 40 decibels; for television, the acceptable ratio is 50 decibels.

significant figures the figures in a number that, by virtue of their place value, express the magnitude of that number to a specified degree of accuracy. The final significant figure is rounded up if the following digit is greater than 5. For example, 5,463,254 to three significant figures is 5,460,000; 3·462891 to four significant figures is 3·463; 0·00347 to two significant figures is 0.0035.

silencer (North American *muffler*) device in the exhaust system of cars and motorbikes. Gases leave the engine at supersonic speeds, and the exhaust system and silencer are designed to slow them down, thereby silencing them.

silica silicon dioxide, SiO_2, the composition of the most common mineral group, of which the most familiar form is quartz. Other silica forms are ◊chalcedony, chert, opal, tridymite, and cristobalite.

Common sand consists largely of silica in the form of quartz.

silicate one of a group of minerals containing silicon and oxygen in tetrahedral units of SiO_4, bound together in various ways to form specific structural types. Silicates are the chief rock-forming minerals. Most rocks are composed, wholly or in part, of silicates (the main exception being limestones). Glass is a manufactured complex polysilicate material in which other elements (boron in borosilicate glass) have been incorporated.

silicon brittle, nonmetallic element, symbol Si, atomic number 14, relative atomic mass 28.086. It is the second-most abundant element (after oxygen) in the Earth's crust and occurs in amorphous and crystalline forms. In nature it is found only in combination with other elements, chiefly with oxygen in silica (silicon dioxide, SiO_2) and the silicates. These form the mineral ◊quartz, which makes up most sands, gravels, and beaches.

silicon chip ◊integrated circuit with microscopically small electrical components on a piece of silicon crystal only a few millimetres square.

sillimanite aluminium silicate, Al_2SiO_5, a mineral that occurs either as white to brownish prismatic crystals or as minute white fibres. It is an indicator of high temperature conditions in metamorphic rocks formed from clay sediments. Andalusite, kyanite, and sillimanite are all polymorphs of Al_2SiO_5.

Silurian period of geological time 439–409 million years ago, the third period of the Palaeozoic era. Silurian sediments are mostly marine and consist of shales and limestone. Luxuriant reefs were built by coral-like organisms. The first land plants began to evolve during this period, and there were many ostracoderms (armoured jawless fishes). The first jawed fishes (called acanthodians) also appeared.

silver white, lustrous, extremely malleable and ductile, metallic element, symbol Ag (from Latin *argentum*), atomic number 47, relative atomic mass 107.868. It occurs in nature in ores and as a free metal; the chief ores are sulphides, from which the metal is extracted by smelting with lead. It is one of the best metallic conductors of both heat and electricity; its most useful compounds are the chloride and bromide, which darken on exposure to light and are the basis of photographic emulsions.

Silver is used ornamentally, for jewellery and tableware, for coinage, in electroplating, electrical contacts, in dentistry, and as a solder. It has been mined since prehistory.

$$\sin(\theta)\ \beta = \frac{\text{opposite}}{\text{hypotenuse}} = \frac{AC}{AB}$$

sine *(left) The sine of an angle. The sine is a function used in the mathematical study of the triangle. If the sine of angle β is known, then the hypotenuse can be found given the length of the opposite side, or the opposite side can be found from the hypotenuse.*

silver plate silverware made by depositing a layer of silver on another metal, usually copper, by the process of ◊electroplating.

sima in geochemistry and geophysics, the substance of the Earth's oceanic ◊crust, as distinct from the ◊sial of the continental crust. The name, now used rarely, is derived from *si*lica and *ma*gnesia, its two main chemical constituents.

simple harmonic motion (SHM) oscillatory or vibrational motion in which an object (or point) moves so that its acceleration towards a central point is proportional to its distance from it. A simple example is a pendulum, which also demonstrates another feature of SHM, that the maximum deflection is the same on each side of the central point.

A graph of the varying distance with respect to time is a sine curve, a characteristic of the oscillating current or voltage of an alternating current (AC), which is another example of SHM.

simulation short for ◊computer simulation.

simultaneous equations in mathematics, one of two or more algebraic equations that contain two or more unknown quantities that may have a unique solution. For example, in the case of two linear equations with two unknown variables, such as (i) $x + 3y = 6$ and (ii) $3y - 2x = 4$, the solution will be those unique values of x and y that are valid for both equations. Linear simultaneous equations can be solved by using algebraic manipulation to eliminate one of the variables, ◊coordinate geometry, or matrices (see ◊matrix).

sine in trigonometry, a function of an angle in a right-angled triangle which is defined as the ratio of the length of the side opposite the angle to the length of the hypotenuse (the longest side).

Various properties in physics vary sinusoidally; that is, they can be represented diagrammatically by a sine wave (a graph obtained by plotting values of angles against the values of their sines). Examples include ◊simple harmonic motion, such as the way alternating current (AC) electricity varies with time.

sine rule in trigonometry, a rule that relates the sides and angles of a triangle, stating that the ratio of the length of each side and the sine of the angle opposite is constant (twice the radius of the circumscribing circle). If the sides of a triangle are *a*, *b*, and *c*, and the angles

skin *The skin of an adult man covers about 1.9 sq m/20 sq ft; a woman's skin covers about 1.6 sq m/17 sq ft. During our lifetime, we shed about 18 kg/40 lb of skin.*

opposite are A, B, and C, respectively, then the sine rule may be expressed as:

$$a/\sin A = b/\sin B = c/\sin C$$

singularity in astrophysics, the point in ◊space–time at which the known laws of physics break down. Singularity is predicted to exist at the centre of a black hole, where infinite gravitational forces compress the infalling mass of a collapsing star to infinite density. It is also thought, according to the Big Bang model of the origin of the universe, to be the point from which the expansion of the universe began.

sinoatrial node the heart's natural ◊pacemaker.

siphon tube in the form of an inverted U with unequal arms. When it is filled with liquid and the shorter arm is placed in a tank or reservoir, liquid flows out of the longer arm provided that its exit is below the level of the surface of the liquid in the tank.

siphonogamy plant reproduction in which a pollen tube grows to enable male gametes to pass to the ovary without leaving the protection of the plant.

Siphonogamous reproduction is found in all angiosperms and most gymnosperms, and has enabled these plants to reduce their dependency on wet conditions for reproduction, unlike the zoidogamous plants (see ◊zoidogamy).

Sirius or *Dog Star* or *Alpha Canis Majoris* the brightest star in the sky, 8.6 light years from Earth in the constellation Canis Major. Sirius is a white star with a mass 2.3 times that of the Sun, a diameter 1.8 times that of the Sun, and a luminosity of 23 Suns. It is orbited every 50 years by a white dwarf, Sirius B, also known as the Pup.

SI units (French *Système International d'Unités*) standard system of scientific units used by scientists worldwide. Originally proposed in 1960, it replaces the ◊m.k.s., ◊c.g.s., and ◊f.p.s. systems. It is based on seven basic units: the metre (m) for length, kilogram (kg) for mass, second (s) for time, ampere (A) for electrical current, kelvin (K) for temperature, mole (mol) for amount of substance, and candela (cd) for luminosity.

skeleton the rigid or semirigid framework

that supports and gives form to an animal's body, protects its internal organs, and provides anchorage points for its muscles. The skeleton may be composed of bone and cartilage (vertebrates), chitin (arthropods), calcium carbonate (molluscs and other invertebrates), or silica (many protists). The human skeleton is composed of 206 bones (see ◊human body for illustration).

It may be internal, forming an ◊*endoskeleton*, or external, forming an ◊*exoskeleton*. Another type of skeleton, found in invertebrates such as earthworms, is the **hydrostatic skeleton**. This gains partial rigidity from fluid enclosed within a body cavity. Because the fluid cannot be compressed, contraction of one part of the body results in extension of another part, giving peristaltic motion.

skew lines straight lines that are not parallel and yet do not meet since they lie in a different plane. Every pair of skew lines has a minimum distance between them, which is the length of their common perpendicular.

skin the covering of the body of a vertebrate. In mammals, the outer layer (epidermis) is dead and its cells are constantly being rubbed away and replaced from below; it helps to protect the body from infection and to prevent dehydration. The lower layer (dermis) contains blood vessels, nerves, hair roots, and sweat and sebaceous glands, and is supported by a network of fibrous and elastic cells.

skull in vertebrates, the collection of flat and irregularly shaped bones (or cartilage) that enclose the brain and the organs of sight, hearing, and smell, and provide support for the jaws. In mammals, the skull consists of 22 bones joined by sutures. The floor of the skull is pierced by a large hole (foramen magnum) for the spinal cord and a number of smaller apertures through which other nerves and blood vessels pass.

Skylab US space station, launched 14 May 1973, made from the adapted upper stage of a Saturn V rocket. At 75 tonnes/82.5 tons, it was the heaviest object ever put into space, and was 25.6 m/84 ft long. *Skylab* contained a workshop for carrying out experiments in weightlessness, an observatory for monitoring the Sun, and cameras for photographing the Earth's surface.

slag in chemistry, the molten mass of impurities that is produced in the smelting or refining of metals.

The slag produced in the manufacture of iron in a ◊blast furnace floats on the surface above the molten iron. It contains mostly silicates, phosphates, and sulphates of calcium. When cooled, the solid is broken up and used

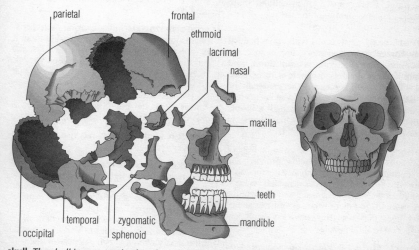

skull *The skull is a protective box for the brain, eyes, and hearing organs. It is also a framework for the teeth and flesh of the face. The cranium has eight bones: occipital, two temporal, two parietal, frontal, sphenoid, ethmoid. The face has 14 bones, the main ones being two maxillae, two nasal, two zygoma, two lacrimal, and the mandible.*

as a core material in the foundations of roads and buildings.

slaked lime $Ca(OH)_2$ (technical name *calcium hydroxide*) substance produced by adding water to quicklime (calcium oxide, CaO). Much heat is given out and the solid crumbles as it absorbs water. A solution of slaked lime is called ◊limewater.

slate fine-grained, usually grey metamorphic rock that splits readily into thin slabs along its ◊cleavage planes. It is the metamorphic equivalent of ◊shale.

sleep state of natural unconsciousness and activity that occurs at regular intervals in most mammals and birds, though there is considerable variation in the amount of time spent sleeping. Sleep differs from hibernation in that it occurs daily rather than seasonally, and involves less drastic reductions in metabolism. The function of sleep is unclear. People deprived of sleep become irritable, uncoordinated, forgetful, hallucinatory, and even psychotic.

slime mould or *myxomycete* extraordinary organism that shows some features of ◊fungus and some of ◊protozoa. Slime moulds are not closely related to any other group, although they are often classed, for convenience, with the fungi. There are two kinds, cellular slime moulds and plasmodial slime moulds, differing in their complex life cycles.

smart card plastic card with an embedded microprocessor and memory. It can store, for example, personal data, identification, and bank-account details, to enable it to be used as a credit or debit card. The card can be loaded with credits, which are then spent electronically, and reloaded as needed. Possible other uses range from hotel door 'keys' to passports.

smart drug any drug or combination of nutrients (vitamins, amino acids, minerals, and sometimes herbs) said to enhance the functioning of the brain, increase mental energy, lengthen the span of attention, and improve the memory. As yet there is no scientific evidence to suggest that these drugs have any significant effect on healthy people.

smart fluid or *electrorheological fluid* liquid suspension that solidifies to form a jelly-like solid when a high-voltage electric field is applied across it and that returns to the liquid state when the field is removed. Most smart fluids are ◊zeolites or metals coated with polymers or oxides.

smell sense that responds to chemical molecules in the air. It works by having receptors for particular chemical groups, into which the airborne chemicals must fit to trigger a message to the brain.

smelling salts or *sal volatile* mixture of ammonium carbonate, bicarbonate, and carbamate together with other strong-smelling substances, formerly used as a restorative for dizziness or fainting.

smelting processing a metallic ore in a furnace to produce the metal. Oxide ores such as iron ore are smelted with coke (carbon), which reduces the ore into metal and also provides fuel for the process.

smokeless fuel fuel that does not give off any smoke when burned, because all the carbon is fully oxidized to carbon dioxide (CO_2). Natural gas, oil, and coke are smokeless fuels.

smoothing capacitor large electronic ◊capacitor connected across the output of a rectifier circuit that has the effect of smoothing out the voltage variations to give a nearly steady DC voltage supply. The voltage and current output from a rectifier circuit fitted with a smoothing capacitor is similar to that provided by a battery.

smooth muscle involuntary muscle capable of slow contraction over a period of time. It is present in hollow organs, such as the intestines, stomach, bladder, and blood vessels. Its presence in the wall of the alimentary canal allows slow rhythmic movements known as ◊peristalsis, which cause food to be mixed and forced along the gut. Smooth muscle has a microscopic structure distinct from other forms.

snellen unit expressing the visual power of the eye.

Snell's law of refraction in optics, the law stating that for light rays passing from one transparent medium to another, the ratio of the sine of the angle of incidence and the sine of the angle of refraction is constant.

The ratio for any pair of transparent media is called its ◊refractive index n. The law is named after its discoverer, Dutch physicist Willebrord Snell (1581–1626).

soap mixture of the sodium salts of various ◊fatty acids: palmitic, stearic, and oleic acid. It is made by the action of sodium hydroxide (caustic soda) or potassium hydroxide (caustic potash) on fats of animal or vegetable origin. Soap makes grease and dirt disperse in water in a similar manner to a ⟡detergent.

soapstone compact, massive form of impure ⟡talc.

social behaviour in zoology, behaviour concerned with altering the behaviour of other individuals of the same species. Social behaviour allows animals to live harmoniously in groups by establishing hierarchies of dominance to discourage disabling fighting. It may be aggressive or submissive (for example, cowering and other signals of appeasement), or designed to establish bonds (such as social grooming or preening).

sociobiology study of the biological basis of all social behaviour, including the application of ⟡population genetics to the evolution of behaviour. It builds on the concept of ⟡inclusive fitness, contained in the notion of the 'selfish gene'. Contrary to some popular interpretations, it does not assume that all behaviour is genetically determined.

soda ash former name for ⟡sodium carbonate.

soda lime powdery mixture of calcium hydroxide and sodium hydroxide or potassium hydroxide, used in medicine and as a drying agent.

sodium soft, waxlike, silver-white, metallic element, symbol Na (from Latin *natrium*), atomic number 11, relative atomic mass 22.898. It is one of the ⟡alkali metals and has a very low density, being light enough to float on water. It is the sixth-most abundant element (the fourth-most abundant metal) in the Earth's crust. Sodium is highly reactive, oxidizing rapidly when exposed to air and reacting violently with water. Its most familiar compound is sodium chloride (common salt), which occurs naturally in the oceans and in salt deposits left by dried-up ancient seas.

Other sodium compounds used industrially include sodium hydroxide (caustic soda, NaOH), sodium carbonate (washing soda, Na_2CO_3) and hydrogencarbonate (sodium bicarbonate, $NaHCO_3$), sodium nitrate (saltpetre, $NaNO_3$, used as a fertilizer), and sodium thiosulphate (hypo, $Na_2S_2O_3$, used as a photographic fixer). Thousands of tons of these are manufactured annually. Sodium metal is used to a limited extent in spectroscopy, in discharge lamps, and alloyed with potassium as a heat-transfer medium in nuclear reactors.

sodium carbonate or *soda ash* Na_2CO_3 anhydrous white solid. The hydrated, crystalline form ($Na_2CO_3.10H_2O$) is also known as washing soda.

sodium chloride or *common salt* or *table salt* NaCl white, crystalline compound found widely in nature. It is a a typical ionic solid with a high melting point (801°C/1,474°F); it is soluble in water, insoluble in organic solvents, and is a strong electrolyte when molten or in aqueous solution. Found in concentrated deposits, it is widely used in the food industry as a flavouring and preservative, and in the chemical industry in the manufacture of sodium, chlorine, and sodium carbonate.

sodium hydrogencarbonate chemical name for ⟡bicarbonate of soda.

sodium hydroxide or *caustic soda* NaOH the commonest ⟡alkali. The solid and the solution are corrosive. It is used to neutralize acids, in the manufacture of soap, and in oven cleaners. It is prepared industrially from sodium chloride by the ⟡electrolysis of concentrated brine.

software in computing, a collection of programs and procedures for making a computer perform a specific task, as opposed to ⟡hardware, the physical components of a computer system. Software is created by programmers and is either distributed on a suitable medium, such as the ⟡floppy disc, or built into the computer in the form of ⟡firmware. Examples of software include ⟡operating systems, ⟡compilers, and applications programs, such as payrolls. No computer can function without some form of software.

software project lifecycle various stages of development in the writing of a major program (software), from the identification of a requirement to the installation, maintenance, and support of the finished program. The process includes ⟡systems analysis and ⟡systems design.

soft water water that contains very few dissolved metal ions such as calcium (Ca^{2+}) or magnesium (Mg^{2+}). It lathers easily with soap, and no ⟡scale is formed inside kettles or boilers. It has been found that the incidence of heart disease is higher in soft-water areas.

soil loose covering of broken rocky material and decaying organic matter overlying the bedrock of the Earth's surface. Various types of soil develop under different conditions: deep soils form in warm wet climates and in valleys; shallow soils form in cool dry areas and on slopes. *Pedology*, the study of soil, is significant because of the relative importance of different soil types to agriculture.

soil mechanics branch of engineering that studies the nature and properties of the soil. Soil is investigated during construction work to ensure that it has the mechanical properties necessary to support the foundations of dams, bridges, and roads.

sol ◊colloid of very small solid particles dispersed in a liquid that retains the physical properties of a liquid.

solar energy energy derived from the Sun's radiation. The amount of energy falling on just 1 sq km/0.3861 sq mi is about 4,000 megawatts, enough to heat and light a small town. In one second the Sun gives off 13 million times more energy than all the electricity used in the USA in one year. *Solar heaters* have industrial or domestic uses. They usually consist of a black (heat-absorbing) panel containing pipes through which air or water, heated by the Sun, is circulated, either by thermal ◊convection or by a pump.

Solar energy may also be harnessed indirectly using *solar cells* (photovoltaic cells) made of panels of ◊semiconductor material (usually silicon), which generate electricity when illuminated by sunlight. Although it is difficult to generate a high output from solar energy compared to sources such as nuclear or fossil fuels, it is a major nonpolluting and renewable energy source used as far north as Scandinavia as well as in the SW USA and in Mediterranean countries.

solar flare eruption from the Sun, see ◊flare, solar.

Solar Maximum Mission satellite launched by the US agency NASA 1980 to study solar activity, which discovered that the Sun's luminosity increases slightly when sunspots are most numerous. It was repaired in orbit by astronauts from the space shuttle 1984 and burned up in the Earth's atmosphere 1989.

solar pond natural or artificial 'pond', such as the Dead Sea, in which salt becomes more soluble in the Sun's heat. Water at the bottom becomes saltier and hotter, and is insulated by the less salty water layer at the top. Temperatures at the bottom reach about 100°C/212°F and can be used to generate electricity.

solar radiation radiation given off by the Sun, consisting mainly of visible light, ◊ultraviolet radiation, and ◊infrared radiation, although the whole spectrum of ◊electromagnetic waves is present, from radio waves to X-rays. High-energy charged particles such as electrons are also emitted, especially from solar ◊flares. When these reach the Earth, they cause magnetic storms (disruptions of the Earth's magnetic field), which interfere with radio communications.

Solar System the Sun (a star) and all the bodies orbiting it: the nine ◊planets (Mercury, Venus, Earth, Mars, Jupiter, Saturn, Uranus, Neptune, and Pluto), their moons, the asteroids, and the comets. It is thought to have formed from a cloud of gas and dust in space about 4.6 billion years ago. The Sun contains 99% of the mass of the Solar System.

solar wind stream of atomic particles, mostly protons and electrons, from the Sun's corona, flowing outwards at speeds of between 300 kps/200 mps and 1,000 kps/600 mps.

solder any of various alloys used when melted for joining metals such as copper, its common alloys (brass and bronze), and tin-plated steel, as used for making food cans.

solenoid coil of wire, usually cylindrical, in which a magnetic field is created by passing an electric current through it (see ◊electromagnet). This field can be used to move an iron rod placed on its axis.

Mechanical valves attached to the rod can be operated by switching the current on or off, so converting electrical energy into mechanical energy. Solenoids are used to relay energy from the battery of a car to the starter motor by means of the ignition switch.

solid in physics, a state of matter that holds its own shape (as opposed to a liquid, which takes up the shape of its container, or a gas, which totally fills its container). According to ◊kinetic theory, the atoms or molecules in a solid are not free to move but merely vibrate about fixed positions, such as those in crystal lattices.

solidification change of state from liquid (or vapour) to solid that occurs at the ◊freezing point of a substance.

solid-state circuit electronic circuit where all the components (resistors, capacitors, transistors, and diodes) and interconnections are made at the same time, and by the same processes, in or on one piece of single-crystal silicon. The small size of this construction accounts for its use in electronics for space vehicles and aircraft.

Solar System *Most of the objects in the Solar System lie close to the plane of the ecliptic. The planets are tiny compared to the Sun. If the Sun were the size of a basketball, the planet closest to the Sun, Mercury, would be the size of a mustard seed 15 m/48 ft from the Sun. The most distant planet, Pluto, would be a pinhead 1.6 km/1 mi away from the Sun. The Earth, which is the third planet out from the Sun, would be the size of a pea 32 m/100 ft from the Sun.*

soliton solitary wave that maintains its shape and velocity, and does not widen and disperse in the normal way. The mathematical equations that sum up the behaviour of solitons are being used to further research in nuclear fusion and superconductivity.

solstice either of the days on which the Sun is farthest north or south of the celestial equator each year. The ***summer solstice***, when the Sun is farthest north, occurs around June 21; the ***winter solstice*** around Dec 22.

solubility in physics, a measure of the amount of solute (usually a solid or gas) that will dissolve in a given amount of solvent (usually a liquid) at a particular temperature. Solubility may be expressed as grams of solute per 100 grams of solvent or, for a gas, in parts per million (ppm) of solvent.

solubility curve graph that indicates how the solubility of a substance varies with temperature. Most salts increase their solubility with an increase in temperature, as they dissolve endothermically. These curves can be used to predict which salt, and how much of it, will crystallize out of a mixture of salts.

solute substance that is dissolved in another substance (see ◊solution).

Solvay process industrial process for the manufacture of sodium carbonate.

solvent substance, usually a liquid, that will dissolve another substance (see ◊solution). Although the commonest solvent is water, in popular use the term refers to low-boiling-point organic liquids, which are harmful if used in a confined space. They can give rise to respiratory problems, liver damage, and neurological complaints.

sonar (acronym for *sound navigation and ranging*) method of locating underwater objects by the reflection of ultrasonic waves. The time taken for an acoustic beam to travel to the object and back to the source enables the distance to be found since the velocity of sound in water is known. Sonar devices, or *echo sounders*, were developed 1920.

sone unit of subjective loudness. A tone of 40 decibels above the threshold of hearing with a frequency of 1,000 hertz is defined as one sone; any sound that seems twice as loud as this has a value of two sones, and so on. A loudness of one sone corresponds to 40 ◊phons.

sonic boom noise like a thunderclap that occurs when an aircraft passes through the ◊sound barrier, or begins to travel faster than the speed of sound. It happens when the cone-shaped shock wave caused by the plane touches the ground.

sonoluminescence emission of light by a liquid that is subjected to high-frequency sound waves. The rapid changes of pressure induced by the sound cause minute bubbles to form in the liquid, which then collapse. Light is emitted at the final stage of the collapse, probably because it squeezes and heats gas inside the bubbles.

sorbic acid $CH_3CH:CHCH:CHCOOH$ tasteless acid found in the fruit of the mountain ash (genus *Sorbus*) and prepared synthetically. It is widely used in the preservation of food – for example, cider, wine, soft drinks, animal feeds, bread, and cheese.

sorting in computing, arranging data in sequence. When sorting a collection, or file, of data made up of several different ◊fields, one must be chosen as the *key field* used to establish the correct sequence. For example, the data in a company's mailing list might include fields for each customer's first names, surname, address, and telephone number. For most purposes the company would wish the records to be sorted alphabetically by surname; therefore, the surname field would be chosen as the key field.

sorus in ferns, a group of sporangia, the reproductive structures that produce ◊spores. They occur on the lower surface of fern fronds.

sound physiological sensation received by the ear, originating in a vibration that communicates itself as a pressure variation in the air and travels in every direction, spreading out as an expanding sphere. All sound waves in air travel with a speed dependent on the temperature; under ordinary conditions, this is about 330 m/1,070 ft per second. The pitch of the sound depends on the number of vibrations imposed on the air per second, but the speed is unaffected. The loudness of a sound is dependent primarily on the amplitude of the vibration of the air.

The lowest note audible to a human being has a frequency of about 20 ◊hertz (vibrations per second), and the highest one of about 15,000 Hz; the lower limit of this range varies little with the person's age, but the upper range falls steadily from adolescence onwards.

sound absorption in acoustics, the conversion of sound energy to heat energy when sound waves strike a surface. The process reduces the amplitude of each reflected sound wave (echo) and thus the degree to which ◊reverberation takes place. Materials with good sound-absorbing properties are often fitted on walls and ceilings in buildings such as offices, factories, and concert halls in order to reduce or control sound levels.

sound barrier concept that the speed of sound, or sonic speed (about 1,220 kph/760 mph at sea level), constitutes a speed limit to flight through the atmosphere, since a badly designed aircraft suffers severe buffeting at near sonic speed owing to the formation of shock waves. US test pilot Chuck Yeager first flew through the 'barrier' in 1947 in a Bell X-1 rocket plane. Now, by careful design, such aircraft as Concorde can fly at supersonic speed with ease, though they create in their wake a ◊sonic boom.

sound wave the longitudinal wave motion with which sound energy travels through a medium. It carries energy away from the source of the sound without carrying the material itself with it. Sound waves are mechanical; unlike electromagnetic waves, they require vibration of their medium's molecules or particles, and this is why sound cannot travel through a vacuum.

source language in computing, the language in which a program is written, as opposed to ◊machine code, which is the form in which the program's instructions are carried out by the computer. Source languages are classified as either ◊high-level languages or ◊low-level languages, according to whether each notation in the source language stands for many or only one instruction in machine code.

source program in computing, a program written in a ◊source language.

source resistance alternative term for ◊internal resistance, the resistance inside an electric power supply.

souring change that occurs to wine on prolonged exposure to air. The ethanol in the wine is oxidized by the air (oxygen) to ethanoic acid. It is the presence of the ethanoic (acetic) acid that produces the sour taste.

$$CH_3CH_2OH_{(aq)} + O_{2(g)} \rightarrow CH_3COOH_{(aq)} + H_2O_{(l)}$$

Southern Cross popular name for the constellation ◊Crux.

southern lights common name for the ◊aurora australis, coloured light in southern skies.

Soyuz Soviet series of spacecraft, capable of carrying up to three cosmonauts. Soyuz spacecraft consist of three parts: a rear section containing engines; the central crew compartment; and a forward compartment that gives additional room for working and living space. They are now used for ferrying crews up to space stations, though they were originally used for independent space flight.

space or *outer space* the void that exists beyond Earth's atmosphere. Above 120 km/75 mi, very little atmosphere remains, so objects can continue to move quickly without extra energy. The space between the planets is not entirely empty, but filled with the tenuous gas of the ◊solar wind as well as dust specks.

Spacelab small space station built by the European Space Agency, carried in the cargo bay of the US space shuttle, in which it remains throughout each flight, returning to Earth with the shuttle. Spacelab consists of a pressurized module in which astronauts can work, and a series of pallets, open to the vacuum of space, on which equipment is mounted.

space probe any instrumented object sent beyond Earth to collect data from other parts of the Solar System and from deep space. The first probe was the Soviet *Lunik 1*, which flew past the Moon 1959. The first successful planetary probe was the US *Mariner 2*, which flew past Venus 1962, using ◊transfer orbit. The first space probe to leave the Solar System was *Pioneer 10* 1983. Space probes include *Galileo, Giotto, Magellan, Mars Observer, Ulysses*, the ◊Moon probes, and the Mariner, Pioneer, Viking, and Voyager series. Japan launched its first space probe 1990.

space shuttle reusable crewed spacecraft. The first was launched 12 April 1981 by the USA. It was developed by NASA to reduce the cost of using space for commercial, scientific, and military purposes. After leaving its payload in space, the space-shuttle orbiter can be flown back to Earth to land on a runway, and is then available for reuse.

Four orbiters were built: *Columbia, Challenger, Discovery*, and *Atlantis. Challenger* was destroyed in a midair explosion just over a minute after its tenth launch 28 Jan 1986, killing all seven crew members, the result of a failure in one of the solid rocket boosters. Flights resumed with redesigned boosters in Sept 1988. A replacement orbiter, *Endeavor*, was built, which had its maiden flight in May 1992. At the end of the 1980s, an average of $375 million had been spent on each space-shuttle mission.

space station any large structure designed for human occupation in space for extended periods of time. Space stations are used for carrying out astronomical observations and surveys of Earth, as well as for biological studies and the processing of materials in weightlessness. The first space station was ◊*Salyut 1*, and the USA has launched ◊*Skylab*.

NASA plans to build a larger space station, to be called *Freedom*, in orbit during the 1990s, in cooperation with other countries, including the European Space Agency, which is building a module called *Columbus*, and Japan, also building a module.

space–time in physics, combination of space and time used in the theory of ◊relativity. When developing relativity, Albert Einstein showed that time was in many respects like an extra dimension (or direction) to space. Space and time can thus be considered as entwined into a single entity, rather than two separate things.

Space–time is considered to have four dimensions: three of space and one of time. In relativity theory, events are described as occurring at points in space–time. The **general theory of relativity** describes how space–time is distorted by the presence of material bodies, an effect that we observe as gravity.

spadix in botany, an ◊inflorescence consisting of a long, fleshy axis bearing many small, stalkless flowers. It is partially enclosed by a large bract or ◊spathe. A spadix is characteristic of plants belonging to the family Araceae, including the arum lily *Zantedeschia aethiopica*.

spark chamber electronic device for recording tracks of charged subatomic ◊particles, decay products, and rays. In combination with a stack of photographic plates, a spark chamber enables the point where an interaction has taken place to be located, to within a cubic centimetre. At its simplest, it consists of two smooth threadlike ◊electrodes that are positioned 1–2 cm/0.5–1 in apart, the space between being filled by an inert gas such as neon. Sparks jump through the gas along the ionized path created by the radiation. See ◊particle detector.

spark plug plug that produces an electric spark in the cylinder of a petrol engine to ignite the fuel mixture. It consists essentially of two electrodes insulated from one another. High-voltage (18,000 V) electricity is fed to a central electrode via the distributor. At the base of the electrode, inside the cylinder, the electricity jumps to another electrode earthed to the engine body, creating a spark. See also ◊ignition coil.

spathe in flowers, the single large bract surrounding the type of inflorescence known as a ◊spadix. It is sometimes brightly coloured and petal-like, as in the brilliant scarlet spathe of the flamingo plant *Anthurium andreanum* from South America; this serves to attract insects.

speciation emergence of a new species during evolutionary history. One cause of speciation is the geographical separation of populations of the parent species, followed by reproductive isolation, so that they no longer produce viable offspring unless they interbreed. Other causes are ◊assortative mating and the establishment of a ◊polyploid population.

species in biology, a distinguishable group of organisms that resemble each other or consist of a few distinctive types (as in ◊polymorphism), and that can all interbreed to produce fertile offspring. Species are the lowest level in the system of biological classification.

Related species are grouped together in a genus. Within a species there are usually two or more separate populations, which may in time become distinctive enough to be designated subspecies or varieties, and could eventually give rise to new species through ◊speciation. Around 1.4 million species have been identified so far, of which 750,000 are insects, 250,000 are plants, and 41,000 are vertebrates. In tropical regions there are roughly two species for each temperate-zone species. It is estimated that one species becomes extinct every day through habitat destruction.

specific gravity alternative term for ◊relative density.

specific heat capacity in physics, quantity of heat required to raise unit mass (1 kg) of a substance by one ◊kelvin (1°C). The unit of specific heat capacity in the SI system is the ◊joule per kilogram kelvin (J kg^{-1} K^{-1}).

specific latent heat in physics, the heat that changes the physical state of a unit mass (one kilogram) of a substance without causing any temperature change.

speckle interferometry technique whereby large telescopes can achieve high resolution of astronomical objects despite the adverse effects of the atmosphere through which light from the object under study must pass. It involves the taking of large numbers of images, each under high magnification and with short exposure times. The pictures are then combined to form the final picture. The technique was introduced by the French astronomer Antoine Labeyrie 1970.

spectator ion in a chemical reaction that takes place in solution, an ion that remains in solution without taking part in the chemical change. For example, in the precipitation of barium sulphate from barium chloride and sodium sulphate, the sodium and chloride ions are spectator ions.

$BaCl_{2\,(aq)} + Na_2SO_{4\,(aq)} = BaSO_{4\,(s)} + 2NaCl_{(aq)}$

spectrometer in physics and astronomy, an instrument used to study the composition of light emitted by a source. The range, or ◊spectrum, of wavelengths emitted by a source depends upon its constituent elements, and may be used to determine its chemical composition.

spectroscopy study of spectra (see ◊spectrum) associated with atoms or molecules in solid, liquid, or gaseous phase. Spectroscopy can be used to identify unknown compounds and is an invaluable tool in science, medicine, and industry (for example, in checking the purity of drugs).

Emission spectroscopy is the study of the characteristic series of sharp lines in the spectrum produced when an ⟡element is heated. Thus an unknown mixture can be analysed for its component elements. Related is *absorption spectroscopy*, dealing with atoms and molecules as they absorb energy in a characteristic way. Again, dark lines can be used for analysis. More detailed structural information can be obtained using *infrared spectroscopy* (concerned with molecular vibrations) or *nuclear magnetic resonance (NMR) spectroscopy* (concerned with interactions between adjacent atomic nuclei).

Supersonic jet laser beam spectroscopy enables the isolation and study of clusters in the gas phase. A laser vaporizes a small sample, which is cooled in helium, and ejected into an evacuated chamber. The jet of clusters expands supersonically, cooling the clusters to near absolute zero, and stabilizing them for study in a ⟡mass spectrometer.

spectrum (plural *spectra*) in physics, an arrangement of frequencies or wavelengths when electromagnetic radiations are separated into their constituent parts. Visible light is part of the ⟡electromagnetic spectrum and most sources emit waves over a range of wavelengths that can be broken up or 'dispersed'; white light can be separated into red, orange, yellow, green, blue, indigo, and violet. The visible spectrum was first studied by Isaac Newton (1642–1727), who showed in 1672 how white light could be broken up into different colours.

There are many types of spectra, both emission and absorption, for radiation and particles, used in ⟡spectroscopy. An incandescent body gives rise to a *continuous spectrum* where the dispersed radiation is distributed uninterruptedly over a range of wavelengths. An element gives a *line spectrum* – one or more bright discrete lines at characteristic wavelengths. Molecular gases give *band spectra* in which there are groups of close-packed lines shaded in one direction of wavelength. In an *absorption spectrum* dark lines or spaces replace the characteristic bright lines of the absorbing medium. The *mass spectrum* of an element is obtained from a mass spectrometer and shows the relative proportions of its constituent ⟡isotopes.

speech recognition or *voice input* in computing, any technique by which a computer can understand ordinary speech. Spoken words are divided into 'frames', each lasting about one-thirtieth of a second, which are converted to a wave form. These are then compared with a series of stored frames to determine the most likely word. Research into speech recognition started in 1938, but the technology did not become sufficiently developed for commercial applications until the late 1980s.

speech synthesis or *voice output* computer-based technology for generating speech. A speech synthesizer is controlled by a computer, which supplies strings of codes representing basic speech sounds (phonemes); together these make up words. Speech-synthesis applications include children's toys, car and aircraft warning systems, and talking books for the blind.

speed the rate at which an object moves. The average speed v of an object may be calculated by dividing the distance s has travelled by the time t taken to do so, and may be expressed as:

$$v = s/t$$

The usual units of speed are metres per second or kilometres per hour.

speed of light speed at which light and other ⟡electromagnetic waves travel through empty space. Its value is 299,792,458 m/186,281 mi per second. The speed of light is the highest speed possible, according to the theory of ⟡relativity, and its value is independent of the motion of its source and of the observer. It is impossible to accelerate any material body to this speed because it would require an infinite amount of energy.

speed of reaction alternative term for ⟡rate of reaction.

speed of sound speed at which sound travels through a medium, such as air or water. In air at a temperature of 0°C/32°F, the speed of sound is 331 m/1,087 ft per second. At higher temperatures, the speed of sound is greater; at 18°C/64°F it is 342 m/1,123 ft per second. It is greater in liquids and solids; for example, in water it is around 1,440 m/4,724 ft per second, depending on the temperature.

speed–time graph graph used to describe the motion of a body by illustrating how its speed or velocity changes with time. The gradient of the graph gives the object's acceleration: if the gradient is zero (the graph is horizontal) then the body is moving with constant speed or uniform velocity; if the gradient is constant, the body is moving with uniform acceleration. The area under the graph gives the total distance travelled by the body.

speleology scientific study of caves, their origin, development, physical structure, flora, fauna, folklore, exploration, mapping, photography, cave-diving, and rescue work. *Potholing*, which involves following the course of underground rivers or streams, has become a popular sport.

sperm or *spermatozoon* in biology, the male ◊gamete of animals. Each sperm cell has a head capsule containing a nucleus, a middle portion containing ◊mitochondria (which provide energy), and a long tail (flagellum). See ◊sexual reproduction.

In most animals, the sperm are motile, and are propelled by a long flagellum, but in some (such as crabs and lobsters) they are non-motile. Sperm cells are produced in the testes (see ◊testis). From there they pass through the sperm ducts via the seminal vesicles and the ◊prostate gland, which produce fluids called semen, that give the sperm cells energy and keep them moving after they leave the body. Hundreds of millions of sperm cells are contained in only a small amount of semen.

Sperm counts have fallen by 50% worldwide since 1940, according to a Danish study 1990. This reduction may be due to increases in pollution as a number of pollutants, including some petroleum by-products and ◊polychlorinated biphenyls, mimic the effects of the female hormone oestrogen on cells.

spermatophore small capsule containing ◊sperm and other nutrients produced in invertebrates, newts, and cephalopods.

spermatophyte in botany, another name for a ◊seed plant.

sperm competition competition between the sperm of rival males within the body of the female. This may involve actual removal of a rival's sperm before mating, for example the penis of a male dragonfly is shaped to scoop out existing sperm; or internal competition between rival sperm cells.

spermicide any cream, jelly, pessary, or other preparation that kills the ◊sperm cells in semen. Spermicides are used for contraceptive purposes, usually in combination with a ◊condom or ◊diaphragm. Sponges impregnated with spermicide have been developed but are not yet in widespread use. Spermicide used alone is only 75% effective in preventing pregnancy.

sperm *Only a single sperm is needed to fertilize a female egg, or ovum. Yet up to 500 million may start the journey towards the egg. Once a sperm has fertilized an egg, the egg's wall cannot be penetrated by other sperm. The unsuccessful sperm die after about three days.*

sphalerite the chief ore of zinc, composed of zinc sulphide with a small proportion of iron, formula $(Zn,Fe)S$. It is brown with a non-metallic lustre unless an appreciable amount of iron is present (up to 26% by weight). Sphalerite usually occurs in ore veins in limestones, where it is often associated with galena. It crystallizes in the cubic system but does not normally form perfect cubes.

sphere in mathematics, a perfectly round object with all points on its surface the same distance from the centre. This distance is the radius of the sphere. For a sphere of radius r, the volume $V = 4/3\pi r^3$ and the surface area $A = 4\pi r^2$.

sphincter ring of muscle found at various points in the alimentary canal, which contracts and relaxes to control the movement of food. The *pyloric sphincter*, at the base of the stomach, controls the release of the gastric contents into the duodenum. After release the sphincter contracts, closing off the stomach.

Spica or *Alpha Virginis* the brightest star in the constellation Virgo and the 16th brightest star in the sky. Spica has a true luminosity of over 1,500 times that of the Sun, and is 140 light years from Earth. It is also a spectroscopic ◊binary star, the components of which orbit each other every four days.

spicules, solar in astronomy, short-lived jets of hot gas in the upper ◊chromosphere of the Sun. Spiky in appearance, they move at high velocities along lines of magnetic force to which they owe their shape, and last for a few minutes each. Spicules appear to disperse material into the ◊corona.

spider any arachnid (eight-legged animal) of the order Araneae. There are about 30,000 known species. Unlike insects, the head and breast are merged to form the cephalothorax, connected to the abdomen by a characteristic narrow waist. There are eight legs, and usually eight simple eyes. On the undersurface of the abdomen are spinnerets, usually six, which exude a viscid fluid. This hardens on exposure to the air to form silky threads, used to make silken egg cases, silk-lined tunnels, or various kinds of webs and snares for catching prey that is then wrapped. The fangs of spiders inject substances to subdue and digest prey, the juices of which are then sucked into the stomach by the spider.

spikelet in botany, one of the units of a grass ◊inflorescence. It comprises a slender axis on which one or more flowers are borne.

spin in physics, the intrinsic ◊angular momentum of a subatomic particle, nucleus, atom, or molecule, which continues to exist even when the particle comes to rest. A particle in a specific energy state has a particular spin, just as it has a particular electric charge and mass. According to ◊quantum theory, this is restricted to discrete and indivisible values, specified by a spin ◊quantum number. Because of its spin, a charged particle acts as a small magnet and is affected by magnetic fields.

spinal cord major component of the ◊central nervous system in vertebrates, encased in the spinal column. It consists of bundles of nerves enveloped in three layers of membrane (the meninges).

spine backbone of vertebrates. In most mammals, it contains 26 small bones called vertebrae, which enclose and protect the spinal

spine *The human spine extends every night during sleep. During the day, the cartilage discs between the vertebrae are squeezed when the body is in a vertical position, standing or sitting, but at night, with pressure released, the discs swell and the spine lengthens by about 8 mm/0.3 in.*

cord (which links the peripheral nervous system to the brain).

The spine articulates with the skull, ribs, and hip bones, and provides attachment for the back muscles. In humans it is made up of individual vertebrae, separated by intervertebral discs. In the adult there are seven cervical *vertebrae* in the neck; 12 thoracic in the upper trunk; five lumbar in the lower back; the sacrum (consisting of five rudimentary vertebrae fused together, joined to the hipbones); and the coccyx (four vertebrae, fused into a tailbone). The human spine has four curves (front to rear), which allow for the increased size of the chest and pelvic cavities, and permit springing, to minimize jolting of the internal organs.

spinel any of a group of 'mixed oxide' minerals consisting mainly of the oxides of magnesium and aluminium, $MgAl_2O_4$ and $FeAl_2O_4$. Spinels crystallize in the cubic system, forming octahedral crystals. They are found in high-temperature igneous and metamorphic rocks. The aluminium oxide spinel contains gem varieties, such as the ruby spinels of Sri Lanka and Myanmar (Burma).

spiracle in insects, the opening of a ◊trachea, through which oxygen enters the body and carbon dioxide is expelled. In cartilaginous fishes (sharks and rays), the same name is given to a circular opening that marks the remains of the first gill slit.

spirits of salts former name for ◊hydrochloric acid.

spirochaete spiral-shaped bacterium. Some spirochaetes are free-living in water, others inhabit the intestines and genital areas of animals. The sexually transmitted disease syphilis is caused by a spirochaete.

spleen organ in vertebrates, part of the reticuloendothelial system, which helps to process ◊lymphocytes. It also regulates the number of red blood cells in circulation by destroying old cells, and stores iron. It is situated on the left side of the body, behind the stomach.

sponge any saclike simple invertebrate of the phylum Porifera, usually marine. A sponge has a hollow body, its cavity lined by cells bearing flagellae, whose whiplike movements keep water circulating, bringing in a stream of food particles. The body walls are strengthened with protein (as in the bath sponge) or small spikes of silica, or a framework of calcium carbonate.

spontaneous combustion burning that is not initiated by the direct application of an external source of heat. A number of materials and chemicals, such as hay and sodium chlorate, can react with their surroundings, usually by oxidation, to produce so much internal heat that combustion results.

spooling in computing, the process in which information to be printed is stored temporarily in a file, the printing being carried out later. It is used to prevent a relatively slow printer from holding up the system at critical times, and to enable several computers or programs to share one printer.

sporangium structure in which spores are produced.

spore small reproductive or resting body, usually consisting of just one cell. Unlike a ◊gamete, it does not need to fuse with another cell in order to develop into a new organism. Spores are produced by the lower plants, most fungi, some bacteria, and certain protozoa. They are generally light and easily dispersed by wind movements.

Plant spores are haploid and are produced by the sporophyte, following ◊meiosis; see ◊alternation of generations.

sporophyte diploid spore-producing generation in the life cycle of a plant that undergoes ◊alternation of generations.

spreadsheet in computing, a program that mimics a sheet of ruled paper, divided into columns and rows. The user enters values in the sheet, then instructs the program to perform some operation on them, such as totalling a column or finding the average of a series of numbers. Highly complex numerical analyses may be built up from these simple steps.

spring device, usually a metal coil, that returns to its original shape after being stretched or compressed. Springs are used in some machines (such as clocks) to store energy, which can be released at a controlled rate. In other machines (such as engines) they are used to close valves.

spring balance instrument for measuring weight that relates the weight of an object to the extent to which it stretches or compresses a vertical spring. According to ◊Hooke's law, the extension or compression will be directly proportional to the weight, providing that the spring is not overstretched. A pointer attached to the spring indicates the weight on a scale, which may be calibrated in newtons (the SI unit of force) for physics experiments, or in grams, kilograms, or pounds (units of mass) for everyday use.

	A	B	C	D	E
1	DESCRIPTION		NUMBER	UNIT COST(£)	COST
2					
3	FLOPPY DISCS		50	0.55	27.5
4	INK RIBBONS		10	2.25	22.5
5	INK JET CARTRIDGES		5	12.5	62.5
6	PAPER (BOX 500 SHEETS)		12	3.75	45
7					
8				TOTAL COST	£ 157.50
9					
10					

spreadsheet *Each cell in a spread may contain an instruction. For instance, the cell E8 (above) contains the instruction to add the figures entered into cells E3–6, and display the results.*

sprite in computing, a graphics object made up of a pattern of ◊pixels (picture elements) defined by a computer programmer. Some ◊high-level languages and ◊applications programs contain routines that allow a user to define the shape, colours, and other characteristics of individual graphics objects. These objects can then be manipulated and combined to produce animated games or graphic screen displays.

Sputnik series of ten Soviet Earth-orbiting satellites. *Sputnik 1* was the first artificial satellite, launched 4 Oct 1957. It weighed 84 kg/185 lb, with a 58 cm/23 in diameter, and carried only a simple radio transmitter which allowed scientists to track it as it orbited Earth. It burned up in the atmosphere 92 days later. Sputniks were superseded in the early 1960s by the Cosmos series.

sq abbreviation for *square* (measure).

SQL (abbreviation of *structured query language*) high-level computer language designed for use with ◊relational databases. Although it can be used by programmers in the same way as other languages, it is often used as a means for programs to communicate with each other. Typically, one program (called the 'client') uses SQL to request data from a database 'server'.

square in geometry, a quadrilateral (four-sided) plane figure with all sides equal and each angle a right angle. Its diagonals bisect each other at right angles. The area A of a square is the length l of one side multiplied by itself ($A = l \times l$).

Also, any quantity multiplied by itself is termed a square, represented by an ◊exponent of power 2; for example, $4 \times 4 = 4^2 = 16$ and $6.8 \times 6.8 = 6.8^2 = 46.24$.

square root in mathematics, a number that when squared (multiplied by itself) equals a given number. For example, the square root of 25 (written $\sqrt{25}$) is $\div 5$, because $5 \times 5 = 25$, and $(-5) \times (-5) = 25$. As an ◊exponent, a square root is represented by $1/2$, for example, $16^{1/2} = 4$.

SRAM (acronym for *s*tatic *r*andom-*a*ccess *m*emory) computer memory device in the form of a silicon chip used to provide ◊immediate access memory. SRAM is faster but more expensive than ◊DRAM (dynamic random-access memory).

stability measure of how difficult it is to move an object from a position of balance or ◊equilibrium with respect to gravity.

An object displaced from equilibrium does not remain in its new position if its weight, acting vertically downwards through its ◊centre of mass, no longer passes through the line of action of the ◊contact force (the force exerted by the surface on which the object is resting), acting vertically upwards through the object's new base. If the lines of action of these two opposite but equal forces do not coincide they will form a couple and create a moment (see ◊moment of a force) that will cause the object either to return to its original rest position or to topple over into another position.

An object in *stable equilibrium* returns to its rest position after being displaced slightly. This form of equilibrium is found in objects that are difficult to topple over; these usually possess a relatively wide base and a low centre of mass – for example, a cone resting on its flat base on a horizontal surface.

An object in **unstable equilibrium** does not remain at rest if displaced, but falls into a new position; it does not return to its original rest position. Objects possessing this form of equilibrium are easily toppled and usually have a relatively small base and a high centre of mass, for example, a cone balancing on its point, or apex, on a horizontal surface.

An object in **neutral equilibrium** stays at rest if it is moved into a new position – neither moving back to its original position nor on any further. This form of equilibrium is found in objects that are able to roll, such as a cone resting on its curved side placed on a horizontal surface.

stabilizer one of a pair of fins fitted to the sides of a ship, especially one governed automatically by a ◊gyroscope mechanism, designed to reduce side-to-side rolling of the ship in rough weather.

stable equilibrium the state of equilibrium possessed by a body that will return to its original rest position if displaced slightly. See ◊stability.

stack in computing, a method of storing data in which the most recent item stored will be the first to be retrieved. The technique is commonly called 'last in, first out'.

stain in chemistry, a coloured compound that will bind to other substances. Stains are used extensively in microbiology to colour microorganisms and in histochemistry to detect the presence and whereabouts in plant and animal tissue of substances such as fats, cellulose, and proteins.

stainless steel widely used ◊alloy of iron, chromium, and nickel that resists rusting. Its chromium content also gives it a high tensile strength. It is used for cutlery and kitchen fittings. Stainless steel was first produced in the UK 1913 and in Germany 1914.

stalactite and stalagmite cave structures formed by the deposition of calcite dissolved in ground water. *Stalactites* grow downwards from the roofs or walls whereas *stalagmites* grow upwards from the cave floor. Growing stalactites and stalagmites may meet to form a continuous column from floor to ceiling.

stamen male reproductive organ of a flower. The stamens are collectively referred to as the ◊androecium. A typical stamen consists of a stalk, or filament, with an anther, the pollen-bearing organ, at its apex, but in some primitive plants, such as *Magnolia*, the stamen may not be markedly differentiated.

standard atmosphere alternative term for ◊atmosphere, a unit of pressure.

standard deviation in statistics, a measure (symbol σ or s) of the spread of data. The deviation (difference) of each of the data items from the mean is found, and their values squared. The mean value of these squares is then calculated. The standard deviation is the square root of this mean.

If n is the number of items of data, x is the value of each item, and $\bar{x}$ is the mean value, the standard deviation $\bar{x}$ may be given by the formula:

$$\sigma = \sqrt{[\Sigma(x-\bar{x})^2/n]}$$

where Σ indicates that the differences between the value of each item of data and the mean should be summed.

standard form method of writing numbers often used by scientists, particularly for very large or very small numbers. The numbers are written with one digit before the decimal point

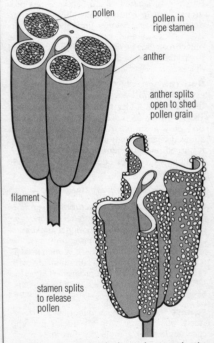

stamen *The stamen is the male reproductive organ of a flower. It has a thin stalk called a filament with an anther at the tip. The anther contains pollen sacs, which split to release tiny grains of pollen.*

and multiplied by a power of 10. The number of digits given after the decimal point depends on the accuracy required. For example, the ◊speed of light is 2.9979×10^8 m/1.8628×10^5 mi per second.

Standard Generalized Markup Language ◊International Standards Organization standard describing how the structure of a text can be identified so that it can be used, probably via ◊filters, in applications such as ◊desktop publishing and ◊electronic publishing.

standard gravity acceleration due to gravity, generally taken as 9.81274 m/32.38204 ft per second per second. See also ◊g scale.

standard illuminant any of three standard light intensities, A, B, and C, used for illumination when phenomena involving colour are measured. A is the light from a filament at 2,848K (2,575°C/4,667°F), B is noon sunlight, and C is normal daylight. B and C are defined with respect to A. Standardization is necessary because colours appear different when viewed in different lights.

standard model in physics, the modern theory of ◊elementary particles and their interactions. According to the standard model, elementary particles are classified as leptons (light particles, such as electrons), ◊hadrons (particles, such as neutrons and protons, that are formed from quarks), and gauge bosons. Leptons and hadrons interact by exchanging ◊gauge bosons, each of which is responsible for a different fundamental force: photons mediate the electromagnetic force, which affects all charged particles; gluons mediate the strong nuclear force, which affects quarks; gravitons mediate the force of gravity; and the weakons (intermediate vector bosons) mediate the weak nuclear force. See also ◊forces, fundamental, ◊quantum electrodynamics, and ◊quantum chromodynamics.

standard temperature and pressure (STP) in chemistry, a standard set of conditions for experimental measurements, to enable comparisons to be made between sets of results. Standard temperature is 0°C and standard pressure 1 atmosphere (101,325 Pa).

standard volume in physics, the volume occupied by one kilogram molecule (the molecular mass in kilograms) of any gas at standard temperature and pressure. Its value is approx 22.414 cubic metres.

standing wave in physics, a wave in which the positions of ◊nodes (positions of zero vibration) and antinodes (positions of maximum vibration) do not move. Standing waves result when two similar waves travel in opposite directions through the same space.

staphylococcus spherical bacterium that occurs in clusters. It is found on the skin and mucous membranes of humans and other animals. It can cause abscesses and systemic infections that may prove fatal.

star luminous globe of gas, mainly hydrogen and helium, which produces its own heat and light by nuclear reactions. Although stars shine for a very long time – many billions of years – they are not eternal, and have been found to change in appearance at different stages in their lives.

Stars are born when nebulae (giant clouds of dust and gas) contract under the influence of gravity. As each new star contracts, the temperature and pressure in its core rises. At about 10 million°C the temperature is hot enough for a nuclear reaction to begin (the fusion of hydrogen nuclei to form helium nuclei); vast amounts of energy are released, contraction stops, and the star begins to shine. Stars at this stage are called *main-sequence stars*; the Sun is such a star and is expected to remain at this stage for the next 5 billion years. Their surface temperatures range from 2,000°C/3,600°F to above 30,000°C/54,000°F and the corresponding colours range from red to blue-white.

The smallest mass possible for a star is about 8% that of the Sun (80 times the mass of the planet Jupiter), otherwise nuclear reactions do not occur. Objects with less than this critical mass shine only dimly, and are termed *brown dwarfs*.

When all the hydrogen at the core of a main-sequence star has been converted into helium, the star swells to become a *red giant*, about 100 times its previous size and with a cooler, redder surface. When, after this brief stage, the star can produce no more nuclear energy, its outer layers drift off into space to form a planetary nebula, and its core collapses in on itself to form a small and very dense body called a *white dwarf*. Eventually the white dwarf fades away, leaving a nonluminous *dark body*.

Some very large main-sequence stars do not end their lives as white dwarfs – they pass through their life cycle quickly, becoming red ◊supergiants that eventually explode into brilliant *supernovae*. Part of the core remaining after such an explosion may collapse to form a small superdense star, which consists almost entirely of neutrons and is therefore called a *neutron star*. Neutron stars, called *pulsars*, spin very quickly, giving off pulses of radio waves (rather as a lighthouse gives off flashes of light). If the collapsing core of the supernova is very massive it does not form a neutron star; instead it forms a ***black hole***, a region so dense that its gravity not only draws in all nearby matter but also all radiation, including its own light. See also ◊binary star, ◊Hertzsprung–Russell diagram, and ◊variable star.

starch widely distributed, high-molecular-mass ◊carbohydrate, produced by plants as a food store; main dietary sources are cereals, legumes, and tubers, including potatoes. It consists of varying proportions of two ◊glucose polymers (◊polysaccharides): straight-chain (amylose) and branched (amylopectin) molecules.

Purified starch is a white powder used to stiffen textiles and paper and as a raw material for making various chemicals. It is used in the food industry as a thickening agent. Chemical treatment of starch gives rise to a range of 'modified starches' with varying properties. Hydrolysis (splitting) of starch by acid or enzymes generates a variety of 'glucose syrups' or 'liquid glucose' for use in the food industry. Complete hydrolysis of starch with acid generates the ◊monosaccharide glucose only. Incomplete hydrolysis or enzymic hydrolysis yields a mixture of glucose, maltose and nonhydrolysed fractions called dextrins.

star cluster group of related stars, usually held together by gravity. Members of a star cluster are thought to form together from one large cloud of gas in space. ***Open clusters*** such as the Pleiades contain from a dozen to many hundreds of young stars, loosely scattered over several light years. ◊Globular clusters are larger and much more densely packed, containing perhaps 100,000 old stars.

state change in science, a change in the physical state (solid, liquid, or gas) of a material. For instance, melting, boiling, evaporation, and their opposites, solidification and condensation, are changes of state. The former set of changes are brought about by heating or decreased pressure; the latter by cooling or increased pressure. In the unusual change of state called ***sublimation***, a solid changes directly to a gas without passing through the liquid state. For example, solid carbon dioxide (dry ice) sublimes to carbon dioxide gas.

states of matter forms (solid, liquid, or gas) in which material can exist. Whether a material is solid, liquid, or gas depends on its temperature and the pressure on it. The transition between states takes place at definite temperatures, called melting point and boiling point.

◊Kinetic theory describes how the state of a material depends on the movement and arrangement of its atoms or molecules. A hot ionized gas or ◊plasma is often called the fourth state of matter, but liquid crystals, ◊colloids, and glass also have a claim to this title.

state symbol symbol used in chemical equations to indicate the physical state of the substances present. The symbols are: (s) for solid, (l) for liquid, (g) for gas, and (aq) for aqueous.

static electricity ◊electric charge that is stationary, usually acquired by a body by means of electrostatic induction or friction. Rubbing different materials can produce static electricity, as seen in the sparks produced on combing one's hair or removing a nylon shirt. In some processes static electricity is useful, as in paint spraying where the parts to be sprayed are charged with electricity of opposite polarity to that on the paint droplets, and in ◊xerography.

statics branch of mechanics concerned with the behaviour of bodies at rest and forces in equilibrium, and distinguished from ◊dynamics.

statistical mechanics branch of physics in which the properties of large collections of particles are predicted by considering the motions of the constituent particles.

statistics branch of mathematics concerned with the collection and interpretation of data. For example, to determine the ◊mean age of the children in a school, a statistically acceptable answer might be obtained by calculating an average based on the ages of a representative sample, consisting, for example, of a random tenth of the pupils from each class. ◊Probability is the branch of statistics dealing with predictions of events.

staurolite silicate mineral, $(Fe,Mg)_2(Al,Fe)_9Si_4O_{20}(OH)_2$. It forms brown crystals that may be twinned in the form of a cross. It is a useful indicator of medium grade (moderate temperature and pressure) in metamorphic rocks formed from clay sediments.

steady-state theory in astronomy, a rival theory to that of the ◊Big Bang, which claims that the universe has no origin but is expanding because new matter is being created continuously throughout the universe. The theory was proposed 1948 by Hermann Bondi (1919–),

Thomas Gold (1920–), and Fred Hoyle (1915–), but was dealt a severe blow in 1965 by the discovery of ♦cosmic background radiation (radiation left over from the formation of the universe) and is now largely rejected.

steam in chemistry, a dry, invisible gas formed by vaporizing water. The visible cloud that normally forms in the air when water is vaporized is due to minute suspended water particles. Steam is widely used in chemical and other industrial processes and for the generation of power.

steam engine engine that uses the power of steam to produce useful work. It was the principal power source during the British Industrial Revolution in the 18th century. The first successful steam engine was built 1712 by English inventor Thomas Newcomen, and it was developed further by Scottish mining engineer James Watt from 1769 and by English mining engineer Richard Trevithick, whose high-pressure steam engine 1802 led to the development of the steam locomotive.

stearic acid $CH_3(CH_2)_{16}COOH$ saturated long-chain ♦fatty acid, soluble in alcohol and ether but not in water. It is found in many fats and oils, and is used to make soap and candles and as a lubricant. The salts of stearic acid are called stearates.

stearin mixture of stearic and palmitic acids, used to make soap.

steel alloy or mixture of iron and up to 1.7% carbon, sometimes with other elements, such as manganese, phosphorus, sulphur, and silicon. The USA, Russia, Ukraine, and Japan are the main steel producers. Steel has innumerable uses, including ship and automobile manufacture, skyscraper frames, and machinery of all kinds.

Steels with only small amounts of other metals are called *carbon steels*. These steels are far stronger than pure iron, with properties varying with the composition. *Alloy steels* contain greater amounts of other metals. Low-alloy steels have less than 5% of the alloying material; high-alloy steels have more. Low-alloy steels containing up to 5% silicon with relatively little carbon have a high electrical resistance and are used in power transformers and motor or generator cores, for example. *Stainless steel* is a high-alloy steel containing at least 11% chromium. Steels with up to 20% tungsten are very hard and are used in high-speed cutting tools. About 50% of the world's steel is now made from scrap.

Steel is produced by removing impurities, such as carbon, from raw or pig iron, produced by a ♦blast furnace.

Stefan–Boltzmann law in physics, a law that relates the energy, E, radiated away from a perfect emitter (a ♦black body), to the temperature, T, of that body. It has the form $M = \sigma T^4$, where M is the energy radiated per unit area per second, T is the temperature, and σ is the *Stefan–Boltzmann constant*. Its value is 5.6697×10^{-8} W m^{-2} K^{-4}. The law was derived by Austrian physicists Joseph Stefan and Ludwig Boltzmann.

stem main supporting axis of a plant that bears the leaves, buds, and reproductive structures; it may be simple or branched. The plant stem usually grows above ground, although some grow underground, including ♦rhizomes, ♦corms, ♦rootstocks, and ♦tubers. Stems contain a continuous vascular system that conducts water and food to and from all parts of the plant.

stepper motor electric motor that can be precisely controlled by signals from a computer. The motor turns through a precise angle each time it receives a signal pulse from the computer. By varying the rate at which signal pulses are produced, the motor can be run at different speeds or turned through an exact angle and then stopped. Switching circuits can be constructed to allow the computer to reverse the direction of the motor.

step rocket another term for ♦multistage rocket.

steradian SI unit (symbol sr) of measure of solid (three-dimensional) angles, the three-dimensional equivalent of the ♦radian. One steradian is the angle at the centre of a sphere when an area on the surface of the sphere equal to the square of the sphere's radius is joined to the centre.

sterilization any surgical operation to terminate the possibility of reproduction. In women, this is normally achieved by sealing or tying off the ♦Fallopian tubes (tubal ligation) so that fertilization can no longer take place. In men, the transmission of sperm is blocked by vasectomy.

sterilization the killing or removal of living organisms such as bacteria and fungi. A sterile environment is necessary in medicine, food processing, and some scientific experiments. Methods include heat treatment (such as boiling), the use of chemicals (such as disinfectants), irradiation with gamma rays, and filtration. See also ♦asepsis.

sterling silver alloy containing 925 parts of silver and 75 parts of copper. The copper hardens the silver, making it more useful.

sternum large flat bone at the front of the chest, joined to the ribs. It gives protection to the heart and lungs. During open heart surgery if must be split to give access to the thorax.

steroid in biology, any of a group of cyclic, unsaturated alcohols (lipids without fatty acid components), which, like sterols, have a complex molecular structure consisting of four carbon rings. Steroids include the sex hormones, such as ♦testosterone, the corticosteroid hormones produced by the ♦adrenal gland, bile acids, and ♦cholesterol. The term is commonly used to refer to ♦anabolic steroid. In medicine, synthetic steroids are used to treat a wide range of conditions.

sterol any of a group of solid, cyclic, unsaturated alcohols, with a complex structure that includes four carbon rings; cholesterol is an example. Steroids are derived from sterols.

stigma in a flower, the surface at the tip of a ♦carpel that receives the ♦pollen. It often has short outgrowths, flaps, or hairs to trap pollen and may produce a sticky secretion to which the grains adhere.

stimulus any agency, such as noise, light, heat, or pressure, that can be detected by an organism's receptors.

stipule outgrowth arising from the base of a leaf or leaf stalk in certain plants. Stipules usually occur in pairs or fused into a single semicircular structure.

stokes c.g.s. unit (symbol St) of kinematic viscosity (a liquid's resistance to flow).

STOL (acronym for *short takeoff and landing*) aircraft fitted with special devices on the wings (such as sucking flaps) that increase aerodynamic lift at low speeds.

stolon in botany, a type of ♦runner.

stoma (plural *stomata*) in botany, a pore in the epidermis of a plant. Each stoma is surrounded by a pair of guard cells that are crescent-shaped when the stoma is open but can collapse to an oval shape, thus closing off the opening between them. Stomata allow the exchange of carbon dioxide and oxygen (needed for ♦photosynthesis and ♦respiration) between the internal tissues of the plant and the outside atmosphere. They are also the main route by which water is lost from the plant, and they can be closed to conserve water, the movements being controlled by changes in turgidity of the guard cells.

stomach the first cavity in the digestive system of animals. In mammals it is a bag of muscle situated just below the diaphragm. Food enters it from the oesophagus, is digested by the acid and ♦enzymes secreted by the stomach lining, and then passes into the duodenum. Some plant-eating mammals have multichambered stomachs that harbour bacteria in one of the chambers to assist in the digestion of ♦cellulose. The gizzard is part of the stomach in birds.

stone (plural *stone*) imperial unit (abbreviation st) of mass. One stone is 14 pounds (6.35 kg).

STP abbreviation for ♦standard temperature and pressure.

strain in the science of materials, the extent to which a body is distorted when a deforming force (stress) is applied to it. It is a ratio of the extension or compression of that body (its length, area, or volume) to its original dimensions (see ♦Hooke's law). For example, linear strain is the ratio of the change in length of a body to its original length.

stoma *The stomata, tiny openings in the epidermis of a plant, are surrounded by pairs of crescent-shaped cells, called guard cells. The guard cells open and close the stoma by changing shape.*

STRING VARIATIONS OF

structure

detail of stomach wall

stomach *The human stomach can hold about 1.5 l/2.6 pt of liquid. The digestive juices are acidic enough to dissolve metal. To avoid damage, the cells of the stomach lining are replaced quickly –500,000 cells are replaced every minute, and the whole stomach lining every three days.*

stratigraphy branch of geology that deals with the sequence of formation of ◊sedimentary rock layers and the conditions under which they were formed. Its basis was developed by William Smith, a British canal engineer. Stratigraphy in the interpretation of archaeological excavations provides a relative chronology for the levels and the artefacts within them. The basic principle of superimposition establishes that upper layers or deposits have accumulated later in time than the lower ones.

stratosphere that part of the atmosphere 10–40 km/6–25 mi from the Earth's surface, where the temperature slowly rises from a low of –55°C/–67°F to around 0°C/32°F. The air is rarefied and at around 25 km/15 mi much ◊ozone is concentrated.

streamlining shaping a body so that it offers the least resistance when travelling through a medium such as air or water. Aircraft, for example, must be carefully streamlined to reduce air resistance, or ◊drag.

strength of acids and bases in chemistry, the ability of ◊acids and ◊bases to dissociate in solution with water, and hence to produce a low or high ◊pH respectively.

A strong acid is fully dissociated in aqueous solution, whereas a weak acid is only partly dissociated. Since the ◊dissociation of acids generates hydrogen ions, a solution of a strong acid will have a high concentration of hydrogen ions and therefore a low pH. A strong base will have a high pH, whereas a weaker base will not dissociate completely and will have a pH of nearer 7.

stress and strain in the science of materials, measures of the deforming force applied to a body (stress) and of the resulting change in its shape (◊strain). For a perfectly elastic material, stress is proportional to strain (◊Hooke's law).

stress protein one of a group of proteins produced by a cell when it is under stress. Stress proteins are created in response to a wide variety of stresses including illness, high temperatures, alcohol, and poisons. They act as a buffer for the cell and repair damage.

stridulatory organs in insects, organs that produce sound when rubbed together. Crickets rub their wings together, but grasshoppers rub a hind leg against a wing. Stridulation is thought to be used for attracting mates, but may also serve to mark territory.

string in computing, a group of characters manipulated as a single object by the computer. In its simplest form a string may consist of a single letter or word – for example, the single word SMITH might be established as a string for processing by a computer. A string can also consist of a combination of words, spaces, and numbers – for example, 33 HIGH STREET ANYTOWN ALLSHIRE could be established as a single string.

string theory mathematical theory developed in the 1980s to explain the behaviour of ◊elementary particles; see ◊superstring theory.

string, vibrations of the standing waves set up in a stretched string or wire when it is plucked, or stroked with a bow. They are formed by the reflection of ◊progressive waves at the fixed ends of the string. Waves of many different ◊frequencies can be established on a string at the same time. Those that match the natural frequencies of the string will, by a process called ◊resonance, produce large-amplitude vibrations. The vibration of lowest frequency is called the ◊fundamental vibration; vibrations of frequencies that are multiples of the fundamental frequency are called harmonics.

strobilus in botany, a reproductive structure found in most ◊gymnosperms and some ◊pteridophytes, notably the club mosses. In conifers the strobilus is commonly known as a ◊cone.

stroboscope instrument for studying continuous periodic motion by using light flashing at the same frequency as that of the motion; for example, rotating machinery can be optically 'stopped' by illuminating it with a stroboscope flashing at the exact rate of rotation.

stromatolite mound produced in shallow water by mats of algae that trap mud particles. Another mat grows on the trapped mud layer and this traps another layer of mud and so on. The stromatolite grows to heights of a metre or so. They are uncommon today but their fossils are among the earliest evidence for living things – over 2,000 million years old.

strong nuclear force one of the four fundamental ◊forces of nature, the other three being the electromagnetic force, gravity, and the weak nuclear force. The strong nuclear force was first described by Japanese physicist Hideki Yukawa 1935. It is the strongest of all the forces, acts only over very small distances (within the nucleus of the atom), and is responsible for binding together ◊quarks to form ◊hadrons, and for binding together protons and neutrons in the atomic nucleus. The particle that is the carrier of the strong nuclear force is the ◊gluon, of which there are eight kinds, each with zero mass and zero charge.

strontium soft, ductile, pale-yellow, metallic element, symbol Sr, atomic number 38, relative atomic mass 87.62. It is one of the ◊alkaline-earth metals, widely distributed in small quantities only as a sulphate or carbonate. Strontium salts burn with a red flame and are used in fireworks and signal flares.

structured programming in computing, the process of writing a program in small, independent parts. This makes it easier to control a program's development and to design and test its individual component parts. Structured programs are built up from units called *modules*, which normally correspond to single ◊procedures or ◊functions. Some programming languages, such as PASCAL and Modula-2, are better suited to structured programming than others.

strychnine $C_{21}H_{22}O_2N_2$ bitter-tasting, poisonous alkaloid. It is a poison that causes violent muscular spasms, and is usually obtained by powdering the seeds of plants of the genus *Strychnos* (for example *S. nux vomica*). Curare is a related drug.

style in flowers, the part of the ◊carpel bearing the ◊stigma at its tip. In some flowers it is very short or completely lacking, while in others it may be long and slender, positioning the stigma in the most effective place to receive the pollen.

subatomic particle in physics, a particle that is smaller than an atom. Such particles may be indivisible ◊elementary particles, such as the ◊electron and ◊quark, or they may be composites, such as the ◊proton, ◊neutron, and ◊alpha particle. See also ◊particle physics.

subduction zone region where two plates of the Earth's rigid lithosphere collide, and one plate descends below the other into the semi-liquid asthenosphere. Subduction occurs along ocean trenches, most of which encircle the Pacific Ocean; portions of the ocean plate slide beneath other plates carrying continents.

sublimation in chemistry, the conversion of a solid to vapour without passing through the liquid phase.

Some substances that do not sublime at atmospheric pressure can be made to do so at low pressures. This is the principle of freeze-drying, during which ice sublimes at low pressure.

subroutine in computing, a small section of a program that is executed ('called') from another part of the program. Subroutines provide a method of performing the same task at more than one point in the program, and also of separating the details of a program from its main logic. In some computer languages, subroutines are similar to ◊functions or ◊procedures.

substitution reaction in chemistry, the replacement of one atom or ◊functional group in an organic molecule by another.

substrate in biochemistry, a compound or mixture of compounds acted on by an enzyme. The term also refers to a substance such as ◊agar that provides the nutrients for the metabolism of microorganisms. Since the enzyme systems of microorganisms regulate their metabolism, the essential meaning is the same.

succession in ecology, a series of changes that occur in the structure and composition of the vegetation in a given area from the time it is first colonized by plants (*primary succession*), or after it has been disturbed by fire, flood, or clearing (*secondary succession*).

succulent plant thick, fleshy plant that stores water in its tissues; for example, cacti and stonecrops *Sedum*. Succulents live either in areas where water is very scarce, such as deserts, or in places where it is not easily obtainable because of the high concentrations of salts in the soil, as in salt marshes. Many desert plants are ◊xerophytes.

suckering in plants, reproduction by new shoots (suckers) arising from an existing root system rather than from seed. Plants that produce suckers include elm, dandelion, and members of the rose family.

sucrase enzyme capable of digesting sucrose into its constituent molecules of glucose and fructose.

sucrose or *cane sugar* or *beet sugar* $C_{12}H_{22}O_{10}$ a sugar found in the pith of sugar cane and in sugar beets. It is popularly known as ◊sugar.

sulfate US spelling of ◊sulphate.

sulfide US spelling of ◊sulphide.

sulfite US spelling of ◊sulphite.

sulfur US spelling of ◊sulphur.

sulphate SO_4^{2-} salt or ester derived from sulphuric acid. Most sulphates are water soluble (the exceptions are lead, calcium, strontium, and barium sulphates), and require a very high temperature to decompose them.

sulphide compound of sulphur and another element in which sulphur is the more electronegative element (see ◊electronegativity). Sulphides occur in a number of minerals. Some of the more volatile sulphides have extremely unpleasant odours (hydrogen sulphide smells of bad eggs).

sulphite SO_3^{2-} salt or ester derived from sulphurous acid.

sulphonamide any of a group of compounds containing the chemical group sulphonamide (SO_2NH_2) or its derivatives, which were, and still are in some cases, used to treat bacterial diseases. Sulphadiazine ($C_{10}H_{10}N_4O_2S$) is an example.

sulphur brittle, pale-yellow, nonmetallic element, symbol S, atomic number 16, relative atomic mass 32.064. It occurs in three allotropic forms: two crystalline (called rhombic and monoclinic, following the arrangements of the atoms within the crystals) and one amorphous. It burns in air with a blue flame and a stifling odour. Insoluble in water but soluble in carbon disulphide, it is a good electrical insulator. Sulphur is widely used in the manufacture of sulphuric acid (used to treat phosphate rock to make fertilizers) and in making paper, matches, gunpowder and fireworks, in vulcanizing rubber, and in medicines.

It is found abundantly in nature in volcanic regions combined with both metals and non-metals, and also in its elemental form as a crystalline solid. insecticides. Between 20 and 50 million tonnes of sulphur are returned from the oceans to the atmosphere every year in the form of dimethyl sulphide (DMS), the gas that gives sea air its bracing smell. DMS is a breakdown product of a salt produced by marine algae to maintain their osmotic balance (see ◊osmosis). Human activity releases about 80 million tonnes of sulphur.

sulphur dioxide SO_2 pungent gas produced by burning sulphur in air or oxygen. It is widely used for disinfecting food vessels and equipment, and as a preservative in some food products. It occurs in industrial flue gases and is a major cause of ◊acid rain.

sulphuric acid or *oil of vitriol* H_2SO_4 a dense, viscous, colourless liquid that is extremely corrosive. It gives out heat when added to water and can cause severe burns. Sulphuric acid is used extensively in the chemical industry, in the refining of petrol, and in the manufacture of fertilizers, detergents, explosives, and dyes. It forms the acid component of car batteries.

sulphurous acid H_2SO_3 solution of sulphur dioxide (SO_2) in water. It is a weak acid.

sulphur trioxide SO_3 colourless solid prepared by reacting sulphur dioxide and oxygen in the presence of a vanadium(V) oxide catalyst in the ◊contact process. It reacts violently with water to give sulphuric acid.

Sun the ◊star at the centre of the Solar System. Its diameter is 1,392,000 km/865,000 mi; its temperature at the surface is about 5,800K (5,530°C/9,980°F), and at the centre 15,000,000K (15,000,000°C/27,000,000°F). It is composed of about 70% hydrogen and 30% helium, with other elements making up less

Sun *The internal structure of the Sun. The visible surface of the Sun is called the photosphere. Above the photosphere lies the chromosphere and the corona. The Sun's energy is generated in the central core, where 4 million tonnes/3.9 million tons of material is converted to energy each second. The energy can take 10 million years to reach the surface from the centre.*

than 1%. The Sun's energy is generated by nuclear fusion reactions that turn hydrogen into helium at its centre. The gas core is far denser than mercury or lead on Earth. The Sun is about 4.7 billion years old, with a predicted lifetime of 10 billion years.

sun-and-planet gear another term for ◊epicyclic gear.

sunspot dark patch on the surface of the Sun, actually an area of cooler gas, thought to be caused by strong magnetic fields that block the outward flow of heat to the Sun's surface. Sunspots consist of a dark central *umbra*, about 4,000K (3,700°C/6,700°F), and a lighter surrounding *penumbra*, about 5,500K (5,200°C/9,400°F). They last from several days to over a month, ranging in size from 2,000 km/1,250 mi to groups stretching for over 100,000 km/62,000 mi. The number of sunspots visible at a given time varies from none to over 100 in a cycle averaging 11 years. They are more common during active periods in the Sun's magnetic cycle, when they are sometimes accompanied by nearby ◊flares.

Sunyaev-Zel'dovich effect slight dip in the ◊cosmic background radiation when observed through the gas that surrounds clusters of galaxies.

superactinide any of a theoretical series of superheavy, radioactive elements, starting with atomic number 113, that extend beyond the ◊transactinide series in the periodic table. They do not occur in nature and none has yet been synthesized.

supercomputer the fastest, most powerful type of computer, capable of performing its basic operations in picoseconds (thousand-billionths of a second), rather than nanoseconds (billionths of a second), like most other computers.

Of the world's 500 most powerful supercomputers 232 are in the USA, 109 in Japan, and 140 in Europe with 23 in Britain. The most powerful is the Thinking Machines CM-5 in Los Alamos National Laboratory, New Mexico. It is capable of making 131 billion calculations per second.

superconductivity in physics, increase in electrical conductivity at low temperatures. The resistance of some metals and metallic compounds decreases uniformly with decreasing temperature until at a critical temperature (the superconducting point), within a few degrees of absolute zero (0 K/–273.16°C/–459.67°F), the resistance suddenly falls to zero. The phenomenon was discovered by Dutch scientist Heike Kamerlingh-Onnes (1853–1926) in 1911.

In this superconducting state, an electric current will continue indefinitely after the magnetic field has been removed, provided that the material remains below the superconducting point. In 1986 IBM researchers achieved superconductivity with some ceramics at –243°C/–405°F; Paul Chu at the University of Houston, Texas, achieved superconductivity at –179°C/–290°F, a temperature that can be sustained using liquid nitrogen. In 1992, two Japanese researchers developed a material which becomes superconducting at around –103°C/–153°F.

supercooling in physics, the lowering in temperature of a ◊saturated solution without crystallization taking place, forming a supersaturated solution. Usually crystallization rapidly follows the introduction of a small (seed) crystal or agitation of the supercooled solution.

supercritical fluid fluid that combines the properties of a gas and a liquid, see ◊fluid, supercritical.

superfluid fluid that flows without viscosity or friction and has a very high thermal conductivity. Liquid helium at temperatures below 2K (−271°C/−456°F) is a superfluid: it shows unexpected behaviour; for instance, it flows uphill in defiance of gravity and, if placed in a container, will flow up the sides and escape.

supergiant the largest and most luminous type of star known, with a diameter of up to 1,000 times that of the Sun and absolute magnitudes of between −5 and −9.

superheterodyne receiver the most widely used type of radio receiver, in which the incoming signal is mixed with a signal of fixed frequency generated within the receiver circuits. The resulting signal, called the intermediate-frequency (i.f.) signal, has a frequency between that of the incoming signal and the internal signal. The intermediate frequency is near the optimum frequency of the amplifier to which the i.f. signal is passed. This arrangement ensures greater gain and selectivity. The superheterodyne system is also used in basic television receivers.

superior planet planet that is farther away from the Sun than the Earth is: that is, Mars, Jupiter, Saturn, Uranus, Neptune, and Pluto.

supernova the explosive death of a star, which temporarily attains a brightness of 100 million Suns or more, so that it can shine as brilliantly as a small galaxy for a few days or weeks. Very approximately, it is thought that a supernova explodes in a large galaxy about once every 100 years. Approximately 50% of supernovae remain undetected because of obscuring by interstellar dust.

Type I supernovae are thought to occur in ◊binary star systems in which gas from one star falls on to a ◊white dwarf, causing it to explode. *Type II* supernovae occur in stars ten or more times as massive as the Sun, which suffer runaway internal nuclear reactions at the ends of their lives, leading to explosions. These are thought to leave behind ◊neutron stars and ◊black holes. Gas ejected by such an explosion causes an expanding radio source, such as the ◊Crab nebula. Supernovae are thought to be the main source of elements heavier than hydrogen and helium.

Supernova 1993J, a Type II supernova, was observed March 1993 in the galaxy M81, 11 million light years away. The explosion occurred in a red ◊supergiant (see ◊red giant) producing one of the 10 brightest supernovae to be seen in modern times.

superphosphate phosphate fertilizer made by treating apatite with sulphuric or phosphoric acid. The commercial mixture contains largely monocalcium phosphate. Single-superphosphate obtained from apatite and sulphuric acid contains 16–20% available phosphorus, as P_2O_5; triple-superphosphate, which contains 45–50% phosphorus, is made by treating apatite with phosphoric acid.

supersaturation in chemistry, the state of a solution that has a higher concentration of ◊solute than would normally be obtained in a ◊saturated solution.

supersonic speed speed greater than that at which sound travels, measured in ◊Mach numbers. In dry air at 0°C/32°F, sound travels at about 1,170 kph/727 mph, but decreases its speed with altitude until, at 12,000 m/39,000 ft, it is only 1,060 kph/658 mph.

superstring theory in physics, a mathematical theory developed in the 1980s to explain the properties of ◊elementary particles and the forces between them (in particular, gravity and the nuclear forces) in a way that combines ◊relativity and ◊quantum theory.

In string theory, the fundamental objects in the universe are not pointlike particles but extremely small stringlike objects. These objects exist in a universe of ten dimensions, although, for reasons not yet understood, only three space dimensions and one dimension of time are discernible.

supersymmetry in physics, a theory that relates the two classes of elementary particle, the ◊fermions and the ◊bosons. According to supersymmetry, each fermion particle has a boson partner particle, and vice versa. It has not been possible to marry up all the known fermions with the known bosons, and so the theory postulates the existence of other, as yet undiscovered fermions, such as the photinos (partners of the photons), gluinos (partners of the gluons), and gravitinos (partners of the gravitons). Using these ideas, it has become possible to develop a theory of gravity – called supergravity that extends Einstein's work and considers the gravitational, nuclear, and electromagnetic forces to be manifestations of an underlying superforce. Supersymmetry has been incorporated into the ◊superstring theory, and appears to be a crucial ingredient in the 'theory of everything' sought by scientists.

support environment in computing, a collection of programs (◊software) used to help people design and write other programs. At its simplest, this includes a ◊text editor (word-processing software) and a ◊compiler for translating programs into executable form; but it can also include interactive debuggers for helping to locate faults, data dictionaries for keeping track of the data used, and rapid prototyping tools for producing quick, experimental mock-ups of programs.

suprarenal gland alternative name for the ◊adrenal gland.

surd expression containing the root of an ◊irrational number that can never be exactly expressed – for example, $\sqrt{3} = 1.732050808...$.

surface area the area of the outside surface of a solid.

surface-area-to-volume ratio the ratio of an animal's surface area (the area covered by its skin) to its total volume. This is high for small animals, but low for large animals such as elephants.

The ratio is important for endothermic (warm-blooded) animals because the amount of heat lost by the body is proportional to its surface area, whereas the amount generated is proportional to its volume. Very small birds and mammals, such as hummingbirds and shrews, lose a lot of heat and need a high intake of food to maintain their body temperature. Elephants, on the other hand, are in danger of overheating, which is why they have no fur.

surface tension in physics, the property that causes the surface of a liquid to behave as if it were covered with a weak elastic skin; this is why a needle can float on water. It is caused by the exposed surface's tendency to contract to the smallest possible area because of unequal cohesive forces between ◊molecules at the surface. Allied phenomena include the formation of droplets, the concave profile of a meniscus, and the ◊capillary action by which water soaks into a sponge.

surfactant (contraction of *surface-active agent*) substance added to a liquid in order to increase its wetting or spreading properties. Detergents are examples.

surgical spirit ◊ethanol to which has been added a small amount of methanol to render it unfit to drink. It is used to sterilize surfaces and to cleanse skin abrasions and sores.

susceptibility in physics, ratio of the intensity of magnetization produced in a material to the intensity of the magnetic field to which the material is exposed. It measures the extent to which a material is magnetized by an applied magnetic field. ◊Diamagnetic materials have small negative susceptibilities; ◊paramagnetic materials have small positive susceptibilities; ◊ferromagnetic materials have large positive susceptibilities. See also ◊magnetism.

suspension mixture consisting of small solid particles dispersed in a liquid or gas, which will settle on standing. An example is milk of magnesia, which is a suspension of magnesium hydroxide in water.

suspensory ligament in the ◊eye, a ring of fibre supporting the lens. The ligaments attach to the ciliary muscles, the circle of muscle mainly responsible for changing the shape of the lens during ◊accommodation. If the ligaments are put under tension, the lens becomes flatter, and therefore able to focus on objects in the far distance.

sweat gland ◊gland within the skin of mammals that produces surface perspiration. In primates, sweat glands are distributed over the whole body, but in most other mammals they are more localized; for example, in cats and dogs, they are restricted to the feet and around the face.

sweetener any chemical that gives sweetness to food. Caloric sweeteners are various forms of ◊sugar; noncaloric, or artificial, sweeteners are used by dieters and diabetics and provide neither energy nor bulk. Questions have been raised about the long-term health effects from several artificial sweeteners.

swim bladder thin-walled, air-filled sac found between the gut and the spine in bony fishes. Air enters the bladder from the gut or from surrounding capillaries (see ◊capillary), and changes of air pressure within the bladder maintain buoyancy whatever the water depth.

syenite grey, crystalline, plutonic (intrusive) ◊igneous rock, consisting of feldspar and hornblende; other minerals may also be present, including small amounts of quartz.

symbiosis any close relationship between two organisms of different species, and one where both partners benefit from the association. A well-known example is the pollination relationship between insects and flowers, where the insects feed on nectar and carry pollen from one flower to another. This is sometimes known as ◊mutualism.

surface area of a **cube**
(faces are identical)
= 6 × area of each surface
= $6l^2$

surface area of a **cuboid**
(opposite faces are identical)
= area of two end faces = area of two sides
 + area of top and base
= $2lh + 2hb + 2lb$
= $2(lh + hb \times lb)$

surface area of a **cylinder**
= area of a curved surface + area of top
 and base
= 2π × (radius of cross-section × height)
 + 2π (radius of cross-section)^2
= $2\pi rh + 2\pi r^2$
= $2\pi r(h + r)$

surface area of a **cone**
= area of a curved surface + area base
= π × (radius of cross-section × slant height)
 + π (radius of cross-section)^2
= $\pi rl + \pi r^2$
= $\pi r(l + r)$

surface area of a **sphere**
= 4π × radius^2
= $4\pi r^2$

surface area *Surface area of common three-dimensional shapes.*

symbol, chemical letter or letters used to represent a chemical element, usually derived from the beginning of its English or Latin name. Symbols derived from English include B, boron; C, carbon; Ba, barium; and Ca, calcium. Those derived from Latin include Na, sodium (Latin *natrium*); Pb, lead (Latin *plumbum*); and Au, gold (Latin *aurum*).

symbolic address in computing, a symbol used in ◊assembly language programming to represent the binary ◊address of a memory location.

symbolic processor computer purpose-built to run so-called symbol-manipulation programs rather than programs involving a great deal of numerical computation. They exist principally for the ◊artificial intelligence language ◊LISP, although some have also been built to run ◊PROLOG.

symmetry exact likeness in shape about a given line (axis), point, or plane. A figure has symmetry if one half can be rotated and/or reflected onto the other. (Symmetry preserves length, angle, but not necessarily orientation.) In a wider sense, symmetry exits if a change in the system leaves the essential features of the system unchanged; for example, reversing the sign of electric charges does not change the electrical behaviour of an arrangement of charges.

synapse junction between two ◊nerve cells, or between a nerve cell and a muscle (a neuromuscular junction), across which a nerve impulse is transmitted. The two cells are separated by a narrow gap called the *synaptic cleft*. The gap is bridged by a chemical ◊neurotransmitter, released by the nerve impulse.

The threadlike extension, or ◊axon, of the transmitting nerve cell has a slightly swollen terminal point, the *synaptic knob*. This forms one half of the synaptic junction and houses membrane-bound vesicles, which contain a chemical neurotransmitter. When nerve impulses reach the knob, the vesicles release the transmitter and this flows across the gap and binds itself to special receptors on the receiving cell's membrane. If the receiving cell is a nerve cell, the other half of the synaptic junction will be one or more extensions called ◊dendrites; these will be stimulated by the neurotransmitter to set up an impulse, which will then be conducted along the length of the nerve cell and on to its own axons. If the receiving cell is a muscle cell, it will be stimulated by the neurotransmitter to contract.

synchrotron particle ◊accelerator in which particles move, at increasing speed, around a hollow ring. The particles are guided around the ring by electromagnets, and accelerated by electric fields at points around the ring. Synchrotrons come in a wide range of sizes, the smallest being about a metre across while the largest is 27 km acros. This huge machine is the ◊Large Electron Positron Collider (LEP) at the ◊CERN laboratories near Geneva in Switzerland. The Tevatron synchrotron, 6 km in circumference, at ◊Fermilab accelerates protons and antiprotons to 1 TeV.

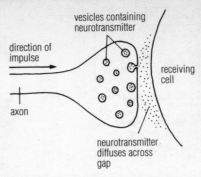

synapse *A synapse, or junction between two nerve cells, is bridged by a chemical neurotransmitter, such as acetylcholine, released by the nerve impulse.*

syncline geological term for a fold in the rocks of the Earth's crust in which the layers or beds dip inwards, thus forming a trough-like structure with a sag in the middle. The opposite structure, with the beds arching upwards, is an ◊anticline.

synodic period the time taken for a planet or moon to return to the same position in its orbit as seen from the Earth; that is, from one ◊opposition to the next. It differs from the ◊sidereal period because the Earth is moving in orbit around the Sun.

synovial fluid viscous colourless fluid that bathes movable joints between the bones of vertebrates. It nourishes and lubricates the ◊cartilage at the end of each bone.

synthesis in chemistry, the formation of a substance or compound from more elementary compounds. The synthesis of a drug can involve several stages from the initial material to the final product; the complexity of these stages is a major factor in the cost of production.

synthesizer device that uses electrical components to produce sounds. In *preset synthesizers*, the sound of various instruments is produced by a built-in computer-type memory. In *programmable synthesizers* any number of new instrumental or other sounds may be produced at the will of the performer. *Speech synthesizers* can break down speech into 128 basic elements (allophones), which are then combined into words and sentences, as in the voices of electronic teaching aids.

synthetic any material made from chemicals. Since the 1900s, more and more of the materials used in everyday life are synthetics, including plastics (polythene, polystyrene), ◊synthetic fibres (nylon, acrylics, polyesters), synthetic resins, and synthetic rubber. Most naturally occurring organic substances are now made synthetically, especially pharmaceuticals.

synthetic fibre fibre made by chemical processes, unknown in nature. There are two kinds. One is made from natural materials that have been chemically processed in some way; ◊rayon, for example, is made by processing the cellulose in wood pulp. The other type is the true synthetic fibre, made entirely from chemicals. ◊Nylon was the original synthetic fibre, made from chemicals obtained from petroleum (crude oil).

syrinx the voice-producing organ of a bird. It is situated where the trachea divides in two and consists of vibrating membranes, a reverberating capsule, and numerous controlling muscles.

Système International d'Unités official French name for ◊SI units.

System Network Architecture (SNA) a set of communication protocols developed by IBM and incorporated in hardware and software implementations.

systems analysis in computing, the investigation of a business activity or clerical procedure, with a view to deciding if and how it can be computerized. The analyst discusses the existing procedures with the people involved, observes the flow of data through the business, and draws up an outline specification of the required computer system (see also ◊systems design).

systems design in computing, the detailed design of an ◊applications package. The designer breaks the system down into component programs, and designs the required input forms, screen layouts, and printouts. Systems design forms a link between systems analysis and ◊programming.

systems program in computing, a program that performs a task related to the operation and performance of the computer system itself. For example, a systems program might control the operation of the display screen, or control and organize backing storage. In contrast, an ◊applications program is designed to carry out tasks for the benefit of the computer user.

System X in communications, a modular, computer-controlled, digital switching system used in telephone exchanges.

t symbol for ◊*tonne*, ◊*ton*.

tachograph combined speedometer and clock that records a vehicle's speed (on a small card disc, magnetic disc, or tape) and the length of time the vehicle is moving or stationary. It is used to monitor a lorry driver's working hours.

tachyon in physics, a hypothetical particle that travels faster than the speed of light. Its existence would explain the emission of ◊Cherenkov radiation.

talc $Mg_3Si_4O_{10}(OH)_2$, mineral, hydrous magnesium silicate. It occurs in tabular crystals, but the massive impure form, known as ***steatite*** or ***soapstone***, is more common. It is formed by the alteration of magnesium compounds and usually found in metamorphic rocks. Talc is very soft, ranked 1 on the Mohs' scale of hardness. It is used in powdered form in cosmetics, lubricants, and as an additive in paper manufacture.

tangent in geometry, a straight line that touches a curve and gives the gradient of the curve at the point of contact. At a maximum, minimum, or point of inflection, the tangent to a curve has zero gradient. Also, in trigonometry, a function of an acute angle in a right-angled triangle, defined as the ratio of the length of the side opposite the angle to the length of the side adjacent to it; a way of expressing the gradient of a line.

tannic acid or *tannin* $C_{14}H_{10}O_9$ yellow astringent substance, composed of several ◊phenol rings, occurring in the bark, wood, roots, fruits, and galls (growths) of certain trees, such as the oak. It precipitates gelatin to give an insoluble compound used in the manufacture of leather from hides (tanning).

tantalum hard, ductile, lustrous, grey-white, metallic element, symbol Ta, atomic number 73, relative atomic mass 180.948. It occurs with niobium in tantalite and other minerals. It can be drawn into wire with a very high melting point and great tenacity, useful for lamp filaments subject to vibration. It is also used in alloys, for corrosion-resistant laboratory apparatus and chemical equipment, as a catalyst in manufacturing synthetic rubber, in tools and instruments, and in rectifiers and capacitors.

tape recording, magnetic method of recording electric signals on a layer of iron oxide, or other magnetic material, coating a thin plastic tape. The electrical signals from the microphone are fed to the electromagnetic recording head, which magnetizes the tape in accordance with the frequency and amplitude of the original signal. The impulses may be audio (for sound recording), video (for television), or data (for computer). For playback, the tape is passed over the same, or another, head to convert magnetic into electrical signals, which are then amplified for reproduction. Tapes are easily demagnetized (erased) for reuse, and come in cassette, cartridge, or reel form.

tape streamer in computing, a backing storage device consisting of a continuous loop of magnetic tape. Tape streamers are largely used to store dumps (rapid backup copies) of important data files.

tapeworm any of various parasitic flatworms of the class Cestoda. They lack digestive and sense organs, can reach 15 m/ 50 ft in length, and attach themselves to the host's intestines by means of hooks and suckers. Tapeworms are made up of hundreds of individual segments, each of which develops into a functional hermaphroditic reproductive unit capable of producing numerous eggs. The larvae of tapeworms usually reach humans in imperfectly cooked meat or fish, causing anaemia and intestinal disorders.

taproot in botany, a single, robust, main ◊root that is derived from the embryonic root, or ◊radicle, and grows vertically downwards, often to considerable depth. Taproots are often modified for food storage and are common in biennial plants such as the carrot *Daucus carota*, where they act as ◊perennating organs.

tar dark brown or black viscous liquid obtained by the destructive distillation of coal, shale, and wood. Tars consist of a mixture of hydrocarbons, acids, and bases. ◊Creosote and ◊paraffin are produced from wood tar. See also ◊coal tar.

tangent to a curve

tangent to a circle

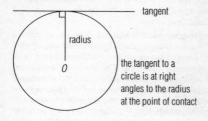

the tangent to a circle is at right angles to the radius at the point of contact

tangent of an angle

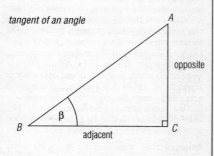

$$\tan\beta = \frac{\sin\beta}{\cos\beta} = \frac{\text{opposite}}{\text{adjacent}} = \frac{AC}{BC}$$

tangent *The tangent of an angle is a mathematical function used in the study of right-angled triangles. If the tangent of an angle β is known, then the length of the opposite side can be found given the length of the adjacent side, or vice versa.*

tartaric acid $HCOO(CHOH)_2COOH$ organic acid present in vegetable tissues and fruit juices in the form of salts of potassium, calcium, and magnesium. It is used in carbonated drinks and baking powders.

tartrazine (E102) yellow food colouring produced synthetically from petroleum. Many people are allergic to foods containing it. Typical effects are skin disorders and respiratory problems. It has been shown to have an adverse effect on hyperactive children.

taste sense that detects some of the chemical constituents of food. The human ◊tongue can distinguish only four basic tastes (sweet, sour, bitter, and salty) but it is supplemented by the sense of smell. What we refer to as taste is really a composite sense made up of both taste and smell.

tau ◊elementary particle with the same electric charge as the electron but a mass nearly double that of a proton. It has a lifetime of around 3×10^{-13} seconds and belongs to the ◊lepton family of particles – those that interact via the electromagnetic, weak nuclear, and gravitational forces, but not the strong nuclear force.

Tau Ceti one of the nearest stars visible to the naked eye, 11.9 light years from Earth in the constellation Cetus. It has a diameter slightly less than that of the Sun, and an actual luminosity of about 45% of the Sun's. Its similarity to the Sun is sufficient to suggest that Tau Ceti may possess a planetary system, although observations have yet to reveal evidence of this.

Taurus zodiacal constellation in the northern hemisphere near Orion, represented as a bull. The Sun passes through Taurus from mid-May to late June. Its brightest star is Aldebaran, seen as the bull's red eye. Taurus contains the Hyades and Pleiades open ◊star clusters, and the Crab nebula. In astrology, the dates for Taurus are between about 20 April and 20 May.

tautomerism form of isomerism in which two interconvertible ◊isomers are in equilibrium. It is often specifically applied to an equilibrium between the keto (–C–C=O) and enol (–C=C–OH) forms of carbonyl compounds.

taxis (plural *taxes*) or *tactic movement* in botany, the movement of a single cell, such as a bacterium, protozoan, single-celled alga, or gamete, in response to an external stimulus. A movement directed towards the stimulus is described as positive taxis, and away from it as negative taxis. The alga *Chlamydomonas*, for example, demonstrates positive *phototaxis* by swimming towards a light source to increase the rate of photosynthesis.

Chemotaxis is a response to a chemical stimulus, as seen in many bacteria that move towards higher concentrations of nutrients.

taxonomy another name for the ◊classification of living organisms.

T cell or *T lymphocyte* immune cell (see ◊immunity and ◊lymphocyte) that plays several roles in the body's defences. T cells are so called because they mature in the ◊thymus.

tear gas any of various volatile gases that produce irritation and watering of the eyes, used by police against crowds and used in chemical warfare. The gas is delivered in pressurized, liquid-filled canisters or grenades, thrown by hand or launched from a specially adapted rifle. Gases (such as Mace) cause violent coughing and blinding tears, which pass when the victim breathes fresh air, and there are no lasting effects.

tears salty fluid exuded by lacrimal glands in the eyes. The fluid contains proteins that are antibacterial, and also absorbs oils and mucus. Apart from cleaning and disinfecting the surface of the eye, the fluid supplies nutrients to the cornea, which does not have a blood supply.

technetium silver-grey, radioactive, metallic element, symbol Tc, atomic number 43, relative atomic mass 98.906. It occurs in nature only in extremely minute amounts, produced as a fission product from uranium in ◊pitchblende and other uranium ores. Its longest-lived isotope, Tc-99, has a half-life of 216,000 years. It is a superconductor and is used as a hardener in steel alloys and as a medical tracer.

technology the use of tools, power, and materials, generally for the purposes of production. Almost every human process for getting food and shelter depends on complex technological systems, which have been developed over a 3-million-year period. Significant milestones include the advent of the ◊steam engine 1712, the introduction of ◊electricity and the ◊internal combustion engine in the mid-1870s, and recent developments in ◊communications, ◊electronics, and the nuclear and space industries. The ***advanced technology*** (highly automated and specialized) on which modern industrialized society depends is frequently contrasted with the ***low technology*** (labour-intensive and unspecialized) that characterizes some developing countries. ◊***Intermediate technology*** is an attempt to adapt scientifically advanced inventions to less developed areas by using local materials and methods of manufacture.

tectonics in geology, the study of the movements of rocks on the Earth's surface. On a small scale tectonics involves the formation of folds and ◊faults, but on a large scale ◊plate tectonics deals with the movement of the Earth's surface as a whole.

Teflon trade name for polytetrafluoroethene (PTFE), a tough, waxlike, heat-resistant plastic used for coating nonstick cookware and in gaskets and bearings.

tektite small, rounded glassy stone, found in certain regions of the Earth, such as Australasia. Tektites are probably the scattered drops of molten rock thrown out by the impact of a large ◊meteorite.

telecommunications communications over a distance, generally by electronic means.

Long-distance voice communication was pioneered 1876 by US inventor Alexander Graham Bell, when he invented the telephone, which was made possible by the discovery of electromagnetism by the English chemist and physicist Michael Faraday. Today it is possible to communicate with most countries by telephone cable, or by satellite or microwave link, with over 100,000 simultaneous conversations and several television channels being carried by the latest satellites. ◊Integrated-Services Digital Network (ISDN) makes videophones and high-quality fax possible; the world's first large-scale centre of ISDN began operating in Japan 1988. ISDN is a system that transmits voice and image data on a single transmission line by changing them into digital signals. The chief method of relaying long-distance calls on land is microwave radio transmission.

telegraphy transmission of coded messages along wires by means of electrical signals. The first modern form of telecommunication, it now uses printers for the transmission and receipt of messages. Telex is an international telegraphy network.

telemetry measurement at a distance, in particular the systems by which information is obtained and sent back by instruments on board a spacecraft. See ◊remote sensing.

telephone instrument for communicating by voice over long distances, invented by US inventor Alexander Graham ◊Bell 1876. The transmitter (mouthpiece) consists of a carbon microphone, with a diaphragm that vibrates when a person speaks into it. The diaphragm vibrations compress grains of carbon to a greater or lesser extent, altering their resistance to an electric current passing through them. This sets up variable electrical signals, which travel along the telephone lines to the receiver of the person being called. There

telephone In the telephone sound vibrations are converted to an electric signal and back again. The mouthpiece contains a carbon microphone that produces an electrical signal which varies in step with the spoken sounds. The signal is routed to the receiver via local or national exchanges. The earpiece contains an electromagnetic loudspeaker which reproduces the sounds by vibrating a diaphragm.

they cause the magnetism of an electromagnet to vary, making a diaphragm above the electromagnet vibrate and give out sound waves, which mirror those that entered the mouthpiece originally.

teleprinter or *teletypewriter* transmitting and receiving device used in telecommunications to handle coded messages. Teleprinters are automatic typewriters keyed telegraphically to convert typed words into electrical signals (using a 5-unit Baudot code, see ◊baud) at the transmitting end, and signals into typed words at the receiving end.

telescope optical instrument that magnifies images of faint and distant objects; any device for collecting and focusing light and other forms of electromagnetic radiation. It is a major research tool in astronomy and is used to sight over land and sea; small telescopes can be attached to cameras and rifles. A telescope with a large aperture, or opening, can distinguish finer detail and fainter objects than one with a small aperture. The *refracting telescope* uses lenses, and the *reflecting telescope* uses mirrors. A third type, the *catadioptric telescope*, with a combination of lenses and mirrors, is used increasingly. See also ◊radio telescope.

The *liquid-mirror telescope* is a reflecting telescope constructed with a rotating mercury mirror. ◊Charge-couple devices are used to compensate for the fact that the a liquid mirror cannot be tilted. In 1994 a small number of liquid-mirror telescopes were in use for research purposes.

telescope *The Schmidt camera telescope uses a corrective plate to achieve a wide field of view and allows a distortion-free image to be formed on a curved photographic plate or film. A second lens is added to compensate for the use of a flat photographic plate. It is one of the most widely used tools of astronomy. There is a Schmidt telescope at Mount Palomar, California, USA.*

teletext broadcast system of displaying information on a television screen. The information – typically about news items, entertainment, sport, and finance – is constantly updated. Teletext is a form of ◊videotext, pioneered in Britain by the British Broadcasting Corporation (BBC) with Ceefax and by Independent Television with Teletext.

television (TV) reproduction at a distance by radio waves of visual images. For transmission, a television camera converts the pattern of light it takes in into a pattern of electrical charges. This is scanned line by line by a beam of electrons from an electron gun, resulting in variable electrical signals that represent the picture. These signals are combined with a radio carrier wave and broadcast as electromagnetic waves. The TV aerial picks up the wave and feeds it to the receiver (TV set). This separates out the vision signals, which pass to a cathode-ray tube where a beam of electrons is made to scan across the screen line by line, mirroring the action of the electron gun in the TV camera. The result is a recreation of the pattern of light that entered the camera. Twenty-five pictures are built up each second with interlaced scanning in Europe (30 in North America), with a total of 625 lines in Europe (525 lines in North America and Japan).

telex (acronym for *teleprinter exchange*) international telecommunications network that handles telegraph messages in the form of coded signals. It uses ◊teleprinters for transmitting and receiving, and makes use of land lines (cables) and radio and satellite links to make connections between subscribers.

tellurium silver-white, semi-metallic (◊metalloid) element, symbol Te, atomic number 52, relative atomic mass 127.60. Chemically it is similar to sulphur and selenium, and it is considered one of the sulphur group. It occurs naturally in telluride minerals, and is used in colouring glass blue-brown, in the electrolytic refining of zinc, in electronics, and as a catalyst in refining petroleum.

Telstar US communications satellite, launched 10 July 1962, which relayed the first live television transmissions between the USA and Europe. *Telstar* orbited the Earth in 158 minutes, and so had to be tracked by ground stations, unlike the geostationary satellites of today.

temperature degree or intensity of heat of a body, and the condition that determines whether or not it will transfer heat to, or receive heat from, another body according to the laws of ◊thermodynamics. It is measured in degrees Celsius (before 1948 called centigrade), kelvin, or Fahrenheit.

temperature regulation the ability of an organism to control its internal body temperature; in warm-blooded animals this is known as ◊homeothermy.

tempering heat treatment for improving the properties of metals, often used for steel alloys. The metal is heated to a certain temperature and then cooled suddenly in a water or oil bath.

temporary hardness hardness of water that is removed by boiling (see ◊hard water).

tendon or *sinew* cord of tough, fibrous connective tissue that joins muscle to bone in vertebrates. Tendons are largely composed of the protein collagen, and because of their inelasticity are very efficient at transforming muscle power into movement.

TELEVISION

television Simplified block diagram of a complete colour television system – transmitting and receiving. The camera separates the picture into three colours – red, blue, green – by using filters and different camera tubes for each colour. The audio signal is produced separately from the video signal. Both signals are transmitted from the same aerial using a special coupling device called a diplexer. There are four sections in the receiver: the aerial, the tuners, the decoders, and the display. As in the transmitter, the audio and video signals are processed separately. The signals are amplified at various points.

tendril in botany, a slender, threadlike structure that supports a climbing plant by coiling around suitable supports, such as the stems and branches of other plants. It may be a modified stem, leaf, leaflet, flower, leaf stalk, or stipule (a small appendage on either side of the leaf stalk), and may be simple or branched. The tendrils of Virginia creeper *Parthenocissus quinquefolia* are modified flower heads with suckerlike pads at the end that stick to walls, while those of the grapevine *Vitis* grow away from the light and thus enter dark crevices where they expand to anchor the plant firmly.

tension reaction force set up in a body that is subjected to stress. In a stretched string or wire it exerts a pull that is equal in magnitude but opposite in direction to the stress being applied at its ends. Tension originates in the net attractive intermolecular force created when a stress causes the mean distance separating a material's molecules to become greater than the equilibrium distance. It is measured in newtons.

terbium soft, silver-grey, metallic element of the ◊lanthanide series, symbol Tb, atomic number 81, relative atomic mass 158.925. It occurs in gadolinite and other ores, with yttrium and ytterbium, and is used in lasers, semiconductors, and television tubes. It was named in 1843 by Swedish chemist Carl Mosander (1797–1858) for the town of Ytterby, Sweden, where it was first found.

terminal in computing, a device consisting of a keyboard and display screen (◊VDU) – or, in older systems, a teleprinter – to enable the operator to communicate with the computer. The terminal may be physically attached to the computer or linked to it by a telephone line (remote terminal). A 'dumb' terminal has no processor of its own, whereas an 'intelligent' terminal has its own processor and takes some of the processing load away from the main computer.

terminal velocity or *terminal speed* the maximum velocity that can be reached by an object moving through a fluid (gas or liquid). As the speed of the object increases so does the total magnitude of the forces resisting its motion. Terminal velocity is reached when the resistive forces exactly balance the applied force that has caused the object to accelerate; because there is now no resultant force, there can be no further acceleration.

terminal voltage the potential difference (pd) or voltage across the terminals of a power supply, such as a battery of cells. When the supply is not connected in circuit its terminal voltage is the same as its ◊electromotive force (emf); however, as soon as it begins to supply current to a circuit its terminal voltage falls because some electric potential energy is lost in driving current against the supply's own ◊internal resistance. As the current flowing in the circuit is increased the terminal voltage of the supply falls.

territorial behaviour in biology, any behaviour that serves to exclude other members of the same species from a fixed area or ◊territory. It may involve aggressively driving out intruders, marking the boundary (with dung piles or secretions from special scent glands), conspicuous visual displays, characteristic songs, or loud calls.

territory in animal behaviour, a fixed area from which an animal or group of animals excludes other members of the same species. Animals may hold territories for many different reasons; for example, to provide a constant food supply, to monopolize potential mates, or to ensure access to refuges or nest sites.

The size of a territory depends in part on its function: some nesting and mating territories may be only a few square metres, whereas feeding territories may be as large as hundreds of square kilometres.

Tertiary period of geological time 65–1.64 million years ago, divided into five epochs: Palaeocene, Eocene, Oligocene, Miocene, and Pliocene. During the Tertiary, mammals took over all the ecological niches left vacant by the extinction of the dinosaurs, and became the prevalent land animals. The continents took on their present positions, and climatic and vegetation zones as we know them became established. Within the geological time column the Tertiary follows the Cretaceous period and is succeeded by the Quaternary period.

Terylene trade name for a polyester synthetic fibre produced by the chemicals company ICI. It is made by polymerizing ethylene glycol and terephthalic acid. Cloth made from Terylene keeps its shape after washing and is hard-wearing.

tesla SI unit (symbol T) of ◊magnetic flux density. One tesla represents a flux density of one ◊weber per square metre, or 10^4 ◊gauss. It is named after the Croatian engineer Nikola Tesla.

testa the outer coat of a seed, formed after fertilization of the ovule. It has a protective function and is usually hard and dry. In some cases the coat is adapted to aid dispersal, for example by being hairy. Humans have found uses for many types of testa, including the fibre of the cotton seed.

test cross in genetics, a breeding experiment used to discover the genotype of an individual organism. By crossing with a double recessive of the same species, the offspring will indicate whether the test individual is homozygous or heterozygous for the characteristic in question. In peas, a tall plant under investigation would be crossed with a double recessive short plant with known genotype tt. The results of the cross will be all tall plants if the test plant is TT. If the individual is in fact Tt then there will be some short plants (genotype tt) among the offspring.

test data data designed to test whether a new computer program is functioning correctly. The test data are carefully chosen to ensure that all possible branches of the program are tested. The expected results of running the data are written down and are then compared with the actual results obtained using the program.

testis (plural *testes*) the organ that produces ◊sperm in male (and hermaphrodite) animals. In vertebrates it is one of a pair of oval structures that are usually internal, but in mammals (other than elephants and marine mammals), the paired testes (or testicles) descend from the body cavity during development, to hang outside the abdomen in a scrotal sac. The testes also secrete the male sex hormone ◊androgen.

testosterone in vertebrates, hormone secreted chiefly by the testes, but also by the ovaries and the cortex of the adrenal glands. It promotes the development of secondary sexual characteristics in males. In animals with a breeding season, the onset of breeding behaviour is accompanied by a rise in the level of testosterone in the blood.

Tethys Sea sea that once separated ◊Laurasia from ◊Gondwanaland. It has now closed up to become the Mediterranean, the Black, the Caspian, and the Aral seas.

tetrachloromethane CCl_4 or *carbon tetrachloride* chlorinated organic compound that is a very efficient solvent for fats and greases, and was at one time the main constituent of household dry-cleaning fluids and of fire extinguishers used with electrical and petrol fires. Its use became restricted after it was discovered to be carcinogenic and it has now been largely removed from educational and industrial laboratories.

tetraethyl lead $Pb(C_2H_5)_4$ compound added to leaded petrol as a component of ◊antiknock to increase the efficiency of combustion in car engines. It is a colourless liquid that is insoluble in water but soluble in organic solvents such as benzene, ethanol, and petrol.

tetrapod type of ◊vertebrate. The group includes mammals, birds, reptiles, and amphibians. Birds are included because they evolved from four-legged ancestors, the forelimbs having become modified to form wings. Even snakes are tetrapods, because they are descended from four-legged reptiles.

T$_E$X (pronounced 'tek') public domain text formatting and typesetting system, developed by Donald Knuth and widely used for producing mathematical and technical documents. Unlike ◊desktop publishing applications, T_EX is not ◊WYSIWYG, although in some implementations screen preview of pages is possible.

text editor in computing, a program that allows the user to edit text on the screen and to store it in a file. Text editors are similar to ◊word processors, except that they lack the ability to format text into paragraphs and pages and to apply different typefaces and styles.

TGV (abbreviation for *train à grande vitesse* – French 'high-speed train') French electrically powered train that provides the world's fastest rail service. Since it began operating 1981, it has carried more than 100 million passengers between Paris and Lyon, at average speeds of 268 kph/167 mph. In 1990 a second service, the Atlantique, was launched, running from Paris to Le Mans and Tours. A third is planned, linking Paris with the Channel Tunnel, Brussels, and Cologne. In 1990, a TGV broke the world speed record, reaching a speed of 515.3 kph/320.2 mph (about half that of a passenger jet aircraft) on a stretch of line near Tours.

thalamus one of two oval masses of grey matter lying in the cerebral hemispheres. It is an integrating centre for sensory information on its way to the ◊cerebral cortex and is also important in motor control.

thallium soft, bluish-white, malleable, metallic element, symbol Tl, atomic number 81, relative atomic mass 204.37. It is a poor conductor of electricity. Its compounds are poisonous and are used as insecticides and rodent poisons; some are used in the optical-glass and infrared-glass industries and in photocells.

thallus any plant body that is not divided into true leaves, stems, and roots. It is often thin and flattened, as in the body of a seaweed, lichen, or liverwort, and the gametophyte generation (◊prothallus) of a fern.

thaumatin naturally occurring, noncarbohydrate sweetener derived from the bacterium *Thaumatococcus danielli*. Its sweetness is not sensed as quickly as that of other sweeteners, and it is not as widely used in the food industry.

thebaine $C_{19}H_{21}NO_3$ highly poisonous extract of ◊opium.

theodolite instrument for the measurement of horizontal and vertical angles, used in surveying. It consists of a small telescope mounted so as to move on two graduated circles, one horizontal and the other vertical, while its axes pass through the centre of the circles. See also ◊triangulation.

theorem mathematical proposition that can be deduced by logic from a set of axioms (basic facts that are taken to be true without proof). Advanced mathematics consists almost entirely of theorems and proofs, but even at a simple level theorems are important.

theory in science, a set of ideas, concepts, principles, or methods used to explain a wide set of observed facts. Among the major theories of science are ◊relativity, ◊quantum theory, ◊evolution, and ◊plate tectonics.

Theory of Everything (ToE) another name for ◊grand unified theory.

therm unit of energy defined as 10^5 British thermal units; equivalent to 1.055×10^8 joules. It is no longer in scientific use.

thermal capacity or *heat capacity* another name for ◊heat capacity

thermal conductivity in physics, the ability of a substance to conduct heat. Good thermal conductors, like good electrical conductors, are generally materials with many free electrons (such as metals).

Thermal conductivity is expressed in units of joules per second per metre per kelvin (J s^{-1} m^{-1} K^{-1}). For a block of material of cross-sectional area a and length l, with temperatures T_1 and T_2 at its end faces, the thermal conductivity λ equals $Hl/at(T_2 - T_1)$, where H is the amount of heat transferred in time t.

thermal decomposition breakdown of a compound into simpler substances by heating it; the reaction is irreversible. The catalytic ◊cracking of hydrocarbons is an example.

thermal dissociation reversible breakdown of a compound into simpler substances by heating it (see ◊dissociation). The splitting of ammonium chloride into ammonia and hydrogen chloride is an example. On cooling, they recombine to form the salt.

thermal expansion in physics, expansion that is due to a rise in temperature. It can be expressed in terms of linear, area, or volume expansion.

The coefficient of linear expansion α is the increase in unit length per degree temperature rise; area, or superficial, expansion β is the increase in unit area per degree; and volume, or cubic, expansion γ is the increase in unit volume per degree. As a close approximation, $\beta = 2\alpha$ and $\gamma = 3\alpha$.

thermal reactor nuclear reactor in which the neutrons released by fission of uranium-235 nuclei are slowed down in order to increase their chances of being captured by other uranium-235 nuclei, and so induce further fission. The material (commonly graphite or heavy water) responsible for doing so is called a *moderator*. When the fast newly-emitted neutrons collide with the nuclei of the moderator's atoms, some of their kinetic energy is lost and their speed is reduced. Those that have been slowed down to a speed that matches the thermal (heat) energy of the surrounding material are called *thermal neutrons*, and it is these that are most likely to induce fission and ensure the continuation of the chain reaction. See ◊nuclear reactor and ◊nuclear energy.

thermic lance cutting tool consisting of a tube of mild steel, enclosing tightly packed small steel rods and fed with oxygen. On ignition, temperatures above 3,000°C/5,400°F are produced and the thermic lance becomes its own sustaining fuel. It rapidly penetrates walls and a 23-cm/9-in steel door can be cut through in less than 30 seconds.

thermionics branch of electronics dealing with the emission of electrons from matter under the influence of heat.

thermistor device whose electrical ◊resistance falls as temperature rises. The current passing through a thermistor increases rapidly as its temperature rises, and so they are used in electrical thermometers.

thermite process method used in incendiary devices and welding operations. It uses a powdered mixture of aluminium and (usually) iron oxide, which, when ignited, gives out enormous heat. The oxide is reduced to iron, which is molten at the high temperatures produced. This can be used to make a weld. The process was discovered 1895 by German chemist Hans Goldschmidt (1861–1923).

thermocouple electric temperature-measuring device consisting of a circuit having two wires made of different metals welded together at their ends. A current flows in the circuit when the two junctions are maintained at different temperatures (◊Seebeck effect). The electromotive force generated – measured by a millivoltmeter – is proportional to the temperature difference.

thermodynamics branch of physics dealing with the transformation of heat into and from other forms of energy. It is the basis of the study of the efficient working of engines, such as the steam and internal-combustion engines. The three laws of thermodynamics are (1) energy can be neither created nor destroyed, heat and mechanical work being mutually convertible; (2) it is impossible for an unaided self-acting machine to convey heat from one body to another at a higher temperature; and (3) it is impossible by any procedure, no matter how idealized, to reduce any system to the ◊absolute zero of temperature (0K/–273°C) in a finite number of operations. Put into mathematical form, these laws have widespread applications in physics and chemistry.

thermography photographic recording of heat patterns. It is used medically as an imaging technique to identify 'hot spots' in the body – for example, tumours, where cells are more active than usual. Thermography was developed in the 1970s and 1980s by the military to assist night vision by detecting the body heat of an enemy or the hot engine of a tank. It uses a photographic method (using infrared radiation) employing infrared-sensitive films.

thermoluminescence release in the form of light of stored energy from a substance heated by ◊irradiation. It occurs with most crystalline substances to some extent. It is used in archaeology to date pottery, and by geologists in studying terrestrial rocks and meteorites.

thermometer instrument for measuring temperature. There are many types, designed to measure different temperature ranges to varying degrees of accuracy. Each makes use of a different physical effect of temperature. Expansion of a liquid is employed in common **liquid-in-glass thermometers**, such as those containing mercury or alcohol. The more accurate *gas thermometer* uses the effect of temperature on the pressure of a gas held at constant volume. A *resistance thermometer* takes advantage of the change in resistance of a conductor (such as a platinum wire) with variation in temperature. Another electrical thermometer is the ◊*thermocouple*. Mechanically, temperature change can be indicated by the change in curvature of a *bimetallic strip* (as commonly used in a ◊thermostat).

thermopile instrument for measuring radiant heat, consisting of a number of ◊thermocouples connected in series with alternate junctions exposed to the radiation. The current generated (measured by an ◊ammeter) is proportional to the radiation falling on the device.

thermoplastic or *thermosoftening plastic* type of ◊plastic that always softens on repeated heating. Thermoplastics include polyethylene (polyethene), polystyrene, nylon, and polyester.

thermoset type of ◊plastic that remains rigid when set, and does not soften with heating. Thermosets have this property because the long-chain polymer molecules cross-link with each other to give a rigid structure. Examples include Bakelite, resins, melamine, and urea-formaldehyde resins.

thermosphere layer in the Earth's ◊atmosphere above the mesosphere and below the exosphere. Its lower level is about 80 km/50 mi above the ground, but its upper level is undefined. The ionosphere is located in the thermosphere. In the thermosphere the temperature rises with increasing height to several thousand degrees Celsius. However, because of the thinness of the air, very little heat is actually present.

thermostat temperature-controlling device that makes use of feedback. It employs a temperature sensor (often a bimetallic strip) to operate a switch or valve to control electricity or fuel supply. Thermostats are used in central heating, ovens, and car engines.

thiamine or *vitamin B_1* a water-soluble vitamin of the B complex. It is found in seeds and grain. Its absence from the diet causes the disease ◊beriberi.

thorax in tetrapod vertebrates, the part of the body containing the heart and lungs, and protected by the rib cage; in arthropods, the middle part of the body, between the head and abdomen.

thorium dark-grey, radioactive, metallic element of the ◊actinide series, symbol Th, atomic number 90, relative atomic mass 232.038. It occurs throughout the world in small quantities in minerals such as thorite and is widely distributed in monazite beach sands. It is one of three fissile elements (the others are uranium and plutonium), and its longest-lived isotope has a half-life of 1.39×10^{10} years.

Thorium is used to strengthen alloys. It was discovered by Jöns Berzelius 1828 and was named by him after the Norse god Thor.

throat in human anatomy, the passage that leads from the back of the nose and mouth to the ◊trachea and ◊oesophagus. It includes the ◊pharynx and the ◊larynx, the latter being at the top of the trachea. The word 'throat' is also used to mean the front part of the neck, both in humans and other vertebrates; for example, in describing the plumage of birds. In engineering, it is any narrowing entry, such as the throat of a carburettor.

thrombin substance that converts the soluble protein fibrinogen to the insoluble fibrin in the final stage of blood clotting. Its precursor in the bloodstream is ◊prothrombin.

thulium soft, silver-white, malleable and ductile, metallic element, of the ◊lanthanide series, symbol Tm, atomic number 69, relative atomic mass 168.94. It is the least abundant of the rare-earth metals, and was first found in gadolinite and various other minerals. It is used in arc lighting.

thymus organ in vertebrates, situated in the upper chest cavity in humans. The thymus processes ◊lymphocyte cells to produce T lymphocytes (T denotes 'thymus-derived'), which are responsible for binding to specific invading organisms and killing them or rendering them harmless.

thyristor type of ◊rectifier, an electronic device that conducts electricity in one direction only. The thyristor is composed of layers of ◊semiconductor material sandwiched between two electrodes called the anode and cathode. The current can be switched on by using a third electrode called the gate.

thyroid ◊endocrine gland of vertebrates, situated in the neck in front of the trachea. It secretes several hormones, principally thyroxine, an iodine-containing hormone that stimulates growth, metabolism, and other functions of the body. The thyroid gland may be thought of as the regulator gland of the body's metabolic rate. If it is overactive, as in ◊hyperthyroidism, the sufferer feels hot and sweaty, has an increased heart rate, diarrhoea, and weight loss. Conversely, an underactive thyroid leads to *myxoedema*, a condition characterized by sensitivity to the cold, constipation, and weight gain. In infants, an underactive thyroid leads to *cretinism*, a form of mental retardation.

tibia the anterior of the pair of bones found between the ankle and the knee. In humans, the tibia is the shinbone.

tidal energy energy derived from the tides. The tides mainly gain their potential energy from the gravitational forces acting between the Earth and the Moon. If water is trapped at a high level during high tide, perhaps by means of a barrage across an estuary, it may then be gradually released and its associated ◊gravitational potential energy exploited to drive turbines and generate electricity. Several schemes have been proposed for the Bristol Channel, in SW England, but environmental concerns as well as construction costs have so far prevented any decision from being taken.

tidal power station ◊hydroelectric power plant that uses the 'head' of water created by the rise and fall of the ocean tides to spin the water turbines. The world's only large tidal power station is located on the estuary of the river Rance in the Gulf of St Malo, Brittany, France, which has been in use since 1966. It produces 240 megawatts and can generate electricity on both the ebb and flow of the tide.

time continuous passage of existence, recorded by division into hours, minutes, and seconds. Formerly the measurement of time was based on the Earth's rotation on its axis, but this was found to be irregular. Therefore the second, the standard ◊SI unit of time, was redefined 1956 in terms of the Earth's annual orbit of the Sun, and 1967 in terms of a radiation pattern of the element caesium.

time-sharing in computing, a way of enabling several users to access the same computer at the same time. The computer rapidly switches between user ◊terminals and programs, allowing each user to work as if he or she had sole use of the system.

tin soft, silver-white, malleable and somewhat ductile, metallic element, symbol Sn (from Latin *stannum*), atomic number 50, relative atomic mass 118.69. Tin exhibits ◊allotropy, having three forms: the familiar lustrous metallic form above 55.8°F/13.2°C; a brittle form above 321.8°F/161°C; and a grey powder form below 55.8°F/13.2°C (commonly called tin pest or tin disease). The metal is quite soft (slightly harder than lead) and can be rolled, pressed, or hammered into extremely thin sheets; it has a low melting point. In nature it occurs rarely as a free metal. It resists corrosion and is therefore used for coating and plating other metals.

tin ore mineral from which tin is extracted, principally cassiterite, SnO_2. The world's chief producers are Malaysia, Thailand, and Bolivia.

tinplate milled steel coated with tin, the metal used for most 'tin' cans. The steel provides the strength, and the tin provides the corrosion resistance, ensuring that the food inside is not contaminated. Tinplate may be made by ♦electroplating or by dipping in a bath of molten tin.

tissue in biology, any kind of cellular fabric that occurs in an organism's body. Several kinds of tissue can usually be distinguished, each consisting of cells of a particular kind bound together by cell walls (in plants) or extracellular matrix (in animals). Thus, nerve and muscle are different kinds of tissue in animals, as are ♦parenchyma and ♦sclerenchyma in plants.

tissue culture process by which cells from a plant or animal are removed from the organism and grown under controlled conditions in a sterile medium containing all the necessary nutrients. Tissue culture can provide information on cell growth and differentiation, and is also used in plant propagation and drug production.

tissue plasminogen activator (tPA) naturally occurring substance in the body tissues that activates the enzyme plasmin that is able to dissolve blood clots. Human tPA, produced in bacteria by genetic engineering, has, like ♦streptokinase, been used to dissolve blood clots in the coronary arteries of heart-attack victims. It has been shown to be more effective than streptokinase when used in conjunction with heparin, but it is much more expensive.

Titan in astronomy, largest moon of the planet Saturn, with a diameter of 5,150 km/3,200 mi and a mean distance from Saturn of 1,222,000 km/759,000 mi. It was discovered 1655 by Dutch mathematician and astronomer Christiaan Huygens (1629–1695), and is the second largest moon in the Solar System (Ganymede, of Jupiter, is larger).

titanium strong, lightweight, silver-grey, metallic element, symbol Ti, atomic number 22, relative atomic mass 47.90. The ninth-most abundant element in the Earth's crust, its compounds occur in practically all igneous rocks and their sedimentary deposits. It is very strong and resistant to corrosion, so it is used in building high-speed aircraft and spacecraft; it is also widely used in making alloys, as it unites with almost every metal except copper and aluminium. Titanium oxide is used in high-grade white pigments.

titanium ore any mineral from which titanium is extracted, principally ilmenite ($FeTiO_3$) and rutile (TiO_2). Brazil, India, and Canada are major producers. Both these ore minerals are found either in rock formations or concentrated in heavy mineral sands.

Titan rocket family of US space rockets, developed from the Titan intercontinental missile. Two-stage Titan rockets launched the ♦Gemini crewed missions. More powerful Titans, with additional stages and strap-on boosters, were used to launch spy satellites and space probes, including the ♦Viking and ♦Voyager probes and ♦*Mars Observer*.

titration in analytical chemistry, a technique to find the concentration of one compound in a solution by determining how much of it will react with a known amount of another compound in solution.

titration *In acid–alkali titration, a standard solution of acid is added little by little to a measured amount of the alkali (with added indicator) until the indicator just changes colour.*

TNT (abbreviation for *trinitrotoluene*) $CH_3C_6H_2(NO_2)_3$, a powerful high explosive. It is a yellow solid, prepared in several isomeric forms from ◊toluene by using sulphuric and nitric acids.

toadstool common name for many umbrella-shaped fruiting bodies of fungi. The term is normally applied to those that are inedible or poisonous.

tocopherol or *vitamin E* fat-soluble chemical found in vegetable oils. Deficiency of tocopherol leads to multiple adverse effects on health. In rats, vitamin E deficiency has been shown to cause sterility.

tog unit of measure of thermal insulation used in the textile trade; a light summer suit provides 1.0 tog.

tokamak acronym for toroidal magnetic chamber, an experimental machine conceived by Soviet physicist Andrei Sakharov and developed in the Soviet Union to investigate controlled nuclear fusion. It consists of a doughnut-shaped chamber surrounded by electromagnets capable of exerting very powerful magnetic fields. The fields are generated to confine a very hot (millions of degrees) ◊plasma of ions and electrons, keeping it away from the chamber walls. See also ◊JET.

toluene or *methyl benzene* $C_6H_5CH_3$ colourless, inflammable liquid, insoluble in water, derived from petroleum. It is used as a solvent, in aircraft fuels, in preparing phenol (carbolic acid, used in making resins for adhesives, pharmaceuticals, and as a disinfectant), and the powerful high explosive ◊TNT.

tomography the technique of using X-rays or ultrasound waves to procure images of structures deep within the body for diagnostic purposes. In modern medical imaging there are several techniques, such as the ◊CAT scan (computerized axial tomography).

ton imperial unit of mass. The *long ton*, used in the UK, is 1,016 kg/2,240 lb; the *short ton*, used in the USA, is 907 kg/2,000 lb. The *metric ton* or *tonne* is 1,000 kg/2,205 lb.

tongue in tetrapod vertebrates, a muscular organ usually attached to the floor of the mouth. It has a thick root attached to a U-shaped bone (hyoid), and is covered with a ◊mucous membrane containing nerves and taste buds. It is the main organ of taste. The tongue directs food to the teeth and into the throat for chewing and swallowing. In humans, it is crucial for speech; in other animals, for lapping up water and for grooming, among other functions.

tonne the metric ton of 1,000 kg/2,204.6 lb; equivalent to 0.9842 of an imperial ◊ton.

tonsils in higher vertebrates, masses of lymphoid tissue situated at the back of the mouth and throat (palatine tonsils), and on the rear surface of the tongue (lingual tonsils). The tonsils contain many ◊lymphocytes and are part of the body's defence system against infection.

tooth in vertebrates, one of a set of hard, bonelike structures in the mouth, used for

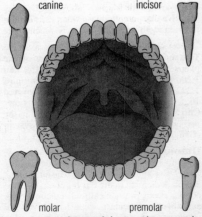

tooth *Adults have 32 teeth: two incisors, one canine, two premolars, and three molars on each side of each jaw. Each tooth has three parts: crown, neck, and root. The crown consists of a dense layer of mineral, the enamel, surrounding hard dentine with a soft centre, the pulp.*

biting and chewing food, and in defence and aggression. In humans, the first set (20 milk teeth) appear from age six months to two and a half years. The permanent ◊dentition replaces these from the sixth year onwards, the wisdom teeth (third molars) sometimes not appearing until the age of 25 or 30. Adults have 32 teeth: two incisors, one canine (eye tooth), two premolars, and three molars on each side of each jaw. Each tooth consists of an enamel coat (hardened calcium deposits), dentine (a thick, bonelike layer), and an inner pulp cavity, housing nerves and blood vessels. Mammalian teeth have roots surrounded by cementum, which fuses them into their sockets in the jawbones.

The neck of the tooth is covered by the ◊gum, while the enamel-covered crown protrudes above the gum line.

topaz mineral, aluminium fluosilicate, $Al_2(F_2SiO_4)$. It is usually yellow, but pink if it has been heated, and is used as a gemstone when transparent. It ranks 8 on the Mohs' scale of hardness.

topography the surface shape and composition of the landscape, comprising both natural and artificial features, and its study. Topographical features include the relief and contours of the land; the distribution of mountains, valleys, and human settlements; and the patterns of rivers, roads, and railways.

topology branch of geometry that deals with those properties of a figure that remain unchanged even when the figure is transformed (bent, stretched) – for example, when a square painted on a rubber sheet is deformed by distorting the sheet.

Topology has scientific applications, as in the study of turbulence in flowing fluids.

torque the turning effect of force on an object. A turbine produces a torque that turns an electricity generator in a power station. Torque is measured by multiplying the force by its perpendicular distance from the turning point.

torque converter device similar to a turbine, filled with oil, used in automatic transmission systems in motor vehicles and locomotives to transmit power (torque) from the engine to the gears.

torr unit of pressure equal to 1/760 of an ◊atmosphere, used mainly in high-vacuum technology.

torsion in physics, the state of strain set up in a twisted material; for example, when a thread, wire, or rod is twisted, the torsion set up in the material tends to return the material to its original state. The **torsion balance**, a sensitive device for measuring small gravitational or magnetic forces, or electric charges, balances these against the restoring force set up by them in a torsion suspension.

torus semicylindrical ring with a D-shaped cross-section used to contain ◊plasma in nuclear fusion reactors such as the Joint European Torus (◊JET) reactor.

total internal reflection the complete reflection of a beam of light that occurs from the surface of an optically 'less dense' material. For example, a beam from an underwater light source can be reflected from the surface of the water, rather than escaping through the surface. Total internal reflection can only happen if a light beam hits a surface at an angle greater than the ◊critical angle for that particular pair of materials.

touch sensation produced by specialized nerve endings in the skin. Some respond to light pressure, others to heavy pressure. Temperature detection may also contribute to the overall sensation of touch. Many animals, such as nocturnal ones, rely on touch more than humans do. Some have specialized organs of touch that project from the body, such as whiskers or antennae.

touch screen in computing, an input device allowing the user to communicate with the computer by touching a display screen with a finger. In this way, the user can point to a required ◊menu option or item of data. Touch screens are used less widely than other pointing devices such as the ◊mouse or ◊joystick.

touch sensor in a computer-controlled ◊robot, a device used to give the robot a sense of touch, allowing it to manipulate delicate objects or move automatically about a room. Touch sensors provide the feedback necessary for the robot to adjust the force of its movements and the pressure of its grip. The main types include the strain gauge and the microswitch.

tourmaline hard, brittle mineral, a complex silicate of various metals, but mainly sodium aluminium borosilicate.

toxicity tests tests carried out on new drugs, cosmetics, food additives, pesticides, and other synthetic chemicals to see whether they are safe for humans to use. They aim to identify potential toxins, carcinogens, teratogens, and mutagens.

toxin any poison produced by another living organism (usually a bacterium) that can damage the living body. In vertebrates, toxins are broken down by ◊enzyme action, mainly in the liver.

trace in computing, a method of checking that a computer program is functioning correctly by causing the changing values of all the ◊variables involved to be displayed while the program is running. In this way it becomes possible to narrow down the search for a bug, or error, in the program to the exact instruction that causes the variables to take unexpected values.

trace element chemical element necessary in minute quantities for the health of a plant or animal. For example, magnesium, which occurs in chlorophyll, is essential to photosynthesis, and iodine is needed by the thyroid gland of mammals for making hormones that control growth and body chemistry.

tracer a small quantity of a radioactive ◊isotope (form of an element) used to follow the path of a chemical reaction or a physical or biological process. The location (and possibly concentration) of the tracer is usually detected by using a Geiger–Muller counter.

trachea tube that forms an airway in air-breathing animals. In land-living ◊vertebrates, including humans, it is also known as the *windpipe* and runs from the larynx to the upper part of the chest. Its diameter is about 1.5 cm/0.6 in and its length 10 cm/4 in. It is strong and flexible, and reinforced by rings of ◊cartilage. In the upper chest, the trachea branches into two tubes: the left and right bronchi, which enter the lungs. Insects have a branching network of tubes called tracheae, which conduct air from holes (◊spiracles) in the body surface to all the body tissues. The finest branches of the tracheae are called tracheoles.

tracheid cell found in the water-conducting tissue (◊xylem) of many plants, including gymnosperms (conifers) and pteridophytes (ferns). It is long and thin with pointed ends.

The cell walls are thickened by ◊lignin, except for numerous small rounded areas, or pits, through which water and dissolved minerals pass from one cell to another. Once mature, the cell itself dies and only its walls remain.

track in computing, part of the magnetic structure created on a disc surface during ◊disc formatting so that data can be stored on it. The disc is first divided into circular tracks and then each circular track is divided into a number of sectors.

transactinide element any of a series of nine radioactive, metallic elements with atomic numbers that extend beyond the ◊actinide series, those from 104 (rutherfordium) to 112 (unnamed). They are grouped because of their expected chemical similarities (all are bivalent), the properties differing only slightly with atomic number. All have half-lives of less than two minutes.

transaction file in computing, a file that contains all the additions, deletions, and amendments required during ◊file updating to produce a new version of a master file.

transcription in living cells, the process by which the information for the synthesis of a protein is transferred from the ◊DNA strand on which it is carried to the messenger ◊RNA strand involved in the actual synthesis.

transducer device that converts one form of energy into another. For example, a thermistor is a transducer that converts heat into an electrical voltage, and an electric motor is a transducer that converts an electrical voltage

trachea *The human trachea, or windpipe. The larynx, or voice box, lies at the entrance to the trachea. The two vocal cords are membranes that normally remain open and still. When they are drawn together, the passage of air makes them vibrate and produce sounds.*

into mechanical energy. Transducers are important components in many types of ◊sensor, converting the physical quantity to be measured into a proportional voltage signal.

transduction the transfer of genetic material between cells by an infectious mobile genetic element such as a virus. Transduction is used in ◊genetic engineering to produce new varieties of bacteria.

transfermium element or *superheavy element* elements with an atomic number above 100. They are all highly unstable and have existed only briefly, and because of this there is a continuing controversy over who should be credited with their discoveries. See also ◊transuranic element.

transfer orbit elliptical path followed by a spacecraft moving from one orbit to another, designed to save fuel although at the expense of a longer journey time.

transformation a mapping or ◊function, especially one which causes a change of shape or position in a geometric figure. Reflection, rotation, enlargement, and ◊translation are the main geometrical transformations.

transformer device in which, by electromagnetic induction, an alternating current (AC) of one voltage is transformed to another voltage, without change of ◊frequency. Transformers are widely used in electrical apparatus of all kinds, and in particular in power transmission where high voltages and low currents are utilized.

A transformer has two coils, a primary for the input and a secondary for the output, wound on a common iron core. The ratio of the primary to the secondary voltages (and currents) is directly (and inversely) proportional to the number of turns in the primary and secondary coils.

transfusion intravenous delivery of blood or blood products (plasma, red cells) into a patient's circulation to make up for deficiencies due to disease, injury, or surgical intervention.

Cross-matching is carried out to ensure the patient receives the right blood group. Because of worries about blood-borne disease, there is a growing interest in autologous transfusion with units of the patient's own blood 'donated' over the weeks before an operation.

transgenic organism plant, animal, bacterium, or other living organism which has had a foreign gene added to it by means of ◊genetic engineering.

transistor solid-state electronic component, made of ◊semiconductor material, with three or more ◊electrodes, that can regulate a current passing through it. A transistor can act as an ◊amplifier, ◊oscillator, ◊photocell, or switch, and (unlike earlier thermionic valves) usually operates on a very small amount of power. Transistors commonly consist of a tiny sandwich of germanium or silicon, alternate layers having different electrical properties because they are impregnated with minute amounts of different impurities.

A crystal of pure germanium or silicon would act as an insulator (nonconductor). By introducing impurities in the form of atoms of other materials (for example, boron, arsenic, or indium) in minute amounts, the layers may be made either *n-type*, having an excess of electrons, or *p-type*, having a deficiency of electrons. This enables electrons to flow from one layer to another in one direction only.

transistor–transistor logic (TTL) in computing, the type of integrated circuit most commonly used in building electronic products. In TTL chips the bipolar transistors are directly connected (usually collector to base). In mass-produced items, large numbers of TTL chips are commonly replaced by a small number of uncommitted logic arrays (ULAs), or logic gate arrays.

transit in astronomy, the passage of a smaller object across the visible disc of a larger one. Transits of the inferior planets occur when they pass directly between the Earth and the Sun, and are seen as tiny dark spots against the Sun's disc.

transition metal any of a group of metallic elements that have incomplete inner electron shells and exhibit variable valency – for example, cobalt, copper, iron, and molybdenum. They are excellent conductors of electricity, and generally form highly coloured compounds.

translation in living cells, the process by which proteins are synthesized. During translation, the information coded as a sequence of nucleotides in messenger ◊RNA is transformed into a sequence of amino acids in a peptide chain. The process involves the 'translation' of the ◊genetic code. See also ◊transcription.

translation program in computing, a program that translates another program written in a high-level language or assembly language into the machine-code instructions that a computer can obey. See ◊assembler, ◊compiler, and ◊interpreter.

transuranic elements

atomic number	name	symbol	year discovered	source of first preparation	isotope identified	half-life of first isotope identified
actinide series						
93	neptunium	Np	1940	irradiation of uranium-238 with neutrons	Np-239	2.35 days
94	plutonium	Pu	1941	bombardment of uranium-238 with deuterons	Pu-238	86.4 years
95	americium	Am	1944	irradiation of plutonium-239 with neutrons	Am-241	458 years
96	curium	Cm	1944	bombardment of plutonium-239 with helium nuclei	Cm-242	162.5 days
97	berkelium	Bk	1949	bombardment of americium-241 with helium nuclei	Bk-243	4.5 hours
98	californium	Cf	1950	bombardment of curium-242 with helium nuclei	Cf-245	44 minutes
99	einsteinium	Es	1952	irradiation of uranium-238 with neutrons in first thermonuclear explosion	Es-253	20 days
100	fermium	Fm	1953	irradiation of uranium-238 with neutrons in first thermonuclear explosion	Fm-235	20 hours
101	mendelevium	Md	1955	bombardment of einsteinium-253 with helium nuclei	Md-256	76 minutes
102	nobelium	No	1958	bombardment of curium-246 with carbon nuclei	No-255	2.3 seconds
103	lawrencium	Lr	1961	bombardment of californium-252 with boron nuclei	Lr-257	4.3 seconds
transactinide elements						
104	unnilquadium (also called rutherfordium or kurchatovium)	Unq	1969	bombardment of californium-249 with carbon-12 nuclei	Unq-257	3.4 seconds
105	unnilpentium (also called hahnium or nielsbohrium)	Unp	1970	bombardment of californium-249 with nitrogen-15 nuclei	Unp-260	1.6 seconds
106	seaborgium	Sg	1974	bombardment of californium-249 with oxygen-18 nuclei	Sg-263	0.9 seconds
107	unnilseptium*	Uns	1977	bombardment of bismuth-209 with nuclei of chromium-54	Uns	102 milliseconds
108	unniloctium*	Uno	1984	bombardment of lead-208 with nuclei of iron-58	Uno-265	1.8 milliseconds
109	unnilennium*	Une	1982	bombardment of bismuth-209 with nuclei of iron-58	Une	3.4 milliseconds

* names for elements are as proposed by the International Union for Pure and Applied Chemistry 1980

transmission electron microscope (TEM) the most powerful type of ◊electron microscope, with a resolving power ten times better than that of a ◊scanning electron microscope and a thousand times better than that of an optical microscope. A fine electron beam passes through the specimen, which must therefore be sliced extremely thinly – typically to about one-thousandth of the thickness of a sheet of paper (100 nanometres). The TEM can resolve objects 0.001 micrometres (0.04 millionth of an inch) apart, a gap that is 100,000 times smaller than the unaided eye can see.

transpiration the loss of water from a plant by evaporation. Most water is lost from the leaves through pores known as ◊stomata, whose primary function is to allow ◊gas exchange between the plant's internal tissues and the atmosphere. Transpiration from the leaf surfaces causes a continuous upward flow of water from the roots via the ◊xylem, which is known as the transpiration stream.

transplant in medicine, the transfer of a tissue or organ from one human being to another or from one part of the body to another (skin grafting). In most organ transplants, the operation is for life-saving purposes, though the immune system tends to reject foreign tissue. Careful matching and immunosuppressive drugs must be used, but these are not always successful.

transposon or *jumping gene* segment of DNA able to move within or between chromosomes. They trigger changes in gene expression by shutting off genes or causing insertion ◊mutations.

transputer in computing, a member of a family of microprocessors designed for parallel

transpiration *Water is taken in by the roots and moves up through the plant.*

processing, developed in the UK by Inmos. In the circuits of a standard computer the processing of data takes place in sequence; in a transputer's circuits processing takes place in parallel, greatly reducing computing time for those programs that have been specifically written for it.

transuranic element or *transuranium element* chemical element with an atomic number of 93 or more – that is, with a greater number of protons in the nucleus than has uranium. All transuranic elements are radioactive. Neptunium and plutonium are found in nature; the others are synthesized in nuclear reactions.

transverse wave ◊wave in which the displacement of the medium's particles is at right-angles to the direction of travel of the wave motion.

trapezius triangular muscle running from high up in the back of the neck, across the back of the shoulder and down the upper spine. It is involved in movements of the ◊scapula (shoulder blade) and also moves the head from side to side.

tree perennial plant with a woody stem, usually a single stem or 'trunk', made up of ◊wood and protected by an outer layer of ◊bark. It absorbs water through a ◊root system. There is no clear dividing line between ◊shrubs and trees, but sometimes a minimum achievable height of 6 m/20 ft is used to define a tree.

tree-and-branch filing system in computing, a filing system where all files are stored within directories, like folders in a filing cabinet. These directories may in turn be stored within further directories. The root directory contains all the other directories and may be thought of as equivalent to the filing cabinet. Another way of picturing the system is as a tree with branches from which grow smaller branches, ending in leaves (individual files).

tree diagram in probability theory, a branching diagram consisting only of arcs and nodes (but not loops curving back on themselves), which is used to establish probabilities.

tree rings rings visible in the wood of a cut tree; see ◊annual rings.

triangle in geometry, a three-sided plane figure, the sum of whose interior angles is 180°. Triangles can be classified by the relative lengths of their sides. A *scalene triangle* has three sides of unequal length; an *isosceles triangle* has at least two equal sides; an *equilateral triangle* has three equal sides (and three equal angles of 60°).

triangle of forces method of calculating the force produced by two other forces (the resultant). It is based on the fact that if three forces acting at a point can be represented by the sides of a triangle, the forces are in equilibrium. See ◊parallelogram of forces.

triangulation technique used in surveying and navigation to determine distances, using the properties of the triangle. To begin, surveyors measure a certain length exactly to provide a base line. From each end of this line they then measure the angle to a distant point, using a ◊theodolite. They now have a triangle in which they know the length of one side and the two adjacent angles. By simple trigonometry they can work out the lengths of the other two sides.

triangle Types of triangles. Triangles are classified by the relative lengths of their sides.

Triassic period of geological time 245–208 million years ago, the first period of the Mesozoic era. The continents were fused together to form the world continent ◊Pangaea. Triassic sediments contain remains of early dinosaurs and other reptiles now extinct. By late Triassic times, the first mammals had evolved.

tribology the science of friction, lubrication, and lubricants. It studies the origin of frictional forces, the wearing of interacting surfaces, and the problems of efficient lubrication.

trichloromethane technical name for ◊chloroform.

tricuspid valve flap of tissue situated on the right side of the ◊heart between the atrium and the ventricle. It prevents blood flowing backwards when the ventricle contracts.

As in all valves, its movements are caused by pressure changes during the beat rather than by any intrinsic muscular activity. As the valve snaps shut, a vibration passes through the chest cavity and is detectable as the first sound of the heartbeat.

triglyceride chemical name for ◊fat.

trigonometry branch of mathematics that solves problems relating to plane and spherical triangles. Its principles are based on the fixed proportions of sides for a particular angle in a right-angled triangle, the simplest of which are known as the ◊sine, ◊cosine, and ◊tangent (so-called trigonometrical ratios). It is of practical importance in navigation, surveying, and simple harmonic motion in physics.

triiodomethane technical name for ◊iodoform.

trilobite any of a large class (Trilobita) of extinct, marine, invertebrate arthropods of the Palaeozoic era, with a flattened, oval body, 1–65 cm/0.4–26 in long. The hard-shelled body was divided by two deep furrows into three lobes.

triode three-electrode thermionic ◊valve containing an anode and a cathode (as does a ◊diode) with an additional negatively biased control grid. Small variations in voltage on the grid bias result in large variations in the current. The triode was commonly used in amplifiers but has now been almost entirely superseded by the ◊transistor.

triple bond three covalent bonds between adjacent atoms, as in the ◊alkynes (–C≡C–).

triploblastic in biology, having a body wall composed of three layers. The outer layer is the *ectoderm*, the middle layer the *mesoderm*, and the inner layer the *endoderm*. This pattern of development is shown by most multicellular animals (including humans).

triticale cereal crop of recent origin that is a cross between wheat *Triticum* and rye *Secale*. It can produce heavy yields of high-protein grain, principally for use as animal feed.

tritium radioactive isotope of hydrogen, three times as heavy as ordinary hydrogen, consisting of one proton and two neutrons. It has a half-life of 12.5 years.

Triton in astronomy, the largest of Neptune's moons. It has a diameter of 2,700 km/1,680 mi, and orbits Neptune every 5.88 days in a retrograde (east to west) direction.

Trojan horse in computing, a ◊virus program that appears to function normally but, while undetected by the normal user, causes damage to other files or circumvents security procedures. The earliest appeared in the UK in about 1988.

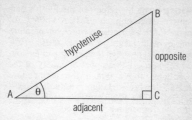

for any right-angled triangle with angle θ as shown the trigonometrical ratios are

$$\sin \theta = \frac{BC}{AB} = \frac{\text{opposite}}{\text{hypotenuse}}$$

$$\cos \theta = \frac{AC}{AB} = \frac{\text{adjacent}}{\text{hypotenuse}}$$

$$\tan \theta = \frac{BC}{AC} = \frac{\text{opposite}}{\text{adjacent}}$$

trigonometry *Trigonometry is the branch of mathematics devoted to solving problems related to triangles.*

trophic level in ecology, the position occupied by a species (or group of species) in a ◊food chain. The main levels are *primary producers* (photosynthetic plants), *primary consumers* (herbivores), *secondary consumers* (carnivores), and *decomposers* (bacteria and fungi).

tropism or *tropic movement* the directional growth of a plant, or part of a plant, in response to an external stimulus such as gravity or light. If the movement is directed towards the stimulus it is described as positive; if away from it, it is negative. *Geotropism* for example, the response of plants to gravity, causes the root (positively geotropic) to grow downwards, and the stem (negatively geotropic) to grow upwards.

Phototropism occurs in response to light, *hydrotropism* to water, *chemotropism* to a chemical stimulus, and *thigmotropism*, or *haptotropism*, to physical contact, as in the tendrils of climbing plants when they touch a support and then grow around it.

Tropic movements are the result of greater rate of growth on one side of the plant organ than the other. Tropism differs from a ◊nastic movement in being influenced by the direction of the stimulus.

troposphere lower part of the Earth's ◊atmosphere extending about 10.5 km/6.5 mi from the Earth's surface, in which temperature decreases with height to about −60°C/−76°F except in local layers of temperature inversion. The *tropopause* is the upper boundary of the troposphere, above which the temperature increases slowly with height within the atmosphere. All of the Earth's weather takes place within the troposphere.

troy system system of units used for precious metals and gems. The pound troy (0.37 kg) consists of 12 ounces (each of 120 carats) or 5,760 grains (each equal to 65 mg).

truncation error in computing, an ◊error that occurs when a decimal result is cut off (truncated) after the maximum number of places allowed by the computer's level of accuracy.

truth table in electronics, a diagram showing the effect of a particular ◊logic gate on every combination of inputs.

Every possible combination of inputs and outputs for a particular gate or combination of gates is described, thereby defining their action in full. When logic value 1 is written in the table, it indicates a 'high' (or 'yes') input of perhaps 5 volts; logic value 0 indicates a 'low' (or 'no') input of 0 volts.

trypsin an enzyme in the vertebrate gut responsible for the digestion of protein molecules. It is secreted by the pancreas but in an inactive form known as trypsinogen. Activation into working trypsin occurs only in the small intestine, owing to the action of another enzyme enterokinase, secreted by the wall of the duodenum. Unlike the digestive enzyme pepsin, found in the stomach, trypsin does not require an acid environment.

TSR abbreviation for ◊*terminate and stay resident*.

tsunami (Japanese 'harbour wave') wave generated by an undersea ◊earthquake or volcanic eruption. In the open ocean it may take the form of several successive waves, rarely in excess of a metre in height but travelling at speeds of 650–800 kph/400–500 mph. In the coastal shallows tsunamis slow down and build up, producing towering waves that can sweep inland and cause great loss of life and property. In 1983 an earthquake in the Pacific caused tsunamis up to 3 m/10 ft high, which killed more than 100 people in Akita, northern Japan.

TTL abbreviation for ◊*transistor–transistor logic*, a family of integrated circuits.

tuber swollen region of an underground stem or root, usually modified for storing food. The potato is a *stem tuber*, as shown by the presence of terminal and lateral buds, the 'eyes' of the potato. *Root tubers*, for example dahlias, developed from ◊adventitious roots (growing from the stem, not from other roots) lack these. Both types of tuber can give rise to new individuals and so provide a means of ◊vegetative reproduction.

Tucana constellation of the southern hemisphere, represented as a toucan. It contains the second most prominent ◊globular cluster in the sky, 47 Tucanae, and the Small ◊Magellanic Cloud.

tufa or *travertine* soft, porous, ◊limestone rock, white in colour, deposited from solution from carbonate-saturated ground water around hot springs and in caves.

tumour overproduction of cells in a specific area of the body, often leading to a swelling or lump. Tumours are classified as **benign** or **malignant** (see ◊cancer). Benign tumours grow more slowly, do not invade surrounding tissues, do not spread to other parts of the body, and do not usually recur after removal. However, benign tumours can be dangerous in areas such as the brain. The most familiar types

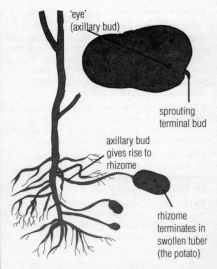

tuber *Tubers are produced underground from stems, as in the potato, or from roots, as in the dahlia. Tubers can grow into new plants.*

of benign tumour are warts on the skin. In some cases, there is no sharp dividing line between benign and malignant tumours.

tung oil oil used in paints and varnishes, obtained from trees of the genus *Aleurites*, family Euphorbiaceae, native to China.

tungsten hard, heavy, grey-white, metallic element, symbol W (from German *Wolfram*), atomic number 74, relative atomic mass 183.85. It occurs in the minerals wolframite, scheelite, and hubertite. It has the highest melting point of any metal (6,170°F/3,410°C) and is added to steel to make it harder, stronger, and more elastic; its other uses include high-speed cutting tools, electrical elements, and thermionic couplings. Its salts are used in the paint and tanning industries.

tungsten ore either of the two main minerals, wolframite (FeMn)WO$_4$ and scheelite, CaWO$_4$, from which tungsten is extracted. Most of the world's tungsten reserves are in China, but the main suppliers are Bolivia, Australia, Canada, and the USA.

Tunguska Event explosion at Tunguska, central Siberia, Russia, in June 1908, which devastated around 6,500 sq km/2,500 sq mi of forest. It is thought to have been caused by either a cometary nucleus or a fragment of ◊Encke's comet about 200 m/220 yards across, or possibly an asteroid. The magnitude of the explosion was equivalent to an atom bomb (10–20 megatons) and produced a colossal shock wave; a bright falling object was seen 600 km/375 mi away and was heard up to 1,000 km/625 mi away.

tunicate any marine ◊chordate of the subphylum Tunicata (Urochordata), for example the sea squirt. Tunicates have transparent or translucent tunics made of cellulose. They vary in size from a few millimetres to 30 cm/1 ft in length, and are cylindrical, circular, or irregular in shape. There are more than a thousand species.

turbine engine in which steam, water, gas, or air is made to spin a rotating shaft by pushing on angled blades, like a fan. Turbines are among the most powerful machines. Steam turbines are used to drive generators in power stations and ships' propellers; water turbines spin the generators in hydroelectric power plants; and gas turbines (as jet engines; see ◊jet propulsion) power most aircraft and drive machines in industry.

turbocharger turbine-driven device fitted to engines to force more air into the cylinders, producing extra power. The turbocharger consists of a 'blower', or ◊compressor, driven by a turbine, which in most units is driven by the exhaust gases leaving the engine.

turbofan jet engine of the type used by most airliners, so called because of its huge front fan. The fan sends air not only into the engine for combustion but also around the engine for additional thrust. This results in a faster and more fuel-efficient propulsive jet (see ◊jet propulsion).

turbojet jet engine that derives its thrust from a jet of hot exhaust gases. Pure turbojets can be very powerful but use a lot of fuel.

turboprop jet engine that derives its thrust partly from a jet of exhaust gases, but mainly from a propeller powered by a turbine in the jet exhaust. Turboprops are more economical than turbojets but can be used only at relatively low speeds.

turbulence irregular fluid (gas or liquid) flow, in which vortices and unpredictable fluctuations and motions occur. ◊Streamlining reduces the turbulence of flow around an object, such as an aircraft, and reduces drag. Turbulent flow of a fluid occurs when the ◊Reynolds number is high.

turgor the rigid condition of a plant caused by the fluid contents of a plant cell exerting a mechanical pressure against the cell wall. Turgor supports plants that do not have woody stems.

turpentine solution of resins distilled from the sap of conifers, used in varnish and as a paint solvent but now largely replaced by ◊white spirit.

turquoise mineral, hydrous basic copper aluminium phosphate. Blue-green, blue, or green, it is a gemstone. Turquoise is found in Australia, Egypt, Ethiopia, France, Germany, Iran, Turkestan, Mexico, and southwestern USA. It was originally introduced into Europe through Turkey, from which its name is derived.

turtle small computer-controlled wheeled robot. The turtle's movements are determined by programs written by a computer user, typically using the high-level programming language ◊LOGO.

TVP (abbreviation for *texturized vegetable protein*) meat substitute usually made from soya beans. In manufacture, the soya-bean solids (what remains after oil has been removed) are ground finely and mixed with a binder to form a sticky mixture. This is forced through a spinneret and extruded into fibres, which are treated with salts and flavourings, wound into hanks, and then chopped up to resemble meat chunks.

twin one of two young produced from a single pregnancy. Human twins may be genetically identical (monozygotic), having been formed from a single fertilized egg that splits into two cells, both of which became implanted. Nonidentical (fraternal or dizygotic) twins are formed when two eggs are fertilized at the same time.

two's complement number system number system, based on the ◊binary number system, that allows both positive and negative numbers to be conveniently represented for manipulation by computer.

two-stroke cycle operating cycle for internal combustion piston engines. The engine cycle is completed after just two strokes (movement up or down) of the piston, which distinguishes it from the more common ◊four-stroke cycle. Power mowers and lightweight motorcycles use two-stroke petrol engines, which are cheaper and simpler than four-strokes.

tympanic membrane another name for the eardrum.

UHF (abbreviation for *ultra high frequency*) referring to radio waves of very short wavelength, used, for example, for television broadcasting.

ulna one of the two bones found in the lower limb of the tetrapod (four-limbed) vertebrate. It articulates with the shorter radius and ◊humerus (upper arm bone) at one end and with the radius and wrist bones at the other.

ultrabasic in geology, an igneous rock with a lower silica content than basic rocks (less than 45% silica).

ultrafiltration process by which substances in solution are separated on the basis of their molecular size. A solution is forced through a membrane with pores large enough to permit the passage of small solute molecules but not large ones.

Ultrafiltration is a vital mechanism in the vertebrate kidney: the cell membranes lining the Bowman's capsule act as semipermeable membranes allowing water and low-molecular-weight substances such as urea and salts to pass through into the urinary tubules but preventing the larger proteins from being lost from the blood.

ultrasonics study and application of the sound and vibrations produced by ultrasonic pressure waves (see ◊ultrasound).

ultrasound pressure waves similar in nature to sound waves but occurring at frequencies above 20,000 Hz (cycles per second), the approximate upper limit of human hearing (15–16 Hz is the lower limit).

Ultrasonics is concerned with the study and practical application of these phenomena.

ultraviolet astronomy study of cosmic ultraviolet emissions using artificial satellites. The USA has launched a series of satellites for this purpose, receiving the first useful data 1968. Only a tiny percentage of solar ultraviolet radiation penetrates the atmosphere, this being the less dangerous longer-wavelength ultraviolet. The dangerous shorter-wavelength radiation is absorbed by gases in the ozone layer high in the Earth's upper atmosphere.

ultraviolet radiation electromagnetic radiation invisible to the human eye, of wavelengths from about 400 to 4 nm (where the ◊X-ray range begins). Physiologically, ultraviolet radiation is extremely powerful, producing sunburn and causing the formation of vitamin D in the skin.

Ulysses space probe to study the Sun's poles, launched 1990 by a US space shuttle. It is a joint project by NASA and the European Space Agency. In Feb 1992, the gravity of Jupiter swung *Ulysses* on to a path that loops it first under the Sun's south pole in 1994 and then over the north pole in 1995 to study the Sun and solar wind at latitudes not observable from the Earth.

umbilical cord connection between the ◊embryo and the ◊placenta of placental mammals. It has one vein and two arteries, transporting oxygen and nutrients to the developing young, and removing waste products. At birth, the connection between the young and the placenta is no longer necessary. The umbilical cord drops off or is severed, leaving a scar called the navel.

umbra region of a ◊shadow that is totally dark because no light reaches it, and from which no part of the light source can be seen (compare ◊penumbra). In astronomy, it is a region of the Earth from which a complete ◊eclipse of the Sun or Moon can be seen.

uncertainty principle or *indeterminacy principle* in quantum mechanics, the principle that it is meaningless to speak of a particle's position, momentum, or other parameters, except as results of measurements; measuring, however, involves an interaction (such as a ◊photon of light bouncing off the particle under scrutiny), which must disturb the particle, though the disturbance is noticeable only at an atomic scale. The principle implies that one cannot, even in theory, predict the moment-to-moment behaviour of such a system.

unconformity in geology, a break in the sequence of ◊sedimentary rocks. It is usually seen as an eroded surface, with the beds above and below lying at different angles. An uncon-

unicellular organism animal or plant consisting of a single cell. Most are invisible without a microscope but a few, such as the giant ◊amoeba, may be visible to the naked eye. The main groups of unicellular organisms are bacteria, protozoa, unicellular algae, and unicellular fungi or yeasts. Some become disease-causing agents, ◊pathogens.

unified field theory in physics, the theory that attempts to explain the four fundamental forces (strong nuclear, weak nuclear, electromagnetic, and gravity) in terms of a single unified force (see ◊particle physics).

Research was begun by Albert Einstein and, by 1971, a theory developed by US physicists Steven Weinberg and Sheldon Glashow, Pakistani physicist Abdus Salam, and others, had demonstrated the link between the weak and electromagnetic forces. The next stage is to develop a theory (called the ◊grand unified theory, or GUT) that combines the strong nuclear force with the electroweak force. The final stage will be to incorporate gravity into the scheme. Work on the ◊superstring theory indicates that this may be the ultimate 'theory of everything'.

uniformitarianism in geology, the principle that processes that can be seen to occur on the Earth's surface today are the same as those that have occurred throughout geological time. For example, desert sandstones containing sand-dune structures must have been formed under conditions similar to those present in deserts today.

unit standard quantity in relation to which other quantities are measured. There have been many systems of units. Some ancient units, such as the day, the foot, and the pound, are still in use. ◊SI units, the latest version of the metric system, are widely used in science.

universal indicator in chemistry, a mixture of ◊pH indicators, used to gauge the acidity or alkalinity of a solution. Each component changes colour at a different pH value, and so the indicator is capable of displaying a range of colours, according to the pH of the test solution, from red (at pH 1) to purple (at pH 13).

universal joint flexible coupling used to join rotating shafts; for example, the propeller shaft in a car. In a typical universal joint the ends of the shafts to be joined end in U-shaped yokes. They dovetail into each other and pivot flexibly about an X-shaped spider. This construction allows side-to-side and up-and-down movement, while still transmitting rotary motion.

universe all of space and its contents, the study of which is called cosmology. The universe is thought to be between 10 billion and 20 billion years old, and is mostly empty space, dotted with ◊galaxies for as far as telescopes can see. The most distant detected galaxies and ◊quasars lie 10 billion light years or more from Earth, and are moving farther apart as the universe expands. Several theories attempt to explain how the universe came into being and evolved, for example, the ◊Big Bang theory of an expanding universe originating in a single explosive event, and the contradictory ◊steady-state theory.

Apart from those galaxies within the ◊Local Group, all the galaxies we see display ◊red shifts in their spectra, indicating that they are moving away from us. The farther we look into space, the greater are the observed red shifts, which implies that the more distant galaxies are receding at ever greater speeds. This observation led to the theory of an expanding universe, first proposed by Edwin Hubble 1929, and to Hubble's law, which states that the speed with which one galaxy moves away from another is proportional to its distance from it. Current data suggest that the galaxies are moving apart at a rate of 50–100 kps/30–60 mps for every million ◊parsecs of distance.

Unix multiuser ◊operating system designed for minicomputers but becoming increasingly popular on large microcomputers, workstations, mainframes, and supercomputers. It was developed by AT&T's Bell Laboratories in the USA during the late 1960s, using the programming language ◊C. It could therefore run on any machine with a C compiler, so ensuring its wide portability. Its wide range of functions and flexibility have made it widely used by universities and in commercial software.

unleaded petrol petrol manufactured without the addition of ◊antiknock. It has a slightly lower octane rating than leaded petrol, but has the advantage of not polluting the atmosphere with lead compounds. Many cars can be converted to running on unleaded petrol by altering the timing of the engine, and most new cars are designed to do so.

Cars fitted with a ◊catalytic converter must use unleaded fuel.

unnilennium synthesized radioactive element of the ◊transactinide series, symbol Une, atomic number 109, relative atomic mass 266. It was first produced in 1982 at the Laboratory for Heavy Ion Research in Darmstadt, Germany, by fusing bismuth and iron nuclei; it took a week to obtain a single new, fused nucleus.

unnilhexium synthesized radioactive element of the ◊transactinide series, atomic number 106, relative atomic mass 263. It was first synthesized in 1974 by two institutions, each of which claims priority. The University of California at Berkeley bombarded californium with oxygen nuclei to get isotope 263; the Joint Institute for Nuclear Research in Dubna, Russia, bombarded lead with chromium nuclei to obtain isotopes 259 and 260. It was named seaborguim (Sg) 1994.

unniloctium synthesized, radioactive element of the ◊transactinide series, symbol Uno, atomic number 108, relative atomic mass 265. It was first synthesized in 1984 by the Laboratory for Heavy Ion Research in Darmstadt, Germany.

unnilpentium synthesized, radioactive, metallic element of the ◊transactinide series, symbol Unp, atomic number 105, relative atomic mass 262. Six isotopes have been synthesized, each with very short (fractions of a second) half-lives. Two institutions claim to have been the first to produce it: the Joint Institute for Nuclear Research in Dubna, Russia, in 1967 (proposed name *nielsbohrium*); and the University of California at Berkeley, USA, who disputed the Soviet claim, in 1970 (proposed name *hahnium*).

unnilquadium synthesized, radioactive, metallic element, the first of the ◊transactinide series, symbol Unq, atomic number 104, relative atomic mass 262. It is produced by bombarding californium with carbon nuclei and has ten isotopes, the longest-lived of which, Unq-262, has a half-life of 70 seconds.

Two institutions claim to be the first to have synthesized it: the Joint Institute for Nuclear Research in Dubna, Russia, in 1964 (proposed name *kurchatovium*); and the University of California at Berkeley, USA, in 1969 (proposed name *rutherfordium*).

unnilseptium synthesized, radioactive element of the ◊transactinide series, symbol Uns, atomic number 107, relative atomic mass 262. It was first synthesized by the Joint Institute for Nuclear Research in Dubna, Russia, in 1976; in 1981 the Laboratory for Heavy Ion Research in Darmstadt, Germany, confirmed its existence.

unsaturated compound chemical compound in which two adjacent atoms are bonded by a double or triple covalent bond.

unsaturated solution solution that is capable of dissolving more solute than it already contains at the same temperature.

unstable equilibrium the state of equilibrium possessed by a body that will not remain at rest if displaced slightly, but will topple over into a new position; it will not return to its original rest position. See ◊stability.

upthrust upward force experienced by all objects that are totally or partially immersed in a fluid (liquid or gas). It acts against the weight of the object, and, according to Archimedes' principle, is always equal to the weight of the fluid displaced by that object. An object will float when the upthrust from the fluid is equal to its weight (see ◊floating).

uraninite uranium oxide, UO_2, an ore mineral of uranium, also known as *pitchblende* when occurring in massive form. It is black or brownish black, very dense, and radioactive. It occurs in veins and as massive crusts, usually associated with granite rocks.

uranium hard, lustrous, silver-white, malleable and ductile, radioactive, metallic element of the ◊actinide series, symbol U, atomic number 92, relative atomic mass 238.029. It is the most abundant radioactive element in the Earth's crust, its decay giving rise to essentially all radioactive elements in nature; its final decay product is the stable element lead. Uranium combines readily with most elements to form compounds that are extremely poisonous. The chief ore is ◊pitchblende, in which the element was discovered by German chemist Martin Klaproth 1789; he named it after the planet Uranus, which had been discovered 1781.

Uranium is one of three fissile elements (the others are thorium and plutonium). It was long considered to be the element with the highest atomic number to occur in nature. The isotopes U-238 and U-235 have been used to help determine the age of the Earth.

Uranium-238, which comprises about 99% of all naturally occurring uranium, has a half-life of 4.51×10^9 years. Because of its abundance, it is the isotope from which fissile plutonium is produced in breeder ◊nuclear reactors. The fissile isotope U-235 has a half-life of 7.13×10^8 years and comprises about 0.7% of naturally occurring uranium; it is used directly as a fuel for nuclear reactors and in the manufacture of nuclear weapons.

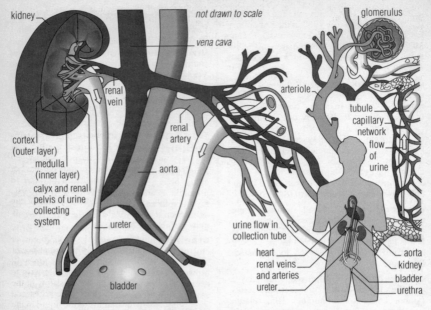

urinary system *The human urinary system. At the bottom right, the complete system in outline; on the left, the arrangement of blood vessels connected to the kidney; at the top right, a detail of the network of vessels within a kidney.*

uranium ore material from which uranium is extracted, often a complex mixture of minerals. The main ore is uraninite (or pitchblende) UO_2, which is commonly found with sulphide minerals. The USA, Canada, and South Africa are the main producers in the West.

Uranus the seventh planet from the Sun, discovered by William Herschel (1738–1822) 1781. It is twice as far out as the sixth planet, Saturn. Uranus has a diameter of 50,800 km/ 31,600 mi and a mass 14.5 times that of Earth. It orbits the Sun in 84 years at an average distance of 2.9 billion km/1.8 billion mi. The spin axis of Uranus is tilted at 98°, so that one pole points towards the Sun, giving extreme seasons. It has 15 moons, and in 1977 was discovered to have thin rings around its equator.

urban ecology study of the ecosystems, animal and plant communities, soils, and microclimates found within an urban landscape.

urea $CO(NH_2)_2$ waste product formed in the mammalian liver when nitrogen compounds are broken down. It is excreted in urine. When purified, it is a white, crystalline solid. In industry it is used to make urea-formaldehyde plastics (or resins), pharmaceuticals, and fertilizers.

ureter tube connecting the kidney to the bladder. Its wall contains fibres of smooth muscle whose contractions aid the movement of urine out of the kidney.

urethra in mammals, a tube connecting the bladder to the exterior. It carries urine and, in males, semen.

uric acid $C_5H_4N_4O_3$ nitrogen-containing waste substance, formed from the breakdown of food and body protein.

It is only slightly soluble in water. Uric acid is the normal means by which most land animals that develop in a shell (birds, reptiles, insects, and land gastropods) deposit their waste products. The young are unable to get rid of their excretory products while in the shell and therefore store them in this insoluble form.

Humans and other primates produce some uric acid as well as urea, the normal nitrogenous waste product of mammals, adult amphibians, and many marine fishes. If formed in excess and not excreted, uric acid may be deposited in sharp crystals in the joints and other tissues, causing gout; or it may form stones (calculi) in the kidneys or bladder.

urinary system system of organs that removes nitrogenous waste products and excess water from the bodies of animals. In vertebrates, it consists of a pair of kidneys, which produce urine; ureters, which drain the kidneys; and (in bony fishes, amphibians, some reptiles, and mammals) a bladder that stores the urine before its discharge. In mammals, the urine is expelled through the urethra; in other vertebrates, the urine drains into a common excretory chamber called a ◊cloaca, and the urine is not discharged separately.

urine amber-coloured fluid filtered out by the kidneys from the blood. It contains excess water, salts, proteins, waste products in the form of urea, a pigment, and some acid.

Ursa Major third-largest constellation in the sky, in the north polar region. Its seven brightest stars make up the familiar shape of the *Big Dipper* or *Plough*. The second star of the 'handle' of the dipper, called Mizar, has a companion star, Alcor. Two stars forming the far side of the 'bowl' act as pointers to the north pole star, Polaris.

Ursa Minor constellation in the northern sky. It is shaped like a dipper, with the north pole star Polaris at the end of the handle.

user-friendly term used to describe the ease of use of a computer system, particularly for those with little understanding or familiarity with computers. Even for experienced users, user-friendly programs are quicker to learn.

user interface in computing, the procedures and methods through which the user operates a program. These might include ◊menus, input forms, error messages, and keyboard procedures. A ◊graphical user interface (GUI or WIMP) is one that makes use of icons (small pictures) and allows the user to make menu selections with a mouse.

UTC abbreviation for *coordinated universal time*, the standard measurement of ◊time.

uterus hollow muscular organ of female mammals, located between the bladder and rectum, and connected to the Fallopian tubes above and the vagina below. The embryo develops within the uterus, and in placental mammals is attached to it after implantation via the ◊placenta and umbilical cord. The lining of the uterus changes during the ◊menstrual cycle. In humans and other higher primates, it is a single structure, but in other mammals it is paired.

utility program in computing, a systems program designed to perform a specific task related to the operation of the computer when requested to do so by the computer user. For example, a utility program might be used to complete a screen dump, format a disc, or convert the format of a data file so that it can be accessed by a different applications program.

U-tube U-shaped tube that may be partly filled with liquid and used as a manometer, or instrument for measuring ◊pressure. Greater pressure on the liquid surface in one arm will force the level of the liquid in the other arm to rise. A difference between the pressures in the spaces in the two arms is therefore registered as a difference in the heights of the liquid in the arms.

UV in physics, abbreviation for *ultraviolet*.

U-value measure of a material's heat-conducting properties. It is used in the building industry to compare the efficiency of insulating products, a good insulator having a low U-value. The U-value of a material is defined as the rate at which heat is conducted through it per unit surface area per unit temperature-difference between its two sides; it is measured in watts per square metre per kelvin (W m^{-2} K^{-1}). In equation terms, it may be calculated as the rate of loss of heat/surface area × temperature difference.

uvula muscular structure descending from the soft palate. Its length is variable (5–20 mm/0.2–0.8 in in humans). It is raised during swallowing and is sometimes used in the production of speech sounds.

V in physics, symbol for *volt*.

v in physics, symbol for *velocity*.

vaccine any preparation of modified pathogens (viruses or bacteria) that is introduced into the body, usually either orally or by a hypodermic syringe, to induce the specific ◊antibody reaction that produces ◊immunity against a particular disease.

vacuole in biology, a fluid-filled, membrane-bound cavity inside a cell. It may be a reservoir for fluids that the cell will secrete to the outside, or may be filled with excretory products or essential nutrients that the cell needs to store. In amoebae (single-cell animals), vacuoles are the sites of digestion of engulfed food particles. Plant cells usually have a large central vacuole for storage.

vacuum in general, a region completely empty of matter; in physics, any enclosure in which the gas pressure is considerably less than atmospheric pressure (101,325 pascals).

vacuum flask or *Dewar flask* or *Thermos flask* container for keeping things either hot or cold. It has two silvered glass walls with a vacuum between them, in a metal or plastic outer case. This design reduces the three forms of heat transfer: radiation (prevented by the silvering), conduction, and convection (both prevented by the vacuum). A vacuum flask is therefore equally efficient at keeping cold liquids cold, or hot liquids hot.

vagina the lower part of the reproductive tract in female mammals, linking the uterus to the exterior. It admits the penis during sexual intercourse, and is the birth canal down which the baby passes during delivery.

valence electron in chemistry, an electron in the outermost shell of an ◊atom. It is the valence electrons that are involved in the formation of ionic and covalent bonds (see ◊molecule). The number of electrons in this outermost shell represents the maximum possible valence for many elements and matches the number of the group that the element occupies in the ◊periodic table of the elements.

valency in chemistry, the measure of an element's ability to combine with other elements, expressed as the number of atoms of hydrogen (or any other standard univalent element) capable of uniting with (or replacing) its atoms. The number of electrons in the outermost shell of the atom dictates the combining ability of an element.

The elements are described as uni-, di-, tri-, and tetravalent when they unite with one, two, three, and four univalent atoms respectively. Some elements have *variable valency*: for example, nitrogen and phosphorus have a valency of both three and five. The valency of oxygen is two: hence the formula for water, H_2O (hydrogen being univalent).

valency shell in chemistry, outermost shell of electrons in an ◊atom. It contains the ◊valence electrons. Elements with four or more electrons in their outermost shell can show variable valency. Chlorine can show valencies of 1, 3, 5, and 7 in different compounds.

validation in computing, the process of checking input data to ensure that it is complete, accurate, and reasonable. Although it would be impossible to guarantee that only valid data are entered into a computer, a suitable combination of validation checks should ensure that most errors are detected.

valve device that controls the flow of a fluid. Inside a valve, a plug moves to widen or close the opening through which the fluid passes. The valve was invented by US radio engineer Lee de Forest (1873–1961).

valve in animals, a structure for controlling the direction of the blood flow. In humans and other vertebrates, the contractions of the beating heart cause the correct blood flow into the arteries because a series of valves prevent back flow. Diseased valves, detected as 'heart murmurs', have decreased efficiency. The tendency for low-pressure venous blood to collect at the base of limbs under the influence of gravity is counteracted by a series of small valves within the veins. It was the existence of these valves that prompted the 17th-century physician William Harvey to suggest that the blood circulated around the body.

valve or *electron tube* in electronics, a glass

tube containing gas at low pressure, which is used to control the flow of electricity in a circuit.

Three or more metal electrodes are inset into the tube. By varying the voltage on one of them, called the **grid electrode**, the current through the valve can be controlled, and the valve can act as an amplifier. Valves have been replaced for most applications by ◊transistors.

vanadium silver-white, malleable and ductile, metallic element, symbol V, atomic number 23, relative atomic mass 50.942. It occurs in certain iron, lead, and uranium ores and is widely distributed in small quantities in igneous and sedimentary rocks. It is used to make steel alloys, to which it adds tensile strength.

vanadium(V) oxide Va_2O_5 or *vanadium pentoxide* crystalline compound used as a catalyst in the ◊contact process for the manufacture of sulphuric acid.

Van Allen radiation belts two zones of charged particles around the Earth's magnetosphere, discovered 1958 by US physicist James Van Allen. The atomic particles come from the Earth's upper atmosphere and the ◊solar wind, and are trapped by the Earth's magnetic field. The inner belt lies 1,000–5,000 km/620–3,100 mi above the equator, and contains ◊protons and ◊electrons. The outer belt lies 15,000–25,000 km/9,300–15,500 mi above the equator, but is lower around the magnetic poles. It contains mostly electrons from the solar wind.

van de Graaff generator electrostatic generator capable of producing a voltage of over a million volts. It consists of a continuous vertical conveyor belt that carries electrostatic charges (resulting from friction) up to a large hollow sphere supported on an insulated stand. The lower end of the belt is earthed, so that charge accumulates on the sphere. The size of the voltage built up in air depends on the radius of the sphere, but can be increased by enclosing the generator in an inert atmosphere, such as nitrogen.

van der Waals' law modified form of the ◊gas laws that includes corrections for the non-ideal behaviour of real gases (the molecules of ideal gases occupy no space and exert no forces on each other). It is named after Dutch physicist J D van der Waals (1837–1923).

The equation derived from the law states that:

$$(P + aV^2)(V - b) = RT$$

where P, V, and T are the pressure, volume, and temperature (in kelvin) of the gas, respectively; R is the ◊gas constant; and a and b are constants for that particular gas.

Vanguard early series of US Earth-orbiting satellites and their associated rocket launcher. *Vanguard 1* was the second US satellite, launched 17 March 1958 by the three-stage Vanguard rocket. Tracking of its orbit revealed that Earth is slightly pear-shaped. The series ended Sept 1959 with *Vanguard 3*.

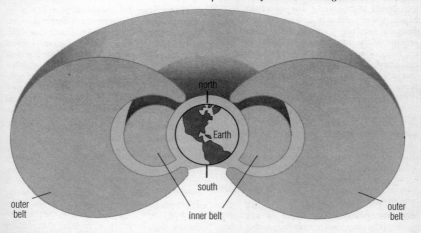

Van Allen radiation belts *The Van Allen radiation belts of trapped charged particles are a hazard to spacecraft, affecting on board electronics and computer systems. Similar belts have been discovered around the planets Mercury, Jupiter, Saturn, Uranus, and Neptune.*

van de Graaff generator *The van de Graaff generator can produce more than a million volts. Such generators are used as particle accelerators for experiments with charged particles.*

vapour one of the three states of matter (see also ◊solid and ◊liquid). The molecules in a vapour move randomly and are far apart, the distance between them, and therefore the volume of the vapour, being limited only by the walls of any vessel in which they might be contained. A vapour differs from a ◊gas only in that a vapour can be liquefied by increased pressure, whereas a gas cannot unless its temperature is lowered below its ◊critical temperature; it then becomes a vapour and may be liquefied.

vapour density density of a gas, expressed as the ◊mass of a given volume of the gas divided by the mass of an equal volume of a reference gas (such as hydrogen or air) at the same temperature and pressure. It is equal approximately to half the relative molecular weight (mass) of the gas.

vapour pressure pressure of a vapour given off by (evaporated from) a liquid or solid, caused by vibrating atoms or molecules continuously escaping from its surface. In an enclosed space, a maximum value is reached when the number of particles leaving the surface is in equilibrium with those returning to it; this is known as the *saturated vapour pressure*.

variable in computing, a quantity that can take different values. Variables can be used to represent different items of data in the course of a program.

variable in mathematics, a changing quantity (one that can take various values), as opposed to a ◊constant. For example, in the algebraic expression $y = 4x^3 + 2$, the variables are x and y, whereas 4 and 2 are constants.

variable star in astronomy, a star whose brightness changes, either regularly or irregularly, over a period ranging from a few hours to months or even years. The ◊Cepheid variables regularly expand and contract in size every few days or weeks.

variation in biology, a difference between individuals of the same species, found in any sexually reproducing population. Variations may be almost unnoticeable in some cases, obvious in others, and can concern many aspects of the organism. Typically, variation in size, behaviour, biochemistry, or colouring may be found. The cause of the variation is genetic (that is, inherited), environmental, or more usually a combination of the two. The origins of variation can be traced to the recombination of the genetic material during the formation of the gametes, and, more rarely, to mutation.

variegation description of plant leaves or stems that exhibit patches of different colours. The term is usually applied to plants that show white, cream, or yellow on their leaves, caused by areas of tissue that lack the green pigment ◊chlorophyll. Variegated plants are bred for their decorative value, but they are often considerably weaker than the normal, uniformly green plant. Many will not breed true and require ◊vegetative reproduction.

vascular bundle in botany, strand of primary conducting tissue (a 'vein') in vascular plants, consisting mainly of water-conducting tissues, metaxylem and protoxylem, which together make up the primary ◊xylem, and nutrient-conducting tissue, ◊phloem. It extends from the roots to the stems and leaves. Typically the phloem is situated nearest to the epidermis and the xylem towards the centre of the bundle. In plants exhibiting ◊secondary growth, the xylem and phloem are separated by a thin layer of vascular ◊cambium, which gives rise to new conducting tissues.

vascular plant plant containing vascular bundles. ◊Pteridophytes (ferns, horsetails, and club mosses), ◊gymnosperms (conifers and cycads), and ◊angiosperms (flowering plants) are all vascular plants.

vas deferens in male vertebrates, a tube conducting sperm from the testis to the urethra. The sperm is carried in a fluid secreted by various glands, and can be transported very rapidly when the smooth muscle in the wall of the vas deferens undergoes rhythmic contraction, as in sexual intercourse.

vasoconstriction narrowing of blood vessels. It is mediated by the vasomotor centre in the ◊brainstem, causing reflex changes in the heart rate with an adjustment of blood pressure. Some drugs act as vasoconstrictors.

vasodilation widening of the blood vessels; the opposite of ◊vasoconstriction. Some drugs act as vasodilators.

vasopressin another name for the antidiuretic hormone, ◊ADH.

VDU abbreviation for ◊visual display unit.

vector disease-carrying agent. The term is usually applied to insects, ticks and the like, that transmit parasites from one person to another or from animals to human beings. The *Anopheles* mosquito is a vector, transmitting the ◊malaria parasite in its salivary glands.

vector graphics computer graphics that are stored in the computer memory by using geometric formulas. Vector graphics can be transformed (enlarged, rotated, stretched, and so on) without loss of picture resolution. It is also possible to select and transform any of the components of a vector-graphics display because each is separately defined in the computer memory. In these respects vector graphics are superior to ◊raster graphics. Vector graphics are typically used for drawing applications, allowing the user to create and modify technical diagrams such as designs for houses or cars.

vector quantity any physical quantity that has both magnitude and direction (such as the velocity or acceleration of an object) as distinct from ◊scalar quantity (such as speed, density, or mass), which has magnitude but no direction. A vector is represented either geometrically by an arrow whose length corresponds to its magnitude and points in an appropriate direction, or by a pair of numbers written vertically and placed within parentheses ($x\ y$).

Vectors can be added graphically by constructing a parallelogram of vectors (such as the ◊parallelogram of forces commonly employed in physics and engineering).

Vega or *Alpha Lyrae* brightest star in the constellation Lyra and the fifth-brightest star in the sky. It is a blue-white star, 25 light years from Earth, with a luminosity 50 times that of the Sun.

vegetative reproduction type of ◊asexual reproduction in plants that relies not on spores, but on multicellular structures formed by the parent plant. Some of the main types are ◊stolons and runners, ◊gemmae, ◊bulbils, sucker shoots produced from roots (such as in the creeping thistle *Cirsium arvense*), ◊tubers, ◊bulbs, ◊corms, and ◊rhizomes.

vein in animals with a circulatory system, any vessel that carries blood from the body to the heart. Veins contain valves that prevent the blood from running back when moving against gravity. They always carry deoxygenated blood, with the exception of the veins leading from the lungs to the heart in birds and mammals, which carry newly oxygenated blood.

Vela constellation of the southern hemisphere near Carina, represented as the sails of a ship. It contains large wisps of gas – called the Gum

vascular bundle *The fluid-carrying tissue of most plants is normally arranged in units called vascular bundles. The vascular tissue is of two types: xylem and phloem. The xylem carries water up through the plant; the phloem distributes food made in the leaves to all parts of the plant.*

nebula after its discoverer, the Australian astronomer Colin Gum (1924–1960) – believed to be the remains of one or more ◊supernovae. Vela also contains the second optical ◊pulsar (a pulsar that flashes at a visible wavelength) to be discovered.

velocity speed of an object in a given direction. Velocity is a ◊vector quantity, since its direction is important as well as its magnitude (or speed).

The velocity at any instant of a particle travelling in a curved path is in the direction of the tangent to the path at the instant considered. The velocity v of an object travelling in a fixed direction may be calculated by dividing the distance s it has travelled by the time t taken to do so, and may be expressed as:

$$v = s/t$$

velocity ratio (VR) or *distance ratio* in a machine, the distance moved by the input force, or effort, divided by the distance moved by the output force, or load in the same time. It follows that the velocities of the effort and the load are in the same ratio. Velocity ratio has no units. See also ◊mechanical advantage and ◊efficiency.

vena cava either of the two great veins of the trunk, returning deoxygenated blood to the right atrium of the ◊heart. The *superior vena cava*, beginning where the arches of the two innominate veins join high in the chest, receives blood from the head, neck, chest, and arms; the *inferior vena cava*, arising from the junction of the right and left common iliac veins, receives blood from all parts of the body below the diaphragm.

Venn diagram in mathematics, a diagram representing a ◊set or sets and the logical relationships between them. The sets are drawn as circles. An area of overlap between two circles (sets) contains elements that are common to both sets, and thus represents a third set. Circles that do not overlap represent sets with no elements in common (disjoint sets). The method is named after the British logician John Venn (1834–1923).

ventral surface the front of an animal. In vertebrates, the side furthest from the backbone; in invertebrates, the side closest to the ground. The positioning of the main nerve pathways on the ventral side is a characteristic of invertebrates.

ventricle in zoology, either of the two lower chambers of the heart that force blood into the circulation by contraction of their muscular walls. The term also refers to any of four cavities within the brain in which cerebrospinal fluid is produced.

Venturi tube device for measuring the rate of fluid flow through a pipe. It consists of a tube with a constriction (narrowing) in the middle of its length. The constriction causes a drop in pressure in the fluid flowing in the pipe. A pressure gauge attached to the constriction measures the pressure drop and this is used to find the rate of fluid flow.

Venus second planet from the Sun. It orbits the Sun every 225 days at an average distance of 108.2 million km/67.2 million mi and can approach the Earth to within 38 million km/24 million mi, closer than any other planet. Its diameter is 12,100 km/7,500 mi and its mass is 0.82 that of Earth. Venus rotates on its axis more slowly than any other planet, once every 243 days and from east to west, the opposite direction to the other planets (except Uranus and possibly Pluto). Venus is shrouded by clouds of sulphuric acid droplets that sweep across the planet from east to west every four days. The atmosphere is almost entirely carbon dioxide, which traps the Sun's heat by the ◊greenhouse effect and raises the planet's surface temperature to 480°C/900°F, with an atmospheric pressure of 90 times that at the surface of the Earth.

verdigris green-blue coating of copper ethanoate that forms naturally on copper, bronze, and brass. It is an irritating, greenish, poisonous compound made by treating copper with ethanoic acid, and was formerly used in wood preservatives, antifouling compositions, and green paints.

verification in computing, the process of checking that data being input to a computer have been accurately copied from a source document.

This may be done visually, by checking the original copy of the data against the copy shown on the VDU screen. A more thorough method is to enter the data twice, using two different keyboard operators, and then to check the two sets of input copies against each other. The checking is normally carried out by the computer itself, any differences between the two copies being reported for correction by one of the the keyboard operators.

vermilion HgS red form of mercuric sulphide; a scarlet that occurs naturally as the crystalline mineral ◊cinnabar.

vernal equinox see spring ◊equinox.

vernalization the stimulation of flowering by exposure to cold. Certain plants will not flower unless subjected to low temperatures during their development. For example, winter wheat will flower in summer only if planted in the previous autumn. However, by placing partially germinated seeds in low temperatures for several days, the cold requirement can be supplied artificially, allowing the wheat to be sown in the spring.

vernier device for taking readings on a graduated scale to a fraction of a division. It consists of a short divided scale that carries an index or pointer and is slid along a main scale. It was invented by Pierre Vernier.

vertebral column the backbone, giving support to an animal and protecting the spinal cord. It is made up of a series of bones or vertebrae running from the skull to the tail with a central canal containing the nerve fibres of the spinal cord. In tetrapods the vertebrae show some specialization with the shape of the bones varying according to position. In the chest region the upper or thoracic vertebrae are shaped to form connections to the ribs. The backbone is only slightly flexible to give adequate rigidity to the animal structure.

vertebrate any animal with a backbone. The 41,000 species of vertebrates include mammals, birds, reptiles, amphibians, and fishes. They include most of the larger animals, but in terms of numbers of species are only a tiny proportion of the world's animals. The zoological taxonomic group Vertebrata is a subgroup of the ◊phylum Chordata.

vertex (plural *vertices*) in geometry, a point shared by three or more sides of a solid figure; the point farthest from a figure's base; or the point of intersection of two sides of a plane figure or the two rays of an angle.

vertical takeoff and landing craft (VTOL) aircraft that can take off and land vertically. ◊Helicopters, airships, and balloons can do this, as can a few fixed-wing aeroplanes, like the convertiplane.

vestigial organ in biology, an organ that remains in diminished form after it has ceased to have any significant function in the adult organism. In humans, the appendix is vestigial, having once had a digestive function in our ancestors.

VHF (abbreviation for *very high frequency*) referring to radio waves that have very short wavelengths (10 m–1 m). They are used for interference-free FM transmissions (see ◊frequency modulation). VHF transmitters have a relatively short range because the waves cannot be reflected over the horizon like longer radio waves.

video camera or *camcorder* portable television camera that takes moving pictures electronically on magnetic tape. It produces an electrical output signal corresponding to rapid line-by-line scanning of the field of view. The output is recorded on video cassette and is played back on a television screen via a video cassette recorder.

video cassette recorder (VCR) device for recording on and playing back video cassettes; see ◊videotape recorder.

video disc disc with pictures and sounds recorded on it, played back by laser. The video disc is a type of ◊compact disc.

video game electronic game played on a visual-display screen or, by means of special additional or built-in components, on the screen of a television set. The first commercially sold was a simple bat-and-ball game developed in the USA 1972, but complex variants are now available in colour and with special sound effects.

videotape recorder (VTR) device for recording pictures and sound on cassettes or spools of magnetic tape. The first commercial VTR was launched 1956 for the television broadcasting industry, but from the late 1970s cheaper models developed for home use, to record broadcast programmes for future viewing and to view rented or owned video cassettes of commercial films.

videotext system in which information (text and simple pictures) is displayed on a television (video) screen. There are two basic systems, known as ◊teletext and ◊viewdata. In the teletext system information is broadcast with the ordinary television signals, whereas in the viewdata system information is relayed to the screen from a central data bank via the telephone network. Both systems require the use of a television receiver (or a connected VTR) with special decoder.

viewdata system of displaying information on a television screen in which the information is extracted from a computer data bank and transmitted via the telephone lines. It is one form of ◊videotext. The British Post Office (now British Telecom) developed the first viewdata system, Prestel, 1975. Similar systems are now in widespread use in other countries. Users have access to a large store of information, presented on the screen in the form of 'pages'.

Viking probes two US space probes to Mars, each one consisting of an orbiter and a lander. They were launched 20 Aug and 9 Sept 1975. They transmitted colour pictures and analysed the soil.

villus plural *villi* small fingerlike projection extending into the interior of the small intestine and increasing the absorptive area of the intestinal wall. Digested nutrients, including sugars and amino acids, pass into the villi and are carried away by the circulating blood.

Virgo zodiacal constellation, the second largest in the sky. It is represented as a maiden holding an ear of wheat. The Sun passes through Virgo from late Sept to the end of Oct. Virgo's brightest star is the first-magnitude ◊Spica. Virgo contains the nearest large cluster of galaxies to us, 50 million light years away, consisting of about 3,000 galaxies centred on the giant elliptical galaxy M87. Also in Virgo is the nearest ◊quasar, 3C 273, an estimated 3 billion light years distant. In astrology, the dates for Virgo are between about 23 Aug and 22 Sept.

virion the smallest unit of a mature ◊virus.

virtual in computing, without physical existence. Some computers have virtual memory, making their immediate-access memory seem larger than it is; some computers can also simulate *virtual devices*. For example, the Acorn A3000 and A5000 computers have only one floppy-disc drive but can behave as if they were equipped with two, using part of the ◊RAM to simulate the second drive. ◊Virtual reality is a computer simulation of a whole physical environment.

virtual memory in computing, a technique whereby a portion of the computer backing storage, or external, ◊memory is used as an extension of its immediate-access, or internal, memory. The contents of an area of the immediate-access memory are stored on, say, a hard disc while they are not needed, and brought back into main memory when required.

virtual reality advanced form of computer simulation, in which a participant has the illusion of being part of an artificial environment. The participant views the environment through two tiny television screens (one for each eye) built into a visor. Sensors detect movements of the participant's head or body, causing the apparent viewing position to change. Gloves (datagloves) fitted with sensors may be worn, which allow the participant seemingly to pick up and move objects in the environment.

virus infectious particle consisting of a core of nucleic acid (DNA or RNA) enclosed in a protein shell. Viruses are acellular and able to function and reproduce only if they can invade a living cell to use the cell's system to replicate themselves. In the process they may disrupt or alter the host cell's own DNA. The healthy human body reacts by producing an antiviral protein, ◊interferon, which prevents the infection spreading to adjacent cells.

Many viruses mutate continuously so that the host's body has little chance of developing permanent resistance; others transfer between species, with the new host similarly unable to develop resistance. The viruses that cause ◊AIDS and Lassa fever are both thought to have 'jumped' to humans from other mammalian hosts.

Among diseases caused by viruses are canine distemper, chickenpox, common cold, herpes, influenza, rabies, smallpox, yellow fever, AIDS, and many plant diseases. Recent evidence implicates viruses in the development of some forms of cancer (see ◊oncogenes). ***Bacteriophages*** are viruses that infect bacterial cells.

virus in computing, a piece of ◊software that can replicate itself and transfer itself from one computer to another, without the user being aware of it. Some viruses are relatively harmless, but others can damage or destroy data. They are written by anonymous programmers, often maliciously, and are spread along telephone lines or on ◊floppy discs. Antivirus software can be used to detect and destroy well-known viruses, but new viruses continually appear and these may bypass existing antivirus programs.

viscera general term for the organs contained in the chest and abdominal cavities.

viscose yellowish, syrupy solution made by treating cellulose with sodium hydroxide and carbon disulphide. The solution is then regenerated as continuous filament for the making of ◊rayon and as cellophane.

viscosity in physics, the resistance of a fluid to flow, caused by its internal friction, which makes it resist flowing past a solid surface or other layers of the fluid. It applies to the motion of an object moving through a fluid as well as the motion of a fluid passing by an object.

visual display unit (VDU) computer terminal consisting of a keyboard for input data and a screen for displaying output. The oldest and the most popular type of VDU screen is the ◊cathode-ray tube (CRT), which uses essentially the same technology as a television screen. Other types use plasma display technology and ◊liquid-crystal displays.

visual field area that can be seen without moving the head or eyes. When both eyes are open and looking straight ahead it is known as the binocular visual field; if only one eye is used it is uni-ocular. Away from the centre of the visual field white objects show up much better than do coloured ones.

vitalism the idea that living organisms derive their characteristic properties from a universal life force. In the present century, this view is associated with the French philosopher Henri ◊Bergson.

vitamin any of various chemically unrelated organic compounds that are necessary in small quantities for the normal functioning of the human body. Many act as coenzymes, small molecules that enable ◊enzymes to function effectively. They are normally present in adequate amounts in a balanced diet. Deficiency of a vitamin may lead to a metabolic disorder ('deficiency disease'), which can be remedied by sufficient intake of the vitamin. They are generally classified as **water-soluble** (B and C) or **fat-soluble** (A, D, E, and K). See separate entries for individual vitamins, also ◊nicotinic acid, ◊folic acid, and ◊pantothenic acid.

main vitamins

Vitamin		Main dietary sources	Major functions in the body	Deficiency symptoms
A	retinol	dairy products, egg yolk, liver; also formed in body from beta-carotene, a pigment present in some leafy vegetables	aids growth, night vision; keeps mucous membranes healthy	
B_1	thiamine	meat, potatoes, cereals, legumes, nuts	essential for carbohydrate metabolism	beriberi
B_2	riboflavin	eggs, liver, milk	involved in energy metabolism	inflammation of tongue and lips; sores in corners of the mouth
B_6	pyridoxine	widely distributed in foods	involved in protein metabolism	neurological problems, kidney stones
B_{12}	cyanocobalamin	meat, fish, eggs, dairy products	involved in synthesis of nucleic acids, maintenance of myelin sheath around nerve fibres; efficient use of folic acid	
	biotin	legumes, lean meat, liver, egg yolk	essential for metabolism of fats	fatigue, muscle pain, dermatitis, nausea, depression
	folic acid	liver, vegetables	involved in synthesis of nucleic acids	megaloblastic anaemia
	nicotinic acid (or niacin)	meat, yeast extract, some cereals; also formed in the body from the amino acid tryptophan	involved in energy metabolism	pellagra
	pantothenic acid	widely distributed in foods	involved in energy metabolism	fatigue, sleep disturbances, poor coordination, nausea
C	ascorbic acid	citrus fruits, green vegetables, tomatoes	important in synthesis of collagen (constituent of connective tissue)	scurvy
D	calciferol, cholecalciferol	liver, fish oil, dairy products, eggs	promotes growth and mineralization of bone	rickets in children; osteomalacia in adults
E	tocopherol	vegetable oils, eggs, butter, some cereals	prevents damage to cell membranes	anaemia
K	phytomenadione, menaquinone	green vegetables, cereals, fruits, meat	essential for blood clotting	haemorrhage

vitamin A another name for ◊retinol.

vitamin B₁ another name for ◊thiamine.

vitamin B₁₂ another name for ◊cyanocobalamin.

vitamin B₂ another name for ◊riboflavin.

vitamin B₆ another name for ◊pyridoxine.

vitamin C another name for ◊ascorbic acid.

vitamin D another name for ◊cholecalciferol.

vitamin E another name for ◊tocopherol.

vitamin H another name for ◊biotin.

vitamin K another name for ◊phytomenadione.

vitreous humour transparent jellylike substance behind the lens of the vertebrate ◊eye. It gives rigidity to the spherical form of the eye and allows light to pass through to the retina.

vitriol any of a number of sulphate salts. Blue, green, and white vitriols are copper, ferrous, and zinc sulphate, respectively. *Oil of vitriol* is sulphuric acid.

vivipary in animals, a method of reproduction in which the embryo develops inside the body of the female from which it gains nourishment (in contrast to ◊ovipary and ◊ovovivipary). Vivipary is best developed in placental mammals, but also occurs in some arthropods, fishes, amphibians, and reptiles that have placentalike structures. In plants, it is the formation of young plantlets or bulbils instead of flowers. The term also describes seeds that germinate prematurely, before falling from the parent plant.

vivisection literally, cutting into a living animal. Used originally to mean experimental surgery or dissection practised on a live subject, the term is often used by antivivisection campaigners to include any experiment on animals, surgical or otherwise.

VLF in physics, the abbreviation for *very low ◊frequency*.

vocal cords the paired folds, ridges, or cords of tissue within a mammal's larynx, and a bird's syrinx. Air constricted between the folds or membranes makes them vibrate, producing sounds. Muscles in the larynx change the pitch of the sounds produced, by adjusting the tension of the vocal cords.

voice sound produced through the mouth and by the passage of air between the ◊vocal cords. In humans the sound is much amplified by the hollow sinuses of the face, and is modified by the movements of the lips, tongue, and cheeks.

vol abbreviation for *volume*.

volatile in chemistry, term describing a substance that readily passes from the liquid to the vapour phase. Volatile substances have a high ◊vapour pressure.

volatile memory in computing, ◊memory that loses its contents when the power supply to the computer is disconnected.

volcano crack in the Earth's crust through which hot magma (molten rock) and gases well up. The magma is termed lava when it reaches the surface. A volcanic mountain, usually cone shaped with a crater on top, is formed around the opening, or vent, by the build-up of solidified lava and ashes (rock fragments). Most volcanoes arise on plate margins (see ◊plate tectonics), where the movements of plates generate magma or allow it to rise from the mantle beneath. However, a number are found far from plate-margin activity, on 'hot spots' where the Earth's crust is thin.

There are two main types of volcano:

Composite volcanoes, such as Stromboli and Vesuvius in Italy, are found at destructive plate margins (areas where plates are being pushed together), usually in association with island arcs and coastal mountain chains. The magma is mostly derived from plate material and is rich in silica. This makes a very stiff lava such as andesite, which solidifies rapidly to form a high, steep-sided volcanic mountain. The magma often clogs the volcanic vent, causing violent eruptions as the blockage is blasted free, as in the eruption of Mount St Helens, USA, in 1980. The crater may collapse to form a caldera.

Shield volcanoes, such as Mauna Loa in Hawaii, are found along the rift valleys and ocean ridges of constructive plate margins (areas where plates are moving apart), and also over hot spots. The magma is derived from the Earth's mantle and is quite free-flowing. The lava formed from this magma – usually basalt – flows for some distance over the surface before it sets and so forms broad low volcanoes. The lava of a shield volcano is not ejected violently but simply flows over the crater rim.

volt SI unit of electromotive force or electric potential, symbol V. A small battery has a potential of 1.5 volts; the domestic electricity supply in the UK is 240 volts (110 volts in the USA); and a high-tension transmission line may carry up to 765,000 volts.

The absolute volt is defined as the potential difference necessary to produce a current of one ampere through an electric circuit with a resistance of one ohm. It can also be defined as the potential difference that requires one joule of work to move a positive charge of one coulomb from the lower to the higher potential. It is named after the Italian scientist Alessandro Volta.

voltage commonly used term for ◊potential difference (pd).

voltage amplifier electronic device that increases an input signal in the form of a voltage or ◊potential difference, delivering an output signal that is larger than the input by a specified ratio.

voltmeter instrument for measuring potential difference (voltage). It has a high internal resistance (so that it passes only a small current), and is connected in parallel with the component across which potential difference is to be measured. A common type is constructed from a sensitive current-detecting moving-coil ◊galvanometer placed in series with a high-value resistor (multiplier). To measure an AC (◊alternating current) voltage, the circuit must usually include a rectifier; however, a moving-iron instrument can be used to measure alternating voltages without the need for such a device.

volume in geometry, the space occupied by a three-dimensional solid object. A prism (such as a cube) or a cylinder has a volume equal to the area of the base multiplied by the height. For a pyramid or cone, the volume is equal to one-third of the area of the base multiplied by the perpendicular height. The volume of a sphere is equal to $4/3 \times \pi r^3$, where r is the radius. Volumes of irregular solids may be calculated by the technique of ◊integration.

volumetric analysis procedure used for determining the concentration of a solution. A known volume of a solution of unknown concentration is reacted with a solution of known concentration (standard). The standard solution is delivered from a ◊burette so the volume added is known. This technique is known as ◊titration. Often an indicator is used to show when the correct proportions have reacted. This procedure is used for acid–base, ◊redox, and certain other reactions involving solutions.

Voskhod Soviet spacecraft used in the mid-1960s; it was modified from the single-seat Vostok, and was the first spacecraft capable of carrying two or three cosmonauts. During *Voskhod 2*'s flight 1965, Aleksi Leonov made the first space walk.

Vostok first Soviet spacecraft, used 1961–63. Vostok was a metal sphere 2.3 m/7.5 ft in diameter, capable of carrying one cosmonaut. It made flights lasting up to five days. *Vostok 1* carried the first person into space, Yuri Gagarin.

Voyager probes two US space probes, originally ◊Mariners. *Voyager 1*, launched 5 Sept 1977, passed Jupiter March 1979, and reached Saturn Nov 1980. *Voyager 2* was launched earlier, 20 Aug 1977, on a slower trajectory that took it past Jupiter July 1979, Saturn Aug 1981, Uranus Jan 1986, and Neptune Aug 1989. Like the ◊Pioneer probes, the *Voyagers* are on their way out of the Solar System. Their tasks now include helping scientists to locate the position of the heliopause, the boundary at which the influence of the Sun gives way to the forces exerted by other stars.

VR abbreviation for ◊*velocity ratio.*

VRAM (abbreviation for *video random-access memory*) form of ◊RAM that allows simultaneous access by two different devices, so that graphics can be handled at the same time as data is updated. VRAM improves graphic display performance.

VSTOL (abbreviation for *vertical/short takeoff and landing*) aircraft capable of taking off and landing either vertically or using a very short length of runway (see ◊STOL). Vertical takeoff requires a vector-control system that permits the thrust of the aircraft engine to be changed from horizontal to vertical for takeoff and back again to horizontal to permit forward flight. An alternative VSTOL technology developed in the USA involves tilting the wings of the aircraft from vertical to horizontal and along with them the aircraft propellers, thus changing from vertical lift to horizontal thrust.

vulcanization technique for hardening rubber by heating and chemically combining it with sulphur. The process also makes the rubber stronger and more elastic. If the sulphur content is increased to as much as 30%, the product is the inelastic solid known as ebonite. More expensive alternatives to sulphur, such as selenium and tellurium, are used to vulcanize rubber for specialized products such as vehicle tyres. The process was discovered accidentally by US inventor Charles Goodyear 1839 and patented 1844.

vulcanology the study of ◊volcanoes and the geological phenomena that cause them.

volume of a **cube**
= length³
= l^3

volume of a **cuboid**
= length × breadth × height
= $l \times b \times h$

volume of a **cylinder**
= π × (radius of cross-section)² × height
= $\pi r^2 h$

volume of a **cone**
= ⅓ π × (radius of cross-section)² × height
= $\frac{1}{3}\pi r^2 h$

volume of a **sphere**
= ⁴⁄₃π radius³
= $\frac{4}{3}\pi r^3$

volume *The volumes of a number of common shapes.*

Vulpecula small constellation in the northern hemisphere of the sky just south of Cygnus, represented as a fox. It contains a major planetary ◊nebula, the Dumbbell, and the first ◊pulsar (pulsating radio source) to be discovered.

vulva female external genitalia. Two folds of fleshy tissue, the labia majora and labia minora, extend downwards from the clitoris to surround the entrances to the ◊urethra and the vagina.

W in physics, symbol for *watt*.

wafer in microelectronics, a 'superchip' some 8–10 cm/3–4 in diameter, for which wafer-scale integration (WSI) is used to link the equivalent of many individual ◊silicon chips, improving reliability, speed, and cooling.

Wallace line imaginary line running down the Lombok Strait in SE Asia, between the island of Bali and the islands of Lombok and Sulawesi. It was identified by English naturalist Alfred Russel Wallace as separating the S Asian (Oriental) and Australian biogeographical regions, each of which has its own distinctive animals.

wall pressure in plants, the mechanical pressure exerted by the cell contents against the cell wall. The rigidity (turgor) of a plant often depends on the level of wall pressure found in the cells of the stem. Wall pressure falls if the plant cell loses water.

WAN abbreviation for *wide area ◊network*.

Wankel engine rotary petrol engine developed by the German engineer Felix Wankel (1902–1988) in the 1950s. It operates according to the same stages as the ◊four-stroke petrol engine cycle, but these stages take place in different sectors of a figure-eight chamber in the space between the chamber walls and a triangular rotor. Power is produced once on every turn of the rotor. The Wankel engine is simpler in construction than the four-stroke piston petrol engine, and produces rotary power directly (instead of via a crankshaft). Problems with rotor seals have prevented its widespread use.

warning coloration in biology, an alternative term for ◊aposematic coloration.

washing soda $Na_2CO_3.10H_2O$ (chemical name *sodium carbonate decahydrate*) substance added to washing water to 'soften' it (see ◊hard water).

water H_2O liquid without colour, taste, or odour. It is an oxide of hydrogen. Water begins to freeze at 0°C or 32°F, and to boil at 100°C or 212°F. When liquid, it is virtually incompressible; frozen, it expands by $1/11$ of its volume. At 4°C/39.2°F, one cubic centimetre of water has a mass of one gram; this is its maximum density, forming the unit of specific gravity. It has the highest known specific heat, and acts as an efficient solvent, particularly when hot. Most of the world's water is in the sea; less than 0.01% is fresh water.

Water covers 70% of the Earth's surface and occurs as standing (oceans, lakes) and running (rivers, streams) water, rain, and vapour and supports all forms of Earth's life.

Water makes up 60–70% of the human body or about 40 litres of which 25 are inside the cells, 15 outside (12 in tissue fluid, and 3 in blood plasma). A loss of 4 litres may cause hallucinations; a loss of 8–10 litres may cause death. About 1.5 litres a day are lost through breathing, perspiration, and faeces, and the additional amount lost in urine is the amount needed to keep the balance between input and output. People cannot survive more than five or six days without water or two or three days in a hot environment.

water-borne disease disease associated with poor water supply. In the Third World four-fifths of all illness is caused by water-borne diseases, with diarrhoea being the leading cause of childhood death. Malaria, carried by mosquitoes dependent on stagnant water for breeding, affects 400 million people every year and kills 5 million. Polluted water is also a problem in industrialized nations, where industrial dumping of chemical, hazardous, and radioactive wastes causes a range of diseases from headache to cancer.

water cycle or *hydrological cycle* the natural circulation of water through the ◊biosphere. Water is lost from the Earth's surface to the atmosphere either by evaporation from the surface of lakes, rivers, and oceans, or through the transpiration of plants. This atmospheric water forms clouds that condense to deposit moisture on the land and sea as rain or snow. The water that collects on land flows to the ocean in streams and rivers.

water gas fuel gas consisting of a mixture of carbon monoxide and hydrogen, made by passing steam over red-hot coke. The gas was

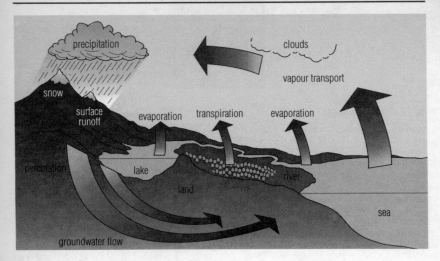

water cycle *About one-third of the solar energy reaching the Earth is used in evaporating water. About 380,000 cubic km/95,000 cubic mi is evaporated each year. The entire contents of the oceans would take about one million years to pass through the water cycle.*

once the chief source of hydrogen for chemical syntheses such as the Haber process for making ammonia, but has been largely superseded in this and other reactions by hydrogen obtained from natural gas.

water glass common name for sodium metasilicate (Na_2SiO_3). It is a colourless, jelly-like substance that dissolves readily in water to give a solution used for preserving eggs and fireproofing porous materials such as cloth, paper, and wood. It is also used as an adhesive for paper and cardboard and in the manufacture of soap and silica gel, a substance that absorbs moisture.

water of crystallization water chemically bonded to a salt in its crystalline state. For example, in copper(II) sulphate, there are five moles of water per mole of copper sulphate: hence its formula is $CuSO_4.5H_2O$. This water is responsible for the colour and shape of the crystalline form. When the crystals are heated gently, the water is driven off as steam and a white powder is formed.

water softener any substance or unit that removes the hardness from water. Hardness is caused by the presence of calcium and magnesium ions, which combine with soap to form an insoluble scum, prevent lathering, and cause deposits to build up in pipes and cookware (kettle fur). A water softener replaces these ions with sodium ions, which are fully soluble and cause no scum.

water supply distribution of water for domestic, municipal, or industrial consumption. Water supply in sparsely populated regions usually comes from underground water rising to the surface in natural springs, supplemented by pumps and wells. Urban sources are deep artesian wells, rivers, and reservoirs, usually formed from enlarged lakes or dammed and flooded valleys, from which water is conveyed by pipes, conduits, and aqueducts to filter beds. As water seeps through layers of shingle, gravel, and sand, harmful organisms are removed and the water is then distributed by pumping or gravitation through mains and pipes. Often other substances are added to the water, such as chlorine and fluorine; aluminium sulphate, a clarifying agent, is the most widely used chemical in water treatment.

In towns, domestic and municipal (road washing, sewage) needs account for about 135 l/30 gal per head each day. In coastal desert areas, such as the Arabian peninsula, desalination plants remove salt from sea water. The Earth's waters, both fresh and saline, have been polluted by industrial and domestic chemicals, some of which are toxic and others radioactive.

water table the upper level of ground water (water collected underground in porous rocks). Water that is above the water table will drain downwards; a spring forms where the water table cuts the surface of the ground. The water table rises and falls in response to rainfall and the rate at which water is extracted, for example, for irrigation and industry.

watt SI unit (symbol W) of power (the rate of expenditure or consumption of energy). A light bulb may use 40, 100, or 150 watts of power; an electric heater will use several kilowatts (thousands of watts). The watt is named after the Scottish engineer James Watt.

The absolute watt is defined as the power used when one joule of work is done in one second. In electrical terms, the flow of one ampere of current through a conductor whose ends are at a potential difference of one volt uses one watt of power (watts = volts × amperes).

wave in physics, a disturbance travelling through a medium (or space). There are two types: in a *longitudinal wave* (such as a ♢sound wave) the disturbance is parallel to the wave's direction of travel; in a *transverse wave* (such as an ♢electromagnetic wave) it is perpendicular. The medium (for example the Earth, for seismic waves) is not permanently displaced by the passage of a wave.

waveguide hollow metallic tube containing a ♢dielectric used to guide a high-frequency electromagnetic wave (microwave) travelling within it. The wave is reflected from the internal surfaces of the guide. Waveguides are extensively used in radar systems.

wavelength the distance between successive crests of a ♢wave. The wavelength of a light wave determines its colour; red light has a wavelength of about 700 nanometres, for example. The complete range of wavelengths of electromagnetic waves is called the electromagnetic ♢spectrum.

wave power power obtained by harnessing the energy of water waves. Various schemes have been advanced since 1973, when oil prices rose dramatically and an energy shortage threatened. In 1974 the British engineer Stephen Salter developed the duck – a floating boom whose segments nod up and down with the waves. The nodding motion can be used to drive pumps and spin generators. Another device, developed in Japan, uses an oscillating water column to harness wave power. However, a major breakthrough will be required if wave power is ever to contribute significantly to the world's energy needs.

wax solid fatty substance of animal, vegetable, or mineral origin. Waxes are composed variously of ♢esters, ♢fatty acids, free ♢alcohols, and solid hydrocarbons.

Mineral waxes are obtained from petroleum and vary in hardness from the soft petroleum jelly (or petrolatum) used in ointments to the hard paraffin wax employed for making candles and waxed paper for drinks cartons.

Animal waxes include beeswax, the wool wax lanolin, and spermaceti from sperm-whale oil; they are used mainly in cosmetics, ointments, and polishes.

Vegetable waxes, which usually occur as a waterproof coating on plants that grow in hot, arid regions, include carnauba wax (from the leaves of the carnauba palm) and candelilla wax, both of which are components of hard polishes such as car waxes.

weak acid acid that only partially ionizes in aqueous solution (see ♢dissociation). Weak acids include ethanoic acid and carbonic acid. The pH of such acids lies between pH 3 and pH 6.

weak base base that only partially ionizes in aqueous solution (see ♢dissociation); for example, ammonia. The pH of such bases lies between pH 8 and pH 10.

weak nuclear force or *weak interaction* one of the four fundamental forces of nature, the other three being gravity, the electromagnetic force, and the strong force. It causes radioactive decay and other subatomic reactions. The particles that carry the weak force are called ♢weakons (or intermediate vector bosons) and comprise the positively and negatively charged W particles and the neutral Z particle.

weakon or *intermediate vector boson* in physics, a ♢gauge boson that carries the weak nuclear force, one of the fundamental forces of nature. There are three types of weakon, the positive and negative W particle and the neutral Z particle.

weapon any implement used for attack and defence, from simple clubs, spears, and bows and arrows in prehistoric times to machine guns and nuclear bombs in modern times. The first revolution in warfare came with the invention of ♢gunpowder and the development of cannons and shoulder-held guns. Many other weapons now exist, such as grenades, shells, torpedoes, rockets, and guided missiles. The ultimate in explosive weapons are the atomic (fission) and hydrogen (fusion) bombs. They release the enormous energy produced when

weathering

physical weathering

temperature changes	weakening rocks by expansion and contraction
frost	wedging rocks apart by the expansion of water on freezing
unloading	the loosening of rock layers by release of pressure after the erosion and removal of those layers above

chemical weathering

carbonation	the breakdown of calcite by reaction with carbonic acid in rainwater
hydrolysis	the breakdown of feldspar into china clay by reaction with carbonic acid in rainwater
oxidation	the breakdown of iron-rich minerals due to rusting
hydration	the expansion of certain minerals due to the uptake of water

atoms split or fuse together. There are also chemical and bacteriological weapons.

weathering process by which exposed rocks are broken down on the spot by the action of rain, frost, wind, and other elements of the weather. It differs from ◊erosion in that no movement or transportation of the broken-down material takes place. Two types of weathering are recognized: physical (or mechanical) and chemical. They usually occur together.

Physical weathering includes such effects as freeze–thaw (the splitting of rocks by the alternate freezing and thawing of water trapped in cracks) and exfoliation, or onion-skin weathering (flaking caused by the alternate expansion and contraction of rocks in response to extreme changes in temperature).

Chemical weathering is brought about by a chemical change in the rocks affected. The most common form is caused by rainwater that has absorbed carbon dioxide from the atmosphere and formed a weak carbonic acid. This then reacts with certain minerals in the rocks and breaks them down. Examples are the solution of caverns in limestone terrains, and the breakdown of feldspars in granite to form china clay or kaolin.

weber SI unit (symbol Wb) of ◊magnetic flux (the magnetic field strength multiplied by the area through which the field passes). It is named after German chemist Wilhelm Weber. One weber equals 10^8 ◊maxwells.

wedge block of triangular cross-section that can be used as a simple machine. An axe is a wedge: it splits wood by redirecting the energy of the downward blow sideways, where it exerts the force needed to split the wood.

weight the force exerted on an object by ◊gravity. The weight of an object depends on its mass – the amount of material in it – and the strength of the Earth's gravitational pull, which decreases with height. Consequently, an object weighs less at the top of a mountain than at sea level. On the Moon, an object has only one-sixth of its weight on Earth, because the pull of the Moon's gravity is one-sixth that of the Earth.

weightlessness condition in which there is no gravitational force acting on a body, either because gravitational force is cancelled out by equal and opposite acceleration, or because the body is so far outside a planet's gravitational field that no force is exerted upon it.

weights and measures see under ◊c.g.s. system, ◊f.p.s. system, ◊m.k.s. system, ◊SI units.

welding joining pieces of metal (or non-metal) at faces rendered plastic or liquid by heat or pressure (or both). The principal processes today are gas and arc welding, in which the heat from a gas flame or an electric arc melts the faces to be joined. Additional 'filler metal' is usually added to the joint.

wheel and axle simple machine with a rope wound round an axle connected to a larger wheel with another rope attached to its rim. Pulling on the wheel rope (applying an effort) lifts a load attached to the axle rope. The velocity ratio of the machine (distance moved by load divided by distance moved by effort) is equal to the ratio of the wheel radius to the axle radius.

white blood cell or *leucocyte* one of a number of different cells that play a part in the body's defences and give immunity against disease. Some (◊phagocytes and ◊macrophages) engulf invading microorganisms, others kill infected cells, while ◊lymphocytes produce more specific immune responses. White blood cells are colourless, with clear or granulated cytoplasm, and are capable of independent amoeboid movement. They occur in the blood, ◊lymph and elsewhere in the body's tissues.

Unlike mammalian red blood cells, they possess a nucleus. Human blood contains about 11,000 leucocytes to the cubic millimetre – about one to every 500 red cells.

White blood cell numbers may be reduced (leucopenia) by starvation, pernicious anaemia, and certain infections, such as typhoid and malaria. An increase in their numbers (leucocytosis) is a reaction to normal events such as digestion, exertion, and pregnancy, and to abnormal ones such as loss of blood, cancer, and most infections.

white dwarf small, hot ◊star, the last stage in the life of a star such as the Sun. White dwarfs have a mass similar to that of the Sun, but only 1% of the Sun's diameter, similar in size to the Earth. Most have surface temperatures of 8,000°C/14,400°F or more, hotter than the Sun. Yet, being so small, their overall luminosities may be less than 1% of that of the Sun. The Milky Way contains an estimated 50 billion white dwarfs.

white spirit colourless liquid derived from petrol; it is used as a solvent and in paints and varnishes.

wide area network in computing, a ◊network that connects computers distributed over a wide geographical area.

Wien's law in physics, a law of radiation stating that the wavelength carrying the maximum energy is inversely proportional to the body's absolute temperature: the hotter a body is, the shorter the wavelength. It has the form $\lambda_{max}T$ = constant, where λ_{max} is the wavelength of maximum intensity and T is the temperature. The law is named after German physicist Wilhelm Wien (1864–1928).

wild type in genetics, the naturally occurring gene for a particular character that is typical of most individuals of a given species, as distinct from new genes that arise by mutation.

wilting the loss of rigidity (◊turgor) in plants, caused by a decreasing wall pressure within the cells making up the supportive tissues. Wilting is most obvious in plants that have little or no wood.

WIMP (acronym for *windows, icons, menus, pointing device*) in computing, another name for ◊graphical user interface (GUI).

Winchester drive in computing, a small hard-disc drive commonly used with microcomputers; *Winchester disc* has become synonymous with ◊hard disc.

wind-chill factor or *wind-chill index* estimate of how much colder it feels when a wind is blowing. It is the sum of the temperature (in °F below zero) and the wind speed (in miles per hour). So for a wind of 15 mph at an air temperature of –5°F, the wind-chill factor is 20.

window in computing, a rectangular area on the screen of a ◊graphical user interface. A window is used to display data and can be manipulated in various ways by the computer user.

Windows ◊graphical user interface (GUI) from Microsoft that has become the standard for IBM PCs and clones using the ◊MS-DOS operating system.

wind power the harnessing of wind energy to produce power. The wind has long been used as a source of energy: sailing ships and windmills are ancient inventions. After the energy crisis of the 1970s ◊wind turbines began to be used to produce electricity on a large scale.

wind tunnel test tunnel in which air is blown over, for example, a stationary model aircraft, motor vehicle, or locomotive to simulate the effects of movement. Lift, drag, and airflow patterns are observed by the use of special cameras and sensitive instruments. Wind-tunnel testing assesses aerodynamic design, preparatory to full-scale construction.

wind turbine windmill of advanced aerodynamic design connected to an electricity generator and used in wind-power installations. Wind turbines can be either large propeller-type rotors mounted on a tall tower, or flexible metal strips fixed to a vertical axle at top and bottom. In 1990, over 20,000 wind turbines were in use throughout the world, generating 1,600 megawatts of power.

wing in biology, the modified forelimb of birds and bats, or the membranous outgrowths of the ◊exoskeleton of insects, which give the power of flight. Birds and bats have two wings. Bird wings have feathers attached to the fused digits ('fingers') and forearm bones, while bat wings consist of skin stretched between the digits. Most insects have four wings, which are strengthened by wing veins.

wolfram alternative name for ◊tungsten.

wolframite iron manganese tungstate, $(Fe,Mn)WO_4$, an ore mineral of tungsten. It is dark grey with a submetallic surface lustre, and often occurs in hydrothermal veins in association with ores of tin.

womb common name for the ◊uterus.

wood the hard tissue beneath the bark of many perennial plants; it is composed of water-conducting cells, or secondary ◊xylem, and gains its hardness and strength from deposits of ◊lignin. *Hardwoods*, such as oak, and *softwoods*, such as pine, have commercial value as structural material and for furniture.

The central wood in a branch or stem is known as *heartwood* and is generally darker and harder than the outer wood; it consists only of dead cells. As well as providing structural support, it often contains gums, tannins, or pigments which may impart a characteristic colour and increased durability. The surrounding *sapwood* is the functional part of the xylem that conducts water.

The *secondary xylem* is laid down by the vascular ◊cambium which forms a new layer of wood annually, on the outside of the existing wood and is visible as an ◊annual ring when the tree is felled; see ◊dendrochronology.

wood pitch by-product of charcoal manufacture, made from *wood tar*, the condensed liquid produced from burning charcoal gases. The wood tar is boiled to produce the correct consistency. It has been used since ancient times for caulking wooden ships (filling in the spaces between the hull planks to make them watertight).

wood pulp wood that has been processed into a pulpy mass of fibres. Its main use is for making paper, but it is also used in making ◊rayon and other cellulose fibres and plastics.

word in computing, a group of bits (binary digits) that a computer's central processing unit treats as a single working unit. The size of a word varies from one computer to another and, in general, increasing the word length leads to a faster and more powerful computer. In the late 1970s and early 1980s, most microcomputers were 8-bit machines. During the 1980s 16-bit microcomputers were introduced and 32-bit microcomputers are now available. Mainframe computers may be 32-bit or 64-bit machines.

word processor in computing, a program that allows the input, amendment, manipulation, storage, and retrieval of text; or a computer system that runs such software. Since word-processing programs became available to microcomputers, the method has been gradually replacing the typewriter for producing letters or other text.

work in physics, a measure of the result of transferring energy from one system to another to cause an object to move. Work should not be confused with ◊energy (the capacity to do work, which is also measured in joules) or with ◊power (the rate of doing work, measured in joules per second).

Work is equal to the product of the force used and the distance moved by the object in the direction of that force. If the force is F newtons and the distance moved is d metres, then the work W is given by: $W=Fd$. For example, the work done when a force of 10 newtons moves an object 5 metres against some sort of resistance is 50 joules (50 newton-metres).

workstation high-performance desktop computer with strong graphics capabilities, traditionally used for engineering (◊CAD and ◊CAM), scientific research, and desktop publishing. Frequently based on fast RISC (reduced instruction-set computer) chips, workstations generally offer more processing power than microcomputers (although the distinction between workstations and the more powerful microcomputer models is becoming increasingly blurred). Most workstations use Unix as their operating system, and have good networking facilities.

worm any of various elongated limbless invertebrates belonging to several phyla. Worms include the ◊flatworms, such as flukes and ◊tapeworms; the roundworms or ◊nematodes, such as the eelworm and the hookworm; the marine ribbon worms or nemerteans; and the segmented worms or ◊annelids.

WORM (acronym for *write once read many times*) in computing, a storage device, similar to ◊CD-ROM. The computer can write to the disc directly, but cannot later erase or overwrite the same area. WORMs are mainly used for archiving and backup copies.

W particle in physics, an ◊elementary particle, one of the weakons responsible for transmitting the ◊weak nuclear force.

wrist joint linking the hand and forearm. It comprises part of the carpus that articulates with the ◊radius and ◊ulna.

write protection device on discs and tapes that provides data security by allowing data to be read but not deleted, altered, or overwritten.

wrought iron fairly pure iron containing some beads of slag, widely used for construction work before the days of cheap steel. It is strong, tough, and easy to machine. It is made in a puddling furnace, invented by Henry Colt in England 1784. Pig iron is remelted and heated strongly in air with iron ore, burning out the carbon in the metal, leaving relatively pure iron and a slag containing impurities. The resulting pasty metal is then hammered to remove as much of the remaining slag as possible. It is still used in fences and grating.

wt abbreviation for *weight*.

WYSIWYG (acronym for *what you see is what you get*) in computing, a program that attempts to display on the screen a faithful representation of the final printed output. For example, a WYSIWYG ◊word processor would show actual page layout – line widths, page breaks, and the sizes and styles of type.

X

xanthophyll yellow pigment in plants that, like ◊chlorophyll, is responsible for the production of carbohydrates by photosynthesis.

X chromosome larger of the two sex chromosomes, the smaller being the ◊Y chromosome. These two chromosomes are involved in sex determination. Females have two X chromosomes, males have an X and a Y. Genes carried on the X chromosome produce the phenomenon of ◊sex linkage.

xenon colourless, odourless, gaseous, nonmetallic element, symbol Xe, atomic number 54, relative atomic mass 131.30. It is grouped with the ◊inert gases and was long believed not to enter into reactions, but is now known to form some compounds, mostly with fluorine. It is a heavy gas present in very small quantities in the air (about one part in 20 million).

xerography dry, electrostatic method of producing images, without the use of negatives or sensitized paper, invented in the USA by Chester Carlson 1938 and applied in the Xerox photocopier.

xerophyte plant adapted to live in dry conditions. Common adaptations to reduce the rate of ◊transpiration include a reduction of leaf size, sometimes to spines or scales; a dense covering of hairs over the leaf to trap a layer of moist air (as in edelweiss); water storage cells; sunken stomata; and permanently rolled leaves or leaves that roll up in dry weather (as in marram grass). Many desert cacti are xerophytes.

X-ray band of electromagnetic radiation in the wavelength range 10^{-11} to 10^{-9} m (between gamma rays and ultraviolet radiation; see ◊electromagnetic waves). Applications of X-rays make use of their short wavelength (as in ◊X-ray diffraction) or their penetrating power (as in medical X-rays of internal body tissues). X-rays are dangerous and can cause cancer.

X-rays with short wavelengths pass through most body tissues, although dense areas such as bone prevent their passage, showing up as white areas on X-ray photographs. The X-rays used in ◊radiotherapy have very short wavelengths that penetrate tissues deeply and destroy them.

X-ray astronomy detection of X-rays from intensely hot gas in the universe. Such X-rays are prevented from reaching the Earth's surface by the atmosphere, so detectors must be placed in rockets and satellites. The first celestial X-ray source, Scorpius X-1, was discovered by a rocket flight 1962.

X-ray diffraction method of studying the atomic and molecular structure of crystalline substances by using ◊X-rays. X-rays directed at such substances spread out as they pass through the crystals owing to ◊diffraction (the slight spreading of waves around the edge of an opaque object) of the rays around the atoms. By using measurements of the position and intensity of the diffracted waves, it is possible to calculate the shape and size of the atoms in the crystal. The method has been used to study substances such as ◊DNA that are found in living material.

X-ray diffraction analysis the use of X-rays to study the atomic and molecular structure of crystalline substances such as

an X-ray imager

X-ray *The X-rays are generated by high-speed electrons impinging on a tungsten target. The rays pass through the specimen and on to a photographic plate or imager.*

ceramics, stone, sediments, and weathering products on metals. The sample is ground to powder, and exposed to X-rays at various angles; the diffraction patterns produced are then compared with reference standards for identification.

X-ray fluorescence spectrometry technique used to determine the major and trace elements in the chemical composition of such materials as ceramics, obsidian, and glass. A sample is bombarded with X-rays, and the wavelengths of the released energy, or fluorescent X-rays, are detected and measured. Different elements have unique wavelengths, and their concentrations can be estimated from the intensity of the released X-rays. This analysis may, for example, help an archaeologist in identifying the source of the material.

xylem tissue found in ◊vascular plants, whose main function is to conduct water and dissolved mineral nutrients from the roots to other parts of the plant. Xylem is composed of a number of different types of cell, and may include long, thin, usually dead cells known as ◊tracheids; fibres (schlerenchyma); thin-walled ◊parenchyma cells; and conducting vessels.

In most ◊angiosperms (flowering plants) water is moved through these vessels. Most ◊gymnosperms and ◊pteridophytes lack vessels and depend on tracheids for water conduction.

Nonwoody plants contain only primary xylem, derived from the procambium, whereas in trees and shrubs this is replaced for the most part by secondary xylem, formed by ◊secondary growth from the actively dividing vascular ◊cambium. The cell walls of the secondary xylem are thickened by a deposit of ◊lignin, providing mechanical support to the plant; see ◊wood.

yard imperial unit (symbol yd) of length, equivalent to three feet (0.9144 m).

Y chromosome smaller of the two sex chromosomes. In male mammals it occurs paired with the other type of sex chromosome (X), which carries far more genes. The Y chromosome is the smallest of all the mammalian chromosomes and is considered to be largely inert (that is, without direct effect on the physical body). See also ◊sex determination.

In humans, about one in 300 males inherits two Y chromosomes at conception, making him an XYY triploid. Few if any differences from normal XY males exist in these individuals, although at one time they were thought to be emotionally unstable and abnormally aggressive. In 1989 the gene determining that a human being is male was found to occur on the X as well as on the Y chromosome; however, it is not activated in the female.

yd abbreviation for *yard*.

year unit of time measurement, based on the orbital period of the Earth around the Sun. The *tropical year*, from one spring ◊equinox to the next, lasts 365.2422 days. It governs the occurrence of the seasons, and is the period on which the calendar year is based. The *sidereal year* is the time taken for the Earth to complete one orbit relative to the fixed stars, and lasts 365.2564 days (about 20 minutes longer than a tropical year). The difference is due to the effect of ◊precession, which slowly moves the position of the equinoxes. The *calendar year* consists of 365 days, with an extra day added at the end of Feb each leap year. *Leap years* occur in every year that is divisible by four, except that a century year is not a leap year unless it is divisible by 400. Hence 1900 was not a leap year, but 2000 will be.

yeast one of various single-celled fungi (especially the genus *Saccharomyces*) that form masses of minute circular or oval cells by budding. When placed in a sugar solution the cells multiply and convert the sugar into alcohol and carbon dioxide. Yeasts are used as fermenting agents in baking, brewing, and the making of wine and spirits. Brewer's yeast *S. cerevisiae* is a rich source of vitamin B.

yeast artificial chromosome (YAC) fragment of ◊DNA from the human genome inserted into a yeast cell. The yeast replicates the fragment along with its own DNA. In this way the fragments are copied to be preserved in a gene library. YACs are characteristically between 250,000 and 1 million base pairs in length. A ◊cosmid works in the same way.

yield point or *elastic limit* the stress beyond which a material deforms by a relatively large amount for a small increase in stretching force. Beyond this stress, the material no longer obeys ◊Hooke's law.

yolk store of food, mostly in the form of fats and proteins, found in the ◊eggs of many animals. It provides nourishment for the growing embryo.

yolk sac sac containing the yolk in the egg of most vertebrates. The term is also used for the membranous sac formed below the developing mammalian embryo and connected with the umbilical cord.

ytterbium soft, lustrous, silvery, malleable, and ductile element of the ◊lanthanide series, symbol Yb, atomic number 70, relative atomic mass 173.04. It occurs with (and resembles) yttrium in gadolinite and other minerals, and is used in making steel and other alloys.

yttrium silver-grey, metallic element, symbol Y, atomic number 39, relative atomic mass 88.905. It is associated with and resembles the rare earth elements (◊lanthanides), occurring in gadolinite, xenotime, and other minerals. It is used in colour-television tubes and to reduce steel corrosion.

Z

Z in physics, the symbol for *impedance* (electricity and magnetism).

zenith uppermost point of the celestial horizon, immediately above the observer; the ◊nadir is below, diametrically opposite. See ◊celestial sphere.

zeolite any of the hydrous aluminium silicates, also containing sodium, calcium, barium, strontium, or potassium, chiefly found in igneous rocks and characterized by a ready loss or gain of water. Zeolites are used as 'molecular sieves' to separate mixtures because they are capable of selective absorption. They have a high ion-exchange capacity and can be used to make petrol, benzene, and toluene from low-grade raw materials, such as coal and methanol.

zidovudine (formerly *AZT*) antiviral drug used in the treatment of ◊AIDS. It is not a cure for AIDS but is effective in prolonging life; it does not, however, delay the onset of AIDS in people carrying the virus.

zinc hard, brittle, bluish-white, metallic element, symbol Zn, atomic number 30, relative atomic mass 65.37. The principal ore is sphalerite or zinc blende (zinc sulphide, ZnS). Zinc is little affected by air or moisture at ordinary temperatures; its chief uses are in alloys such as brass and in coating metals (for example, galvanized iron). Its compounds include zinc oxide, used in ointments (as an astringent) and cosmetics, paints, glass, and printing ink.

zinc chloride $ZnCl_2$ white, crystalline compound that is deliquescent and sublimes easily. It is used as a catalyst, as a dehydrating agent, and as a flux in soldering.

zinc ore mineral from which zinc is extracted, principally sphalerite (Zn,Fe)S, but also zincite, ZnO_2, and smithsonite, $ZnCO_3$, all of which occur in mineralized veins. Ores of lead and zinc often occur together, and are common worldwide; Canada, the USA, and Australia are major producers.

zinc oxide ZnO white powder, yellow when hot, that occurs in nature as the mineral zincite. It is used in paints and as an antiseptic in zinc ointment; it is the main ingredient of calamine lotion.

zinc sulphide ZnS yellow-white solid that occurs in nature as the mineral sphalerite (also called zinc blende). It is the principal ore of zinc, and is used in the manufacture of fluorescent paints.

zircon zirconium silicate, $ZrSiO_4$, a mineral that occurs in small quantities in a wide range of igneous, sedimentary, and metamorphic rocks. It is very durable and is resistant to erosion and weathering. It is usually coloured brown, but can be other colours, and when transparent may be used as a gemstone.

zirconium lustrous, greyish-white, strong, ductile, metallic element, symbol Zr, atomic number 40, relative atomic mass 91.22. It occurs in nature as the mineral zircon (zirconium silicate), from which it is obtained commercially. It is used in some ceramics, alloys for wire and filaments, steel manufacture, and nuclear reactors, where its low neutron absorption is advantageous.

zodiac zone of the heavens containing the paths of the Sun, Moon, and planets. When this was devised by the ancient Greeks, only five planets were known, making the zodiac about 16° wide. In astrology, the zodiac is divided into 12 signs, each 30° in extent: Aries, Taurus, Gemini, Cancer, Leo, Virgo, Libra, Scorpio, Sagittarius, Capricorn, Aquarius, and Pisces. These do not cover the same areas of sky as the astronomical constellations.

zodiacal light cone-shaped light sometimes seen extending from the Sun along the ◊ecliptic, visible after sunset or before sunrise. It is due to thinly spread dust particles in the central plane of the Solar System. It is very faint, and requires a dark, clear sky to be seen.

zoidogamy type of plant reproduction in which male gametes (antherozoids) swim in a film of water to the female gametes. Zoidogamy is found in algae, bryophytes, pteridophytes, and some gymnosperms (others use ◊siphonogamy).

zoo (abbreviation for *zoological gardens*), a place where animals are kept in captivity. Originally created purely for visitor entertainment and education, zoos have become major centres for the breeding of endangered species of animals; a 1984 report identified 2,000 vertebrate species in need of such maintenance. The Arabian oryx was preserved in this way; it was captured 1962, bred in captivity, and released again in the desert 1972, where it has flourished. In 1991 the number of animals in Britain's zoos totalled 35,000.

zoology branch of biology concerned with the study of animals. It includes description of present-day animals, the study of evolution of animal forms, anatomy, physiology, embryology, behaviour, and geographical distribution.

Z particle in physics, an ◊elementary particle, one of the weakons responsible for carrying the ◊weak nuclear force.

zwitterion ion that has both a positive and a negative charge, such as an ◊amino acid in neutral solution. For example, glycine contains both a basic amino group (NH^2) and an acidic carboxyl group ($-COOH$); when both these are ionized in aqueous solution, the acid group loses a proton to the amino group, and the molecule is positively charged at one end and negatively charged at the other.

zygomatic arch arch of bone on each side of the face, just below the eyes.

zygote ◊ovum (egg) after ◊fertilization but before it undergoes cleavage to begin embryonic development.

SI units

quantity	SI unit	symbol
absorbed radiation dose	gray	Gy
amount of substance	mole*	mol
electric capacitance	farad	F
electric charge	coulomb	C
electric conductance	siemens	S
electric current	ampere*	A
energy or work	joule	J
force	newton	N
frequency	hertz	Hz
illuminance	lux	lx
inductance	henry	H
length	metre*	m
luminous flux	lumen	lm
luminous intensity	candela*	cd
magnetic flux	weber	Wb
magnetic flux density	tesla	T
mass	kilogram*	kg
plane angle	radian	rad
potential difference	volt	V
power	watt	W
pressure	pascal	Pa
radiation dose equivalent	sievert	Sv
radiation exposure	roentgen	r
radioactivity	becquerel	Bq
resistance	ohm	Ω
solid angle	steradian	sr
sound intensity	decibel	dB
temperature	°Celsius	°C
temperature, thermodynamic	kelvin*	K
time	second*	s

*SI base unit

SI prefixes

multiple	prefix	symbol	example
1,000,000,000,000,000,000 (10^{18})	exa-	E	Eg (exagram)
1,000,000,000,000,000 (10^{15})	peta-	P	PJ (petajoule)
1,000,000,000,000 (10^{12})	tera-	T	TV (teravolt)
1,000,000,000 (10^9)	giga-	G	GW (gigawatt)
1,000,000 (10^6)	mega-	M	MHz (megahertz)
1,000 (10^3)	kilo-	k	kg (kilogram)
100 (10^2)	hecto-	h	hm (hectometre)
10	deca-	da	daN (decanewton)
1/10 (10^{-1})	deci-	d	dC (decicoulomb)
1/100 (10^{-2})	centi-	c	cm (centimetre)
1/1,000 (10^{-3})	milli-	m	mA (milliampere)
1/1,000,000 (10^{-6})	micro-	μ	μF (microfarad)
1/1,000,000,000 (10^{-9})	nano-	n	nm (nanometre)
1/1,000,000,000,000 (10^{-12})	pico-	p	ps (picosecond)
1/1,000,000,000,000,000 (10^{-15})	femto-	f	frad (femtoradian)
1/1,000,000,000,000,000,000 (10^{-18})	atto-	a	aT (attotesla)

WEIGHTS AND MEASURES

area

imperial/metric
1 sq inch (in^2)			=	6.4516 cm^2
1 sq yard (yd^2)	=	9 ft^2	=	0.8361 m^2
1 acre	=	4840 yd^2	=	4046.9 m^2
1 sq mile (mile ^{2})	=	640 acres	=	2.59 km^2

metric/imperial
1 sq cm (cm^2)	=	100 mm^2	=	0.1550 in^2
1 sq metre (m^2)	=	10,000 cm^2	=	1.1960 yd^2
1 hectare (ha)	=	10,000 m^2	=	2.4711 acres
1 sq km (km^2)	=	100 ha	=	0.3861 mile^2

length

imperial/metric
1 inch (in)			=	2.54 cm
1 foot (ft)	=	12 in	=	0.3048 m
1 yard (yd)	=	3 ft	=	0.9144 m
1 mile	=	1760 yd	=	1.6093 km

metric/imperial
1 millimetre (mm)			=	0.0394 in
1 centimetre (cm)	=	10 mm	=	0.3937 in
1 metre (m)	=	100 cm	=	1.0936 yd
1 kilometre (km)	=	1000 m	=	0.6214 mile

mass/weight

imperial/metric
1 ounce (oz)	=	437.5 grain	=	28.35 g
1 pound (lb)	=	16 oz	=	0.4536 kg
1 hundredweight (cwt)	=	112 lb	=	50.802 kg
1 (long) ton	=	20 cwt	=	1.0161 t
1 US (short) ton	=	2000 lb	=	0.9072 t

metric/imperial
1 milligram (mg)			=	0.0154 gm
1 gram (g)	=	1000 mg	=	0.0353 oz
1 kilogram (kg)	=	1000 g	=	2.2046 lb
1 tonne (t)	=	1000 kg	=	0.9842 ton

temperature

°Celsius	=	$\frac{5}{9}$(°F − 32)	°Fahrenheit	=	$\frac{9}{5}$ °C (+ 32)

volume

imperial/metric
1 cu inch (in^3)			=	16.3877 cm^3
1 cu foot (ft^3)	=	1728 in^3	=	0.0283 m^3
1 fluid ounce (fl oz)			=	28.413 ml
1 pint (pt)	=	20 fl oz	=	0.5683 l
1 gallon (gal)	=	8 pt	=	4.5461 l
1 US gallon	=	8 US pints	=	3.7854 l

metric/imperial
1 cu cm (cm^3)			=	0.061 in^3
1 cu decimetre (dm^3)	=	1000 cm^3	=	0.0353 ft^3
1 cu metre (m^3)	=	1000 dm^3	=	1.3080 yd^3
1 litre (l)	=	1 dm^3	=	1.76 pt
1 hectolitre (hl)	=	100 l	=	21.997 gal